Interdisciplinary Applied Mathematics

Volume 8/I

Editors
S.S. Antman   J.E. Marsden
L. Sirovich

Geophysics and Planetary Sciences

Mathematical Biology
L. Glass, J.D. Murray

Mechanics and Materials
R.V. Kohn

Systems and Control
S.S. Sastry, P.S. Krishnaprasad

Problems in engineering, computational science, and the physical and biological sciences are using increasingly sophisticated mathematical techniques. Thus, the bridge between the mathematical sciences and other disciplines is heavily traveled. The correspondingly increased dialog between the disciplines has led to the establishment of the series: *Interdisciplinary Applied Mathematics.*

The purpose of this series is to meet the current and future needs for the interaction between various science and technology areas on the one hand and mathematics on the other. This is done, firstly, by encouraging the ways that mathematics may be applied in traditional areas, as well as point towards new and innovative areas of applications; and, secondly, by encouraging other scientific disciplines to engage in a dialog with mathematicians outlining their problems to both access new methods and suggest innovative developments within mathematics itself.

The series will consist of monographs and high-level texts from researchers working on the interplay between mathematics and other fields of science and technology.

Interdisciplinary Applied Mathematics

James Keener   James Sneyd

Mathematical Physiology
I: Cellular Physiology

Second Edition

Springer

James Keener
Department of Mathematics
University of Utah
Salt Lake City, 84112
USA
keener@math.utah.edu

James Sneyd
Department of Mathematics
University of Auckland
Private Bag 92019
Auckland, New Zealand
sneyd@math.auckland.ac.nz

Series Editors
S.S. Antman
Department of Mathematics and
    Institute for Physical Science and
    Technology
University of Maryland
College Park, MD 20742
USA
ssa@math.umd.edu

J.E. Marsden
Control and Dynamical Systems
Mail Code 107-81
California Institute of Technology
Pasadena, CA 91125
USA
marsden@cds.caltech.edu

L. Sirovich
Laboratory of Applied Mathematics
Department of Biomathematics
Mt. Sinai School of Medicine
Box 1012
NYC 10029
USA
Lawrence.Sirovich@mssm.edu

ISBN 978-0-387-75846-6   e-ISBN 978-0-387-75847-3

DOI 10.1007/978-0-387-75847-3

Library of Congress Control Number: 2008931057

Printed on acid-free paper.

springer.com

To Monique,

and

To Kristine, patience personified.

Preface to the Second Edition

If, in 1998, it was presumptuous to attempt to summarize the field of mathematical physiology in a single book, it is even more so now. In the last ten years, the number of applications of mathematics to physiology has grown enormously, so that the field, large then, is now completely beyond the reach of two people, no matter how many volumes they might write.

Nevertheless, although the bulk of the field can be addressed only briefly, there are certain fundamental models on which stands a great deal of subsequent work. We believe strongly that a prerequisite for understanding modern work in mathematical physiology is an understanding of these basic models, and thus books such as this one serve a useful purpose.

With this second edition we had two major goals. The first was to expand our discussion of many of the fundamental models and principles. For example, the connection between Gibbs free energy, the equilibrium constant, and kinetic rate theory is now discussed briefly, Markov models of ion exchangers and ATPase pumps are discussed at greater length, and agonist-controlled ion channels make an appearance. We also now include some of the older models of fluid transport, respiration/perfusion, blood diseases, molecular motors, smooth muscle, neuroendocrine cells, the baroreceptor loop, tubuloglomerular oscillations, blood clotting, and the retina. In addition, we have expanded our discussion of stochastic processes to include an introduction to Markov models, the Fokker–Planck equation, the Langevin equation, and applications to such things as diffusion, and single-channel data.

Our second goal was to provide a pointer to recent work in as many areas as we can. Some chapters, such as those on calcium dynamics or the heart, close to our own fields of expertise, provide more extensive references to recent work, while in other chapters, dealing with areas in which we are less expert, the pointers are neither complete nor extensive. Nevertheless, we hope that in each chapter, enough information is given to enable the interested reader to pursue the topic further.

Of course, our survey has unavoidable omissions, some intentional, others not. We can only apologize, yet again, for these, and beg the reader's indulgence. As with the first edition, ignorance and exhaustion are the cause, although not the excuse.

Since the publication of the first edition, we have received many comments (some even polite) about mistakes and omissions, and a number of people have devoted considerable amounts of time to help us improve the book. Our particular thanks are due to Richard Bertram, Robin Callard, Erol Cerasi, Martin Falcke, Russ Hamer, Harold Layton, Ian Parker, Les Satin, Jim Selgrade and John Tyson, all of whom assisted above and beyond the call of duty. We also thank Peter Bates, Dan Beard, Andrea Ciliberto, Silvina Ponce Dawson, Charles Doering, Elan Gin, Erin Higgins, Peter Jung, Yue Xian Li, Mike Mackey, Robert Miura, Kim Montgomery, Bela Novak, Sasha Panfilov, Ed Pate, Antonio Politi, Tilak Ratnanather, Timothy Secomb, Eduardo Sontag, Mike Steel, and Wilbert van Meerwijk for their help and comments.

Finally, we thank the University of Auckland and the University of Utah for continuing to pay our salaries while we devoted large fractions of our time to writing, and we thank the Royal Society of New Zealand for the James Cook Fellowship to James Sneyd that has made it possible to complete this book in a reasonable time.

University of Utah   James Keener
University of Auckland   James Sneyd
2008

Preface to the First Edition

It can be argued that of all the biological sciences, physiology is the one in which mathematics has played the greatest role. From the work of Helmholtz and Frank in the last century through to that of Hodgkin, Huxley, and many others in this century, physiologists have repeatedly used mathematical methods and models to help their understanding of physiological processes. It might thus be expected that a close connection between applied mathematics and physiology would have developed naturally, but unfortunately, until recently, such has not been the case.

There are always barriers to communication between disciplines. Despite the quantitative nature of their subject, many physiologists seek only verbal descriptions, naming and learning the functions of an incredibly complicated array of components; often the complexity of the problem appears to preclude a mathematical description. Others want to become physicians, and so have little time for mathematics other than to learn about drug dosages, office accounting practices, and malpractice liability. Still others choose to study physiology precisely because thereby they hope not to study more mathematics, and that in itself is a significant benefit. On the other hand, many applied mathematicians are concerned with theoretical results, proving theorems and such, and prefer not to pay attention to real data or the applications of their results. Others hesitate to jump into a new discipline, with all its required background reading and its own history of modeling that must be learned.

But times are changing, and it is rapidly becoming apparent that applied mathematics and physiology have a great deal to offer one another. It is our view that teaching physiology without a mathematical description of the underlying dynamical processes is like teaching planetary motion to physicists without mentioning or using Kepler's laws; you can observe that there is a full moon every 28 days, but without Kepler's laws you cannot determine when the next total lunar or solar eclipse will be nor when

Halley's comet will return. Your head will be full of interesting and important facts, but it is difficult to organize those facts unless they are given a quantitative description. Similarly, if applied mathematicians were to ignore physiology, they would be losing the opportunity to study an extremely rich and interesting field of science.

To explain the goals of this book, it is most convenient to begin by emphasizing what this book is not; it is not a physiology book, and neither is it a mathematics book. Any reader who is seriously interested in learning physiology would be well advised to consult an introductory physiology book such as Guyton and Hall (1996) or Berne and Levy (1993), as, indeed, we ourselves have done many times. We give only a brief background for each physiological problem we discuss, certainly not enough to satisfy a real physiologist. Neither is this a book for learning mathematics. Of course, a great deal of mathematics is used throughout, but any reader who is not already familiar with the basic techniques would again be well advised to learn the material elsewhere.

Instead, this book describes work that lies on the border between mathematics and physiology; it describes ways in which mathematics may be used to give insight into physiological questions, and how physiological questions can, in turn, lead to new mathematical problems. In this sense, it is truly an interdisciplinary text, which, we hope, will be appreciated by physiologists interested in theoretical approaches to their subject as well as by mathematicians interested in learning new areas of application.

It is also an introductory survey of what a host of other people have done in employing mathematical models to describe physiological processes. It is necessarily brief, incomplete, and outdated (even before it was written), but we hope it will serve as an introduction to, and overview of, some of the most important contributions to the field. Perhaps some of the references will provide a starting point for more in-depth investigations.

Unfortunately, because of the nature of the respective disciplines, applied mathematicians who know little physiology will have an easier time with this material than will physiologists with little mathematical training. A complete understanding of all of the mathematics in this book will require a solid undergraduate training in mathematics, a fact for which we make no apology. We have made no attempt whatever to water down the models so that a lower level of mathematics could be used, but have instead used whatever mathematics the physiology demands. It would be misleading to imply that physiological modeling uses only trivial mathematics, or vice versa; the essential richness of the field results from the incorporation of complexities from both disciplines.

At the least, one needs a solid understanding of differential equations, including phase plane analysis and stability theory. To follow everything will also require an understanding of basic bifurcation theory, linear transform theory (Fourier and Laplace transforms), linear systems theory, complex variable techniques (the residue theorem), and some understanding of partial differential equations and their numerical simulation. However, for those whose mathematical background does not include all of these topics, we have included references that should help to fill the gap. We also make extensive use of asymptotic methods and perturbation theory, but include explanatory material to help the novice understand the calculations.

This book can be used in several ways. It could be used to teach a full-year course in mathematical physiology, and we have used this material in that way. The book includes enough exercises to keep even the most diligent student busy. It could also be used as a supplement to other applied mathematics, bioengineering, or physiology courses. The models and exercises given here can add considerable interest and challenge to an otherwise traditional course.

The book is divided into two parts, the first dealing with the fundamental principles of cell physiology, and the second with the physiology of systems. After an introduction to basic biochemistry and enzyme reactions, we move on to a discussion of various aspects of cell physiology, including the problem of volume control, the membrane potential, ionic flow through channels, and excitability. Chapter 5 is devoted to calcium dynamics, emphasizing the two important ways that calcium is released from stores, while cells that exhibit electrical bursting are the subject of Chapter 6. This book is not intentionally organized around mathematical techniques, but it is a happy coincidence that there is no use of partial differential equations throughout these beginning chapters.

Spatial aspects, such as synaptic transmission, gap junctions, the linear cable equation, nonlinear wave propagation in neurons, and calcium waves, are the subject of the next few chapters, and it is here that the reader first meets partial differential equations. The most mathematical sections of the book arise in the discussion of signaling in two- and three-dimensional media—readers who are less mathematically inclined may wish to skip over these sections. This section on basic physiological mechanisms ends with a discussion of the biochemistry of RNA and DNA and the biochemical regulation of cell function.

The second part of the book gives an overview of organ physiology, mostly from the human body, beginning with an introduction to electrocardiology, followed by the physiology of the circulatory system, blood, muscle, hormones, and the kidneys. Finally, we examine the digestive system, the visual system, ending with the inner ear.

While this may seem to be an enormous amount of material (and it is!), there are many physiological topics that are not discussed here. For example, there is almost no discussion of the immune system and the immune response, and so the work of Perelson, Goldstein, Wofsy, Kirschner, and others of their persuasion is absent. Another glaring omission is the wonderful work of Michael Reed and his collaborators on axonal transport; this work is discussed in detail by Edelstein-Keshet (1988). The study of the central nervous system, including fascinating topics like nervous control, learning, cognition, and memory, is touched upon only very lightly, and the field of pharmacokinetics and compartmental modeling, including the work of John Jacquez, Elliot Landaw, and others, appears not at all. Neither does the wound-healing work of Maini, Sherratt, Murray, and others, or the tumor modeling of Chaplain and his colleagues. The list could continue indefinitely. Please accept our apologies if your favorite topic (or life's work) was omitted; the reason is exhaustion, not lack of interest.

As well as noticing the omission of a number of important areas of mathematical physiology, the reader may also notice that our view of what "mathematical" means appears to be somewhat narrow as well. For example, we include very little discussion of statistical methods, stochastic models, or discrete equations, but concentrate almost wholly on continuous, deterministic approaches. We emphasize that this is not from any inherent belief in the superiority of continuous differential equations. It results rather from the unpleasant fact that choices had to be made, and when push came to shove, we chose to include work with which we were most familiar. Again, apologies are offered.

Finally, with a project of this size there is credit to be given and blame to be cast; credit to the many people, like the pioneers in the field whose work we freely borrowed, and many reviewers and coworkers (Andrew LeBeau, Matthew Wilkins, Richard Bertram, Lee Segel, Bruce Knight, John Tyson, Eric Cytrunbaum, Eric Marland, Tim Lewis, J.G.T. Sneyd, Craig Marshall) who have given invaluable advice. Particular thanks are also due to the University of Canterbury, New Zealand, where a significant portion of this book was written. Of course, as authors we accept all the blame for not getting it right, or not doing it better.

University of Utah   James Keener
University of Michigan   James Sneyd
1998

Acknowledgments

With a project of this size it is impossible to give adequate acknowledgment to everyone who contributed: My family, whose patience with me is herculean; my students, who had to tolerate my rantings, ravings, and frequent mistakes; my colleagues, from whom I learned so much and often failed to give adequate attribution. Certainly the most profound contribution to this project was from the Creator who made it all possible in the first place. I don't know how He did it, but it was a truly astounding achievement. To all involved, thanks.

University of Utah   James Keener

Between the three of them, Jim Murray, Charlie Peskin and Dan Tranchina have taught me almost everything I know about mathematical physiology. This book could not have been written without them, and I thank them particularly for their, albeit unaware, contributions. Neither could this book have been written without many years of support from my parents and my wife, to whom I owe the greatest of debts.

University of Auckland   James Sneyd

Table of Contents

CONTENTS, I: Cellular Physiology

CONTENTS, II: Systems Physiology

CHAPTER 1

Biochemical Reactions

Cells can do lots of wonderful things. Individually they can move, contract, excrete, reproduce, signal or respond to signals, and carry out the energy transactions necessary for this activity. Collectively they perform all of the numerous functions of any living organism necessary to sustain life. Yet, remarkably, all of what cells do can be described in terms of a few basic natural laws. The fascination with cells is that although the rules of behavior are relatively simple, they are applied to an enormously complex network of interacting chemicals and substrates. The effort of many lifetimes has been consumed in unraveling just a few of these reaction schemes, and there are many more mysteries yet to be uncovered.

1.1 The Law of Mass Action

The fundamental "law" of a chemical reaction is the law of mass action. This law describes the rate at which chemicals, whether large macromolecules or simple ions, collide and interact to form different chemical combinations. Suppose that two chemicals, say A and B, react upon collision with each other to form product C,

$$A + B \xrightarrow{k} C. \tag{1.1}$$

The rate of this reaction is the rate of accumulation of product, $\frac{d[C]}{dt}$. This rate is the product of the number of collisions per unit time between the two reactants and the probability that a collision is sufficiently energetic to overcome the free energy of activation of the reaction. The number of collisions per unit time is taken to be proportional to the product of the concentrations of A and B with a factor of proportionality that depends on the geometrical shapes and sizes of the reactant molecules and on the temperature of the mixture. Combining these factors, we have

$$\frac{d[\text{C}]}{dt} = k[\text{A}][\text{B}]. \tag{1.2}$$

The identification of (1.2) with the reaction (1.1) is called the *law of mass action*, and the constant k is called the *rate constant* for the reaction. However, the law of mass action is not a law in the sense that it is inviolable, but rather it is a useful model, much like Ohm's law or Newton's law of cooling. As a model, there are situations in which it is not valid. For example, at high concentrations, doubling the concentration of one reactant need not double the overall reaction rate, and at extremely low concentrations, it may not be appropriate to represent concentration as a continuous variable.

For thermodynamic reasons all reactions proceed in both directions. Thus, the reaction scheme for A, B, and C should have been written as

$$\text{A} + \text{B} \underset{k_-}{\overset{k_+}{\rightleftharpoons}} \text{C}, \tag{1.3}$$

with k_+ and k_- denoting, respectively, the forward and reverse rate constants of reaction. If the reverse reaction is slow compared to the forward reaction, it is often ignored, and only the primary direction is displayed. Since the quantity A is consumed by the forward reaction and produced by the reverse reaction, the rate of change of [A] for this bidirectional reaction is

$$\frac{d[\text{A}]}{dt} = k_-[\text{C}] - k_+[\text{A}][\text{B}]. \tag{1.4}$$

At equilibrium, concentrations are not changing, so that

$$\frac{k_-}{k_+} \equiv K_{\text{eq}} = \frac{[\text{A}]_{\text{eq}}[\text{B}]_{\text{eq}}}{[\text{C}]_{\text{eq}}}. \tag{1.5}$$

The ratio k_-/k_+, denoted by K_{eq}, is called the *equilibrium constant* of the reaction. It describes the relative preference for the chemicals to be in the combined state C compared to the dissociated state. If K_{eq} is small, then at steady state most of A and B are combined to give C.

If there are no other reactions involving A and C, then $[\text{A}] + [\text{C}] = A_0$ is constant, and

$$[\text{C}]_{\text{eq}} = A_0 \frac{[\text{B}]_{\text{eq}}}{K_{\text{eq}} + [\text{B}]_{\text{eq}}}, \qquad [\text{A}]_{\text{eq}} = A_0 \frac{K_{\text{eq}}}{K_{\text{eq}} + [\text{B}]_{\text{eq}}}. \tag{1.6}$$

Thus, when $[\text{B}]_{\text{eq}} = K_{\text{eq}}$, half of A is in the bound state at equilibrium.

There are several other features of the law of mass action that need to be mentioned. Suppose that the reaction involves the dimerization of two monomers of the same species A to produce species C,

$$A + A \underset{k_-}{\overset{k_+}{\rightleftarrows}} C. \tag{1.7}$$

For every C that is made, two of A are used, and every time C degrades, two copies of A are produced. As a result, the rate of reaction for A is

$$\frac{d[A]}{dt} = 2k_-[C] - 2k_+[A]^2. \tag{1.8}$$

However, the rate of production of C is half that of A,

$$\frac{d[C]}{dt} = -\frac{1}{2}\frac{d[A]}{dt}, \tag{1.9}$$

and the quantity [A]+2[C] is conserved (provided there are no other reactions).

In a similar way, with a trimolecular reaction, the rate at which the reaction takes place is proportional to the product of three concentrations, and three molecules are consumed in the process, or released in the degradation of product. In real life, there are probably no truly trimolecular reactions. Nevertheless, there are some situations in which a reaction might be effectively modeled as trimolecular (Exercise 2).

Unfortunately, the law of mass action cannot be used in all situations, because not all chemical reaction mechanisms are known with sufficient detail. In fact, a vast number of chemical reactions cannot be described by mass action kinetics. Those reactions that follow mass action kinetics are called *elementary reactions* because presumably, they proceed directly from collision of the reactants. Reactions that do not follow mass action kinetics usually proceed by a complex mechanism consisting of several elementary reaction steps. It is often the case with biochemical reactions that the elementary reaction steps are not known or are very complicated to write down.

1.2 Thermodynamics and Rate Constants

There is a close relationship between the rate constants of a reaction and thermodynamics. The fundamental concept is that of *chemical potential*, which is the Gibbs free energy, G, per mole of a substance. Often, the Gibbs free energy per mole is denoted by μ rather than G. However, because μ has many other uses in this text, we retain the notation G for the Gibbs free energy.

For a mixture of ideal gases, X_i, the chemical potential of gas i is a function of temperature, pressure, and concentration,

$$G_i = G_i^0(T, P) + RT \ln(x_i), \tag{1.10}$$

where x_i is the mole fraction of X_i, R is the universal gas constant, T is the absolute temperature, and P is the pressure of the gas (in atmospheres); values of these constants, and their units, are given in the appendix. The quantity $G_i^0(T, P)$ is the standard free energy per mole of the pure ideal gas, i.e., when the mole fraction of the gas is 1. Note that, since $x_i \leq 1$, the free energy of an ideal gas in a mixture is always less than that of the pure ideal gas. The total Gibbs free energy of the mixture is

$$G = \sum_i n_i G_i, \tag{1.11}$$

where n_i is the number of moles of gas i.

The theory of Gibbs free energy in ideal gases can be extended to ideal dilute solutions. By redefining the standard Gibbs free energy to be the free energy at a concentration of 1 M, i.e., 1 mole per liter, we obtain

$$G = G^0 + RT \ln(c), \tag{1.12}$$

where the concentration, c, is in units of moles per liter. The standard free energy, G^0, is obtained by measuring the free energy for a dilute solution and then extrapolating to $c = 1$ M. For biochemical applications, the dependence of free energy on pressure is ignored, and the pressure is assumed to be 1 atm, while the temperature is taken to be 25°C. Derivations of these formulas can be found in physical chemistry textbooks such as Levine (2002) and Castellan (1971).

For nonideal solutions, such as are typical in cells, the free energy formula (1.12) should use the chemical activity of the solute rather than its concentration. The relationship between chemical activity a and concentration is nontrivial. However, for dilute concentrations, they are approximately equal.

Since the free energy is a potential, it denotes the preference of one state compared to another. Consider, for example, the simple reaction

$$A \longrightarrow B. \tag{1.13}$$

The change in chemical potential ΔG is defined as the difference between the chemical potential for state B (the product), denoted by G_B, and the chemical potential for state A (the reactant), denoted by G_A,

$$\begin{aligned}\Delta G &= G_B - G_A \\ &= G_B^0 - G_A^0 + RT \ln([B]) - RT \ln([A]) \\ &= \Delta G^0 + RT \ln([B]/[A]). \end{aligned} \tag{1.14}$$

The sign of ΔG is important, which is why it is defined with only one reaction direction shown, even though we know that the back reaction also occurs. In fact, there is a wonderful opportunity for confusion here, since there is no obvious way to decide which is the forward and which is the backward direction for a given reaction. If $\Delta G < 0$, then state B is preferred to state A, and the reaction tends to convert A into B, whereas, if $\Delta G > 0$, then state A is preferred to state B, and the reaction tends to convert B into A. Equilibrium occurs when neither state is preferred, so that $\Delta G = 0$, in which case

$$\frac{[B]_{eq}}{[A]_{eq}} = e^{\frac{-\Delta G^0}{RT}}. \tag{1.15}$$

Expressing this reaction in terms of forward and backward reaction rates,

$$A \underset{k_-}{\overset{k_+}{\rightleftarrows}} B, \qquad (1.16)$$

we find that in steady state, $k_+[A]_{eq} = k_-[B]_{eq}$, so that

$$\frac{[A]_{eq}}{[B]_{eq}} = \frac{k_-}{k_+} = K_{eq}. \qquad (1.17)$$

Combining this with (1.15), we observe that

$$K_{eq} = e^{\frac{\Delta G^0}{RT}}. \qquad (1.18)$$

In other words, the more negative the difference in standard free energy, the greater the propensity for the reaction to proceed from left to right, and the smaller is K_{eq}. Notice, however, that this gives only the ratio of rate constants, and not their individual amplitudes. We learn nothing about whether a reaction is fast or slow from the change in free energy.

Similar relationships hold when there are multiple components in the reaction. Consider, for example, the more complex reaction

$$\alpha A + \beta B \longrightarrow \gamma C + \delta D. \qquad (1.19)$$

The change of free energy for this reaction is defined as

$$\begin{aligned}
\Delta G &= \gamma G_C + \delta G_D - \alpha G_A - \beta G_B \\
&= \gamma G_C^0 + \delta G_D^0 - \alpha G_A^0 - \beta G_B^0 + RT \ln\left(\frac{[C]^\gamma [D]^\delta}{[A]^\alpha [B]^\beta}\right) \\
&= \Delta G^0 + RT \ln\left(\frac{[C]^\gamma [D]^\delta}{[A]^\alpha [B]^\beta}\right),
\end{aligned} \qquad (1.20)$$

and at equilibrium,

$$\Delta G^0 = RT \ln\left(\frac{[A]_{eq}^\alpha [B]_{eq}^\beta}{[C]_{eq}^\gamma [D]_{eq}^\delta}\right) = RT \ln(K_{eq}). \qquad (1.21)$$

An important example of such a reaction is the hydrolysis of adenosine triphosphate (ATP) to adenosine diphosphate (ADP) and inorganic phosphate P_i, represented by the reaction

$$ATP \longrightarrow ADP + P_i. \qquad (1.22)$$

The standard free energy change for this reaction is

$$\Delta G^0 = G_{ADP}^0 + G_{P_i}^0 - G_{ATP}^0 = -31.0 \text{ kJ mol}^{-1}, \qquad (1.23)$$

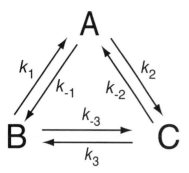

Figure 1.1   Schematic diagram of a reaction loop.

and from this we could calculate the equilibrium constant for this reaction. However, the primary significance of this is not the size of the equilibrium constant, but rather the fact that ATP has free energy that can be used to drive other less favorable reactions. For example, in all living cells, ATP is used to pump ions against their concentration gradient, a process called free energy transduction. In fact, if the equilibrium constant of this reaction is achieved, then one can confidently assert that the system is dead. In living systems, the ratio of [ATP] to [ADP][P_i] is held well above the equilibrium value.

1.3   Detailed Balance

Suppose that a set of reactions forms a loop, as shown in Fig. 1.1. By applying the law of mass action and setting the derivatives to zero we can find the steady-state concentrations of A, B and C. However, for the system to be in thermodynamic equilibrium a stronger condition must hold. Thermodynamic equilibrium requires that the free energy of each state be the same so that each individual reaction is in equilibrium. In other words, at equilibrium there is not only, say, no net change in [B], there is also no net conversion of B to C or B to A. This condition means that, at equilibrium, $k_1[\text{B}] = k_{-1}[\text{A}]$, $k_2[\text{A}] = k_{-2}[\text{C}]$ and $k_3[\text{C}] = k_{-3}[\text{B}]$. Thus, it must be that

$$k_1 k_2 k_3 = k_{-1} k_{-2} k_{-3}, \tag{1.24}$$

or

$$K_1 K_2 K_3 = 1, \tag{1.25}$$

where $K_i = k_{-i}/k_i$. Since this condition does not depend on the concentrations of A, B or C, it must hold in general, not only at equilibrium.

For a more general reaction loop, the principle of detailed balance requires that the product of rates in one direction around the loop must equal the product of rates in the other direction. If any of the rates are dependent on concentrations of other chemicals, those concentrations must also be included.

1.4 Enzyme Kinetics

To see where some of the more complicated reaction schemes come from, we consider a reaction that is catalyzed by an enzyme. Enzymes are catalysts (generally proteins) that help convert other molecules called *substrates* into products, but they themselves are not changed by the reaction. Their most important features are catalytic power, specificity, and regulation. Enzymes accelerate the conversion of substrate into product by lowering the free energy of activation of the reaction. For example, enzymes may aid in overcoming charge repulsions and allowing reacting molecules to come into contact for the formation of new chemical bonds. Or, if the reaction requires breaking of an existing bond, the enzyme may exert a stress on a substrate molecule, rendering a particular bond more easily broken. Enzymes are particularly efficient at speeding up biological reactions, giving increases in speed of up to 10 million times or more. They are also highly specific, usually catalyzing the reaction of only one particular substrate or closely related substrates. Finally, they are typically regulated by an enormously complicated set of positive and negative feedbacks, thus allowing precise control over the rate of reaction. A detailed presentation of enzyme kinetics, including many different kinds of models, can be found in Dixon and Webb (1979), the encyclopedic Segel (1975) or Kernevez (1980). Here, we discuss only some of the simplest models.

One of the first things one learns about enzyme reactions is that they do not follow the law of mass action directly. For, as the concentration of substrate (S) is increased, the rate of the reaction increases only to a certain extent, reaching a maximal reaction velocity at high substrate concentrations. This is in contrast to the law of mass action, which, when applied directly to the reaction of S with the enzyme E

$$S + E \longrightarrow P + E$$

predicts that the reaction velocity increases linearly as [S] increases.

A model to explain the deviation from the law of mass action was first proposed by Michaelis and Menten (1913). In their reaction scheme, the enzyme E converts the substrate S into the product P through a two-step process. First E combines with S to form a complex C which then breaks down into the product P releasing E in the process. The reaction scheme is represented schematically by

$$S + E \underset{k_{-1}}{\overset{k_1}{\rightleftharpoons}} C \underset{k_{-2}}{\overset{k_2}{\rightleftharpoons}} P + E.$$

Although all reactions must be reversible, as shown here, reaction rates are typically measured under conditions where P is continually removed, which effectively prevents the reverse reaction from occurring. Thus, it often suffices to assume that no reverse reaction occurs. For this reason, the reaction is usually written as

$$S + E \underset{k_{-1}}{\overset{k_1}{\rightleftharpoons}} C \overset{k_2}{\longrightarrow} P + E.$$

The reversible case is considered in Section 1.4.5.

There are two similar, but not identical, ways to analyze this equation; the equilibrium approximation, and the quasi-steady-state approximation. Because these methods give similar results it is easy to confuse them, so it is important to understand their differences.

We begin by defining $s = [S]$, $c = [C]$, $e = [E]$, and $p = [P]$. The law of mass action applied to this reaction mechanism gives four differential equations for the rates of change of $s, c, e,$ and p,

$$\frac{ds}{dt} = k_{-1}c - k_1 se, \tag{1.26}$$

$$\frac{de}{dt} = (k_{-1} + k_2)c - k_1 se, \tag{1.27}$$

$$\frac{dc}{dt} = k_1 se - (k_2 + k_{-1})c, \tag{1.28}$$

$$\frac{dp}{dt} = k_2 c. \tag{1.29}$$

Note that p can be found by direct integration, and that there is a conserved quantity since $\frac{de}{dt} + \frac{dc}{dt} = 0$, so that $e + c = e_0$, where e_0 is the total amount of available enzyme.

1.4.1   The Equilibrium Approximation

In their original analysis, Michaelis and Menten assumed that the substrate is in instantaneous equilibrium with the complex, and thus

$$k_1 se = k_{-1}c. \tag{1.30}$$

Since $e + c = e_0$, we find that

$$c = \frac{e_0 s}{K_1 + s}, \tag{1.31}$$

where $K_1 = k_{-1}/k_1$. Hence, the velocity, V, of the reaction, i.e., the rate at which the product is formed, is given by

$$V = \frac{dp}{dt} = k_2 c = \frac{k_2 e_0 s}{K_1 + s} = \frac{V_{\max}s}{K_1 + s}, \tag{1.32}$$

where $V_{\max} = k_2 e_0$ is the maximum reaction velocity, attained when all the enzyme is complexed with the substrate.

At small substrate concentrations, the reaction rate is linear, at a rate proportional to the amount of available enzyme e_0. At large concentrations, however, the reaction rate saturates to $V_{\max}$, so that the maximum rate of the reaction is limited by the amount of enzyme present and the dissociation rate constant k_2. For this reason, the dissociation reaction $C \xrightarrow{k_2} P + E$ is said to be *rate limiting* for this reaction. At $s = K_1$, the reaction rate is half that of the maximum.

It is important to note that (1.30) cannot be exactly correct at all times; if it were, then according to (1.26) substrate would not be used up, and product would not be formed. This points out the fact that (1.30) is an approximation. It also illustrates the need for a systematic way to make approximate statements, so that one has an idea of the magnitude and nature of the errors introduced in making such an approximation.

It is a common mistake with the equilibrium approximation to conclude that since (1.30) holds, it must be that $\frac{ds}{dt} = 0$, which if this is true, implies that no substrate is being used up, nor product produced. Furthermore, it appears that if (1.30) holds, then it must be (from (1.28)) that $\frac{dc}{dt} = -k_2c$, which is also false. Where is the error here?

The answer lies with the fact that the equilibrium approximation is equivalent to the assumption that the reaction (1.26) is a very fast reaction, faster than others, or more precisely, that $k_{-1} \gg k_2$. Adding (1.26) and (1.28), we find that

$$\frac{ds}{dt} + \frac{dc}{dt} = -k_2c, \tag{1.33}$$

expressing the fact that the total quantity $s + c$ changes on a slower time scale. Now when we use that $c = \frac{e_0 s}{K_s + s}$, we learn that

$$\frac{d}{dt}\left(s + \frac{e_0 s}{K_1 + s}\right) = -k_2\frac{e_0 s}{K_1 + s}, \tag{1.34}$$

and thus,

$$\frac{ds}{dt}\left(1 + \frac{e_0 K_1}{(K_1 + s)^2}\right) = -k_2\frac{e_0 s}{K_1 + s}, \tag{1.35}$$

which specifies the rate at which s is consumed.

This way of simplifying reactions by using an equilibrium approximation is used many times throughout this book, and is an extremely important technique, particularly in the analysis of Markov models of ion channels, pumps and exchangers (Chapters 2 and 3). A more mathematically systematic description of this approach is left for Exercise 20.

1.4.2   The Quasi-Steady-State Approximation

An alternative analysis of an enzymatic reaction was proposed by Briggs and Haldane (1925) who assumed that the rates of formation and breakdown of the complex were essentially equal at all times (except perhaps at the beginning of the reaction, as the complex is "filling up"). Thus, dc/dt should be approximately zero.

To give this approximation a systematic mathematical basis, it is useful to introduce dimensionless variables

$$\sigma = \frac{s}{s_0}, \quad x = \frac{c}{e_0}, \quad \tau = k_1 e_0 t, \quad \kappa = \frac{k_{-1} + k_2}{k_1 s_0}, \quad \epsilon = \frac{e_0}{s_0}, \quad \alpha = \frac{k_{-1}}{k_1 s_0}, \tag{1.36}$$

in terms of which we obtain the system of two differential equations

$$\frac{d\sigma}{d\tau} = -\sigma + x(\sigma + \alpha),\tag{1.37}$$

$$\epsilon\frac{dx}{d\tau} = \sigma - x(\sigma + \kappa).\tag{1.38}$$

There are usually a number of ways that a system of differential equations can be nondimensionalized. This nonuniqueness is often a source of great confusion, as it is often not obvious which choice of dimensionless variables and parameters is "best." In Section 1.6 we discuss this difficult problem briefly.

The remarkable effectiveness of enzymes as catalysts of biochemical reactions is reflected by their small concentrations needed compared to the concentrations of the substrates. For this model, this means that ϵ is small, typically in the range of 10^{-2} to 10^{-7}. Therefore, the reaction (1.38) is fast, equilibrates rapidly and remains in near-equilibrium even as the variable σ changes. Thus, we take the *quasi-steady-state approximation* $\epsilon\frac{dx}{d\tau} = 0$. Notice that this is *not* the same as taking $\frac{dx}{d\tau} = 0$. However, because of the different scaling of x and c, it is equivalent to taking $\frac{dc}{dt} = 0$ as suggested in the introductory paragraph.

One useful way of looking at this system is as follows; since

$$\frac{dx}{d\tau} = \frac{\sigma - x(\sigma + \kappa)}{\epsilon},\tag{1.39}$$

$dx/d\tau$ is large everywhere, except where $\sigma - x(\sigma + \kappa)$ is small, of approximately the same size as ϵ. Now, note that $\sigma - x(\sigma + \kappa) = 0$ defines a curve in the σ, x phase plane, called the *slow manifold* (as illustrated in the right panel of Fig. 1.14). If the solution starts away from the slow manifold, $dx/d\tau$ is initially large, and the solution moves rapidly to the vicinity of the slow manifold. The solution then moves along the slow manifold in the direction defined by the equation for σ; in this case, σ is decreasing, and so the solution moves to the left along the slow manifold.

Another way of looking at this model is to notice that the reaction of x is an exponential process with time constant at least as large as $\frac{\kappa}{\epsilon}$. To see this we write (1.38) as

$$\epsilon\frac{dx}{d\tau} + \kappa x = \sigma(1 - x).\tag{1.40}$$

Thus, the variable x "tracks" the steady state with a short delay.

It follows from the quasi-steady-state approximation that

$$x = \frac{\sigma}{\sigma + \kappa},\tag{1.41}$$

$$\frac{d\sigma}{d\tau} = -\frac{q\sigma}{\sigma + \kappa},\tag{1.42}$$

where $q = \kappa - \alpha = \frac{k_2}{k_1 s_0}$. Equation (1.42) describes the rate of uptake of the substrate and is called a *Michaelis–Menten law*. In terms of the original variables, this law is

$$V = \frac{dp}{dt} = -\frac{ds}{dt} = \frac{k_2 e_0 s}{s + K_m} = \frac{V_{\max} s}{s + K_m}, \tag{1.43}$$

where $K_m = \frac{k_{-1} + k_2}{k_1}$. In quasi-steady state, the concentration of the complex satisfies

$$c = \frac{e_0 s}{s + K_m}. \tag{1.44}$$

Note the similarity between (1.32) and (1.43), the only difference being that the equilibrium approximation uses K_1, while the quasi-steady-state approximation uses K_m. Despite this similarity of form, it is important to keep in mind that the two results are based on different approximations. The equilibrium approximation assumes that $k_{-1} \gg k_2$ whereas the quasi-steady-state approximation assumes that $\epsilon \ll 1$. Notice, that if $k_{-1} \gg k_2$, then $K_m \approx K_1$, so that the two approximations give similar results.

As with the law of mass action, the Michaelis–Menten law (1.43) is not universally applicable but is a useful approximation. It may be applicable even if $\epsilon = e_0/s_0$ is not small (see, for example, Exercise 14), and in model building it is often invoked without regard to the underlying assumptions.

While the individual rate constants are difficult to measure experimentally, the ratio K_m is relatively easy to measure because of the simple observation that (1.43) can be written in the form

$$\frac{1}{V} = \frac{1}{V_{\max}} + \frac{K_m}{V_{\max}} \frac{1}{s}. \tag{1.45}$$

In other words, $1/V$ is a linear function of $1/s$. Plots of this double reciprocal curve are called *Lineweaver–Burk plots*, and from such (experimentally determined) plots, $V_{\max}$ and K_m can be estimated.

Although a Lineweaver–Burk plot makes it easy to determine $V_{\max}$ and K_m from reaction rate measurements, it is not a simple matter to determine the reaction rate as a function of substrate concentration during the course of a single experiment. Substrate concentrations usually cannot be measured with sufficient accuracy or time resolution to permit the calculation of a reliable derivative. In practice, since it is more easily measured, the initial reaction rate is determined for a range of different initial substrate concentrations.

An alternative method to determine K_m and $V_{\max}$ from experimental data is the direct linear plot (Eisenthal and Cornish-Bowden, 1974; Cornish-Bowden and Eisenthal, 1974). First we write (1.43) in the form

$$V_{\max} = V + \frac{V}{s} K_m, \tag{1.46}$$

and then treat V_{max} and K_m as variables for each experimental measurement of V and s. (Recall that typically only the initial substrate concentration and initial velocity are used.) Then a plot of the straight line of V_{max} against K_m can be made. Repeating this for a number of different initial substrate concentrations and velocities gives a family of straight lines, which, in an ideal world free from experimental error, intersect at the single point V_{max} and K_m for that reaction. In reality, experimental error precludes an exact intersection, but V_{max} and K_m can be estimated from the median of the pairwise intersections.

1.4.3 Enzyme Inhibition

An enzyme inhibitor is a substance that inhibits the catalytic action of the enzyme. Enzyme inhibition is a common feature of enzyme reactions, and is an important means by which the activity of enzymes is controlled. Inhibitors come in many different types. For example, *irreversible inhibitors*, or *catalytic poisons*, decrease the activity of the enzyme to zero. This is the method of action of cyanide and many nerve gases. For this discussion, we restrict our attention to *competitive* inhibitors and *allosteric* inhibitors.

To understand the distinction between competitive and allosteric inhibition, it is useful to keep in mind that an enzyme molecule is usually a large protein, considerably larger than the substrate molecule whose reaction is catalyzed. Embedded in the large enzyme protein are one or more *active sites*, to which the substrate can bind to form the complex. In general, an enzyme catalyzes a single reaction of substrates with similar structures. This is believed to be a steric property of the enzyme that results from the three-dimensional shape of the enzyme allowing it to fit in a "lock-and-key" fashion with a corresponding substrate molecule.

If another molecule has a shape similar enough to that of the substrate molecule, it may also bind to the active site, preventing the binding of a substrate molecule, thus inhibiting the reaction. Because the inhibitor competes with the substrate molecule for the active site, it is called a competitive inhibitor.

However, because the enzyme molecule is large, it often has other binding sites, distinct from the active site, the binding of which affects the activity of the enzyme at the active site. These binding sites are called *allosteric* sites (from the Greek for "another solid") to emphasize that they are structurally different from the catalytic active sites. They are also called *regulatory sites* to emphasize that the catalytic activity of the protein is regulated by binding at this allosteric site. The ligand (any molecule that binds to a specific site on a protein, from Latin *ligare*, to bind) that binds at the allosteric site is called an *effector* or *modifier*, which, if it increases the activity of the enzyme, is called an allosteric activator, while if it decreases the activity of the enzyme, is called an allosteric inhibitor. The allosteric effect is presumed to arise because of a conformational change of the enzyme, that is, a change in the folding of the polypeptide chain, called an *allosteric transition*.

Competitive Inhibition

In the simplest example of a competitive inhibitor, the reaction is stopped when the inhibitor is bound to the active site of the enzyme. Thus,

$$S + E \underset{k_{-1}}{\overset{k_1}{\rightleftharpoons}} C_1 \overset{k_2}{\longrightarrow} E + P,$$

$$E + I \underset{k_{-3}}{\overset{k_3}{\rightleftharpoons}} C_2.$$

Using the law of mass action we find

$$\frac{ds}{dt} = -k_1 se + k_{-1} c_1, \tag{1.47}$$

$$\frac{di}{dt} = -k_3 ie + k_{-3} c_2, \tag{1.48}$$

$$\frac{dc_1}{dt} = k_1 se - (k_{-1} + k_2) c_1, \tag{1.49}$$

$$\frac{dc_2}{dt} = k_3 ie - k_{-3} c_2. \tag{1.50}$$

where $s = [S]$, $c_1 = [C_1]$, and $c_2 = [C_2]$. We know that $e + c_1 + c_2 = e_0$, so an equation for the dynamics of e is superfluous. As before, it is not necessary to write an equation for the accumulation of the product. To be systematic, the next step is to introduce dimensionless variables, and identify those reactions that are rapid and equilibrate rapidly to their quasi-steady states. However, from our previous experience (or from a calculation on a piece of scratch paper), we know, assuming the enzyme-to-substrate ratios are small, that the fast equations are those for c_1 and c_2. Hence, the quasi-steady states are found by (formally) setting $dc_1/dt = dc_2/dt = 0$ and solving for c_1 and c_2. Recall that this does *not* mean that c_1 and c_2 are unchanging, rather that they are changing in quasi-steady-state fashion, keeping the right-hand sides of these equations nearly zero. This gives

$$c_1 = \frac{K_i e_0 s}{K_m i + K_i s + K_m K_i}, \tag{1.51}$$

$$c_2 = \frac{K_m e_0 i}{K_m i + K_i s + K_m K_i}, \tag{1.52}$$

where $K_m = \frac{k_{-1} + k_2}{k_1}$, $K_i = k_{-3}/k_3$. Thus, the velocity of the reaction is

$$V = k_2 c_1 = \frac{k_2 e_0 s K_i}{K_m i + K_i s + K_m K_i} = \frac{V_{\max} s}{s + K_m (1 + i/K_i)}. \tag{1.53}$$

Notice that the effect of the inhibitor is to increase the effective equilibrium constant of the enzyme by the factor $1 + i/K_i$, from K_m to $K_m(1 + i/K_i)$, thus decreasing the velocity of reaction, while leaving the maximum velocity unchanged.

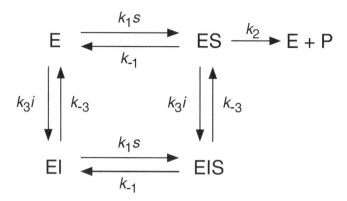

Figure 1.2   Diagram of the possible states of an enzyme with one allosteric and one catalytic binding site.

Allosteric Inhibitors

If the inhibitor can bind at an allosteric site, we have the possibility that the enzyme could bind both the inhibitor and the substrate simultaneously. In this case, there are four possible binding states for the enzyme, and transitions between them, as demonstrated graphically in Fig. 1.2.

The simplest analysis of this reaction scheme is the equilibrium analysis. (The more complicated quasi-steady-state analysis is left for Exercise 6.) We define $K_s = k_{-1}/k_1$, $K_i = k_{-3}/k_3$, and let x, y, and z denote, respectively, the concentrations of ES, EI and EIS. Then, it follows from the law of mass action that at equilibrium (take each of the 4 transitions to be at equilibrium),

$$(e_0 - x - y - z)s - K_s x = 0, \tag{1.54}$$

$$(e_0 - x - y - z)i - K_i y = 0, \tag{1.55}$$

$$ys - K_s z = 0, \tag{1.56}$$

$$xi - K_i z = 0, \tag{1.57}$$

where $e_0 = e + x + y + z$ is the total amount of enzyme. Notice that this is a linear system of equations for x, y, and z. Although there are four equations, one is a linear combination of the other three (the system is of rank three), so that we can determine x, y, and z as functions of i and s, finding

$$x = \frac{e_0 K_i}{K_i + i} \frac{s}{K_s + s}. \tag{1.58}$$

It follows that the reaction rate, $V = k_2 x$, is given by

$$V = \frac{V_{max}}{1 + i/K_i} \frac{s}{K_s + s}, \tag{1.59}$$

where $V_{max} = k_2 e_0$. Thus, in contrast to the competitive inhibitor, the allosteric inhibitor decreases the maximum velocity of the reaction, while leaving K_s unchanged.

(The situation is more complicated if the quasi-steady-state approximation is used, and no such simple conclusion follows.)

1.4.4 Cooperativity

For many enzymes, the reaction velocity is not a simple hyperbolic curve, as predicted by the Michaelis–Menten model, but often has a sigmoidal character. This can result from cooperative effects, in which the enzyme can bind more than one substrate molecule but the binding of one substrate molecule affects the binding of subsequent ones.

Much of the original theoretical work on cooperative behavior was stimulated by the properties of hemoglobin, and this is often the context in which cooperativity is discussed. A detailed discussion of hemoglobin and oxygen binding is given in Chapter 13, while here cooperativity is discussed in more general terms.

Suppose that an enzyme can bind two substrate molecules, so it can exist in one of three states, namely as a free molecule E, as a complex with one occupied binding site, C_1, and as a complex with two occupied binding sites, C_2. The reaction mechanism is then

$$S + E \underset{k_{-1}}{\overset{k_1}{\rightleftharpoons}} C_1 \overset{k_2}{\rightarrow} E + P, \tag{1.60}$$

$$S + C_1 \underset{k_{-3}}{\overset{k_3}{\rightleftharpoons}} C_2 \overset{k_4}{\rightarrow} C_1 + P. \tag{1.61}$$

Using the law of mass action, one can write the rate equations for the 5 concentrations [S], [E], [C_1], [C_2], and [P]. However, because the amount of product [P] can be determined by quadrature, and because the total amount of enzyme molecule is conserved, we only need three equations for the three quantities [S], [C_1], and [C_2]. These are

$$\frac{ds}{dt} = -k_1 se + k_{-1} c_1 - k_3 sc_1 + k_{-3} c_2, \tag{1.62}$$

$$\frac{dc_1}{dt} = k_1 se - (k_{-1} + k_2)c_1 - k_3 sc_1 + (k_4 + k_{-3})c_2, \tag{1.63}$$

$$\frac{dc_2}{dt} = k_3 sc_1 - (k_4 + k_{-3})c_2, \tag{1.64}$$

where $s = [S], c_1 = [C_1], c_2 = [C_2]$, and $e + c_1 + c_2 = e_0$.

Proceeding as before, we invoke the quasi-steady-state assumption that $dc_1/dt = dc_2/dt = 0$, and solve for c_1 and c_2 to get

$$c_1 = \frac{K_2 e_0 s}{K_1 K_2 + K_2 s + s^2}, \tag{1.65}$$

$$c_2 = \frac{e_0 s^2}{K_1 K_2 + K_2 s + s^2}, \tag{1.66}$$

where $K_1 = \frac{k_{-1}+k_2}{k_1}$ and $K_2 = \frac{k_4+k_{-3}}{k_3}$. The reaction velocity is thus given by

$$V = k_2 c_1 + k_4 c_2 = \frac{(k_2 K_2 + k_4 s)e_0 s}{K_1 K_2 + K_2 s + s^2}. \tag{1.67}$$

Use of the equilibrium approximation to simplify this reaction scheme gives, as expected, similar results, in which the formula looks the same, but with different definitions of K_1 and K_2 (Exercise 10).

It is instructive to examine two extreme cases. First, if the binding sites act independently and identically, then $k_1 = 2k_3 = 2k_+$, $2k_{-1} = k_{-3} = 2k_-$ and $2k_2 = k_4$, where k_+ and k_- are the forward and backward reaction rates for the individual binding sites. The factors of 2 occur because two identical binding sites are involved in the reaction, doubling the amount of the reactant. In this case,

$$V = \frac{2k_2 e_0 (K+s)s}{K^2 + 2Ks + s^2} = 2\frac{k_2 e_0 s}{K+s}, \tag{1.68}$$

where $K = \frac{k_-+k_2}{k_+}$ is the K_m of the individual binding site. As expected, the rate of reaction is exactly twice that for the individual binding site.

In the opposite extreme, suppose that the binding of the first substrate molecule is slow, but that with one site bound, binding of the second is fast (this is large positive cooperativity). This can be modeled by letting $k_3 \to \infty$ and $k_1 \to 0$, while keeping $k_1 k_3$ constant, in which case $K_2 \to 0$ and $K_1 \to \infty$ while $K_1 K_2$ is constant. In this limit, the velocity of the reaction is

$$V = \frac{k_4 e_0 s^2}{K_m^2 + s^2} = \frac{V_{max} s^2}{K_m^2 + s^2}, \tag{1.69}$$

where $K_m^2 = K_1 K_2$, and $V_{max} = k_4 e_0$.

In general, if n substrate molecules can bind to the enzyme, there are n equilibrium constants, K_1 through K_n. In the limit as $K_n \to 0$ and $K_1 \to \infty$ while keeping $K_1 K_n$ fixed, the rate of reaction is

$$V = \frac{V_{max} s^n}{K_m^n + s^n}, \tag{1.70}$$

where $K_m^n = \Pi_{i=1}^n K_i$. This rate equation is known as the *Hill equation*. Typically, the Hill equation is used for reactions whose detailed intermediate steps are not known but for which cooperative behavior is suspected. The exponent n and the parameters V_{max} and K_m are usually determined from experimental data. Observe that

$$n \ln s = n \ln K_m + \ln \left(\frac{V}{V_{max} - V} \right), \tag{1.71}$$

so that a plot of $\ln(\frac{V}{V_{max}-V})$ against $\ln s$ (called a *Hill plot*) should be a straight line of slope n. Although the exponent n suggests an n-step process (with n binding sites), in practice it is not unusual for the best fit for n to be noninteger.

An enzyme can also exhibit negative cooperativity (Koshland and Hamadani, 2002), in which the binding of the first substrate molecule *decreases* the rate of subsequent

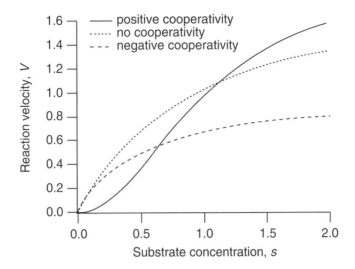

Figure 1.3   Reaction velocity plotted against substrate concentration, for three different cases. Positive cooperativity, $K_1 = 1000$, $K_2 = 0.001$; independent binding sites, $K_1 = 0.5$, $K_2 = 2$; and negative cooperativity, $K_1 = 0.5$, $K_2 = 100$. The other parameters are $e_0 = 1$, $k_2 = 1$, $k_4 = 2$. Concentration and time units are arbitrary.

binding. This can be modeled by decreasing k_3. In Fig. 1.3 we plot the reaction velocity against the substrate concentration for the cases of independent binding sites (no cooperativity), extreme positive cooperativity (the Hill equation), and negative cooperativity. From this figure it can be seen that with positive cooperativity, the reaction velocity is a sigmoidal function of the substrate concentration, while negative cooperativity primarily decreases the overall reaction velocity.

The Monod–Wyman–Changeux Model

Cooperative effects occur when the binding of one substrate molecule alters the rate of binding of subsequent ones. However, the above models give no explanation of how such alterations in the binding rate occur. The earliest model proposed to account for cooperative effects in terms of the enzyme's conformation was that of Monod, Wyman, and Changeux (1965). Their model is based on the following assumptions about the structure and behavior of enzymes.

1. Cooperative proteins are composed of several identical reacting units, called *protomers*, or subunits, each containing one binding site, that occupy equivalent positions within the protein.
2. The protein has two conformational states, usually denoted by R and T, which differ in their ability to bind ligands.

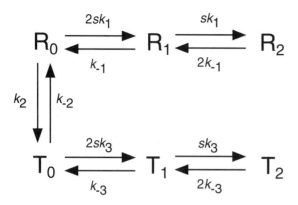

Figure 1.4 Diagram of the states of the protein, and the possible transitions, in a six-state Monod–Wyman–Changeux model.

3. If the binding of a ligand to one protomer induces a conformational change in that protomer, an identical conformational change is induced in all protomers. Because of this assumption, Monod–Wyman–Changeux (MWC) models are often called *concerted* models, as each subunit acts in concert with the others.

    To illustrate how these assumptions can be quantified, we consider a protein with two binding sites. Thus, the protein can exist in one of six states: $R_i, i = 0, 1, 2$, or $T_i, i = 0, 1, 2$, where the subscript i is the number of bound ligands. (In the original model of Monod, Wyman and Changeux, R denoted a *relaxed* state, while T denoted a *tense* state.) For simplicity, we also assume that R_1 cannot convert directly to T_1, or vice versa, and similarly for R_2 and T_2. The general case is left for Exercise 7. The states of the protein and the allowable transitions are illustrated in Fig. 1.4. As with other enzyme models, we assume that the production rate of product is proportional to the amount of substrate that is bound to the enzyme.

    We now assume that all the reactions are in equilibrium. We let a lowercase letter denote a concentration, and thus r_i and t_i denote the concentrations of chemical species R_i and T_i respectively. Also, as before, we let s denote the concentration of the substrate. Then, the fraction Y of occupied sites (also called the *saturation function*) is

$$Y = \frac{r_1 + 2r_2 + t_1 + 2t_2}{2(r_0 + r_1 + r_2 + t_0 + t_1 + t_2)}. \tag{1.72}$$

(This is also proportional to the production rate of product.) Furthermore, with $K_i = k_{-i}/k_i$, for $i = 1, 2, 3$, we find that

$$r_1 = 2sK_1^{-1}r_0, \qquad r_2 = s^2 K_1^{-2} r_0, \tag{1.73}$$

$$t_1 = 2sK_3^{-1}t_0, \qquad t_2 = s^2 K_3^{-2} t_0. \tag{1.74}$$

Substituting these into (1.72) gives

$$Y = \frac{sK_1^{-1}(1 + sK_1^{-1}) + K_2^{-1}[sK_3^{-1}(1 + sK_3^{-1})]}{(1 + sK_1^{-1})^2 + K_2^{-1}(1 + sK_3^{-1})^2}, \tag{1.75}$$

where we have used that $r_0/t_0 = K_2$. More generally, if there are n binding sites, then

$$Y = \frac{sK_1^{-1}(1 + sK_1^{-1})^{n-1} + K_2^{-1}[sK_3^{-1}(1 + sK_3^{-1})^{n-1}]}{(1 + sK_1^{-1})^n + K_2^{-1}(1 + sK_3^{-1})^n}. \tag{1.76}$$

In general, Y is a sigmoidal function of s.

It is not immediately apparent how cooperative binding kinetics arises from this model. After all, each binding site in the R conformation is identical, as is each binding site in the T conformation. In order to get cooperativity it is necessary that the binding affinity of the R conformation be different from that of the T conformation. In the special case that the binding affinities of the R and T conformations are equal (i.e., $K_1 = K_3 = K$, say) the binding curve (1.76) reduces to

$$Y = \frac{s}{K + s}, \tag{1.77}$$

which is simply noncooperative Michaelis–Menten kinetics.

Suppose that one conformation, T say, binds the substrate with a higher affinity than does R. Then, when the substrate concentration increases, T_0 is pushed through to T_1 faster than R_0 is pushed to R_1, resulting in an increase in the amount of substrate bound to the T state, and thus increased overall binding of substrate. Hence the cooperative behavior of the model.

If $K_2 = \infty$, so that only one conformation exists, then once again the saturation curve reduces to the Michaelis–Menten equation, $Y = s/(s + K_1)$. Hence each conformation, by itself, has noncooperative Michaelis–Menten binding kinetics. It is only when the overall substrate binding can be biased to one conformation or the other that cooperativity appears.

Interestingly, MWC models cannot exhibit negative cooperativity. No matter whether $K_1 > K_3$ or *vice versa*, the binding curve always exhibits positive cooperativity.

The Koshland–Nemethy–Filmer model

One alternative to the MWC model is that proposed by Koshland, Nemethy and Filmer in 1966 (the KNF model). Instead of requiring that all subunit transitions occur in concert, as in the MWC model, the KNF model assumes that substrate binding to one subunit causes a conformational change in that subunit only, and that this conformational change causes a change in the binding affinity of the neighboring subunits. Thus, in the KNF model, each subunit can be in a different conformational state, and transitions from one state to the other occur sequentially as more substrate is bound. For this reason KNF models are often called *sequential* models. The increased generality of the KNF model allows for the possibility of negative cooperativity, as the binding to one subunit can *decrease* the binding affinity of its neighbors.

When binding shows positive cooperativity, it has proven difficult to distinguish between the MWC and KNF models on the basis of experimental data. In one of the most intensely studied cooperative mechanisms, that of oxygen binding to hemoglobin, there is experimental evidence for both models, and the actual mechanism is probably a combination of both.

There are many other models of enzyme cooperativity, and the interested reader is referred to Dixon and Webb (1979) for a comprehensive discussion and comparison of other models in the literature.

1.4.5 Reversible Enzyme Reactions

Since all enzyme reactions are reversible, a general understanding of enzyme kinetics must take this reversibility into account. In this case, the reaction scheme is

$$S + E \underset{k_{-1}}{\overset{k_1}{\rightleftarrows}} C \underset{k_{-2}}{\overset{k_2}{\rightleftarrows}} P + E.$$

Proceeding as usual, we let $e + c = e_0$ and make the quasi-steady-state assumption

$$0 = \frac{dc}{dt} = k_1 s(e_0 - c) - (k_{-1} + k_2)c + k_{-2}p(e_0 - c), \tag{1.78}$$

from which it follows that

$$c = \frac{e_0(k_1 s + k_{-2}p)}{k_1 s + k_{-2}p + k_{-1} + k_2}. \tag{1.79}$$

The reaction velocity, $V = \frac{dP}{dt} = k_2 c - k_{-2}pe$, can then be calculated to be

$$V = e_0 \frac{k_1 k_2 s - k_{-1}k_{-2}p}{k_1 s + k_{-2}p + k_{-1} + k_2}. \tag{1.80}$$

When p is small (e.g., if product is continually removed), the reverse reaction is negligible and we get the previous answer (1.43).

In contrast to the irreversible case, the equilibrium and quasi-steady-state assumptions for reversible enzyme kinetics give qualitatively different answers. If we assume that S, E, and C are in fast equilibrium (instead of assuming that C is at quasi-steady state) we get

$$k_1 s(e_0 - c) = k_{-1}c, \tag{1.81}$$

from which it follows that

$$V = k_2 c - k_{-2}p(e_0 - c) = e_0 \frac{k_1 k_2 s - k_{-1}k_{-2}p}{k_1 s + k_{-1}}. \tag{1.82}$$

Comparing this to (1.80), we see that the quasi-steady-state assumption gives additional terms in the denominator involving the product p. These differences result from the assumption underlying the fast-equilibrium assumption, that k_{-1} and k_1 are both substantially larger than k_{-2} and k_2, respectively. Which of these approximations is best depends, of course, on the details of the reaction.

Calculation of the equations for a reversible enzyme reaction in which the enzyme has multiple binding sites is left for the exercises (Exercise 11).

1.4.6 The Goldbeter–Koshland Function

As is seen in many ways in this book, cooperativity is an important ingredient in the construction of biochemical switches. However, highly sensitive switching behavior requires large Hill coefficients, which would seem to require multiple interacting enzymes or binding sites, making these unlikely to occur. An alternative mechanism by which highly sensitive switching behavior is possible, suggested by Goldbeter and Koshland (1981), uses only two enzymatic transitions. In this model reaction, a substrate can be in one of two forms, say W and W*, and transferred from state W to W* by one enzyme, say E_1, and transferred from state W* to W by another enzyme, say E_2. For example, W* could be a phosphorylated state of some enzyme, E_1 could be the kinase that phosphorylates W, and E_2 could be the phosphatase that dephosphorylates W*. Numerous reactions of this type are described in Chapter 10, where W is itself an enzyme whose activity is determined by its phosphorylation state. Thus, the reaction scheme is

$$W + E_1 \underset{k_{-1}}{\overset{k_1}{\rightleftharpoons}} C_1 \overset{k_2}{\longrightarrow} E_1 + W^*,$$

$$W^* + E_2 \underset{k_{-3}}{\overset{k_3}{\rightleftharpoons}} C_2 \overset{k_4}{\longrightarrow} E_2 + W.$$

Although the full analysis of this reaction scheme is not particularly difficult, a simplified analysis quickly shows the salient features. If we suppose that the enzyme reactions take place at Michaelis–Menten rates, the reaction simplifies to

$$W \underset{r_{-1}}{\overset{r_1}{\rightleftharpoons}} W^*, \tag{1.83}$$

where

$$r_1 = \frac{V_1 E_1}{K_1 + W}, \qquad r_{-1} = \frac{V_2 E_2}{K_2 + W^*}, \tag{1.84}$$

and the concentration of W is governed by the differential equation

$$\frac{dW}{dt} = r_{-1}(W_t - W) - r_1 W, \tag{1.85}$$

where $W + W^* = W_t$. In steady state, the forward and backward reaction rates are the same, leading to the equation

$$\frac{V_1 E_1}{V_2 E_2} = \frac{W^*(K_1 + W)}{W(K_2 + W^*)}. \tag{1.86}$$

This can be rewritten as

$$\frac{v_1}{v_2} = \frac{(1 - y)(\hat{K}_1 + y)}{y(\hat{K}_2 + 1 - y)}, \tag{1.87}$$

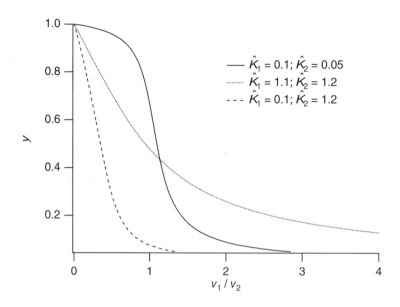

Figure 1.5 Plots of y as a function of the ratio $\frac{v_1}{v_2}$.

where $y = \frac{W}{W_t}$, $\hat{K}_i = K_i/W_t$, $v_i = V_i E_i$, for $i = 1, 2$. Plots of y as a function of the ratio $\frac{v_1}{v_2}$ are easy to draw. One simply plots $\frac{v_1}{v_2}$ as a function of y and then reverses the axes. Examples of these are shown in Fig. 1.5. As is seen in this figure, the ratio $\frac{v_1}{v_2}$ controls the relative abundance of y in a switch-like fashion. In particular, the switch becomes quite sharp when the equilibrium constants $\hat{K}_1$ and $\hat{K}_2$ are small compared to 1. In other words, if the enzyme reactions are running at highly saturated levels, then there is sensitive switch-like dependence on the enzyme velocity ratio $\frac{v_1}{v_2}$.

Equation (1.87) is a quadratic polynomial in y, with explicit solution

$$y = \frac{\beta - \sqrt{\beta^2 - 4\alpha\gamma}}{2\alpha}, \tag{1.88}$$

where

$$\alpha = \frac{v_1}{v_2} - 1, \tag{1.89}$$

$$\beta = (1 - \hat{K}_1) - \frac{v_1}{v_2}(\hat{K}_2 + 1), \tag{1.90}$$

$$\gamma = \hat{K}_1. \tag{1.91}$$

The function

$$G(v_1, v_2, \hat{K}_1, \hat{K}_2) = \frac{\beta - \sqrt{\beta^2 - 4\alpha\gamma}}{2\alpha} \tag{1.92}$$

is called the *Goldbeter–Koshland function*. The Goldbeter–Koshland function is often used in descriptions of biochemical networks (Chapter 10). For example, V_1 and V_2

could depend on the concentration of another enzyme, $\tilde{E}$ say, leading to switch-like regulation of the concentration of W as a function of the concentration of $\tilde{E}$. In this way, networks of biochemical reactions can be constructed in which some of the components are switched on or switched off, relatively abruptly, by other components.

1.5 Glycolysis and Glycolytic Oscillations

Metabolism is the process of extracting useful energy from chemical bonds. A metabolic pathway is the sequence of enzymatic reactions that take place in order to transfer chemical energy from one form to another. The common carrier of energy in the cell is the chemical adenosine triphosphate (ATP). ATP is formed by the addition of an inorganic phosphate group (HPO_4^{2-}) to adenosine diphosphate (ADP), or by the addition of two inorganic phosphate groups to adenosine monophosphate (AMP). The process of adding an inorganic phosphate group to a molecule is called *phosphorylation*. Since the three phosphate groups on ATP carry negative charges, considerable energy is required to overcome the natural repulsion of like-charged phosphates as additional groups are added to AMP. Thus, the hydrolysis (the cleavage of a bond by water) of ATP to ADP releases large amounts of energy.

Energy to perform chemical work is made available to the cell by the oxidation of glucose to carbon dioxide and water, with a net release of energy. The overall chemical reaction for the oxidation of glucose can be written as

$$C_6H_{12}O_6 + 6O_2 \longrightarrow 6CO_2 + 6H_2O + \text{energy}, \tag{1.93}$$

but of course, this is not an elementary reaction. Instead, this reaction takes place in a series of enzymatic reactions, with three major reaction stages, *glycolysis*, the *Krebs cycle*, and the *electron transport* (or *cytochrome*) *system*.

The oxidation of glucose is associated with a large negative free energy, $\Delta G^0 = -2878.41$ kJ/mol, some of which is dissipated as heat. However, in living cells much of this free energy in stored in ATP, with one molecule of glucose resulting in 38 molecules of ATP.

Glycolysis involves 11 elementary reaction steps, each of which is an enzymatic reaction. Here we consider a simplified model of the initial steps. (To understand more of the labyrinthine complexity of glycolysis, interested readers are encouraged to consult a specialized book on biochemistry, such as Stryer, 1988.) The first three steps of glycolysis are (Fig. 1.6)

1. the phosphorylation of glucose to glucose 6-phosphate;
2. the isomerization of glucose 6-phosphate to fructose 6-phosphate; and
3. the phosphorylation of fructose 6-phosphate to fructose 1,6-bisphosphate.

The direct reaction of glucose with phosphate to form glucose 6-phosphate has a relatively large positive standard free energy change ($\Delta G^0 = 14.3$ kJ/mol) and so

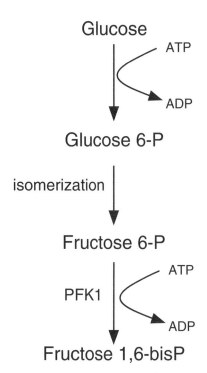

Figure 1.6 The first three reactions in the glycolytic pathway.

does not occur significantly under physiological conditions. However, the first step of metabolism is coupled with the hydrolysis of ATP to ADP (catalyzed by the enzyme hexokinase), giving this step a net negative standard free energy change and making the reaction strongly spontaneous. This feature turns out to be important for the efficient operation of glucose membrane transporters, which are described in the next chapter.

The second step of glycolysis has a relatively small positive standard free energy change ($\Delta G^0 = 1.7$ kJ/mol), with an equilibrium constant of 0.5. This means that significant amounts of product are formed under normal conditions.

The third step is, like the first step, energetically unfavorable, were it not coupled with the hydrolysis of ATP. However, the net standard free energy change ($\Delta G^0 = -14.2$ kJ/mol) means that not only is this reaction strongly favored, but also that it augments the reaction in the second step by depleting the product of the second step.

This third reaction is catalyzed by the enzyme phosphofructokinase (PFK1). PFK1 is an example of an allosteric enzyme as it is allosterically inhibited by ATP. Note that ATP is both a substrate of PFK1, binding at a catalytic site, and an allosteric inhibitor, binding at a regulatory site. The inhibition due to ATP is removed by AMP, and thus the activity of PFK1 increases as the ratio of ATP to AMP decreases. This feedback enables PFK1 to regulate the rate of glycolysis based on the availability of ATP. If ATP levels fall, PFK1 activity increases thereby increasing the rate of production of ATP, whereas, if ATP levels become high, PFK1 activity drops shutting down the production of ATP.

As PFK1 phosphorylates fructose 6-P, ATP is converted to ADP. ADP, in turn, is converted back to ATP and AMP by the reaction

$$2\text{ADP} \rightleftharpoons \text{ATP} + \text{AMP},$$

which is catalyzed by the enzyme adenylate kinase. Since there is normally little AMP in cells, the conversion of ADP to ATP and AMP serves to significantly decrease the ATP/AMP ratio, thus activating PFK1. This is an example of a positive feedback loop; the greater the activity of PFK1, the lower the ATP/AMP ratio, thus further increasing PFK1 activity.

It was discovered in 1980 that in some cell types, another important allosteric activator of PFK1 is fructose 2,6-bisphosphate (Stryer, 1988), which is formed from fructose 6-phosphate in a reaction catalyzed by phosphofructokinase 2 (PFK2), a different enzyme from phosphofructokinase (PFK1) (you were given fair warning about the labyrinthine nature of this process!). Of particular significance is that an abundance of fructose 6-phosphate leads to a corresponding abundance of fructose 2,6-bisphosphate, and thus a corresponding increase in the activity of PFK1. This is an example of a negative feedback loop, where an increase in the substrate concentration leads to a greater rate of substrate reaction and consumption. Clearly, PFK1 activity is controlled by an intricate system of reactions, the collective behavior of which is not obvious a priori.

Under certain conditions the rate of glycolysis is known to be oscillatory, or even chaotic (Nielsen et al., 1997). This biochemical oscillator has been known and studied experimentally for some time. For example, Hess and Boiteux (1973) devised a flow reactor containing yeast cells into which a controlled amount of substrate (either glucose or fructose) was continuously added. They measured the pH and fluorescence of the reactants, thereby monitoring the glycolytic activity, and they found ranges of continuous input under which glycolysis was periodic.

Interestingly, the oscillatory behavior is different in intact yeast cells and in yeast extracts. In intact cells the oscillations are sinusoidal in shape, and there is strong evidence that they occur close to a Hopf bifurcation (Danø et al., 1999). In yeast extract the oscillations are of relaxation type, with widely differing time scales (Madsen et al., 2005).

Feedback on PFK is one, but not the only, mechanism that has been proposed as causing glycolytic oscillations. For example, hexose transport kinetics and autocatalysis of ATP have both been proposed as possible mechanisms (Madsen et al., 2005), while some authors have claimed that the oscillations arise as part of the entire network of reactions, with no single feedback being of paramount importance (Bier et al., 1996; Reijenga et al., 2002). Here we focus only on PFK regulation as the oscillatory mechanism.

A mathematical model describing glycolytic oscillations was proposed by Sel'kov (1968) and later modified by Goldbeter and Lefever (1972). It is designed to capture only the positive feedback of ADP on PFK1 activity. In the Sel'kov model, PFK1 is inactive in its unbound state but is activated by binding with several ADP molecules. Note that, for simplicity, the model does not take into account the conversion of ADP to AMP and ATP, but assumes that ADP activates PFK1 directly, since the overall effect is similar. In the active state, the enzyme catalyzes the production of ADP from ATP as fructose-6-P is phosphorylated. Sel'kov's reaction scheme for this process is as follows: PFK1 (denoted by E) is activated or deactivated by binding or unbinding with γ molecules of ADP (denoted by S_2)

$$\gamma S_2 + E \underset{k_{-3}}{\overset{k_3}{\rightleftharpoons}} ES_2^{\gamma},$$

and ATP (denoted S_1) can bind with the activated form of enzyme to produce a product molecule of ADP. In addition, there is assumed to be a steady supply rate of S_1, while product S_2 is irreversibly removed. Thus,

$$\overset{v_1}{\longrightarrow} S_1, \tag{1.94}$$

$$S_1 + ES_2^{\gamma} \underset{k_{-1}}{\overset{k_1}{\rightleftharpoons}} S_1 ES_2^{\gamma} \overset{k_2}{\longrightarrow} ES_2^{\gamma} + S_2, \tag{1.95}$$

$$S_2 \overset{v_2}{\longrightarrow}. \tag{1.96}$$

Note that (1.95) is an enzymatic reaction of exactly Michaelis–Menten form so we should expect a similar reduction of the governing equations.

Applying the law of mass action to the Sel'kov kinetic scheme, we find five differential equations for the production of the five species $s_1 = [S_1], s_2 = [S_2], e = [E], x_1 = [ES_2^{\gamma}], x_2 = [S_1 ES_2^{\gamma}]$:

$$\frac{ds_1}{dt} = v_1 - k_1 s_1 x_1 + k_{-1} x_2, \tag{1.97}$$

$$\frac{ds_2}{dt} = k_2 x_2 - \gamma k_3 s_2^{\gamma} e + \gamma k_{-3} x_1 - v_2 s_2, \tag{1.98}$$

$$\frac{dx_1}{dt} = -k_1 s_1 x_1 + (k_{-1} + k_2) x_2 + k_3 s_2^{\gamma} e - k_{-3} x_1, \tag{1.99}$$

$$\frac{dx_2}{dt} = k_1 s_1 x_1 - (k_{-1} + k_2) x_2. \tag{1.100}$$

The fifth differential equation is not necessary, because the total available enzyme is conserved, $e + x_1 + x_2 = e_0$. Now we introduce dimensionless variables $\sigma_1 = \frac{k_1 s_1}{k_2 + k_{-1}}, \sigma_2 = (\frac{k_3}{k_{-3}})^{1/\gamma} s_2, u_1 = x_1/e_0, u_2 = x_2/e_0, t = \frac{k_2 + k_{-1}}{e_0 k_1 k_2} \tau$ and find

$$\frac{d\sigma_1}{d\tau} = v - \frac{k_2 + k_{-1}}{k_2} u_1 \sigma_1 + \frac{k_{-1}}{k_2} u_2, \tag{1.101}$$

$$\frac{d\sigma_2}{d\tau} = \alpha \left[u_2 - \frac{\gamma k_{-3}}{k_2} \sigma_2^{\gamma} (1 - u_1 - u_2) + \frac{\gamma k_{-3}}{k_2} u_1 \right] - \eta \sigma_2, \tag{1.102}$$

$$\epsilon \frac{du_1}{d\tau} = u_2 - \sigma_1 u_1 + \frac{k_{-3}}{k_2 + k_{-1}} \left[\sigma_2^\gamma (1 - u_1 - u_2) - u_1 \right], \tag{1.103}$$

$$\epsilon \frac{du_2}{d\tau} = \sigma_1 u_1 - u_2, \tag{1.104}$$

where $\epsilon = \frac{e_0 k_1 k_2}{(k_2 + k_{-1})^2}$, $\nu = \frac{v_1}{k_2 e_0}$, $\eta = \frac{v_2(k_2 + k_{-1})}{k_1 k_2 e_0}$, $\alpha = \frac{k_2 + k_{-1}}{k_1} (\frac{k_3}{k_{-3}})^{1/\gamma}$. If we assume that ϵ is a small number, then both u_1 and u_2 are fast variables and can be set to their quasi-steady values,

$$u_1 = \frac{\sigma_2^\gamma}{\sigma_2^\gamma \sigma_1 + \sigma_2^\gamma + 1}, \tag{1.105}$$

$$u_2 = \frac{\sigma_1 \sigma_2^\gamma}{\sigma_2^\gamma \sigma_1 + \sigma_2^\gamma + 1} = f(\sigma_1, \sigma_2), \tag{1.106}$$

and with these quasi-steady values, the evolution of σ_1 and σ_2 is governed by

$$\frac{d\sigma_1}{d\tau} = \nu - f(\sigma_1, \sigma_2), \tag{1.107}$$

$$\frac{d\sigma_2}{d\tau} = \alpha f(\sigma_1, \sigma_2) - \eta \sigma_2. \tag{1.108}$$

The goal of the following analysis is to demonstrate that this system of equations has oscillatory solutions for some range of the supply rate ν. First observe that because of saturation, the function $f(\sigma_1, \sigma_2)$ is bounded by 1. Thus, if $\nu > 1$, the solutions of the differential equations are not bounded. For this reason we consider only $0 < \nu < 1$. The nullclines of the flow are given by the equations

$$\sigma_1 = \frac{\nu}{1 - \nu} \frac{1 + \sigma_2^\gamma}{\sigma_2^\gamma} \qquad \left(\frac{d\sigma_1}{d\tau} = 0 \right), \tag{1.109}$$

$$\sigma_1 = \frac{1 + \sigma_2^\gamma}{\sigma_2^{\gamma - 1}(p - \sigma_2)} \qquad \left(\frac{d\sigma_2}{d\tau} = 0 \right), \tag{1.110}$$

where $p = \alpha/\eta$. These two nullclines are shown plotted as dotted and dashed curves respectively in Fig. 1.7.

The steady-state solution is unique and satisfies

$$\sigma_2 = p\nu, \tag{1.111}$$

$$\sigma_1 = \frac{\nu(1 + \sigma_2^\gamma)}{(1 - \nu)\sigma_2^\gamma}. \tag{1.112}$$

The stability of the steady solution is found by linearizing the differential equations about the steady-state solution and examining the eigenvalues of the linearized system. The linearized system has the form

$$\frac{d\tilde{\sigma}_1}{d\tau} = -f_1 \tilde{\sigma}_1 - f_2 \tilde{\sigma}_2, \tag{1.113}$$

$$\frac{d\tilde{\sigma}_2}{d\tau} = \alpha f_1 \tilde{\sigma}_1 + (\alpha f_2 - \eta) \tilde{\sigma}_2, \tag{1.114}$$

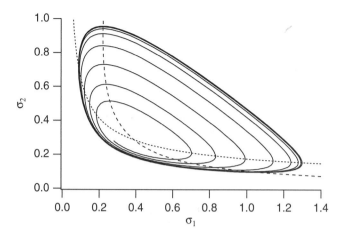

Figure 1.7 Phase portrait of the Sel'kov glycolysis model with $\nu = 0.0285, \eta = 0.1, \alpha = 1.0$, and $\gamma = 2$. Dotted curve: $\frac{d\sigma_1}{d\tau} = 0$. Dashed curve: $\frac{d\sigma_2}{d\tau} = 0$.

where $f_j = \frac{\partial f}{\partial \sigma_j}, j = 1, 2$, evaluated at the steady-state solution, and where $\tilde{\sigma}_i$ denotes the deviation from the steady-state value of σ_i. The characteristic equation for the eigenvalues λ of the linear system (1.113)–(1.114) is

$$\lambda^2 - (\alpha f_2 - \eta - f_1)\lambda + f_1 \eta = 0. \tag{1.115}$$

Since f_1 is always positive, the stability of the linear system is determined by the sign of $H = \alpha f_2 - \eta - f_1$, being stable if $H < 0$ and unstable if $H > 0$. Changes of stability, if they exist, occur at $H = 0$, and are Hopf bifurcations to periodic solutions with approximate frequency $\omega = \sqrt{f_1 \eta}$.

The function $H(\nu)$ is given by

$$H(\nu) = \frac{(1 - \nu)}{(1 + y)}(\eta\gamma + (\nu - 1)y) - \eta, \tag{1.116}$$

$$y = (p\nu)^\gamma. \tag{1.117}$$

Clearly, $H(0) = \eta(\gamma - 1), H(1) = -\eta$, so for $\gamma > 1$, there must be at least one Hopf bifurcation point, below which the steady solution is unstable. Additional computations show that this Hopf bifurcation is supercritical, so that for ν slightly below the bifurcation point, there is a stable periodic orbit.

An example of this periodic orbit is shown in Fig. 1.7 with coefficients $\nu = 0.0285, \eta = 0.1, \alpha = 1.0$, and $\gamma = 2$. The evolution of σ_1 and σ_2 are shown plotted as functions of time in Fig. 1.8.

A periodic orbit exists only in a very small region of parameter space, rapidly expanding until it becomes infinitely large in amplitude as ν decreases. For still smaller values of ν, there are no stable trajectories. This information is summarized in a bifurcation diagram (Fig. 1.9), where we plot the steady state, σ_1, against one of the

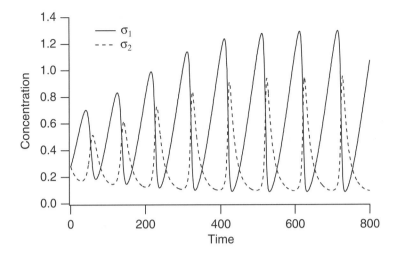

Figure 1.8 Evolution of σ_1 and σ_2 for the Sel'kov glycolysis model toward a periodic solution. Parameters are the same as in Fig. 1.7.

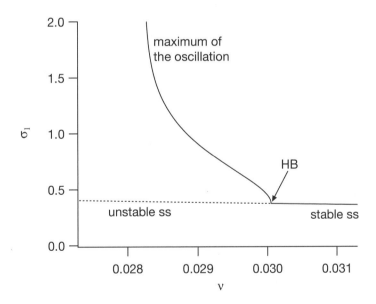

Figure 1.9 Bifurcation diagram for the Sel'kov glycolysis model.

parameters, in this case ν. Thus, ν is called the *bifurcation parameter*. The dashed line labeled "unstable ss" is the curve of unstable steady states as a function of ν, while the solid line labeled "stable ss" is the curve of stable steady states as a function of ν. As is typical in such bifurcation diagrams, we also include the maximum of the oscillation (when it exists) as a function of ν. We could equally have chosen to plot the minimum of the oscillation (or both the maximum and the minimum). Since the oscillation is stable, the maximum of the oscillation is plotted with a solid line.

From the bifurcation diagram we see that the stable branch of oscillations originates at a supercritical Hopf bifurcation (labeled HB), and that the periodic orbits only exist for a narrow range of values of ν. The question of how this branch of periodic orbits terminates is not important for the discussion here, so we ignore this important point for now.

We use bifurcation diagrams throughout this book, and many are considerably more complicated than that shown in Fig. 1.9. Readers who are unfamiliar with the basic theory of nonlinear bifurcations, and their representation in bifurcation diagrams, are urged to consult an elementary book such as Strogatz (1994).

While the Sel'kov model has certain features that are qualitatively correct, it fails to agree with the experimental results at a number of points. Hess and Boiteux (1973) report that for high and low substrate injection rates, there is a stable steady-state solution. There are two Hopf bifurcation points, one at the flow rate of 20 mM/hr and another at 160 mM/hr. The period of oscillation at the low flow rate is about 8 minutes and decreases as a function of flow rate to about 3 minutes at the upper Hopf bifurcation point. In contrast, the Sel'kov model has but one Hopf bifurcation point.

To reproduce these additional experimental features we consider a more detailed model of the reaction. In 1972, Goldbeter and Lefever proposed a model of Monod–Wyman–Changeux type that provided a more accurate description of the oscillations. More recently, by fitting a simpler model to experimental data on PFK1 kinetics in skeletal muscle, Smolen (1995) has shown that this level of complexity is not necessary; his model assumes that PFK1 consists of four independent, identical subunits, and reproduces the observed oscillations well. Despite this, we describe only the Goldbeter–Lefever model in detail, as it provides an excellent example of the use of Monod–Wyman–Changeux models.

In the Goldbeter–Lefever model of the phosphorylation of fructose-6-P, the enzyme PFK1 is assumed to be a dimer that exists in two states, an active state R and an inactive state T. The substrate, S_1, can bind to both forms, but the product, S_2, which is an activator, or positive effector, of the enzyme, binds only to the active form. The enzymatic forms of R carrying substrate decompose irreversibly to yield the product ADP. In addition, substrate is supplied to the system at a constant rate, while product is removed at a rate proportional to its concentration. The reaction scheme for this is as follows: let T_j represent the inactive T form of the enzyme bound to j molecules of substrate and let R_{ij} represent the active form R of the enzyme bound to i substrate molecules and j product molecules. This gives the reaction diagram shown in Fig. 1.10. In this system, the substrate S_1 holds the enzyme in the inactive state by binding with T_0 to produce T_1 and T_2, while product S_2 holds the enzyme in the active state by binding with R_{00} to produce R_{01} and binding with R_{01} to produce R_{02}. There is a factor of two in the rates of reaction because a dimer with two available binding sites reacts like twice the same amount of monomer.

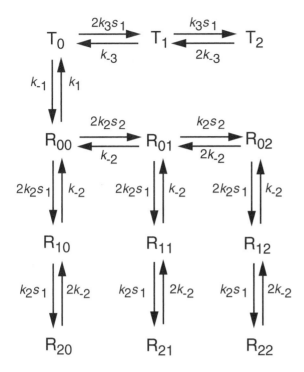

Figure 1.10 Possible states of the enzyme PFK1 in the Goldbeter–Lefever model of glycolytic oscillations.

In addition to the reactions shown in Fig. 1.10, the enzyme complex can disassociate to produce product via the reaction

$$R_{ij} \xrightarrow{k} R_{i-1,j} + S_2, \tag{1.118}$$

provided $i \geq 1$.

The analysis of this reaction scheme is substantially more complicated than that of the Sel'kov scheme, although the idea is the same. We use the law of mass action to write differential equations for the fourteen chemical species. For example, the equation for $s_1 = [S_1]$ is

$$\frac{ds_1}{dt} = v_1 - F, \tag{1.119}$$

where

$$\begin{aligned}
F = {} & k_{-2}(r_{10} + r_{11} + r_{12}) + 2k_{-2}(r_{20} + r_{21} + r_{22}) \\
& - 2k_2 s_1 (r_{00} + r_{01} + r_{02}) - k_2 s_1 (r_{10} + r_{11} + r_{12}) \\
& - 2k_3 s_1 t_0 - k_3 s_1 t_1 + k_{-3} t_1 + 2k_{-3} t_2,
\end{aligned} \tag{1.120}$$

and the equation for $r_{00} = [R_{00}]$ is

$$\frac{dr_{00}}{dt} = -(k_1 + 2k_2 s_1 + 2k_2 s_2) r_{00} + (k_{-2} + k) r_{10} + k_{-2} r_{01} + k_{-1} t_0. \tag{1.121}$$

We then assume that all twelve of the intermediates are in quasi-steady state. This leads to a 12 by 12 linear system of equations, which, if we take the total amount of enzyme to be e_0, can be solved. We substitute this solution into the differential equations for s_1 and s_2 with the result that

$$\frac{ds_1}{dt} = v_1 - F(s_1, s_2), \tag{1.122}$$

$$\frac{ds_2}{dt} = F(s_1, s_2) - v_2 s_2, \tag{1.123}$$

where

$$F(s_1, s_2) = \left(\frac{2k_2 k_{-1} k e_0}{k + k_{-2}} \right) \left(\frac{s_1 \left(1 + \frac{k_2}{k + k_{-2}} s_1 \right) (s_2 + K_2)^2}{K_2^2 k_1 \left(\frac{k_3}{k_{-3}} s_1 + 1 \right)^2 + k_{-1} \left(1 + \frac{k_2}{k + k_{-2}} s_1 \right)^2 (K_2 + s_2)^2} \right), \tag{1.124}$$

where $K_2 = \frac{k_{-2}}{k_2}$. Now we introduce dimensionless variables $\sigma_1 = \frac{s_1}{K_2}, \sigma_2 = \frac{s_2}{K_2}, t = \frac{\tau}{\tau_c}$ and parameters $v = \frac{k_2 v_1}{k_{-2} \tau_c}, \eta = \frac{v_2}{\tau_c}$, where $\tau_c = \frac{2k_2 k_{-1} k e_0}{k_1 (k + k_{-2})}$, and arrive at the system (1.107)–(1.108), but with a different function $f(\sigma_1, \sigma_2)$, and with $\alpha = 1$. If, in addition, we assume that

1. the substrate does not bind to the T form ($k_3 = 0$, T is completely inactive),
2. T_0 is preferred over R_{00} ($k_1 \gg k_{-1}$), and
3. if the substrate S_1 binds to the R form, then formation of product S_2 is preferred to dissociation ($k \gg k_{-2}$),

then we can simplify the equations substantially to obtain

$$f(\sigma_1, \sigma_2) = \sigma_1 (1 + \sigma_2)^2. \tag{1.125}$$

The nullclines for this system of equations are somewhat different from the Sel'kov system, being

$$\sigma_1 = \frac{v}{(1 + \sigma_2)^2} \qquad \left(\frac{d\sigma_1}{d\tau} = 0 \right), \tag{1.126}$$

$$\sigma_1 = \frac{\eta \sigma_2}{(1 + \sigma_2)^2} \qquad \left(\frac{d\sigma_2}{d\tau} = 0 \right), \tag{1.127}$$

and the unique steady-state solution is given by

$$\sigma_1 = \frac{v}{(1 + \sigma_2)^2}, \tag{1.128}$$

$$\sigma_2 = \frac{v}{\eta}. \tag{1.129}$$

The stability of the steady-state solution is again determined by the characteristic equation (1.115), and the sign of the real part of the eigenvalues is the same as the sign of

$$H = f_2 - f_1 - \eta = 2\sigma_1 (1 + \sigma_2) - (1 + \sigma_2)^2 - \eta, \tag{1.130}$$

evaluated at the steady state (1.126)–(1.127). Equation (1.130) can be written as the cubic polynomial

$$\frac{1}{\eta}y^3 - y + 2 = 0, \qquad y = 1 + \frac{\nu}{\eta}. \qquad (1.131)$$

For η sufficiently large, the polynomial (1.131) has two roots greater than 2, say, y_1 and y_2. Recall that ν is the nondimensional flow rate of substrate ATP. To make some correspondence with the experimental data, we assume that the flow rate ν is proportional to the experimental supply rate of glucose. This is not strictly correct, although ATP is produced at about the same rate that glucose is supplied. Accepting this caveat, we see that to match experimental data, we require

$$\frac{y_2 - 1}{y_1 - 1} = \frac{\nu_2}{\nu_1} = \frac{160}{20} = 8. \qquad (1.132)$$

Requiring (1.131) to hold at y_1 and y_2 and requiring (1.132) to hold as well, we find numerical values

$$y_1 = 2.08, \quad y_2 = 9.61, \quad \eta = 116.7, \qquad (1.133)$$

corresponding to $\nu_1 = 126$ and $\nu_2 = 1005$.

At the Hopf bifurcation point, the period of oscillation is

$$T_i = \frac{2\pi}{\omega_i} = \frac{2\pi}{\sqrt{\eta}(1 + \sigma_2)} = \frac{2\pi}{\sqrt{\eta}y_i}. \qquad (1.134)$$

For the numbers (1.133), we obtain a ratio of periods $T_1/T_2 = 4.6$, which is acceptably close to the experimentally observed ratio $T_1/T_2 = 2.7$.

The behavior of the solution as a function of the parameter ν is summarized in the bifurcation diagram, Fig. 1.11, shown here for $\eta = 120$. The steady-state solution is stable below $\eta = 129$ and above $\eta = 1052$. Between these values of η the steady-state solution is unstable, but there is a branch of stable periodic solutions which terminates and collapses into the steady-state solution at the two points where the stability changes, the Hopf bifurcation points.

A typical phase portrait for the periodic solution that exists between the Hopf bifurcation points is shown in Fig. 1.12, and the concentrations of the two species are shown as functions of time in Fig. 1.13.

1.6 Appendix: Math Background

It is certain that some of the mathematical concepts and tools that we routinely invoke here are not familiar to all of our readers. In this first chapter alone, we have used nondimensionalization, phase-plane analysis, linear stability analysis, bifurcation theory, and asymptotic analysis, all the while assuming that these are familiar to the reader.

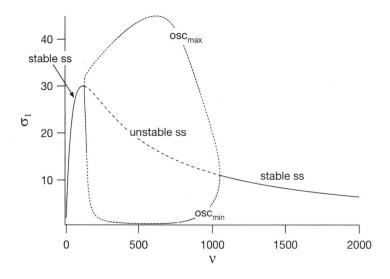

Figure 1.11 Bifurcation diagram for the reduced Goldbeter–Lefever glycolysis model, with $\eta = 120$.

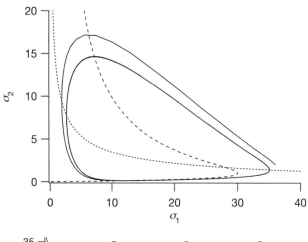

Figure 1.12 Phase portrait of the Goldbeter–Lefever model with $\nu = 200$, $\eta = 120$. Dotted curve: $\frac{d\sigma_1}{d\tau} = 0$. Dashed curve: $\frac{d\sigma_2}{d\tau} = 0$.

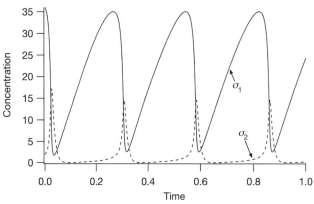

Figure 1.13 Solution of the Goldbeter–Lefever model with $\nu = 200$, $\eta = 120$.

The purpose of this appendix is to give a brief guide to those techniques that are a basic part of the applied mathematician's toolbox.

1.6.1   Basic Techniques

In any problem, there are a number of parameters that are dictated by the problem. However, it often happens that not all parameter variations are *independent*; that is, different variations in different parameters may lead to identical changes in the behavior of the model. Second, there may be parameters whose influence on a behavior is negligible and can be safely ignored for a given context.

The way to identify independent parameters and to determine their relative magnitudes is to nondimensionalize the problem. Unfortunately, there is not a unique algorithm for nondimensionalization; nondimensionalization is as much art as it is science.

There are, however, rules of thumb to apply. In any system of equations, there are a number of independent variables (time, space, etc.), dependent variables (concentrations, etc.) and parameters (rates of reaction, sizes of containers, etc.). Nondimensionalization begins by rescaling the independent and dependent variables by "typical" units, rendering them thereby dimensionless. One goal may be to ensure that the dimensionless variables remain of a fixed order of magnitude, not becoming too large or negligibly small. This usually requires some a priori knowledge about the solution, as it can be difficult to choose typical scales unless something is already known about typical solutions. Time and space scales can be vastly different depending on the context.

Once this selection of scales has been made, the governing equations are written in terms of the rescaled variables and dimensionless combinations of the remaining parameters are identified. The number of remaining free dimensionless parameters is usually less than the original number of physical parameters. The primary difficulty (at least to understand and apply the process) is that there is not necessarily a single way to scale and nondimensionalize the equations. Some scalings may highlight certain features of the solution, while other scalings may emphasize others. Nonetheless, nondimensionalization often (but not always) provides a good starting point for the analysis of a model system.

An excellent discussion of scaling and nondimensionalization can be found in Lin and Segel (1988, Chapter 6). A great deal of more advanced work has also been done on this subject, particularly its application to the quasi-steady-state approximation, by Segel and his collaborators (Segel, 1988; Segel and Slemrod, 1989; Segel and Perelson, 1992; Segel and Goldbeter, 1994; Borghans et al., 1996; see also Frenzen and Maini, 1988).

Phase-plane analysis and linear stability analysis are standard fare in introductory courses on differential equations. A nice introduction to these topics for the biologically inclined can be found in Edelstein-Keshet (1988, Chapter 5) or Braun (1993, Chapter 4). A large number of books discuss the qualitative theory of differential equations, for example, Boyce and Diprima (1997), or at a more advanced level, Hale and Koçak (1991), or Hirsch and Smale (1974).

Bifurcation Theory

Bifurcation theory is a topic that is gradually finding its way into introductory literature. The most important terms to understand are those of *steady-state* bifurcations, *Hopf* bifurcations, *homoclinic* bifurcations, and *saddle-node* bifurcations, all of which appear in this book. An excellent introduction to these concepts is found in Strogatz (1994, Chapters 3, 6, 7, 8), and an elementary treatment, with particular application to biological systems, is given by Beuter et al. (2003, Chapters 2, 3). More advanced treatments include those in Guckenheimer and Holmes (1983), Arnold (1983) or Wiggins (2003).

One way to summarize the behavior of the model is with a bifurcation diagram (examples of which are shown in Figs. 1.9 and 1.11), which shows how certain features of the model, such as steady states or limit cycles, vary as a parameter is varied. When models have many parameters there is a wide choice for which parameter to vary. Often, however, there are compelling physiological or experimental reasons for the choice of parameter. Bifurcation diagrams are important in a number of chapters of this book, and are widely used in the analysis of nonlinear systems. Thus, it is worth the time to become familiar with their properties and how they are constructed. Nowadays, most bifurcation diagrams of realistic models are constructed numerically, the most popular choice of software being AUTO (Doedel, 1986; Doedel et al., 1997, 2001). The bifurcation diagrams in this book were all prepared with XPPAUT (Ermentrout, 2002), a convenient implementation of AUTO.

In this text, the bifurcation that is seen most often is the Hopf bifurcation. The Hopf bifurcation theorem describes conditions for the appearance of small periodic solutions of a differential equation, say

$$\frac{du}{dt} = f(u, \lambda), \tag{1.135}$$

as a function of the parameter λ. Suppose that there is a steady-state solution, $u = u_0(\lambda)$, and that the system linearized about u_0,

$$\frac{dU}{dt} = \frac{\partial f(u_0(\lambda), \lambda)}{\partial u} U, \tag{1.136}$$

has a pair of complex eigenvalues $\mu(\lambda) = \alpha(\lambda) \pm i\beta(\lambda)$. Suppose further that $\alpha(\lambda_0) = 0$, $\alpha'(\lambda_0) \neq 0$, and $\beta(\lambda_0) \neq 0$, and that at $\lambda = \lambda_0$ no other eigenvalues of the system have zero real part. Then λ_0 is a Hopf bifurcation point, and there is a branch of periodic solutions emanating from the point $\lambda = \lambda_0$. The periodic solutions could exist (locally) for $\lambda > \lambda_0$, for $\lambda < \lambda_0$, or in the degenerate (nongeneric) case, for $\lambda = \lambda_0$. If the periodic solutions occur in the region of λ for which $\alpha(\lambda) > 0$, then the periodic solutions are stable (provided all other eigenvalues of the system have negative real part), and this branch of solutions is said to be supercritical. On the other hand, if the periodic solutions occur in the region of λ for which $\alpha(\lambda) < 0$, then the periodic solutions are unstable, and this branch of solutions is said to be subcritical.

The Hopf bifurcation theorem applies to ordinary differential equations and delay differential equations. For partial differential equations, there are some technical issues having to do with the nature of the spectrum of the linearized operator that complicate matters, but we do not concern ourselves with these here. Instead, rather than checking all the conditions of the theorem, we find periodic solutions by looking only for a change of the sign of the real part of an eigenvalue, using numerical computations to verify the existence of periodic solutions, and calling it good.

1.6.2 Asymptotic Analysis

Applied mathematicians love small parameters, because of the hope that the solution of a problem with a small parameter might be approximated by an *asymptotic representation*. A commonplace notation has emerged in which ϵ is often the small parameter. An asymptotic representation has a precise mathematical meaning. Suppose that $G(\epsilon)$ is claimed to be an asymptotic representation of $g(\epsilon)$, expressed as

$$g(\epsilon) = G(\epsilon) + O(\phi(\epsilon)). \tag{1.137}$$

The precise meaning of this statement is that there is a constant A such that

$$\left| \frac{g(\epsilon) - G(\epsilon)}{\phi(\epsilon)} \right| \leq A \tag{1.138}$$

for all ϵ with $|\epsilon| \leq \epsilon_0$ and $\epsilon > 0$. The function $\phi(\epsilon)$ is called a *gauge function*, a typical example of which is a power of ϵ.

Perturbation Expansions

It is often the case that an asymptotic representation can be found as a power series in powers of the small parameter ϵ. Such representations are called *perturbation expansions*. Usually, a few terms of this power series representation suffice to give a good approximation to the solution. It should be kept in mind that under no circumstances does this power series development imply that a complete power series (with an infinite number of terms) exists or is convergent. Terminating the series at one or two terms is deliberate.

However, there are times when a full power series could be found and would be convergent in some nontrivial ϵ domain. Such problems are called *regular perturbation problems* because their solutions are regular, or analytic, in the parameter ϵ.

There are numerous examples of regular perturbation problems, including all of those related to bifurcation theory. These problems are regular because their solutions can be developed in a convergent power series of some parameter.

There are, however, many problems with small parameters whose solutions are not regular, called *singular perturbation problems*. Singular perturbation problems are characterized by the fact that their dependence on the small parameter is not regular, but *singular*, and their convergence as a function of ϵ is not uniform.

Singular problems come in two basic varieties. Characteristic of the first type is a small region of width ϵ somewhere in the domain of interest (either space or time) in which the solution changes rapidly. For example, the solution of the boundary value problem

$$\epsilon u'' + u' + u = 0 \tag{1.139}$$

subject to boundary conditions $u(0) = u(1) = 1$ is approximated by the asymptotic representation

$$u(x; \epsilon) = (1 - e)e^{-x/\epsilon} + e^{1-x} + O(\epsilon). \tag{1.140}$$

Notice the nonuniform nature of this solution, as

$$e = \lim_{x \to 0^+} \left(\lim_{\epsilon \to 0^+} u(x; \epsilon) \right) \neq \lim_{\epsilon \to 0^+} \left(\lim_{x \to 0^+} u(x; \epsilon) \right) = 1.$$

Here the term $e^{-x/\epsilon}$ is a *boundary layer correction*, as it is important only in a small region near the boundary at $x = 0$.

Other terms that are typical in singular perturbation problems are *interior layers* or *transition layers*, typified by expressions of the form $\tan(\frac{x-x_0}{\epsilon})$, and *corner layers*, locations where the derivative changes rapidly but the solution itself changes little. Transition layers are of great significance in the study of excitable systems (Chapter 5). While corner layers show up in this book, we do not study or use them in any detail.

Singular problems of this type can often be identified by the fact that the order of the system decreases if ϵ is set to zero. An example that we have already seen is the quasi-steady-state analysis used to simplify reaction schemes in which some reactions are significantly faster than others. Setting ϵ to zero in these examples reduces the order of the system of equations, signaling a possible problem. Indeed, solutions of these equations typically have *initial layers* near time $t = 0$. We take a closer look at this example below.

The second class of singular perturbation problems is that in which there are two scales in operation everywhere in the domain of interest. Problems of this type show up throughout this book. For example, action potential propagation in cardiac tissue is through a cellular medium whose detailed structure varies rapidly compared to the length scale of the action potential wave front. Physical properties of the cochlear membrane in the inner ear vary slowly compared to the wavelength of waves that propagate along it. For problems of this type, one must make explicit the dependence on multiple scales, and so solutions are often expressed as functions of two variables, say x and x/ϵ, which are treated as independent variables. Solution techniques that exploit the multiple-scale nature of the solution are called *multiscale methods* or *averaging methods*.

Detailed discussions of these asymptotic methods may be found in Murray (1984), Kevorkian and Cole (1996), and Holmes (1995).

1.6.3   Enzyme Kinetics and Singular Perturbation Theory

In most of the examples of enzyme kinetics discussed in this chapter, extensive use was made of the quasi-steady-state approximation (1.44), according to which the concentration of the complex remains constant during the course of the reaction. Although this assumption gives the right answers (which, some might argue, is justification enough), mathematicians have sought for ways to justify this approximation rigorously. Bowen et al. (1963) and Heineken et al. (1967) were the first to show that the quasi-steady-state approximation can be derived as the lowest-order term in an asymptotic expansion of the solution. This has since become one of the standard examples of the application of singular perturbation theory to biological systems, and it is discussed in detail by Rubinow (1973), Lin and Segel (1988), and Murray (2002), among others.

Starting with (1.37) and (1.38),

$$\frac{d\sigma}{d\tau} = -\sigma + x(\sigma + \alpha), \tag{1.141}$$

$$\epsilon \frac{dx}{d\tau} = \sigma - x(\sigma + \kappa), \tag{1.142}$$

with initial conditions

$$\sigma(0) = 1, \tag{1.143}$$

$$x(0) = 0, \tag{1.144}$$

we begin by looking for solutions of the form

$$\sigma = \sigma_0 + \epsilon \sigma_1 + \epsilon^2 \sigma_2 + \cdots, \tag{1.145}$$

$$x = x_0 + \epsilon x_1 + \epsilon^2 x_2 + \cdots. \tag{1.146}$$

We substitute these solutions into the differential equations and equate coefficients of powers of ϵ. To lowest order (i.e., equating all those terms with no ϵ) we get

$$\frac{d\sigma_0}{d\tau} = -\sigma_0 + x_0(\sigma_0 + \alpha), \tag{1.147}$$

$$0 = \sigma_0 - x_0(\sigma_0 + \kappa). \tag{1.148}$$

Note that, because we are matching powers of ϵ, the differential equation for x has been converted into an algebraic equation for x_0, which can be solved to give

$$x_0 = \frac{\sigma_0}{\sigma_0 + \kappa}. \tag{1.149}$$

It follows that

$$\frac{d\sigma_0}{d\tau} = -\sigma_0 + x_0(\sigma_0 + \alpha) = -\sigma_0 \left(\frac{\kappa - \alpha}{\sigma_0 + \kappa} \right). \tag{1.150}$$

These solutions for x_0 and σ_0 (i.e., for the lowest-order terms in the power series expansion) are the quasi-steady-state approximation of Section 1.4.2. We could carry on and solve for σ_1 and x_1, but since the calculations rapidly become quite tedious, with little to no benefit, the lowest-order solution suffices.

However, it is important to notice that this lowest-order solution cannot be correct for all times. For, clearly, the initial conditions $\sigma(0) = 1, x(0) = 0$ are inconsistent with (1.149). In fact, by setting ϵ to be zero, we have decreased the order of the differential equations system, making it impossible to satisfy the initial conditions.

There must therefore be a brief period of time at the start of the reaction during which the quasi-steady-state approximation does not hold. It is not that ϵ is not small, but rather that $\epsilon \frac{dx}{dt}$ is not small during this initial period, since dx/dt is large. Indeed, it is during this initial time period that the enzyme is "filling up" with substrate, until the concentration of complexed enzyme reaches the value given by the quasi-steady-state approximation. Since there is little enzyme compared to the total amount of substrate, the concentration of substrate remains essentially constant during this period.

For most biochemical reactions this transition to the quasi-steady state happens so fast that it is not physiologically important, but for mathematical reasons, it is interesting to understand these kinetics for early times as well. To see how the reaction behaves for early times, we make a change of time scale, $\eta = \tau/\epsilon$. This change of variables expands the time scale on which we look at the reaction and allows us to study events that happen on a fast time scale. To be more precise, we also denote the solution on this fast time scale by a tilde. In the new time scale, (1.37)–(1.38) become

$$\frac{d\tilde{\sigma}}{d\eta} = \epsilon(-\tilde{\sigma} + \tilde{x}(\tilde{\sigma} + \alpha)), \tag{1.151}$$

$$\frac{d\tilde{x}}{d\eta} = \tilde{\sigma} - \tilde{x}(\tilde{\sigma} + \kappa). \tag{1.152}$$

The initial conditions are $\tilde{\sigma}(0) = 1, \tilde{x}(0) = 0$.

As before, we expand $\tilde{\sigma}$ and $\tilde{x}$ in power series in ϵ, substitute into the differential equations, and equate coefficients of powers of ϵ. To lowest order in ϵ this gives

$$\frac{d\tilde{\sigma}_0}{d\eta} = 0, \tag{1.153}$$

$$\frac{d\tilde{x}_0}{d\eta} = \tilde{\sigma}_0 - \tilde{x}_0(\tilde{\sigma}_0 + \kappa). \tag{1.154}$$

Simply stated, this means that $\tilde{\sigma}_0$ does not change on this time scale, so that $\tilde{\sigma}_0 = 1$. Furthermore, we can solve for $\tilde{x}_0$ as

$$\tilde{x}_0 = \frac{1}{1 + \kappa}(1 - e^{-(1+\kappa)\eta}), \tag{1.155}$$

where we have used the initial condition $\tilde{x}_0(0) = 0$.

Once again, we could go on to solve for $\tilde{\sigma}_1$ and $\tilde{x}_1$, but such calculations, being long and of little use, are rarely done. Thus, from now on, we omit the subscript 0, since it plays no essential role.

One important thing to notice about this solution for $\tilde{\sigma}$ and $\tilde{x}$ is that it cannot be valid at large times. After all, σ cannot possibly be a constant for all times. Thus, $\tilde{\sigma}$ and

$\tilde{x}$ are valid for small times (since they satisfy the initial conditions), but not for large times.

At first sight, it looks as if we are at an impasse. We have a solution, σ and x, that works for large times but not for small times, and we have another solution, $\tilde{\sigma}$ and $\tilde{x}$, that works for small times, but not for large ones. The goal now is to match them to obtain a single solution that is valid for all times. Fortunately, this is relatively simple to do for this example.

In terms of the original time variable τ, the solution for $\tilde{x}$ is

$$\tilde{x}(\tau) = \frac{\tilde{\sigma}}{\tilde{\sigma} + \kappa}(1 - e^{-(1+\kappa)\frac{\tau}{\epsilon}}). \tag{1.156}$$

As τ gets larger than order ϵ, the exponential term disappears, leaving only

$$\tilde{x}(\tau) = \frac{\tilde{\sigma}}{\tilde{\sigma} + \kappa}, \tag{1.157}$$

which has the same form as (1.149). It thus follows that the solution

$$x(\tau) = \frac{\sigma}{\sigma + \kappa}(1 - e^{-(1+\kappa)\frac{\tau}{\epsilon}}) \tag{1.158}$$

is valid for all times.

The solution for σ is obtained by direct solution of (1.150), which gives

$$\sigma + \kappa \log \sigma = (\alpha - \kappa)t + 1, \tag{1.159}$$

where we have used the initial condition $\sigma(0) = 1$. Since σ does not change on the short time scale, this solution is valid for both small and large times.

This simple analysis shows that there is first a time span during which the enzyme products rapidly equilibrate, consuming little substrate, and after this initial "layer" the reaction proceeds according to Michaelis–Menten kinetics along the quasi-steady-state curve. This is shown in Fig. 1.14. In the phase plane one can see clearly how the solution moves quickly until it reaches the quasi-steady-state curve (the slow manifold) and

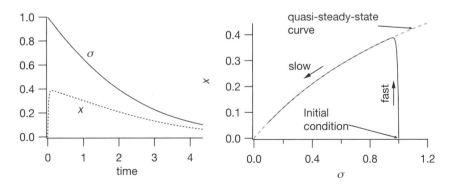

Figure 1.14 The solution to the quasi-steady-state approximation, plotted as functions of time (left panel) and in the phase plane (right panel). Calculated using $\kappa = 1.5$, $\alpha = 0.5$, $\epsilon = 0.05$.

then moves slowly along that curve toward the steady state. Note that the movement to the quasi-steady-state curve is almost vertical, since during that time σ remains approximately unchanged from its initial value. A similar procedure can be followed for the equilibrium approximation (Exercise 20). In this case, the fast movement to the slow manifold is not along lines of constant σ, but along lines of constant $\sigma + \alpha x$, where $\alpha = e_0/s_0$.

In this problem, the analysis of the initial layer is relatively easy and not particularly revealing. However, this type of analysis is of much greater importance later in this book when we discuss the behavior of excitable systems.

1.7   EXERCISES

1.   Consider the simple chemical reaction in which two monomers of A combine to form a dimer B, according to

$$A + A \underset{k_-}{\overset{k_+}{\rightleftarrows}} B.$$

(a)   Use the law of mass action to find the differential equations governing the rates of production of A and B.

(b)   What quantity is conserved? Use this conserved quantity to find an equation governing the rate of production of A that depends only on the concentration of A.

(c)   Nondimensionalize this equation and show that these dynamics depend on only one dimensionless parameter.

2.   In the real world trimolecular reactions are rare, although trimerizations are not. Consider the following trimerization reaction in which three monomers of A combine to form the trimer C,

$$A + A \underset{k_{-1}}{\overset{k_1}{\rightleftarrows}} B,$$

$$A + B \underset{k_{-2}}{\overset{k_2}{\rightleftarrows}} C.$$

(a)   Use the law of mass action to find the rate of production of the trimer C.

(b)   Suppose $k_{-1} \gg k_{-2}, k_2 A$. Use the appropriate quasi-steady-state approximation to find the rates of production of A and C, and show that the rate of production of C is proportional to $[A]^3$. Explain in words why this is so.

3.   The length of microtubules changes by a process called treadmilling, in which monomer is added to one end of the microtubule and taken off at the other end. To model this process, suppose that monomer A_1 is self-polymerizing in that it can form dimer A_2 via

$$A_1 + A_1 \overset{k_+}{\longrightarrow} A_2.$$

Furthermore, suppose A_1 can polymerize an n-polymer A_n at one end making an $n + 1$-polymer A_{n+1}

$$A_1 + A_n \xrightarrow{k_+} A_{n+1}.$$

Finally, degradation can occur one monomer at a time from the opposite end at rate k_-. Find the steady-state distribution of polymer lengths after an initial amount of monomer A_0 has fully polymerized.

4. Suppose that the reaction rates for the three reactant loop of Fig. 1.1 do not satisfy detailed balance. What is the net rate of conversion of A into B when the reaction is at steady state?

5. Consider an enzymatic reaction in which an enzyme can be activated or inactivated by the same chemical substance, as follows:

$$E + X \underset{k_{-1}}{\overset{k_1}{\rightleftharpoons}} E_1, \tag{1.160}$$

$$E_1 + X \underset{k_{-2}}{\overset{k_2}{\rightleftharpoons}} E_2, \tag{1.161}$$

$$E_1 + S \xrightarrow{k_3} P + Q + E. \tag{1.162}$$

Suppose further that X is supplied at a constant rate and removed at a rate proportional to its concentration. Use quasi-steady-state analysis to find the nondimensional equation describing the degradation of X,

$$\frac{dx}{dt} = \gamma - x - \frac{\beta xy}{1 + x + y + \frac{\alpha}{\delta}x^2}. \tag{1.163}$$

Identify all the parameters and variables, and the conditions under which the quasi-steady state approximation is valid.

6. Using the quasi-steady-state approximation, show that the velocity of the reaction for an enzyme with an allosteric inhibitor (Section 1.4.3) is given by

$$V = \left(\frac{V_{\max}K_3}{i + K_3}\right)\left(\frac{s(k_{-1} + k_3i + k_1s + k_{-3})}{k_1(s + K_1)^2 + (s + K_1)(k_3i + k_{-3} + k_2) + k_2k_{-3}/k_1}\right). \tag{1.164}$$

Identify all parameters. Under what conditions on the rate constants is this a valid approximation? Show that this reduces to (1.59) in the case $K_1 = \kappa_1$.

7. (a) Derive the expression (1.76) for the fraction of occupied sites in a Monod–Wyman–Changeux model with n binding sites.

   (b) Modify the Monod–Wyman–Changeux model shown in Fig. 1.4 to include transitions between states R_1 and T_1, and between states R_2 and T_2. Use the principle of detailed balance to derive an expression for the equilibrium constant of each of these transitions. Find the expression for Y, the fraction of occupied sites, and compare it to (1.72).

8. An enzyme-substrate system is believed to proceed at a Michaelis–Menten rate. Data for the (initial) rate of reaction at different concentrations is shown in Table 1.1.

   (a) Plot the data V vs. s. Is there evidence that this is a Michaelis–Menten type reaction?

   (b) Plot V vs. V/s. Are these data well approximated by a straight line?

Table 1.1 Data for Problem 8.

Substrate Concentration (mM)	Reaction Velocity (mM/s)
0.1	0.04
0.2	0.08
0.5	0.17
1.0	0.24
2.0	0.32
3.5	0.39
5.0	0.42

Table 1.2 Data for Problem 9.

Substrate Concentration (mM)	Reaction Velocity (mM/s)
0.2	0.01
0.5	0.06
1.0	0.27
1.5	0.50
2.0	0.67
2.5	0.78
3.5	0.89
4.0	0.92
4.5	0.94
5.0	0.95

(c) Use linear regression and (1.46) to estimate K_m and V_{max}. Compare the data to the Michaelis–Menten rate function using these parameters. Does this provide a reasonable fit to the data?

9. Suppose the maximum velocity of a chemical reaction is known to be 1 mM/s, and the measured velocity V of the reaction at different concentrations s is shown in Table 1.2.

(a) Plot the data V vs. s. Is there evidence that this is a Hill type reaction?

(b) Plot $\ln\left(\frac{V}{V_{max}-V}\right)$ vs. $\ln(s)$. Is this approximately a straight line, and if so, what is its slope?

(c) Use linear regression and (1.71) to estimate K_m and the Hill exponent n. Compare the data to the Hill rate function with these parameters. Does this provide a reasonable fit to the data?

10. Use the equilibrium approximation to derive an expression for the reaction velocity of the scheme (1.60)–(1.61).

Answer:

$$V = \frac{(k_2 K_3 + k_4 s)e_0 s}{K_1 K_3 + K_3 s + s^2},$$ (1.165)

where $K_1 = k_{-1}/k_1$ and $K_3 = k_{-3}/k_3$.

11. (a) Find the velocity of reaction for an enzyme with three active sites.

    (b) Under what conditions does the velocity reduce to a Hill function with exponent three? Identify all parameters.

    (c) What is the relationship between rate constants when the three sites are independent? What is the velocity when the three sites are independent?

12. The Goldbeter–Koshland function (1.92) is defined using the solution of the quadratic equation with a negative square root. Why?

13. Suppose that a substrate can be broken down by two different enzymes with different kinetics. (This happens, for example, in the case of cAMP or cGMP, which can be hydrolyzed by two different forms of phosphodiesterase—see Chapter 19).

    (a) Write the reaction scheme and differential equations, and nondimensionalize, to get the system of equations

$$\frac{d\sigma}{dt} = -\sigma + \alpha_1(\mu_1 + \sigma)x + \alpha_2(\mu_2 + \sigma)y,$$ (1.166)

$$\epsilon_1 \frac{dx}{dt} = \frac{1}{\lambda_1}\sigma(1 - x) - x,$$ (1.167)

$$\epsilon_2 \frac{dy}{dt} = \frac{1}{\lambda_2}\sigma(1 - y) - y.$$ (1.168)

where x and y are the nondimensional concentrations of the two complexes. Identify all parameters.

    (b) Apply the quasi-steady-state approximation to find the equation governing the dynamics of substrate σ. Under what conditions is the quasi-steady-state approximation valid?

    (c) Solve the differential equation governing σ.

    (d) For this system of equations, show that the solution can never leave the positive octant $\sigma, x, y \geq 0$. By showing that $\sigma + \epsilon_1 x + \epsilon_2 y$ is decreasing everywhere in the positive octant, show that the solution approaches the origin for large time.

14. For some enzyme reactions (for example, the hydrolysis of cAMP by phosphodiesterase in vertebrate retinal cones) the enzyme is present in large quantities, so that e_0/s_0 is not a small number. Fortunately, there is an alternate derivation of the Michaelis–Menten rate equation that does not require that $\epsilon = \frac{e_0}{s_0}$ be small. Instead, if one or both of k_{-1} and k_2 are much larger than $k_1 e_0$, then the formation of complex c is a rapid exponential process, and can be taken to be in quasi-steady state. Make this argument systematic by introducing appropriate nondimensional variables and then find the resulting quasi-steady-state dynamics. (Segel, 1988; Frenzen and Maini, 1988; Segel and Slemrod, 1989; Sneyd and Tranchina, 1989).

15. ATP is known to inhibit its own dephosphorylation. One possible way for this to occur is if ATP binds with the enzyme, holding it in an inactive state, via

$$S_1 + E \underset{k_{-4}}{\overset{k_4}{\rightleftharpoons}} S_1 E.$$

Add this reaction to the Sel'kov model of glycolysis and derive the equations governing glycolysis of the form (1.107)–(1.108). Explain from the model why this additional reaction is inhibitory.

16. In the case of noncompetitive inhibition, the inhibitor combines with the enzyme-substrate complex to give an inactive enzyme-substrate-inhibitor complex which cannot undergo further reaction, but the inhibitor does not combine directly with free enzyme or affect its reaction with substrate. Use the quasi-steady-state approximation to show that the velocity of this reaction is

$$V = V_{\max} \frac{s}{K_m + s + \frac{i}{K_i}s}. \tag{1.169}$$

Identify all parameters. Compare this velocity with the velocity for other types of inhibition discussed in the text.

17. The following reaction scheme is a simplified version of the Goldbeter–Lefever reaction scheme for glycolytic oscillations:

$$R_0 \underset{k_{-1}}{\overset{k_1}{\rightleftharpoons}} T_0,$$

$$S_1 + R_j \underset{k_{-2}}{\overset{k_2}{\rightleftharpoons}} C_j \overset{k}{\rightarrow} R_j + S_2, \qquad j = 0, 1, 2,$$

$$S_2 + R_0 \underset{k_{-3}}{\overset{2k_3}{\rightleftharpoons}} R_1,$$

$$S_2 + R_1 \underset{2k_{-3}}{\overset{k_3}{\rightleftharpoons}} R_2.$$

Show that, under appropriate assumptions about the ratios k_1/k_{-1} and $\frac{k_{-2}+k_3}{k_2}$ the equations describing this reaction are of the form (1.107)–(1.108) with $f(\sigma_1, \sigma_2)$ given by (1.125).

18. Use the law of mass action and the quasi-steady-state assumption for the enzymatic reactions to derive a system of equations of the form (1.107)–(1.108) for the Goldbeter–Lefever model of glycolytic oscillations. Verify (1.124).

19. When much of the ATP is depleted in a cell, a considerable amount of cAMP is formed as a product of ATP degradation. This cAMP activates an enzyme phosphorylase that splits glycogen, releasing glucose that is rapidly metabolized, replenishing the ATP supply.

    Devise a model of this control loop and determine conditions under which the production of ATP is oscillatory.

20. (a) Nondimensionalize (1.26)–(1.29) in a way appropriate for the equilibrium approximation (rather than the quasi-steady-state approximation of Section 1.4.2). Hint: Recall that for the equilibrium approximation, the assumption is that $k_1 e_0$ and k_{-1} are large

compared to k_2. You should end up with equations that look something like

$$\epsilon \frac{d\sigma}{d\tau} = \alpha x - \beta\alpha\sigma(1-x),\tag{1.170}$$

$$\epsilon \frac{dx}{d\tau} = \beta\sigma(1-x) - x - \epsilon x,\tag{1.171}$$

where $\epsilon = k_2/k_{-1}$, $\alpha = e_0/s_0$ and $\beta = s_0/K_1$.

(b) Find the behavior, to lowest order in ϵ for this system. (Notice that the slow variable is $\sigma + \alpha x$.)

(c) To lowest order in ϵ, what is the differential equation for σ on this time scale?

(d) Rescale time to find equations valid for small times.

(e) Show that, to lowest order in ϵ, $\sigma + \alpha x =$ constant for small times.

(f) Without calculating the exact solution, sketch the solution in the phase plane, showing the initial fast movement to the slow manifold, and then the movement along the slow manifold.

CHAPTER 2

Cellular Homeostasis

2.1 The Cell Membrane

The cell membrane provides a boundary separating the internal workings of the cell from its external environment. More importantly, it is selectively permeable, permitting the free passage of some materials and restricting the passage of others, thus regulating the passage of materials into and out of the cell. It consists of a double layer (a *bilayer*) of phospholipid molecules about 7.5 nm (75 angstroms) thick (Fig. 2.1). The term *lipid* is used to specify a category of water-insoluble, energy rich macromolecules, typical of fats, waxes, and oils. Irregularly dispersed throughout the phospholipid bilayer are aggregates of globular proteins, which are free to move within the layer, giving the membrane a fluid-like appearance. The membrane also contains water-filled pores with diameters of about 0.8 nm, as well as protein-lined pores, called *channels*, which allow passage of specific molecules. Both the intracellular and extracellular environments consist of, among many other things, a dilute aqueous solution of dissolved salts, primarily NaCl and KCl, which dissociate into Na^+, K^+, and Cl^- ions. The cell membrane acts as a barrier to the free flow of these ions and maintains concentration differences of these ions. In addition, the cell membrane acts as a barrier to the flow of water.

Molecules can be transported across the cell membrane by passive transport or active processes. An active process is one that requires the expenditure of energy, while a passive process results solely from the inherent, random movement of molecules. *Osmosis*, i.e., the diffusion of water down its concentration gradient, is the most important process by which water moves through the cell membrane. Simple diffusion

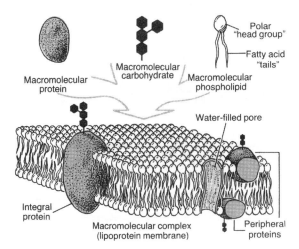

Figure 2.1 Schematic diagram of the cell membrane. (Davis et al., 1985, Fig. 3-1, p. 41.)

accounts for the passage of small molecules through pores and of lipid-soluble molecules through the lipid bilayer. For example, water, urea (a nitrogenous waste product of metabolism), and hydrated Cl^- ions diffuse through membrane pores. Oxygen and carbon dioxide diffuse through the membrane readily because they are soluble in lipids. Sodium and K^+ ions pass through ion-specific channels, driven by diffusion and electrical forces. Some other mechanism must account for the transport of larger sugar molecules such as galactose, glucose, and sucrose, as they are too large to pass through membrane pores.

Concentration differences are set up and maintained by active mechanisms that use energy to pump ions against their concentration gradient. One of the most important of these pumps is the Na^+–K^+ ATPase, which uses the energy stored in ATP molecules to pump Na^+ out of the cell and K^+ in. Another pump, the Ca^{2+} ATPase, pumps Ca^{2+} out of the cell or into the endoplasmic reticulum. There are also a variety of exchange pumps that use the energy inherent in the concentration gradient of one ion type to pump another ion type against its concentration gradient. For example, the Na^+–Ca^{2+} exchanger removes Ca^{2+} from the cell at the expense of Na^+ entry, and similarly for the Na^+–H^+ exchanger. Typical values for intracellular and extracellular ionic concentrations are given in Table 2.1.

Differences in ionic concentrations create a potential difference across the cell membrane that drives ionic currents. Water is also absorbed into the cell because of concentration differences of these ions and also because of other large molecules contained in the cell, whose presence provides an osmotic pressure for the absorption of water. It is the balance of these forces that regulates both the cell volume and the membrane potential.

Table 2.1 Typical values for intracellular and extracellular ionic concentrations, Nernst potentials and resting potentials, from three different cell types. Concentrations are given in units of mM, and potentials are in units of mV. Extracellular concentrations for the squid giant axon are for seawater, while those for frog muscle and red blood cells are for plasma. (Adapted from Mountcastle, 1974, Table 1-1.)

	Squid Giant Axon	Frog Sartorius Muscle	Human Red Blood Cell
Intracellular concentrations			
Na^+	50	13	19
K^+	397	138	136
Cl^-	40	3	78
Mg^{2+}	80	14	5.5
Extracellular concentrations			
Na^+	437	110	155
K^+	20	2.5	5
Cl^-	556	90	112
Mg^{2+}	53	1	2.2
Nernst potentials			
V_{Na}	+56	+55	+55
V_K	−77	−101	−86
V_{Cl}	−68	−86	−9
Resting potentials	−65	−99	−6 to −10

2.2 Diffusion

To keep track of a chemical concentration or any other measurable entity, we must track where it comes from and where it goes; that is, we must write a *conservation law*. If U is some chemical species in some region, then the appropriate conservation law takes the following form (in words):

rate of change of U = rate of production of U + accumulation of U due to transport.

If Ω is a region of space, this conservation law can be written symbolically as

$$\frac{d}{dt} \int_\Omega u \, dV = \int_\Omega f \, dV - \int_{\partial\Omega} \mathbf{J} \cdot \mathbf{n} \, dA, \tag{2.1}$$

where u is the concentration of the chemical species U, $\partial\Omega$ is the boundary of the region Ω, $\mathbf{n}$ is the outward unit normal to the boundary of Ω, f represents the local production density of U per unit volume, and $\mathbf{J}$ is the flux density of U. According to the divergence theorem, if $\mathbf{J}$ is sufficiently smooth, then

$$\int_{\partial\Omega} \mathbf{J} \cdot \mathbf{n}\, dA = \int_{\Omega} \nabla \cdot \mathbf{J}\, dV, \qquad (2.2)$$

so that if the volume in which u is being measured is fixed but arbitrary, the integrals can be dropped, with the result that

$$\frac{\partial u}{\partial t} = f - \nabla \cdot \mathbf{J}. \qquad (2.3)$$

This, being a conservation law, is inviolable. However, there are many ways in which the production term f and the flux $\mathbf{J}$ can vary. Indeed, much of our study in this book is involved in determining appropriate models of production and flux.

2.2.1 Fick's Law

Suppose that u is a function of a single spatial variable, x, and consider the two situations shown in Fig. 2.2, one where u has a steep gradient, the other with a shallow gradient. It is intuitively reasonable that the flux of u should be greater in magnitude in the first case than in the second, and this is indeed what is found experimentally, provided u is not too large. Thus

$$J = -D\frac{du}{dx}. \qquad (2.4)$$

Note the sign of J. By definition, a flux of u from left to right is identified as a positive flux, and thus the flux is opposite in sign to the gradient.

In higher dimensions

$$\mathbf{J} = -D\nabla u. \qquad (2.5)$$

Equation (2.5) is called a *constitutive relationship*, and for chemical species it is called *Fick's law*. The scalar D is the *diffusion coefficient* and is characteristic of the solute and the fluid in which it is dissolved. If u represents the heat content of the volume, (2.5) is called *Newton's law of cooling*. Fick's law is not really a law, but is a reasonable

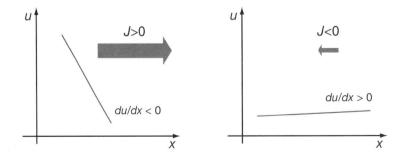

Figure 2.2   Fick's Law. The flux is proportional to the gradient, and opposite in sign.

approximation to reality if the concentration of the chemical species is not too high. When Fick's law applies, the conservation equation becomes the reaction–diffusion equation

$$\frac{\partial u}{\partial t} = \nabla \cdot (D\nabla u) + f, \tag{2.6}$$

or, if D is a constant,

$$\frac{\partial u}{\partial t} = D\nabla^2 u + f. \tag{2.7}$$

The diffusion equation can also be derived from a random walk (Section 2.9.1).

There is a vast literature on reaction–diffusion equations. To mention but a very few, Aronson and Weinberger (1975), Britton (1986) and Grindrod (1991) are biologically oriented, as is Murray (2002), while Smoller (1994) and Fife (1979) are more theoretical presentations.

2.2.2 Diffusion Coefficients

A quantitative understanding of diffusion was given by Einstein (1906) in his theory of Brownian motion. He showed that if a spherical solute molecule is large compared to the solvent molecule, then

$$D = \frac{kT}{6\pi \mu a}, \tag{2.8}$$

where $k = \frac{R}{N_A}$ is Boltzmann's constant, N_A is Avogadro's number, T is the absolute temperature of the solution, μ is the coefficient of viscosity for the solute, and a is the radius of the solute molecule. For nonspherical molecules, Einstein's formula generalizes to

$$D = \frac{kT}{f}, \tag{2.9}$$

where f is the Stokes frictional coefficient of the particle and $f = 6\pi \mu a$ for a sphere of radius a. The molecular weight of a spherical molecule is

$$M = \frac{4}{3}\pi a^3 \rho, \tag{2.10}$$

where ρ is the molecular density, so that, in terms of molecular weight,

$$D = \frac{kT}{3\mu}\left(\frac{\rho}{6\pi^2 M}\right)^{1/3}. \tag{2.11}$$

The density of most large protein molecules is nearly constant (about 1.3–1.4 g/cm^3), so that $DM^{1/3}$ is nearly the same for spherical molecules at a fixed temperature. The diffusion of small molecules, such as the respiratory gases, is different, being proportional to $M^{-1/2}$.

Table 2.2 Molecular weight and diffusion coefficients of some biochemical substances in dilute aqueous solution.

Substance	Molecular Weight	$D(\text{cm}^2/\text{s})$
hydrogen	1	4.5×10^{-5}
oxygen	32	2.1×10^{-5}
carbon dioxide	48	1.92×10^{-5}
glucose	192	6.60×10^{-6}
insulin	5,734	2.10×10^{-6}
Cytochrome c	13,370	1.14×10^{-6}
Myoglobin	16,900	5.1×10^{-7}
Serum albumin	66,500	6.03×10^{-7}
hemoglobin	64,500	6.9×10^{-7}
Catalase	247,500	4.1×10^{-7}
Urease	482,700	3.46×10^{-7}
Fibrinogen	330,000	1.97×10^{-7}
Myosin	524,800	1.05×10^{-7}
Tobacco mosaic virus	40,590,000	5.3×10^{-8}

2.2.3 Diffusion Through a Membrane: Ohm's Law

We can use Fick's law to derive the chemical analogue of Ohm's law for a membrane of thickness L. Suppose that a membrane separates two large reservoirs of a dilute chemical, with concentration c_l on the left (at $x = 0$), and concentration c_r on the right (at $x = L$). According to the diffusion equation, in the membrane (assuming that the only gradients are transverse to the membrane)

$$\frac{\partial c}{\partial t} = D\frac{\partial^2 c}{\partial x^2}, \tag{2.12}$$

subject to boundary conditions $c(0, t) = c_l, c(L, t) = c_r$.

The full time-dependent solution can be found using separation of variables, but for our purposes here, the steady-state solution is sufficient. At steady state, $\frac{\partial c}{\partial t} = 0$, so that $\frac{\partial J}{\partial x} = -D\frac{\partial^2 c}{\partial x^2} = 0$, from which it follows that $J = -D\frac{\partial c}{\partial x} = $ constant, or that $c(x) = ax + b$, for some constants a and b. Applying the boundary conditions, we obtain

$$c(x) = c_l + (c_r - c_l)\frac{x}{L}. \tag{2.13}$$

From Fick's law it follows that the flux of chemical is constant, independent of x, and is given by

$$J = \frac{D}{L}(c_l - c_r). \tag{2.14}$$

The ratio L/D is the effective "resistance" of the membrane, and so D/L is called the *conductance*, or *permeability*, per unit area.

2.2.4 Diffusion into a Capillary

Suppose that a long capillary, open at one end, with uniform cross-sectional area A and filled with water, is inserted into a solution of known chemical concentration C_0, and the chemical species is free to diffuse into the capillary through the open end. Since the concentration of the chemical species depends only on the distance along the tube and time, it is governed by the diffusion equation

$$\frac{\partial c}{\partial t} = D\frac{\partial^2 c}{\partial x^2}, \qquad 0 < x < \infty, \qquad t > 0, \tag{2.15}$$

where for convenience we assume that the capillary is infinitely long. Because the solute bath in which the capillary sits is large, it is reasonable to assume that the chemical concentration at the tip is fixed at $C(0,t) = C_0$, and since the tube is initially filled with pure water we set $C(x,0) = 0$.

The solution of this problem is given by

$$C(x,t) = 2C_0\left(1 - \frac{1}{\sqrt{2\pi}}\int_{-\infty}^{z}\exp\left(-\frac{s^2}{2}\right)ds\right), \qquad z = \frac{x}{\sqrt{2Dt}}. \tag{2.16}$$

From this, one can easily calculate that the total number of molecules that enter the capillary in a fixed time T is

$$N = A\int_0^{\infty} C(x,T)\,dx = 2C_0 A\sqrt{\frac{TD}{\pi}}. \tag{2.17}$$

From this equation it is possible to determine the diffusion coefficient by solving (2.17) for D, yielding

$$D = \frac{\pi N^2}{4C_0^2 A^2 T}. \tag{2.18}$$

A second useful piece of information is found from (2.16) by observing that $C(x,t)/C_0$ is constant on any curve for which z is constant. Thus, the curve $t = x^2/D$ is a level curve for the concentration, and gives a measure of how fast the substance is moving into the capillary. The time $t = x^2/D$ is called the *diffusion time* for the process. To give some idea of the effectiveness of diffusion in various cellular contexts, in Table 2.3 is shown typical diffusion times for a variety of cellular structures. Clearly, diffusion is effective for transport when distances are short, but totally inadequate for longer distances, such as along a nerve axon. Obviously, biological systems must employ other transport mechanisms in these situations in order to survive.

2.2.5 Buffered Diffusion

It is often the case that reactants in an enzymatic reaction (as in Chapter 1) are free to diffuse, so that one must keep track of the effects of both diffusion and reaction. Such problems, called *reaction–diffusion systems*, are of fundamental significance in physiology and are also important and difficult mathematically.

Table 2.3 Estimates of diffusion times for cellular structures of typical dimensions, computed from the relation $t = x^2/D$ using $D = 10^{-5} \text{cm}^2/\text{s}$ (typical for molecules the size of oxygen or carbon dioxide).

x	t	Example
10 nm	100 ns	Thickness of cell membrane
1 μm	1 ms	Size of mitochondrion
10 μm	100 ms	Radius of small mammalian cell
100 μm	10 s	Diameter of a large muscle fiber
250 μm	60 s	Radius of squid giant axon
1 mm	16.7 min	Half-thickness of frog sartorius muscle
2 mm	1.1 h	Half-thickness of lens in the eye
5 mm	6.9 h	Radius of mature ovarian follicle
2 cm	2.6 d	Thickness of ventricular myocardium
1 m	31.7 yrs	Length of a (long!) nerve axon

An important situation, in which reaction and diffusion interact to modify the behavior, occurs when a diffusing species is buffered by a larger diffusing molecule. This occurs, for example, with oxygen in muscle (which we discuss below), or Ca^{2+}, or H^+. The earliest studies of the buffered diffusion equation were those of Irving et al. (1990) and Wagner and Keizer (1994), while Neher and his colleagues (Zhou and Neher, 1993; Naraghi and Neher, 1997; Naraghi et al. 1998) have done a great deal of work on Ca^{2+} buffering. More theoretical analyses have been performed by Sneyd et al. (1998), Smith et al. (2001), and Tsai and Sneyd (2005, 2007a,b).

Consider a "one-dimensional" cell is which there are hydrogen ions (for example) and buffer, B. We assume that the buffering reaction follows

$$\text{H}^+ + \text{B} \underset{k_-}{\overset{k_+}{\rightleftharpoons}} \text{HB}. \tag{2.19}$$

Conservation implies

$$\frac{\partial u}{\partial t} = D_h \frac{\partial^2 u}{\partial x^2} + k_- w - k_+ uv + f(t, x, u), \tag{2.20}$$

$$\frac{\partial v}{\partial t} = D_b \frac{\partial^2 v}{\partial x^2} + k_- w - k_+ uv, \tag{2.21}$$

$$\frac{\partial w}{\partial t} = D_b \frac{\partial^2 w}{\partial x^2} - k_- w + k_+ uv, \tag{2.22}$$

where $u = [\text{H}^+]$, $v = [\text{B}]$, and $w = [\text{HB}]$. Since the buffer is a large molecule, we assume that the diffusion of B is the same as that of HB. We impose no-flux boundary conditions at the ends of the cell and assume that v and w are initially uniform (for example, if w is initially zero, and the buffer is uniformly distributed). The reaction term $f(t, x, u)$ denotes all the other reactions of u apart from the buffering.

Adding (2.21) and (2.22) we obtain

$$\frac{\partial(v + w)}{\partial t} = D_b \frac{\partial^2(v + w)}{\partial x^2}. \tag{2.23}$$

Since $v + w$ is initially uniform, it remains uniform for all time, so that $v + w = w_0$, where w_0 is the total amount of buffer.

If the buffering reaction is fast compared to the other reactions (i.e., those described by $f(t, x, u)$), then we can assume u and v to be in quasi-equilibrium, so that

$$k_-(w_0 - v) - k_+ uv = 0, \tag{2.24}$$

which implies that

$$v = \frac{K_{eq} w_0}{K_{eq} + u}, \qquad \text{where } K_{eq} = \frac{k_-}{k_+}. \tag{2.25}$$

Subtracting (2.21) from (2.20) yields

$$\frac{\partial(u - v)}{\partial t} = D_h \frac{\partial^2 u}{\partial x^2} - D_b \frac{\partial^2 v}{\partial x^2} + f(t, x, u). \tag{2.26}$$

However, since we know v as a function of u, we can eliminate v to find a nonlinear reaction–diffusion equation for u alone,

$$\frac{\partial}{\partial t}\left(u - \frac{K_{eq} w_0}{K_{eq} + u}\right) = D_h \frac{\partial^2 u}{\partial x^2} - D_b \frac{\partial^2}{\partial x^2}\left(\frac{K_{eq} w_0}{K_{eq} + u}\right) + f(t, x, u). \tag{2.27}$$

We expand some of the derivatives and find

$$\left(1 + \frac{K_{eq} w_0}{(K_{eq} + u)^2}\right) u_t = D_h \frac{\partial^2 u}{\partial x^2} + D_b \frac{\partial}{\partial x}\left(\frac{K_{eq} w_0}{(K_{eq} + u)^2} u_x\right) + f(t, x, u). \tag{2.28}$$

Letting

$$\theta(u) = \frac{K_{eq} w_0}{(K_{eq} + u)^2} \tag{2.29}$$

then gives

$$u_t = \frac{D_h + \theta(u) D_b}{1 + \theta(u)} u_{xx} + \frac{D_b \theta'(u)}{1 + \theta(u)} (u_x)^2 + \frac{f(t, x, u)}{1 + \theta(u)}. \tag{2.30}$$

Thus, buffering gives rise to a nonlinear transport equation with a diffusion coefficient that is a nonlinear function of u.

In some cases it is reasonable to assume that $u \ll K_{eq}$. In this limit we find that u has an effective diffusion coefficient

$$D_{eff} = \frac{D_h + D_b \frac{w_0}{K_{eq}}}{1 + \frac{w_0}{K_{eq}}}, \tag{2.31}$$

a convex linear combination of the two diffusion coefficients, D_h and D_b. In addition, the reaction terms are scaled by the constant factor $\frac{1}{1 + w_0/K_{eq}}$. If, additionally, $D_b = 0$, we recover the usual diffusion equation, but one for which both the diffusion coefficient and the reaction terms are scaled by the same constant factor.

2.3 Facilitated Diffusion

A second important example in which both diffusion and reaction play a role is known as *facilitated diffusion*. Facilitated diffusion occurs when the flux of a chemical is amplified by a reaction that takes place in the diffusing medium. An example of facilitated diffusion occurs with the flux of oxygen in muscle fibers. In muscle fibers, oxygen is bound to myoglobin and is transported as oxymyoglobin, and this transport is greatly enhanced above the flow of oxygen in the absence of myoglobin (Wyman, 1966; Murray, 1971; Murray and Wyman, 1971; Rubinow and Dembo, 1977).

This well-documented observation needs further explanation, because at first glance it seems counterintuitive. Myoglobin molecules are much larger (molecular weight $M = 16{,}890$) than oxygen molecules (molecular weight $M = 32$) and therefore have a much smaller diffusion coefficient ($D = 4.4 \times 10^{-7}$ and $D = 1.2 \times 10^{-5} \text{cm}^2/\text{s}$ for myoglobin and oxygen, respectively). The diffusion of oxymyoglobin would therefore seem to be much slower than the diffusion of free oxygen. Further, from the calculation in the last section, the diffusion of free oxygen is much slower when it is buffered by myoglobin since the effective diffusion coefficient of oxygen is lowered substantially by diffusion.

To anticipate slightly, the answer is that, at steady state, the total transport of oxygen is the sum of the free oxygen transport and additional oxygen that is transported by the diffusing buffer. If there is a lot of buffer, with a lot of oxygen bound, this additional transport due to the buffer can be substantial.

A simple model of this phenomenon is as follows. Suppose we have a slab reactor containing diffusing myoglobin. On the left (at $x = 0$) the oxygen concentration is held fixed at s_0, and on the right (at $x = L$) it is held fixed at s_L, which is assumed to be less than s_0.

If f is the rate of uptake of oxygen into oxymyoglobin, then equations governing the concentrations of $s = [\text{O}_2], e = [\text{Mb}], c = [\text{MbO}_2]$ are

$$\frac{\partial s}{\partial t} = D_s \frac{\partial^2 s}{\partial x^2} - f, \tag{2.32}$$

$$\frac{\partial e}{\partial t} = D_e \frac{\partial^2 e}{\partial x^2} - f, \tag{2.33}$$

$$\frac{\partial c}{\partial t} = D_c \frac{\partial^2 c}{\partial x^2} + f. \tag{2.34}$$

It is reasonable to take $D_e = D_c$, since myoglobin and oxymyoglobin are nearly identical in molecular weight and structure. Since myoglobin and oxymyoglobin remain inside the slab, it is also reasonable to specify the boundary conditions $\partial e/\partial x = \partial c/\partial x = 0$ at $x = 0$ and $x = L$. Because it reproduces the oxygen saturation curve (discussed in

Chapter 13), we assume that the reaction of oxygen with myoglobin is governed by the elementary reaction

$$O_2 + Mb \underset{k_-}{\overset{k_+}{\rightleftharpoons}} MbO_2,$$

so that (from the law of mass action) $f = -k_-c + k_+se$. The total amount of myoglobin is conserved by the reaction, so that at steady state $e + c = e_0$ and (2.33) is superfluous.

At steady state,

$$0 = s_t + c_t = D_s s_{xx} + D_c c_{xx}, \tag{2.35}$$

and thus there is a second conserved quantity, namely

$$D_s \frac{ds}{dx} + D_c \frac{dc}{dx} = -J, \tag{2.36}$$

which follows by integrating (2.35) once with respect to x. The constant J (which is yet to be determined) is the sum of the flux of free oxygen and the flux of oxygen in the complex oxymyoglobin, and therefore represents the total flux of oxygen. Integrating (2.36) with respect to x between $x = 0$ and $x = L$, we can express the total flux J in terms of boundary values of the two concentrations as

$$J = \frac{D_s}{L}(s_0 - s_L) + \frac{D_c}{L}(c_0 - c_L), \tag{2.37}$$

although the values c_0 and c_L are as yet unknown.

To further understand this system of equations, we introduce dimensionless variables, $\sigma = \frac{k_+}{k_-}s$, $u = c/e_0$, and $x = Ly$, in terms of which (2.32) and (2.34) become

$$\epsilon_1 \sigma_{yy} = \sigma(1 - u) - u = -\epsilon_2 u_{yy}, \tag{2.38}$$

where $\epsilon_1 = \frac{D_s}{e_0 k_+ L^2}$, $\epsilon_2 = \frac{D_c}{k_- L^2}$.

Reasonable numbers for the uptake of oxygen by myoglobin (Wittenberg, 1966) are $k_+ = 1.4 \times 10^{10} \text{cm}^3 \text{ M}^{-1}\text{s}^{-1}$, $k_- = 11 \text{ s}^{-1}$, and $L = 0.022$ cm in a solution with $e_0 = 1.2 \times 10^{-5}$ M/cm^3. (These numbers are for an experimental setup in which the concentration of myoglobin was substantially higher than what naturally occurs in living tissue.) With these numbers we estimate that $\epsilon_1 = 1.5 \times 10^{-7}$, and $\epsilon_2 = 8.2 \times 10^{-5}$. Clearly, both of these numbers are small, suggesting that oxygen and myoglobin are at quasi-steady state throughout the medium, with

$$c = e_0 \frac{s}{K + s}, \tag{2.39}$$

where $K = k_-/k_+$. Now we substitute (2.39) into (2.37) to find the flux

$$
\begin{aligned}
J &= \frac{D_s}{L}(s_0 - s_L) + \frac{D_c}{L}e_0\left(\frac{s_0}{K + s_0} - \frac{s_L}{K + s_L}\right) \\
&= \frac{D_s}{L}(s_0 - s_L)\left(1 + \frac{D_c}{D_s}\frac{e_0 K}{(s_0 + K)(s_L + K)}\right) \\
&= \frac{D_s}{L}(1 + \mu\rho)(s_0 - s_L),
\end{aligned} \tag{2.40}
$$

where $\rho = \frac{D_c}{D_s}\frac{e_0}{K}, \mu = \frac{K^2}{(s_0+K)(s_L+K)}$.

In terms of dimensionless variables the full solution is given by

$$
\sigma(y) + \rho u(y) = y[\sigma(1) + \rho u(1)] + (1 - y)[\sigma(0) + \rho u(0)], \tag{2.41}
$$

$$
u(y) = \frac{\sigma(y)}{1 + \sigma(y)}. \tag{2.42}
$$

Now we see how diffusion can be facilitated by an enzymatic reaction. In the absence of a diffusing carrier, $\rho = 0$ and the flux is purely Fickian, as in (2.14). However, in the presence of carrier, diffusion is enhanced by the factor $\mu\rho$. The maximum enhancement possible is at zero concentration, when $\mu = 1$. With the above numbers for myoglobin, this maximum enhancement is substantial, being $\rho = 560$. If the oxygen supply is sufficiently high on the left side (near $x = 0$), then oxygen is stored as oxymyoglobin. Moving to the right, as the total oxygen content drops, oxygen is released by the myoglobin. Thus, even though the bound oxygen diffuses slowly compared to free oxygen, the quantity of bound oxygen is high (provided that e_0 is large compared to the half saturation level K), so that lots of oxygen is transported. We can also understand that to take advantage of the myoglobin-bound oxygen, the concentration of oxygen must drop to sufficiently low levels so that myoglobin releases its stored oxygen.

To explain it another way, note from (2.40) that J is the sum of two terms, the usual ohmic flux term and an additional term that depends on the diffusion coefficient of MbO$_2$. The total oxygen flux is the sum of the flux of free oxygen and the flux of oxygen bound to myoglobin. Clearly, if oxymyoglobin is free to diffuse, the total flux is thereby increased. But since oxymyoglobin can only diffuse down its gradient, the concentration of oxymyoglobin must be higher on one side than the other.

In Fig. 2.3A are shown the dimensionless free oxygen concentration σ and the dimensionless bound oxygen concentration u plotted as functions of position. Notice that the free oxygen content falls at first, indicating higher free oxygen flux, and the bound oxygen decreases more rapidly at larger y. Perhaps easier to interpret is Fig. 2.3B, where the dimensionless flux of free oxygen and the dimensionless flux of bound oxygen are shown as functions of position. Here we can see that as the free oxygen concentration drops, the flux of free oxygen also drops, but the flux of bound oxygen increases. For large y, most of the flux is due to the bound oxygen. For these figures, $\rho = 10, \sigma(0) = 2.0$, $\sigma(1) = 0.1$.

One mathematical detail that was ignored in this discussion is the validity of the quasi-steady-state solution (2.39) as an approximation of (2.38). Usually, when one

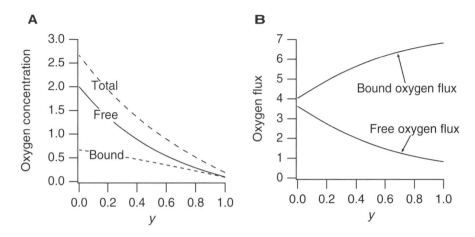

Figure 2.3 A: Free oxygen content $\sigma(y)$ and bound oxygen content $u(y)$ as functions of y. B: Free oxygen flux $-\sigma'(y)$ and bound oxygen flux $-\rho u'(y)$ as functions of y.

makes an approximation to boundary value problems in which the order of the system is reduced (as here where the order is four, and drops by two when ϵ_1 and ϵ_2 are ignored), there are difficulties with the solution at the boundary, because the boundary conditions cannot, in general, be met. Such problems, discussed briefly in Chapter 1 in the context of enzyme kinetics, are called *singular perturbation problems*, because the behavior of the solutions as functions of the small parameters is not regular, but singular (certain derivatives become infinitely large as the parameters approach zero). In this problem, however, there are no boundary layers, and the quasi-steady-state solution is a uniformly valid approximation to the solution. This occurs because the boundary conditions on c are of no-flux (Neumann) type, rather than of fixed (Dirichlet) type. That is, since the value of c is not specified by the boundary conditions, c is readily adjusted so that there are no boundary layers. Only a slight correction to the quasi-steady-state solution is needed to meet the no-flux boundary conditions, but this correction affects only the derivative, not the value, of c in a small region near the boundaries.

2.3.1 Facilitated Diffusion in Muscle Respiration

Even at rest, muscle fibers consume oxygen. This is because ATP is constantly consumed to maintain a nonzero membrane potential across a muscle cell wall, and this consumption of energy requires constant metabolizing of sugar, which consumes oxygen. Although sugar can be metabolized anaerobically, the waste product of this reaction is lactic acid, which is toxic to the cell. In humans, the oxygen consumption of live muscle tissue at rest is about 5×10^{-8} mol/cm³s, and the concentration of myoglobin is about 2.8×10^{-7} mol/cm³. Thus, when myoglobin is fully saturated, it contains only about a 5 s supply of oxygen. Further, the oxygen at the exterior of the muscle cell must penetrate to the center of the cell to prevent the oxygen concentration at the center falling to zero, a condition called *oxygen debt*.

To explain how myoglobin aids in providing oxygen to a muscle cell and helps to prevent oxygen debt, we examine a model of oxygen consumption that includes the effects of diffusion of oxygen and myoglobin. We suppose that a muscle fiber is a long circular cylinder (radius $a = 2.5 \times 10^{-3}$ cm) and that diffusion takes place only in the radial direction. We suppose that the oxygen concentration at the boundary of the fiber is a fixed constant and that the distribution of chemical species is radially symmetric. With these assumptions, the steady-state equations governing the diffusion of oxygen and oxymyoglobin are

$$D_s \frac{1}{r} \frac{d}{dr}\left(r \frac{ds}{dr}\right) - f - g = 0, \tag{2.43}$$

$$D_c \frac{1}{r} \frac{d}{dr}\left(r \frac{dc}{dr}\right) + f = 0, \tag{2.44}$$

where, as before, $s = [O_2]$, $c = [MbO_2]$, and $f = -k_- c + k_+ se$. The coordinate r is in the radial direction. The new term in these equations is the constant g, corresponding to the constant consumption of oxygen. The boundary conditions are $s = s_a, dc/dr = 0$ at $r = a$, and $ds/dr = dc/dr = 0$ at $r = 0$. For muscle, s_a is typically 3.5×10^{-8} mol/cm^3 (corresponding to the partial pressure 20 mm Hg). Numerical values for the parameters in this model are difficult to obtain, but reasonable numbers are $D_s = 10^{-5}$ cm^2/s, $D_c = 5 \times 10^{-7}$ cm^2/s, $k_+ = 2.4 \times 10^{10}$ cm^3/mol $\cdot$ s, and $k_- = 65$/s (Wyman, 1966).

Introducing nondimensional variables $\sigma = \frac{k_+}{k_-}s, u = c/e_0$, and $r = ay$, we obtain the differential equations

$$\epsilon_1 \frac{1}{y} \frac{d}{dy}\left(y \frac{d\sigma}{dy}\right) - \gamma = \sigma(1-u) - u = -\epsilon_2 \frac{1}{y} \frac{d}{dy}\left(y \frac{du}{dy}\right), \tag{2.45}$$

where $\epsilon_1 = \frac{D_s}{e_0 k_+ a^2}, \epsilon_2 = \frac{D_c}{k_- a^2}, \gamma = g/k_-$. Using the parameters appropriate for muscle, we estimate that $\epsilon_1 = 2.3 \times 10^{-4}, \epsilon_2 = 1.2 \times 10^{-3}, \gamma = 3.3 \times 10^{-3}$. While these numbers are not as small as for the experimental slab described earlier, they are small enough to warrant the assumption that the quasi-steady-state approximation (2.39) holds in the interior of the muscle fiber.

It also follows from (2.45) that

$$\epsilon_1 \frac{1}{y} \frac{d}{dy}\left(y \frac{d\sigma}{dy}\right) + \epsilon_2 \frac{1}{y} \frac{d}{dy}\left(y \frac{du}{dy}\right) = \gamma. \tag{2.46}$$

We integrate (2.46) twice with respect to y to find

$$\epsilon_1 \sigma + \epsilon_2 u = A \ln y + B + \frac{\gamma}{4} y^2. \tag{2.47}$$

The constants A and B are determined by boundary conditions. Since we want the solution to be bounded at the origin, $A = 0$, and B is related to the concentration at the origin.

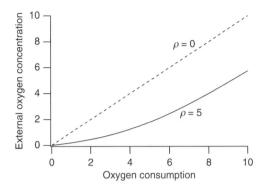

Figure 2.4 Critical concentration σ_0 plotted as a function of oxygen consumption $\frac{\gamma}{4\epsilon_1}$. The dashed curve is the critical concentration with no facilitated diffusion.

Now suppose that there is just enough oxygen at the boundary to prevent oxygen debt. In this model, oxygen debt occurs if σ falls to zero. Marginal oxygen debt occurs if $\sigma = u = 0$ at $y = 0$. For this boundary condition, we take $A = B = 0$. Then the concentration at the boundary must be at least as large as σ_0, where, using the quasi-steady state $\sigma(1 - u) = u$,

$$\sigma_0 + \rho\frac{\sigma_0}{\sigma_0 + 1} = \frac{\gamma}{4\epsilon_1}, \tag{2.48}$$

and where $\rho = \epsilon_2/\epsilon_1$. Otherwise, the center of the muscle is in oxygen debt. Note also that σ_0 is a decreasing function of ρ, indicating a reduced need for external oxygen because of facilitated diffusion.

A plot of this critical concentration σ_0 as a function of the scaled consumption $\frac{\gamma}{4\epsilon_1}$ is shown in Fig. 2.4. For this plot $\rho = 5$, which is a reasonable estimate for muscle. The dashed curve is the critical concentration when there is no facilitated diffusion ($\rho = 0$). The easy lesson from this plot is that facilitated diffusion decreases the likelihood of oxygen debt, since the external oxygen concentration necessary to prevent oxygen debt is smaller in the presence of myoglobin than without.

A similar lesson comes from Fig. 2.5, where the internal free oxygen content σ is shown, plotted as a function of radius y. The solid curves show the internal free oxygen with facilitated diffusion, and the dashed curve is without. The smaller of the two solid curves and the dashed curve have exactly the critical external oxygen concentration, showing clearly that in the presence of myoglobin, oxygen debt is less likely at a given external oxygen concentration. The larger of the two solid curves has the same external oxygen concentration as the dashed curve, showing again the contribution of facilitation toward preventing oxygen debt. For this figure, $\rho = 5$, $\gamma/\epsilon_1 = 14$.

2.4 Carrier-Mediated Transport

Some substances are insoluble in the cell membrane and yet pass through by a process called *carrier-mediated transport*. It is also called *facilitated diffusion* in many physiology books, although we prefer to reserve this expression for the process described in the

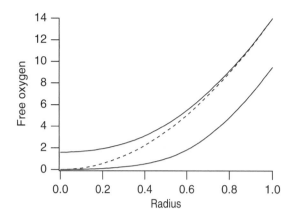

Figure 2.5 Free oxygen σ as a function of radius y. Solid curves show oxygen concentration in the presence of myoglobin ($\rho = 5$), the lower of the two having the critical external oxygen concentration. The dashed curve shows the oxygen concentration without facilitation at the critical external concentration level.

previous section. Carrier-mediated transport is the means by which some sugars cross the cell membrane to provide an energy source for the cell. For example, glucose, the most important of the sugars, combines with a carrier protein at the outer boundary of the membrane, and by means of a conformational change is released from the inner boundary of the membrane.

There are three types of carrier-mediated transport. Carrier proteins that transport a single solute from one side of the membrane to the other are called *uniports*. Other proteins function as coupled transporters by which the simultaneous transport of two solute molecules is accomplished, either in the same direction (called a *symport*) or in the opposite direction (called an *antiport*).

2.4.1 Glucose Transport

Although the details are not certain, the transport of glucose across the lipid bilayer of the cell membrane is thought to occur when the carrier molecule alternately exposes the solute binding site first on one side and then on the other side of the membrane. It is considered highly unlikely that the carrier molecule actually diffuses back and forth through the membrane.

We can model the process of glucose transport as follows: We suppose that the population of enzymatic carrier proteins C has two conformational states, C_i and C_e, with its glucose binding site exposed on the cell interior (subscript i) or exterior (subscript e) of the membrane, respectively. The glucose substrate on the interior S_i can bind with C_i and the glucose substrate on the exterior can bind with enzyme C_e to form the complex P_i or P_e, respectively. Finally, a conformational change transforms P_i into P_e and vice versa. These statements are summarized in Fig. 2.6.

The equations describing this model are

$$\frac{dp_i}{dt} = kp_e - kp_i + k_{+}s_i c_i - k_{-}p_i,$$ (2.49)

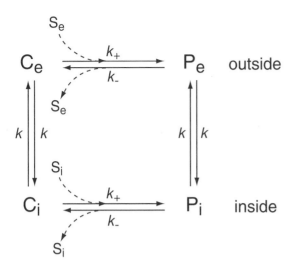

Figure 2.6 Schematic diagram of the glucose transporter described by (2.49)–(2.52).

$$\frac{dp_e}{dt} = kp_i - kp_e + k_+s_ec_e - k_-p_e, \tag{2.50}$$

$$\frac{dc_i}{dt} = kc_e - kc_i + k_-p_i - k_+s_ic_i, \tag{2.51}$$

$$\frac{dc_e}{dt} = kc_i - kc_e + k_-p_e - k_+s_ec_e. \tag{2.52}$$

where $s_i = [S_i]$, $p_i = [P_i]$, etc. Since the total amount of receptor is conserved, we have $p_i + p_e + c_i + c_e = C_0$, where C_0 is a constant (the total transporter concentration). Hence there are only three independent equations, not four. The flux, J, is

$$J = k_-p_i - k_+s_ic_i = k_+s_ec_e - k_-p_e, \tag{2.53}$$

where we have defined a flux from outside to inside to be positive.

We find the steady-state flux by setting all derivatives to zero and solving the resulting algebraic system. It follows that

$$J = \frac{1}{2}KkC_0 \frac{s_e - s_i}{(s_i + K + K_d)(s_e + K + K_d) - K_d^2}, \tag{2.54}$$

where $K = k_-/k_+$ and $K_d = k/k_+$. Since k is the rate at which conformational change takes place, it acts like a diffusion coefficient in that it reflects the effect of random thermal activity at the molecular level.

The nondimensional flux is

$$j = \frac{\sigma_e - \sigma_i}{(\sigma_i + 1 + \kappa)(\sigma_e + 1 + \kappa) - \kappa^2}, \tag{2.55}$$

where $\sigma_i = s_i/K, \sigma_e = s_e/K, \kappa = K_d/K$. A plot of this nondimensional flux is shown in Fig. 2.7, plotted as a function of extracellular glucose σ_e, with fixed intracellular glucose and fixed κ. We can see that the rate of transport is limited by saturation of the enzyme kinetics (this saturation is observed experimentally) and thermal conformational

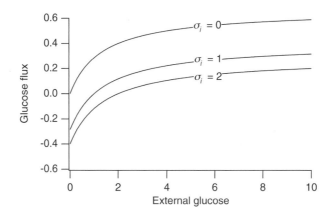

Figure 2.7 Plot of the (nondimensional) flux of glucose as a function of extracellular glucose, for three fixed intracellular glucose concentrations (σ_i), with $\kappa = K_d/K = 0.5$.

change is crucial to the transport process, as transport J drops to zero if $K_d = 0$. The binding affinity of the carrier protein for glucose (k_+), and hence the flux of glucose, is controlled by insulin.

It is important to recognize that the above expression for J is for the steady-state flux only. If the system is not at steady state, then the flux from the outside to the transporter, $J_{on} = k_+s_ec_e - k_-p_e$, need not be the same as the flux off the transporter to the inside, $J_{off} = k_-p_i - k_+s_ic_i$. Obviously, in this case the differential equations must be solved to obtain J_{on} and J_{off}.

It should be noted that there are two ways that the model of Fig. 2.6 can be understood. First, as we did here, we can let each variable represent the concentration of transporters in each of the four possible states. In this case, the conservation relationship is $s_i + p_i + s_e + p_e = C_0$. If each of the variables is scaled by C_0, the conservation relationship becomes $s_i + p_i + s_e + p_e = 1$, and each variable is then the fraction of the population in each state.

However, there is another way to interpret this second conservation relationship. If $s_i + p_i + s_e + p_e = 1$ we can interpret the model as referring to the behavior of a single exchanger, in which case the variables are probabilities of being in a given state, and the exchanger is modeled as a Markov process (see the Appendix to this chapter).

Markov models such as that shown in Fig. 2.6 can often be simplified by assuming that some of the transitions are much faster than others. The technique of reduction using a fast time scale is used in many places throughout this book; indeed, it is used in Chapter 1, in the equilibrium and quasi-steady-state approximations of enzyme kinetics; even though the technique is described in Chapter 1, it is sufficiently important that it warrants repeating.

The procedure can be simply illustrated with this model of the glucose transporter. Suppose that the binding and release of glucose is much faster than the change in conformation, i.e., that the transitions between C_e and P_e, and between C_i and P_i, are much faster than those between C_i and C_e, or between P_e and P_i, so that $K_d \ll 1$. Assuming fast equilibrium between C_e and P_e, and between C_i and P_i, gives

$$s_e c_e = K p_e, \qquad s_i c_i = K p_i. \qquad (2.56)$$

Now, we introduce two variables, $x = c_e + p_e$, $y = c_i + p_i = 1 - x$ (taking $C_0 = 1$). The differential equation for x is found by adding (2.50) and (2.52) to be

$$\begin{aligned} \frac{dx}{dt} &= k c_i + k p_i - k c_e - k p_e \\ &= ky - kx \\ &= k(1 - 2x), \end{aligned} \qquad (2.57)$$

from which it follows that the steady value of x is $x_0 = 1/2$.

Next, from (2.56) we find that

$$x = c_e(1 + s_e/K) = p_e(1 + K/s_e), \qquad (2.58)$$

with similar equations for y. Hence, at steady state, the flux through the transporter is given by

$$\begin{aligned} J = k_+ s_e c_e - k_- p_e &= k p_i - k p_e \\ &= \frac{k s_i x_0}{s_i + K} - \frac{k s_e x_0}{s_e + K} \\ &= \frac{k s_i \frac{1}{2}}{s_i + K} - \frac{k s_e \frac{1}{2}}{s_e + K} \\ &= k K \frac{1}{2} \frac{s_i - s_e}{(s_i + K)(s_e + K)}, \end{aligned} \qquad (2.59)$$

where we have used (2.58) to replace p_e, and the analogous equation to replace p_i.

Notice that this answer is the same as found by letting $K_d \to 0$ in (2.54). However, while the two approaches give the same answer, the quasi-steady-state reduction of the full model is often preferable, especially when the solution of the full model is difficult to obtain.

Other examples of how to simplify Markov models with a fast time scale reduction are given in Exercises 12 and 13.

2.4.2 Symports and Antiports

Models of symport and antiport transporters follow in similar fashion. For a symport the protein carrier has multiple binding sites, which can be exposed to the intracellular or extracellular space. A change of conformation exchanges the location of all of the participating binding sites, from inside to outside, or vice versa. An example of a symport is the Na^+-driven glucose symport that transports glucose and Na^+ from the lumen of the gut to the intestinal epithelium. A similar process occurs in epithelial cells lining the proximal tubules in the kidney, to remove glucose and amino acids from the filtrate (discussed in Chapter 17). Five different amino acid cotransporters have been identified.

If there are k binding sites that participate in the exchange, then there are 2^k possible combinations of bound and unbound sites. The key assumption that makes this model of transport work is that only the completely unbound or completely bound carrier participates in a conformational change. Thus, there is a carrier molecule, say C, with two conformations, C_i and C_e, and a fully bound complex P, also with two conformations, P_i and P_e, and possible transformation between the two conformations,

$$C_i \underset{k_{-c}}{\overset{k_c}{\rightleftharpoons}} C_e, \qquad P_i \underset{k_{-p}}{\overset{k_p}{\rightleftharpoons}} P_e. \tag{2.60}$$

In addition, there are 2^k possible combinations of binding and unbinding in each of the two conformations. For example, with two substrates S and T, and one binding site for each, we have the complexes C, SC, CT, and SCT = P. The possible reactions are summarized in Fig. 2.8.

Unfortunately, the analysis of this fully general reaction scheme is quite complicated. However, it simplifies significantly if we assume that the intermediates can be safely ignored and postulate the multi-molecular reaction scheme

$$m\text{S} + n\text{T} + \text{C} \underset{k_-}{\overset{k_+}{\rightleftharpoons}} \text{P}. \tag{2.61}$$

Now the result for a symport is strikingly similar to the uniport flux, with

$$J = \frac{1}{2} K_d K k_+ C_0 \frac{s_e^m t_e^n - s_i^m t_i^n}{(s_i^m t_i^n + K + K_d)(s_e^m t_e^n + K + K_d) - K_d^2}, \tag{2.62}$$

where the flux of s is mJ and the flux of t is nJ. Here we have set $k_c = k_{-c} = k_p = k_{-p} = k$ and then $K = k_-/k_+$ and $K_d = k/k_+$.

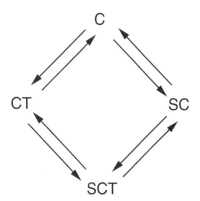

Figure 2.8 States and possible transitions of a transporter with two substrates, S and T, and one binding site for each.

For an antiport, the subscripts on one of the substances must be exchanged, to give

$$J = \frac{1}{2} K_d K k_+ C_0 \frac{s_e^m t_i^n - s_i^m t_e^n}{(s_i^m t_e^n + K + K_d)(s_e^m t_i^n + K + K_d) - K_d^2}. \tag{2.63}$$

The effectiveness of this type of exchanger is determined by the coefficients m and n. For this antiport, flux is positive (S flows inward and T flows outward) if

$$\left(\frac{s_e}{s_i}\right)^m > \left(\frac{t_e}{t_i}\right)^n. \tag{2.64}$$

For example, for the Na^+–Ca^{2+} exchanger (discussed in more detail in the next section) which exchanges three Na^+ ions for one Ca^{2+} ion, a ratio of extracellular to intracellular Na^+ of about 8 can be used to effectively pump Ca^{2+} out of a cell even when the ratio of extracellular to intracellular Ca^{2+} is 500.

2.4.3 Sodium–Calcium Exchange

For the glucose transporter described above, membrane flux is driven by a concentration difference of glucose across the membrane, and if glucose concentrations equilibrate, the transmembrane flux becomes zero. However, because it relies on two concentration differences, an antiport transporter such as the Na^+–Ca^{2+} exchanger can act as a pump. Although this transporter is a passive pump (because it consumes no chemical energy directly), it is often described as a secondarily active pump; it uses the Na^+ gradient to pump Ca^{2+} out of the cell against its concentration gradient, but energy is required to establish and maintain the Na^+ gradient. Na^+–Ca^{2+} exchange is an important mechanism for Ca^{2+} removal in a number of cell types, particularly cardiac ventricular cells, in which much of the Ca^{2+} that enters the cell during an action potential is removed from the cell by the Na^+–Ca^{2+} exchanger (Chapter 12). It has therefore been studied extensively, and a number of highly detailed models have been constructed (Hilgemann, 2004; Kang and Hilgemann, 2004). Here we describe a simple model of this important transporter.

In our model (see Fig. 2.9), E_i is the exchanger protein in the conformation for which the binding sites are exposed to the interior of the cell, and E_e is the conformation for which the binding sites are exposed to the exterior. Starting at state X_1 in the top left of the figure, the exchanger can bind Ca^{2+} inside the cell, simultaneously releasing three Na^+ ions to the interior. A change of conformation to E_e then allows the exchanger to release the Ca^{2+} to the outside and bind three external Na^+. A return to the E_i conformation completes the cycle. Of course, it is a crude approximation to assume that one Ca^{2+} and three Na^+ ions bind or unbind the exchanger simultaneously.

It is now straightforward to calculate the steady flux for this model. As with the previous transporter models, we first solve for the steady-state values of x_1, x_2, y_1, and y_2, the fraction of exchangers in the state X_1, X_2, Y_1, and Y_2, respectively. There are four equations: three differential equations for exchanger states and one conservation

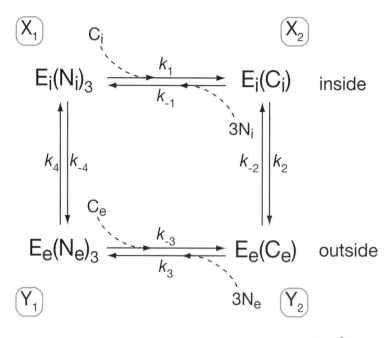

Figure 2.9   Schematic diagram of a simple model of the Na$^+$–Ca^{2+} exchanger.

equation. These are

$$\frac{dx_1}{dt} = k_{-1}n_i^3 x_2 + k_4 y_1 - (k_1 c_i + k_{-4})x_1, \tag{2.65}$$

$$\frac{dx_2}{dt} = k_{-2}y_2 + k_1 c_i x_1 - (k_2 + k_{-1}n_i^3)x_2, \tag{2.66}$$

$$\frac{dy_1}{dt} = k_{-4}x_1 + k_3 n_e^3 y_2 - (k_4 + k_{-3}c_e)y_1, \tag{2.67}$$

$$1 = x_1 + x_2 + y_1 + y_2. \tag{2.68}$$

Here c and n denote, respectively, Ca^{2+} and Na$^+$ concentration, and the subscripts e and i represent external and internal concentrations.

Using a symbolic package such as Maple, the steady-state solution of these equations is easily calculated. The flux, J, is found to be

$$J = k_4 y_1 - k_{-4}x_1 = \frac{k_1 k_2 k_3 k_4 (c_i n_e^3 - K_1 K_2 K_3 K_4 c_e n_i^3)}{16 \text{ positive terms}}, \tag{2.69}$$

where, as usual, $K_i = k_{-i}/k_i$.

Notice that the units of the flux J here (1/time) are different from those in the previous examples (concentration/time), because here the variables x_i and y_i are fractions of exchangers in a particular state (or probabilities) rather than concentrations of exchangers in a particular state. Hence, the flux in this model is a turnover rate, i.e., the number of times the exchanger goes around the cycle per unit time. This can easily be converted to a concentration per time if the concentration of the exchangers is known.

An Electrogenic Exchanger

An important difference between the Na^+–Ca^{2+} exchange process and the transport processes discussed previously is that Na^+ and Ca^{2+} are ions. Since each cycle of the Na^+–Ca^{2+} exchanger transports two positive charges out and three positive charges in, it generates an electric current. Such exchangers are said to be *electrogenic*.

As is discussed in Section 2.6, all cells have an electrical potential difference across their membranes. Clearly, additional work is necessary for the exchanger to move electric current against a potential difference. To take this into account, consider a ligand, L, with a charge z, and suppose that there is a process that moves L from the cell interior with potential V_i to the cell exterior with potential V_e, i.e.,

$$L_i \to L_e. \tag{2.70}$$

The change in chemical potential (cf. Section 1.2) for this reaction is

$$\begin{aligned}\Delta G &= G^0_{L_e} + RT \ln([L_e]) + zFV_e - G^0_{L_i} - RT \ln([L_i]) - zFV_i \\ &= RT \ln\left(\frac{[L_e]}{[L_i]}\right) - zFV, \end{aligned} \tag{2.71}$$

where $V = V_i - V_e$ is the transmembrane potential. (The standard convention is to define the potential difference across the membrane as the internal potential minus the external potential, as discussed further in Section 2.6.1.) The standard free energy for L is the same on both sides of the membrane, so $G^0_{L_e} = G^0_{L_i}$. At equilibrium, $\Delta G = 0$, so that

$$K = \frac{[L_i]_{eq}}{[L_e]_{eq}} = \exp\left(\frac{-zFV}{RT}\right), \tag{2.72}$$

where K is the equilibrium constant for the reaction.

For the Na^+–Ca^{2+} exchanger, the overall reaction begins with three Na^+ outside the cell and one Ca^{2+} inside the cell, and ends with three Na^+ inside the cell and one Ca^{2+} outside. We can write this as

$$3Na^+_e + Ca^{2+}_i \longrightarrow 3Na^+_i + Ca^{2+}_e. \tag{2.73}$$

The change in chemical potential for this reaction is

$$\Delta G = RT \ln\left(\frac{n_i^3 c_e}{n_e^3 c_i}\right) + FV. \tag{2.74}$$

At equilibrium we must have $\Delta G = 0$, in which case

$$\frac{n_{i,eq}^3 c_{e,eq}}{n_{e,eq}^3 c_{i,eq}} = \exp\left(-\frac{FV}{RT}\right). \tag{2.75}$$

Recall that detailed balance requires that around any closed reaction loop the product of the forward rates must be the same as the product of the reverse rates. It follows that

$$k_1 c_{i,eq} k_2 k_3 n_{e,eq}^3 k_4 = n_{i,eq}^3 k_{-1} k_{-4} c_{e,eq} k_{-3} k_{-2}, \qquad (2.76)$$

and thus

$$K_1 K_2 K_3 K_4 = \frac{c_{i,eq}}{c_{e,eq}} \frac{n_{e,eq}^3}{n_{i,eq}^3}. \qquad (2.77)$$

Combining (2.76) and (2.77), we get

$$K_1 K_2 K_3 K_4 = \exp\left(\frac{FV}{RT}\right), \qquad (2.78)$$

which, being independent of the concentrations, must hold in general.

It follows from (2.69) that the flux is given by

$$J = \frac{k_1 k_2 k_3 k_4 (c_i n_e^3 - e^{\frac{FV}{RT}} c_e n_i^3)}{16 \text{ positive terms}}. \qquad (2.79)$$

All of the terms in the denominator are cubic products of rate constants, so that the flux J has units of inverse time. In general, the denominator of this expression also contains terms that depend on the membrane potential difference.

In writing (2.78), no assumption was made about where the charge transfer takes place. From Fig. 2.9 it might appear that the charge transfer takes place during the transitions $Y_1 \to X_1$ and $X_2 \to Y_2$. However, this is not necessarily the case. It could be that those conformational changes are accompanied by no charge transfer, but that the charge transfer occurs during other transitions. However, if we assume that one Ca^{2+} ion is transferred from inside to outside during the $X_2 \to Y_2$ transition, and three Na^+ ions are transferred during the $Y_1 \to X_1$ transition, free energy arguments yield the additional constraints

$$\frac{k_{-2}}{k_2} = \tilde{K}_2 \exp\left(\frac{-2FV}{RT}\right), \qquad \frac{k_4}{k_{-4}} = \tilde{K}_4^{-1} \exp\left(\frac{-3FV}{RT}\right), \qquad (2.80)$$

where $\tilde{K}_2$ and $\tilde{K}_4$ are independent of voltage, and where $K_1 \tilde{K}_2 K_3 \tilde{K}_4 = 1$.

The most important observation is that for given n_i and n_e (set by other mechanisms such as the Na^+–K^+ ATPase discussed in the next section), a negative V enhances the rate at which the Na^+–Ca^{2+} exchanger removes Ca^{2+} from the cell. This makes sense; if V is negative, the potential inside the cell is negative compared to the outside and thus it is easier for the exchanger to move one positive charge into the cell. Since cells typically have a negative resting potential (Section 2.6), the electrogenic nature of the exchanger increases its ability to remove Ca^{2+} in resting conditions. To be specific, if the ratio of extracellular to intracellular Na^+ is 8, and the potential difference is

$V = -85$ mV (which is typical), then Ca^{2+} is removed, provided

$$\frac{c_i}{c_e} > \frac{n_i^3}{n_e^3} e^{\frac{FV}{RT}} = 7 \times 10^{-5}. \tag{2.81}$$

Notice that the difference in potential gives an improvement in the capability of the exchanger by a factor of 27 over an exchanger that is not electrogenic.

2.5 Active Transport

The carrier-mediated transport described above is always down electrochemical gradients, and so is identified with diffusion. Any process that works against gradients requires the expenditure of energy.

There are three primary means by which cells use energy to pump chemical species. The first is to keep the concentration of the cargo in the downstream domain small by binding or modifying it in some way. A binding protein in one compartment could sequester the transported cargo, or the cargo could be covalently modified in one compartment so that it no longer interacts with the transporter. For example, the flux of glucose is inward because intracellular glucose is quickly phosphorylated, thereby keeping the concentration of intracellular glucose low. However, phosphorylation of intracellular glucose requires the hydrolysis of an ATP molecule, from which the needed energy is extracted.

The second means is to use the gradient of one species to pump another species against its gradient. This is the mechanism of the Na^+–Ca^{2+} exchanger as well as numerous other exchangers that use to advantage the energy stored in the Na^+ gradient.

The third means is to regulate the binding of the cargo to the transporter in such a way that binding to the transporter is favored in one compartment and unbinding is favored in the other compartment. This change in affinity is driven by the hydrolysis of ATP or GTP. One important example of such an active (energy-consuming) exchanger is the Na^+–K^+ ATPase. This pump acts as an antiport, actively pumping Na^+ ions out of the cell and pumping K^+ ions in, each against a steep electrochemical gradient. It accomplishes this by using the energy released by the hydrolysis of ATP, and thus is called an ATPase. As is described later in this chapter, the Na^+–K^+ ATPase is important for regulating the cell volume and maintaining a membrane potential. Indeed, almost a third of the energy requirement of a typical animal cell is consumed in fueling this pump; in electrically active nerve cells, this figure approaches two-thirds of the cell's energy requirement. Other important ATPases are the sarco/endoplasmic reticulum calcium ATPase pumps (SERCA pumps) that pump Ca^{2+} into the endoplasmic or sarcoplasmic reticulum, or the plasma membrane Ca^{2+} ATPases, which pump Ca^{2+} out of the cell.

2.5.1   A Simple ATPase

We begin by considering a model of an ATPase that pumps a ligand, L, up its concentration gradient (Fig. 2.10). This hypothetical ATPase exists in six states. E is the base state; ATP can bind to E, followed by binding of the ligand L, to form the top line of states in Fig. 2.10. In each of these states the L binding site is exposed to the inside of the cell. Once ATP and L are bound, the ATPase changes conformation, exposing the L binding site to the outside, and at the same time decreasing the affinity of L for its binding site. Thus, L leaves the ATPase, followed by the hydrolysis of ATP and eventual return of the ATPase to its base state to complete the cycle. The overall cycle results in the transport of one L molecule from inside to outside. Although a realistic ATPase cycle is considerably more complicated than this one, this simple model serves to illustrate the basic principles.

If there is also a transition from $E \cdot ATP$ to $E_e \cdot ATP$, as shown by the dashed line, then the overall cycle can break into two separate subcycles as indicated. This is called *slippage*, since each of the subcycles accomplishes nothing toward the goal of pumping L. The subcycle on the left goes naturally in a clockwise direction and hydrolyzes ATP to ADP and inorganic phosphate, P_i, without using this energy to pump L. Similarly,

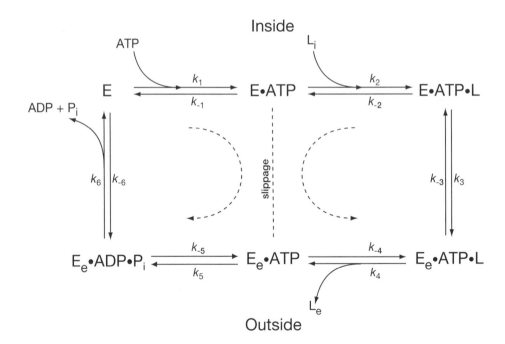

Figure 2.10   Schematic diagram of an ATPase pump that transports one ligand, L, from the inside to the outside against its concentration gradient. For each L transported, one molecule of ATP is hydrolyzed. A subscript e denotes the ATPase conformation in which the L binding sites are exposed to the exterior of the cell.

the subcycle on the right goes naturally in the direction that allows L to flow down its concentration gradient. The energy of the ATP is used to pump L against its gradient only when the ATPase proceeds around the entire cycle.

We use the law of mass action to write the differential equations for the six ATPase states. For example,

$$\frac{d[E]}{dt} = k_{-1}[E \cdot ATP] + k_6[E_e \cdot ADP \cdot P_i] - (k_1[ATP] + k_{-6}[P_i][ADP])[E], \qquad (2.82)$$

with similar equations for each of the other states. The steady-state flux, J, is given by

$$J = k_1[ATP][E] - k_{-1}[E \cdot ATP]. \qquad (2.83)$$

Even a relatively simple model of six states gives a long expression for the steady-state flux. In this case (with no slippage),

$$J = \frac{\frac{[ATP][L_i]}{[ADP][P_i][L_e]} - K_1 K_2 K_3 K_4 K_5 K_6}{\phi}, \qquad (2.84)$$

where $\phi > 0$ is a complicated function of rate constants and concentrations, and where, as usual, $K_i = k_{-i}/k_i$. (Even though it is not obvious from the way it is written, the flux J has, as before, units of inverse time.)

Since detailed balance requires

$$\prod_{i=1}^{6} K_i = \frac{[L_i]_{eq}}{[L_e]_{eq}} \frac{[ATP]_{eq}}{[ADP]_{eq}[P_i]_{eq}}, \qquad (2.85)$$

it follows that

$$J = \frac{\frac{[L_i]}{[L_e]} \frac{[ATP]}{[ADP][P_i]} - \frac{[L_i]_{eq}}{[L_e]_{eq}} \frac{[ATP]_{eq}}{[ADP]_{eq}[P_i]_{eq}}}{\phi}. \qquad (2.86)$$

We see from the numerator that the flux is either positive or negative depending on how far the concentrations of L, ATP, ADP, and P_i are from their equilibrium concentrations. In general, [ATP] is much higher than its equilibrium concentration (due to other processes in the cell that are continuously generating ATP), and it is this that causes a positive pump flux, pumping L against its gradient. However, if $[L_e]$ is high enough it can force the pump to work in reverse, allowing L to move from the outside to the inside of the cell, generating ATP in the process.

To relate the rate constants to the change in free energy we use that the overall reaction is

$$L_i + ATP \longrightarrow L_e + ADP + P_i. \qquad (2.87)$$

The change in free energy is given by

$$\Delta G = G_{\text{ADP}} + G_{\text{P}_i} + G_{\text{L}_e} - G_{\text{ATP}} - G_{\text{L}_i} \tag{2.88}$$

$$= G^0_{\text{ADP}} + G^0_{\text{P}_i} - G^0_{\text{ATP}} - RT \ln \left(\frac{[\text{ATP}][\text{L}_i]}{[\text{ADP}][\text{P}_i][\text{L}_e]} \right) \tag{2.89}$$

$$= \Delta G^0_{\text{ATP}} - RT \ln \left(\frac{[\text{ATP}][\text{L}_i]}{[\text{ADP}][\text{P}_i][\text{L}_e]} \right). \tag{2.90}$$

Note that the standard free energy of L is the same inside and outside the cell. At equilibrium $\Delta G = 0$ and thus

$$\Delta G^0_{\text{ATP}} = RT \ln \left(\frac{[\text{L}_i]_{\text{eq}}}{[\text{L}_e]_{\text{eq}}} \frac{[\text{ATP}]_{\text{eq}}}{[\text{ADP}]_{\text{eq}}[\text{P}_i]_{\text{eq}}} \right). \tag{2.91}$$

Combining this with (2.85) gives

$$\prod_{i=1}^{6} K_i = e^{\frac{\Delta G^0_{\text{ATP}}}{RT}} \text{M}^{-1}, \tag{2.92}$$

which, because it is independent of concentrations, must hold in general. Notice from (2.91) that both sides of (2.92) must have units of liters per mole, or M^{-1}. In fact, since the free energy released by the hydrolysis of ATP is well known to be -31 kJ mole^{-1}, it follows that

$$e^{\frac{\Delta G^0_{\text{ATP}}}{RT}} = 3.73 \times 10^{-6}. \tag{2.93}$$

Now from (2.84) it follows that

$$J = \frac{\frac{[\text{L}_i]}{[\text{L}_e]} \frac{[\text{ATP}]}{[\text{ADP}][\text{P}_i]} - e^{\frac{\Delta G^0_{\text{ATP}}}{RT}} \text{M}^{-1}}{\phi}. \tag{2.94}$$

2.5.2 Active Transport of Charged Ions

Suppose that the interior of the cell has an electric potential of V_i while the exterior has a potential of V_e, and suppose further that L has a charge z. Then the change in potential of the ATPase cycle (2.87) is

$$\Delta G = \Delta G^0_{\text{ATP}} - RT \ln \left(\frac{[\text{ATP}][\text{L}_i]}{[\text{ADP}][\text{P}_i][\text{L}_e]} \right) - zFV, \tag{2.95}$$

where $V = V_i - V_e$ is, as usual, the membrane potential difference.

An identical argument to before shows that

$$\prod_{i=1}^{6} K_i = e^{\frac{\Delta^0 G_{\text{ATP}}}{RT}} e^{\frac{-zFV}{RT}} 1 \text{ mole}^{-1}, \tag{2.96}$$

and thus

$$J = \frac{\frac{[L_i]}{[L_e]} \frac{[ATP]}{[ADP][P_i]} - e^{\frac{\Delta^0 G_{ATP}}{RT}} e^{\frac{-zFV}{RT}} M^{-1}}{\phi(V)}. \tag{2.97}$$

Note that the denominator is now a function of V also, since ϕ depends on the rate constants, which are themselves functions of V, and thus the precise dependence of J on V is not immediately clear. However, if $z > 0$ and $V > 0$, the flux is zero at lower concentrations of L_i, while if $V < 0$, the flux is zero at higher concentrations of L_i. Thus we conclude that a positive membrane potential makes it easier for the pump to move positive ions from inside to outside, while a negative membrane potential makes this more difficult. Although this is not a rigorous argument, a more detailed calculation shows that this result holds in general.

Although this thermodynamic argument shows that there must be some voltage-dependence in the rate constants, it does not tell us in which step (or steps) the voltage-dependence occurs. For example, in this model, the transition from $E \cdot ATP \cdot L$ to $E_e \cdot ATP \cdot L$ involves the net movement of the charge across the cell membrane, so that $\frac{k_{-3}}{k_3} = e^{\frac{-zFV}{RT}}$. (The argument here is identical to the argument used for the voltage-dependence of the Na^+–Ca^{2+} exchanger.) However, there are other possibilities. Although each model must have the same solution when $J = 0$, and the expressions for J must have the same sign, the models can have significantly different expressions for $\phi(V)$.

2.5.3 A Model of the Na⁺–K⁺ ATPase

One of the best-known ATPases is the Na^+–K^+ ATPase, which pumps K^+ into the cell and Na^+ out of the cell through the overall reaction scheme

$$ATP + 3Na_i^+ + 2K_e^+ \longrightarrow ADP + P_i + 3Na_e^+ + 2K_i^+. \tag{2.98}$$

It is an electrogenic pump (each pump cycle transfers one positive charge from inside to out) and a member of the family of P-type active cation transporters which includes the SERCA ATPases that are discussed at length in Chapter 7. A great deal of work has been done to determine the mechanisms that underlie Na^+ and K^+ transport by this ATPase; the most widely accepted model is the Post–Albers model which was developed by two independent groups in the 1960s (Albers et al., 1963; Charnock and Post, 1963). A more recent review is Apell (2004), while a history of investigations into the Na^+–K^+ ATPase is given by Glynn (2002). An excellent mathematical implementation of the Post–Albers scheme is that of Smith and Crampin (2004), and this is the model that we follow here.

In the Post–Albers scheme, phosphorylation of the pump (i.e., exchange of ATP for ADP) is associated with Na^+ efflux, while hydrolysis (i.e., loss of the additional phosphate group) is associated with K^+ influx. During the transition across the membrane each ion type is occluded, i.e., bound to the pump in a conformation in which

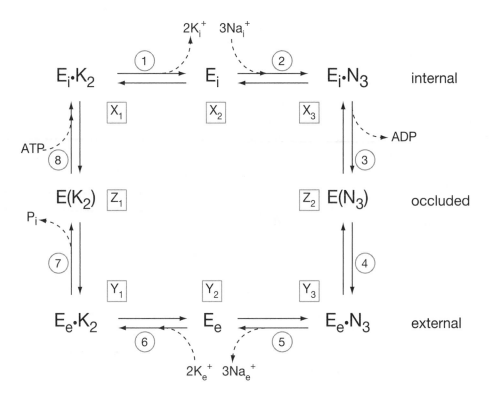

Figure 2.11   Model of the Na^+–K^+ ATPase based on the Post–Albers scheme.

it is accessible from neither side of the membrane. Occlusion prevents slippage, thus increasing the efficiency of the pump.

This scenario is illustrated in Fig. 2.11. Starting at the top left of the figure (state X_1), the ATPase begins in the conformation E_i, in which the binding sites for Na^+ and K^+ are exposed to the inside of the cell. The ATPase then loses two K^+ ions (which is assumed to occur in a single step) and gains three Na^+ ions, again in a single step, to move through states X_2 and X_3. ATP remains bound to the pump in each of the states X_1, X_2, and X_3, although this is not shown explicitly in the diagram. Loss of ADP then drives the ATPase to the occluded state Z_2, in which the three Na^+ ions are inaccessible to both the inside and outside of the cell. After another conformational change to the E_e state, in which the Na^+ and K^+ binding sites are exposed to the outside of the cell, the ATPase loses its three Na^+ to the outside, picks up another two K^+, and loses its extra phosphate to move through to the occluded state Z_1, in which the K^+ ions are shielded. Binding of ATP then returns the ATPase to the E_i conformation to complete the cycle. The rate constants are not shown explicitly, but each transition between states is labeled by a circled number. For each $i = 1, \ldots, 8$, transition i has two rate constants, k_i in the clockwise direction and k_{-i} in the counterclockwise direction.

From this diagram we can easily write the differential equations for each of the ATPase states. For example, letting a lowercase letter denote the fraction of the ATPase in that state, we have

$$\frac{dx_1}{dt} = k_{-1}[K_i^+]^2 x_2 + k_8[ATP]z_1 - (k_{-8} + k_1)x_1, \tag{2.99}$$

and so on, subject to the constraint $x_1 + x_2 + x_3 + y_1 + y_2 + y_3 + z_1 + z_2 = 1$. The resultant expression for the flux is long and unwieldy, of the form

$$J = \frac{[ATP]n_i^3\kappa_e^2 - [ADP][P_i]\kappa_i^2 n_e^3(\prod_{i=1}^{8} K_i)}{\phi}, \tag{2.100}$$

where ϕ is the sum of a large number of terms involving products of the rate constants and concentrations. Here, n denotes the Na^+ concentration, and κ denotes the K^+ concentration. This expression for the flux is similar to that derived in the simpler model of Section 2.5.1. The same thermodynamic constraints apply, and so some of the rate constants are functions of the membrane potential. Smith and Crampin (2004), following the ideas of Apell (1989), incorporate voltage dependence into the rate constants for Na^+ binding and unbinding, i.e., K_2 and K_5 in this model.

A simplified version of this model is discussed in Exercise 13.

2.5.4 Nuclear Transport

The transport of proteins from the cytoplasm to the nucleus (or the reverse) is accomplished by means that combine features of each of the above transport mechanisms. The nuclear membrane contains protein structures called nuclear pore complexes (NPCs) that allow free diffusion of soluble carrier proteins. However, these carrier proteins can pass through the pore complex only when they are bound. These carrier proteins (called importins) recognize and readily bind cargo destined for translocation. The energy to transport cargo against its gradient is provided by the hydrolysis of GTP via a GTPase enzyme called Ran. Ran-GTP has a very high binding affinity for the carrier protein ($\Delta G^0 = -51$ kJ mol^{-1}), effectively excluding the cargo from binding. The transportin/Ran-GTP complex is disassembled by the hydrolysis of Ran-GTP ($\Delta G^0 = -33$ kJ mol^{-1}) to Ran-GDP, which has a binding affinity for the carrier protein that is 10,000-fold lower than that of Ran-GTP. The endogenous GTPase activity rate is extremely slow ($k_{cat} = 1.5 \times 10^{-5}$s^{-1}). However, the hydrolysis of GTP to GDP on Ran is catalyzed by a cytoplasmic GTPase-activating protein called RanGAP, which accelerates this rate by as much as 500,000-fold.

One cycle of transport works as follows. Cargo in the cytoplasm that is targeted for transport binds to the carrier molecule and moves via diffusion through the NPC. In the nucleus, when the cargo unbinds, Ran-GTP quickly binds to the carrier, preventing the cargo from rebinding. Ran-GTP is kept at high concentration in the nucleus by another mechanism, so the Ran-GTP carrier complex diffuses into the cytoplasm through the NPC. On the cytoplasmic side of the membrane, Ran-GTP is quickly hydrolyzed to

Ran-GDP, which because of its much lower binding affinity, unbinds from the carrier molecule, completing the cycle. Although all the reactions are reversible, the directionality is maintained by the free energy of GTP hydrolysis and the high concentration of GTP in the nucleus.

We leave the development of a model of this transport mechanism to the interested reader. In the absence of RanGAP, the model is similar to that of the Na^+–Ca^{2+} transporter described above. On the other hand, if the hydrolysis of Ran-GTP to RanGDP is assumed to be so fast that there is no unbinding of RanGTP in the cytoplasm, then the model is similar to that of the simple ATPase described above.

2.6   The Membrane Potential

The principal function of the active ATPase transport processes described above is to regulate the intracellular ionic composition of the cell. For example, the operation of the Na^+–K^+ ATPase results in high intracellular K^+ concentrations and low intracellular Na^+ concentrations. This is necessary for a cell's survival, as without such regulation, cells swell and burst. However, before we consider models of cell volume regulation, we consider the effects of ionic separation.

2.6.1   The Nernst Equilibrium Potential

One of the most important equations in electrophysiology is the Nernst equation, which describes how a difference in ionic concentration can result in a potential difference across the membrane separating the two concentrations.

Suppose there are two reservoirs containing the same ion S, but at different concentrations, as shown schematically in Fig. 2.12. The reservoirs are separated by a

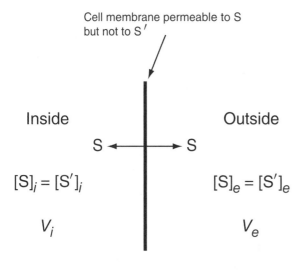

Cell membrane permeable to S but not to S$'$

Inside

Outside

S $\longleftarrow$ | $\longrightarrow$ S

$[S]_i = [S']_i$

$[S]_e = [S']_e$

V_i

V_e

Figure 2.12 Schematic diagram of a membrane separating two solutions with different ionic concentrations. By convention, the potential difference, V, across the membrane is defined to be $V = V_i - V_e$.

semipermeable membrane. The solutions on each side of the membrane are assumed to be electrically neutral (at least initially), and thus each ion S is balanced by another ion, S', with opposite charge. For example, S could be Na^+, while S' could be Cl^-. Because we ultimately wish to apply the Nernst equation to cellular membranes, we call the left of the membrane the inside and the right the outside of the cell.

If the membrane is permeable to S but not to S', the concentration difference across the membrane results in a net flow of S from one side to another, down its concentration gradient. However, because S' cannot diffuse through the membrane, the diffusion of S causes a buildup of charge across the membrane. This charge imbalance, in turn, sets up an electric field that opposes the further net movement of S through the membrane. Equilibrium is reached when the electric field exactly balances the diffusion of S. Note that at steady state there are more S ions than S' ions on one side and fewer S ions than S' ions on the other, and thus neither side of the membrane is exactly electrically neutral. However, because the force from the charge buildup is so strong, only a small amount of S moves across the membrane. To a good approximation, the concentrations of S on either side of the membrane remain unchanged, the solutions on either side of the membrane remain electrically neutral, and the small excess charge accumulates near the interface. The region in which there is a charge imbalance is called the Debye layer, and is on the order of a few nanometers thick.

The chemical potential of S on the inside of the membrane is

$$G_{S,i} = G_S^0 + RT \ln([S]_i) + zFV_i, \tag{2.101}$$

while on the outside it is

$$G_{S,e} = G_S^0 + RT \ln([S]_e) + zFV_e. \tag{2.102}$$

The chemical potential difference is

$$\Delta G_S = G_{S,i} - G_{S,e} = RT \ln\left(\frac{[S]_i}{[S]_e}\right) + zFV. \tag{2.103}$$

At equilibrium, it must be that $\Delta G_S = 0$, and thus the equilibrium potential difference, V_S, across the membrane must be

$$V_S = \frac{RT}{zF} \ln\left(\frac{[S]_e}{[S]_i}\right) = \frac{kT}{zq} \ln\left(\frac{[S]_e}{[S]_i}\right), \tag{2.104}$$

called the *Nernst potential*. Here k is Boltzmann's constant $k = \frac{R}{N_A}$, N_A is Avogadro's number, q is the charge on a proton, and z is the charge on the ion S. When $V = V_S$, there is no net current of S across the membrane, as the tendency of ions to move down their gradient is exactly balanced by the electric potential difference.

Throughout this book we follow the usual convention and define the potential difference, V, across the membrane to be

$$V = V_i - V_e, \tag{2.105}$$

i.e., the intracellular potential minus the extracellular potential.

Typical concentrations (in this case, for squid axon) are 397, 50, and 40 mM for K^+, Na^+, and Cl^-, respectively, in the intracellular space, and 20, 437, and 556 mM in the extracellular space. With these concentrations, the Nernst potentials for squid nerve axon are $V_{Na} = 56$ mV, $V_K = -77$ mV, $V_{Cl} = -68$ mV (using $RT/F = 25.8$ mV at 27°C. See Table 2.1).

The Nernst equation is independent of how the ions move through the membrane and depends only on the ratio of concentrations. In this sense, it is a universal law (although because it was derived from an ideal, yet approximate, law, it too is approximate). Any equation that expresses the transmembrane current of S in terms of the membrane potential, no matter what its form, must have the reversal potential of V_S; i.e., the current must be zero at the Nernst potential $V = V_S$. However, although this is true when a single ion species crosses the membrane, the situation is considerably more complicated when more than one type of ion can cross the membrane. In this case, the membrane potential that generates zero total current does not necessarily have zero current for each individual ion. For example, a current of S in one direction might be balanced by a current of S' in the same direction. Hence, when multiple ion types can diffuse through the membrane, the concentrations are not, in general, at equilibrium, even when there is no total current. Therefore, the arguments of chemical equilibrium used to derive the Nernst equation cannot be used, and there is no universal expression for the reversal potential in the multiple ion case. In this case, the reversal potential depends on the model used to describe the individual transmembrane ionic flows (see Chapter 3).

2.6.2 Gibbs–Donnan Equilibrium

Suppose one side of the membrane contains large charged macromolecules that cannot cross the membrane, but that both of the ion species S and S' freely diffuse across the membrane. To be specific, suppose that the macromolecules are negatively charged with valence $-z_x$, S is positively charged with valence z, and S' is negatively charged with valence $-z$. Note that both z and z_x are defined to be positive.

Outside the cell, S and S' must have the same concentration, to maintain charge neutrality. Inside, charge neutrality requires more S than S', in order to balance the negative charge on the macromolecules. At equilibrium, the membrane potential must be the Nernst potential for both S and S', namely

$$V_S = \frac{RT}{zF} \ln\left(\frac{[S]_e}{[S]_i}\right) = -\frac{RT}{zF} \ln\left(\frac{[S']_e}{[S']_i}\right), \tag{2.106}$$

where

$$z_x[X] + z[S']_i = z[S]_i \quad \text{and} \quad [S']_e = [S]_e. \tag{2.107}$$

It follows that

$$[S']_e[S]_e = [S']_i[S]_i, \tag{2.108}$$

and thus

$$[S]_i \left([S]_i - \frac{z_x}{z}[X] \right) - ([S]_e)^2 = 0. \tag{2.109}$$

Now, we assume that the external concentration is fixed, and treat $[S]_e$ as a known parameter. In this case, (2.109) can be solved to find a unique positive value for $[S]_i$,

$$[S]_i = \sigma[S]_e, \qquad \sigma = \frac{1}{2}(Z + \sqrt{Z^2 + 4}), \tag{2.110}$$

where $Z = \frac{z_x[X]}{z[S]_e}$, and from this the transmembrane potential can be determined using (2.106).

If, instead, we have a fixed volume and a fixed total amount of S, say, then we use the constraint

$$v_i[S]_i + v_e[S]_e = [S]_{\text{tot}}, \tag{2.111}$$

where $[S]_{\text{tot}}$ is a constant, and v_i and v_e are the internal and external volumes, respectively. We can now solve (2.109) subject to this constraint to find $[S]_i$ and the transmembrane potential. Note, however, that a physical solution is not always possible in this case, as there may be insufficient S or S′ to reach equilibrium.

This equilibrium is called the *Gibbs–Donnan equilibrium* (Exercise 15). The potential difference generated in this way is known to occur across cell membranes and also across the edge of a gel in aqueous solution. This potential drop occurs across the edge of a gel if the charged macromolecules are immobilized in the gel, and therefore unable to diffuse out of the gel.

2.6.3 Electrodiffusion: The Goldman–Hodgkin–Katz Equations

In general, the flow of ions through the membrane is driven by concentration gradients and also by the electric field. The contribution to the flow from the electric field is given by *Planck's equation*

$$\mathbf{J} = -u\frac{z}{|z|}c\nabla\phi, \tag{2.112}$$

where u is the *mobility* of the ion, defined as the velocity of the ion under a constant unit electric field; z is the valence of the ion, so that $z/|z|$ is the sign of the force on the ion; c is the concentration of S; and ϕ is the electrical potential, so that $-\nabla\phi$ is the electrical field.

There is a relationship, determined by Einstein, between the ionic mobility u and Fick's diffusion constant:

$$D = \frac{uRT}{|z|F}. \tag{2.113}$$

When the effects of concentration gradients and electrical gradients are combined, we obtain the *Nernst–Planck equation*

$$\mathbf{J} = -D\left(\nabla c + \frac{zF}{RT}c\nabla\phi\right). \tag{2.114}$$

If the flow of ions and the electric field are transverse to the membrane, (2.114) can be viewed as the one-dimensional relation

$$J = -D\left(\frac{dc}{dx} + \frac{zF}{RT}c\frac{d\phi}{dx}\right). \tag{2.115}$$

The Nernst Equation

The Nernst equation can also be derived from the Nernst–Planck equation (2.115). When the flux J is zero, we obtain

$$-D\left(\frac{dc}{dx} + \frac{zF}{RT}c\frac{d\phi}{dx}\right) = 0, \tag{2.116}$$

so that

$$\frac{1}{c}\frac{dc}{dx} + \frac{zF}{RT}\frac{d\phi}{dx} = 0. \tag{2.117}$$

Now suppose that the cell membrane extends from $x = 0$ (the inside) to $x = L$ (the outside), and let subscripts i and e denote internal and external quantities respectively. Then, integrating (2.117) from $x = 0$ to $x = L$ we get

$$\ln(c)\big|_{c_i}^{c_e} = \frac{zF}{RT}(\phi_i - \phi_e), \tag{2.118}$$

and thus the potential difference across the membrane, $V = \phi_i - \phi_e$, is given by

$$V = \frac{RT}{zF}\ln\left(\frac{c_e}{c_i}\right), \tag{2.119}$$

which is the Nernst equation.

The Constant Field Approximation

In general, the electric potential ϕ is determined by the local charge density, and so, if it is not zero, J must be found by solving a coupled system of equations (discussed in detail in Chapter 3). However, a useful result is obtained by assuming that the electric field in the membrane is constant, and thus decoupled from the effects of charges moving through the membrane. Suppose two reservoirs are separated by a semipermeable membrane of thickness L, such that the potential difference across the membrane is V. On the left of the membrane (the inside) $[S] = c_i$, and on the right (the outside) $[S] = c_e$. If the electric field is constant through the membrane, $\partial\phi/\partial x = -V/L$, where $V = \phi(0) - \phi(L)$ is the membrane potential.

At steady state and with no production of ions, the flux must be constant. In this case, the Nernst–Planck equation (2.114) is an ordinary differential equation for the concentration c,

$$\frac{dc}{dx} - \frac{zFV}{RTL}c + \frac{J}{D} = 0, \tag{2.120}$$

whose solution is

$$\exp\left(\frac{-zVFx}{RTL}\right)c(x) = \frac{JRTL}{DzVF}\left[\exp\left(\frac{-zVFx}{RTL}\right) - 1\right] + c_i, \tag{2.121}$$

where we have used the left boundary condition $c(0) = c_i$. To satisfy the boundary condition $c(L) = c_e$, it must be that

$$J = \frac{D}{L}\frac{zFV}{RT}\frac{c_i - c_e\exp\left(\frac{-zVF}{RT}\right)}{1 - \exp\left(\frac{-zVF}{RT}\right)}, \tag{2.122}$$

where J is the flux density with units (typically) of moles per area per unit time. Note that these units are equivalent to units of concentration × speed. This flux density becomes an electrical current density (current per unit area) when multiplied by zF, the amount of charge carried per mole, and thus

$$I_S = P_S\frac{z^2F^2}{RT}V\frac{c_i - c_e\exp\left(\frac{-zFV}{RT}\right)}{1 - \exp\left(\frac{-zFV}{RT}\right)}, \tag{2.123}$$

where $P_S = D/L$ is the permeability of the membrane to S. This is the famous Goldman–Hodgkin–Katz (GHK) current equation, and plays an important role in models of cellular electrical activity. Notice that the GHK flux (2.122) reduces to Fick's law (2.14) in the limit $V \to 0$.

The current is zero if the diffusively driven flow and the electrically driven flow are in balance, which occurs, provided that $z \neq 0$, if

$$V = V_S = \frac{RT}{zF}\ln\left(\frac{c_e}{c_i}\right), \tag{2.124}$$

which is, as expected, the Nernst potential.

If there are several ions that are separated by the same membrane, then the flow of each of these is governed separately by its own current–voltage relationship. In general there is no potential at which these currents are all individually zero. However, the potential at which the net electrical current is zero is called the Goldman–Hodgkin–Katz potential. For a collection of ions all with valence $z = \pm1$, we can calculate the GHK potential directly. For zero net electrical current, it must be that

$$0 = \sum_{z=1}P_j\frac{c_i^j - c_e^j\exp\left(\frac{-VF}{RT}\right)}{1 - \exp\left(\frac{-VF}{RT}\right)} + \sum_{z=-1}P_j\frac{c_i^j - c_e^j\exp\left(\frac{VF}{RT}\right)}{1 - \exp\left(\frac{VF}{RT}\right)}, \tag{2.125}$$

where $P_j = D_j/L$. This expression can be solved for V, to get

$$V = -\frac{RT}{F} \ln \left(\frac{\sum_{z=-1} P_j c_e^j + \sum_{z=1} P_j c_i^j}{\sum_{z=-1} P_j c_i^j + \sum_{z=1} P_j c_e^j} \right). \qquad (2.126)$$

For example, if the membrane separates Na^+ ($z = 1$), K^+ ($z = 1$), and Cl^- ($z = -1$) ions, then the GHK potential is

$$V_r = -\frac{RT}{F} \ln \left(\frac{P_{Na}[Na^+]_i + P_K[K^+]_i + P_{Cl}[Cl^-]_e}{P_{Na}[Na^+]_e + P_K[K^+]_e + P_{Cl}[Cl^-]_i} \right). \qquad (2.127)$$

It is important to emphasize that neither the GHK potential nor the GHK current equation are universal expressions like the Nernst equation. Both depend on the assumption of a constant electric field, and other models give different expressions for the transmembrane current and reversal potential. In Chapter 3 we discuss other models of ionic current, and compare them to the GHK equations. However, the importance of the GHK equations is so great, and their use so widespread, that their separate presentation here is justified.

2.6.4 Electrical Circuit Model of the Cell Membrane

Since the cell membrane separates charge, it can be viewed as a capacitor. The capacitance of any insulator is defined as the ratio of the charge across the capacitor to the voltage potential necessary to hold that charge, and is denoted by

$$C_m = \frac{Q}{V}. \qquad (2.128)$$

From standard electrostatics (Coulomb's law), one can derive the fact that for two parallel conducting plates separated by an insulator of thickness d, the capacitance is

$$C_m = \frac{k\epsilon_0}{d}, \qquad (2.129)$$

where k is the dielectric constant for the insulator and ϵ_0 is the permittivity of free space. The capacitance of cell membrane is typically $1.0 \ \mu F/cm^2$. Using that $\epsilon_0 = (10^{-9}/(36\pi))F/m$, we calculate that the dielectric constant for cell membrane is about 8.5, compared to $k = 3$ for oil.

A simple electrical circuit model of the cell membrane is shown in Fig. 2.13. It is assumed that the membrane acts like a capacitor in parallel with a resistor (although not necessarily ohmic). Since the current is dQ/dt, it follows from (2.128) that the capacitive current is $C_m dV/dt$, provided that C_m is constant. Since there can be no net buildup of charge on either side of the membrane, the sum of the ionic and capacitive currents must be zero, and so

$$C_m \frac{dV}{dt} + I_{ion} = 0, \qquad (2.130)$$

where, as usual, $V = V_i - V_e$.

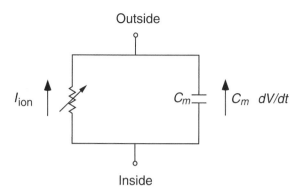

Outside

Inside

I_{ion} C_m $C_m\ dV/dt$

Figure 2.13 Electrical circuit model of the cell membrane.

This equation appears many times in this book, as it is the basis for much of theoretical electrophysiology. A significant challenge is to determine the form of I_{ion}. We have already derived one possible choice, the GHK current equation (2.123), and others are discussed in Chapter 3.

Another common model describes I_{ion} as a linear function of the membrane potential. In Chapter 3 we show how a linear I–V curve can be derived from more realistic models; however, because it is used so widely, we present a brief, heuristic, derivation here. Consider the movement of an ion S across a membrane. We assume that the potential drop across the membrane has two components. First, the potential drop due to concentration differences is given by the Nernst equation

$$V_S = \frac{RT}{zF} \ln\left(\frac{[S]_e}{[S]_i}\right), \tag{2.131}$$

and, second, if the channel is ohmic, the potential drop due to an electrical current is rI_S, where r is the channel resistance and I_S is the transmembrane current (positive outward) of S. Summing these two contributions we obtain

$$V = rI_S + V_S, \tag{2.132}$$

and solving for the current, we get the current–voltage relationship

$$I_S = g(V - V_S), \tag{2.133}$$

where $g = 1/r$ is the *membrane conductance*. The current I_S and conductance g are usually specified per unit area of membrane, being the product of the single channel conductance times the number of channels per unit area of membrane.

Notice that this current–voltage relationship also has zero current when $V = V_S$, the Nernst potential, as it must.

2.7 Osmosis

Suppose two chambers of water are separated by a rigid porous membrane. Because it is porous, water can flow between the two chambers. If the two chambers are topped by pistons, then water can be driven between the two chambers by applying different pressures to the two pistons. In general there is a linear relationship between the pressure difference and the flux of water through the membrane, given by

$$rQ = P_1 - P_2, \tag{2.134}$$

where Q is the flux (volume per unit time) of water from chamber one to chamber two, P_1 and P_2 are the applied pressures for chambers one and two, respectively, and r is the flow resistance of the membrane (not the same as the resistance to flow of ions). The expression (2.134) is actually a definition of the flow resistance r, and this linear relationship is analogous to Ohm's law relating current and voltage in a conductor. It is useful but not universally correct.

Suppose that a solute is added to chamber one, and that the membrane is impermeable to the solute. The difference in free energy per mole (or chemical potential) of solvent (i.e., water) between the two chambers is

$$\Delta G = RT \ln \frac{S_1}{S_2}, \tag{2.135}$$

where S_i is the mole fraction of solvent in the ith chamber. Note that because this expression involves the ratio of S_1 to S_2, we can use whatever units are most convenient. Hence we use mole fraction rather than concentration, which is standard. Because it dilutes the solvent ($S_1 < S_2$), the presence of a solute lowers the chemical potential of the solvent and induces a flow of solvent from chamber two to chamber one. In other words, the solvent diffuses from a region of higher concentration to one of lower concentration.

At constant temperature, equilibrium can be attained either by diluting the solution until it is pure solvent, or by increasing the pressure on the solution. The *osmotic pressure* π_s is defined to be the pressure that must be applied to chamber 1 to bring the free energy back to the free energy of the pure solvent. It follows that

$$RT \ln \frac{S_1}{S_2} + \pi_s v_s = 0, \tag{2.136}$$

where v_s is the molar volume (liters per mole) of the solvent. Note that, from the ideal gas law $PV = nRT$, we see that $\pi_s v_s$ has the same units as RT.

Since $S_2 = 1$ it now follows that

$$\pi_s = -\frac{RT}{v_s} \ln(S_1) = -\frac{RT}{v_s} \ln(1 - N) \approx \frac{RT}{v_s} N, \tag{2.137}$$

where N is the mole fraction of solvent. Also, since $N = \frac{n}{n+n_s} \approx \frac{n}{n_s}$, where n and n_s are the number of moles of solute and solvent, respectively, we have that

$$\pi_s = \frac{RT}{v_s}\frac{n}{n_s} \approx RcT, \qquad (2.138)$$

since $n_s v_s$ is quite close to the volume, v, of solution. Here c is the concentration of solvent in units of moles per liter. Using that $c = n/v$, (2.138) becomes

$$\pi_s v = nRT, \qquad (2.139)$$

which is the same as the ideal gas law. Equation (2.138) was first found empirically by van't Hoff.

If n has the units of numbers of molecules per liter, rather than moles per liter, as above, then (2.139) becomes

$$\pi_s v = nkT. \qquad (2.140)$$

As with all things derived from ideal properties, the expression (2.139) is an approximation, and for real solutions at physiological concentrations, the deviation can be significant. The formula

$$\pi_s v = \phi nRT, \qquad (2.141)$$

works much better, where ϕ is a concentration-dependent correction factor found experimentally. For all solutes, ϕ approaches one for sufficiently small concentrations. At concentrations typical of extracellular fluids in mammals, $\phi = 1.01$ for glucose and lactose, whereas for NaCl and KCl, $\phi = 0.93$ and 0.92, respectively. Deviation from ideality is even more significant for proteins, and is typically more concentration dependent as well. In spite of this, in the remainder of this book we use van't Hoff's law (2.138) to calculate osmotic pressure.

Notice that π_s is not the pressure of the solute but rather the pressure that must be applied to the solution to prevent solvent from flowing in through the semipermeable membrane. Thus, the flow rate of solvent is modified by osmotic pressure to be

$$rQ = P_1 - \pi_s - P_2, \qquad (2.142)$$

The flux of water due to osmotic pressure is called *osmosis*. The effect of the osmotic pressure is to draw water into chamber one, causing an increase in its volume and thereby to decrease the concentration of solute.

Osmotic pressure is determined by the number of particles per unit volume of fluid, and not the mass of the particles. The unit that expresses the concentration in terms of number of particles is called the *osmole*. One osmole is 1 gram molecular weight (that is, one mole) of an undissociated solute. Thus, 180 grams of glucose (1 gram molecular weight) is 1 osmole of glucose, since glucose does not dissociate in water. On the other hand, 1 gram molecular weight of sodium chloride, 58.5 grams, is 2 osmoles, since it dissociates into 2 moles of osmotically active ions in water.

A solution with 1 osmole of solute dissolved in a kilogram of water is said to have osmolality of 1 osmole per kilogram. Since it is difficult to measure the amount of water in a solution, a more common unit of measure is osmolarity, which is the osmoles per liter of aqueous solution. In dilute conditions, such as in the human body, osmolarity and osmolality differ by less than one percent. At body temperature, 37° C, a concentration of 1 osmole per liter of water has an osmotic pressure of 19,300 mm Hg, which corresponds to a column of water over 250 meters high. Clearly, osmotic pressures can be very large. It is for this reason that red blood cells burst when the blood serum is diluted with pure water, and this is known to have been the cause of death in hospital patients when pure water was accidentally injected into the veins.

Suppose two columns (of equal cross-section) of water are separated at the bottom by a rigid porous membrane. If n molecules of sugar are dissolved in column one, what will be the height difference between the two columns after they achieve steady state? At steady state there is no flux between the two columns, so at the level of the membrane, $P_1 - \pi_s = P_2$. Since P_1 and P_2 are related to the height of the column of water through $P = \rho g h$, where ρ is the density of the fluid, g is the gravitational constant, and h is the height of the column. We suppose that the density of the two columns is the same, unaffected by the presence of the dissolved molecule, so we have

$$\rho g h_2 = \rho g h_1 - \frac{nkT}{h_1 A}, \tag{2.143}$$

where A is the cross-sectional area of the columns. Since fluid is conserved, $h_1 + h_2 = 2h_0$, where h_0 is the height of the two columns of water before the sugar was added. From these, we find a single quadratic equation for h_1:

$$h_1^2 - h_0 h_1 - \frac{nkT}{2\rho g A} = 0. \tag{2.144}$$

The positive root of this equation is $h_1 = h_0/2 + \frac{1}{2}\sqrt{h_0^2 + \frac{2nkT}{\rho g A}}$, so that

$$h_1 - h_2 = \sqrt{h_0^2 + \frac{2nkT}{\rho g A}} - h_0. \tag{2.145}$$

When the solute is at a high enough concentration, physical solutions of (2.145) are not possible. Specifically, if the solute is too concentrated with $\frac{nkT}{\rho g A} > 4h_0^2$, the weight of a column of water of height $2h_0$ is insufficient to balance the osmotic pressure, in which case there is not enough water to reach equilibrium.

2.8 Control of Cell Volume

The principal function of the ionic pumps is to set up and maintain concentration differences across the cell membrane, concentration differences that are necessary for the cell to control its volume. In this section we describe how this works by means of a simple model in which the volume of the cell is regulated by the balance between ionic pumping and ionic flow down concentration gradients (Tosteson and Hoffman, 1960; Jakobsson, 1980; Hoppensteadt and Peskin, 2001).

Because the cell membrane is a thin lipid bilayer, it is incapable of withstanding any hydrostatic pressure differences. This is a potentially fatal weakness. For a cell to survive, it must contain a large number of intracellular proteins and ions, but if their concentrations become too large, osmosis causes the entry of water into the cell, causing it to swell and burst (this is what happens to many cells when their pumping machinery is disabled). Thus, for cells to survive, they must regulate their intracellular ionic composition (Macknight, 1988).

An even more difficult problem for some cells is to transport large quantities of water, ions, or other molecules while maintaining a steady volume. For example, Na^+-transporting epithelial cells, found (among other places) in the bladder, the colon, and nephrons of the kidney, are designed to transport large quantities of Na^+ from the lumen of the gut or the nephron to the blood. Indeed, these cells can transport an amount of Na^+ equal to their entire intracellular contents in one minute. However, the rate of transport varies widely, depending on the concentration of Na^+ on the mucosal side. Thus, these cells must regulate their volume and ionic composition under a wide variety of conditions and transport rates (Schultz, 1981).

2.8.1 A Pump–Leak Model

We begin by modeling the active and passive transport of ionic species across the cell membrane. We have already derived two equations for ionic current as a function of membrane potential: the GHK current equation (2.123) and the linear relationship (2.133). For our present purposes it is convenient to use the linear expression for ionic currents. Active transport of Na^+ and K^+ is performed primarily by the Na^+–K^+ ATPase (see Section 2.5.3).

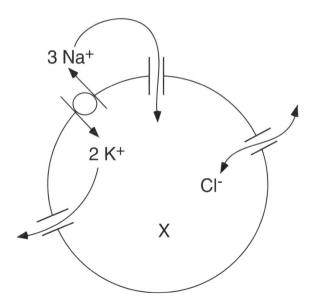

Figure 2.14 Schematic diagram of the pump–leak model.

Combining the expressions for active and passive ion transport, we find that the Na^+, K^+, and Cl^- currents are given by

$$I_{Na} = g_{Na}\left[V - \frac{RT}{F}\ln\left(\frac{[Na^+]_e}{[Na^+]_i}\right)\right] + 3pq, \qquad (2.146)$$

$$I_K = g_K\left[V - \frac{RT}{F}\ln\left(\frac{[K^+]_e}{[K^+]_i}\right)\right] - 2pq, \qquad (2.147)$$

$$I_{Cl} = g_{Cl}\left[V + \frac{RT}{F}\ln\left(\frac{[Cl^-]_e}{[Cl^-]_i}\right)\right], \qquad (2.148)$$

where p is the rate at which the ion exchange pump works and q is the charge of a single ion.

We can express these current–voltage equations as differential equations by noting that an outward ionic current of ion A^{z+} affects the intracellular concentration of that ion through

$$I_A = -\frac{d}{dt}(zFw[A^{z+}]), \qquad (2.149)$$

with w denoting the cell volume. (We use w rather than v to denote the cell volume to prevent confusion with V, the membrane potential.) Thus we have

$$-\frac{d}{dt}(Fw[Na^+]_i) = g_{Na}\left[V - \frac{RT}{F}\ln\left(\frac{[Na^+]_e}{[Na^+]_i}\right)\right] + 3pq, \qquad (2.150)$$

$$-\frac{d}{dt}(Fw[K^+]_i) = g_K\left[V - \frac{RT}{F}\ln\left(\frac{[K^+]_e}{[K^+]_i}\right)\right] - 2pq, \qquad (2.151)$$

$$\frac{d}{dt}(Fw[Cl^-]_i) = g_{Cl}\left[V + \frac{RT}{F}\ln\left(\frac{[Cl^-]_e}{[Cl^-]_i}\right)\right]. \qquad (2.152)$$

Next, we let X denote the number of moles of large negatively charged molecules (with valence $z_x \leq -1$) that are trapped inside the cell. The flow of water across the membrane is driven by osmotic pressure, so that the change of cell volume is given by

$$r\frac{dw}{dt} = RT\left([Na^+]_i - [Na^+]_e + [K^+]_i - [K^+]_e + [Cl^-]_i - [Cl^-]_e + \frac{X}{w}\right). \qquad (2.153)$$

Here we have assumed that the mechanical (hydrostatic) pressure difference across the membrane is zero, and we have also assumed that the elastic restoring force for the membrane is negligible.

Now to determine the membrane potential, we could use the electrical circuit model of the cell membrane, and write

$$C_m\frac{dV}{dt} + I_{Na} + I_K + I_{Cl} = 0. \qquad (2.154)$$

However, the system of equations (2.150)–(2.154) has an infinite number of steady states. This can be seen from the fact that, at steady state, we must have $I_{Na} = I_K = I_{Cl} = 0$, from which it follows that dV/dt must also necessarily be zero. Since the solution is thus determined by the choice of initial condition, it is better to use the integrated form of (2.154), i.e.,

$$C_m V = Q_i - Q_e, \tag{2.155}$$

where Q_i and Q_e are the total positive charge in the intracellular and extracellular spaces, respectively. (Note that here C_m refers to the total cell capacitance, assumed to be independent of cell volume. In other chapters of this book, C_m refers to membrane capacitance per unit area.) Since total charge is the difference between total number of positive and negative charges, we take

$$Q_i = qw([\text{Na}^+]_i + [\text{K}^+]_i - [\text{Cl}^-]_i) + z_x qX, \tag{2.156}$$

$$Q_e = qw_e([\text{Na}^+]_e + [\text{K}^+]_e - [\text{Cl}^-]_e). \tag{2.157}$$

This expression would be correct if the concentration of an ion was defined as the total number of ions in a region divided by the volume of that region, or if the distribution of ions was spatially homogeneous. But such is not the case here. This is because excess charge always accumulates in a thin region at the boundary of the domain (the Debye layer). However, this excess charge is quite small. To see this, consider a cylindrical piece of squid axon of typical radius 500 μm. With a capacitance of 1 μF/cm^2 and a typical membrane potential of 100 mV, the total charge is $Q = C_m V = \pi \times 10^{-8}$ C/cm. In comparison, the charge associated with intracellular K$^+$ ions at 400 mM is about 0.1 π C/cm, showing a relative charge deflection of about 10^{-7}.

Thus, since Q_i and Q_e are so small compared to the charges of each ion, it is a excellent approximation to assume that both the extracellular and intracellular media are electroneutral. Thus, Na$^+$, K$^+$, and Cl$^-$ are assumed to be in electrical balance in the extracellular space. In view of the numbers for squid axon, this assumption is not quite correct, indicating that there must be other ions around to maintain electrical balance. In the intracellular region, Na$^+$, K$^+$, and Cl$^-$ are not even close to being in electrical balance, but here, electroneutrality is maintained by the large negatively charged proteins trapped within the cell's interior. It is, of course, precisely the presence of these proteins in the interior of the cell that makes this whole exercise necessary. If a cell were not full of proteins (negatively charged or otherwise), it could avoid excessive osmotic pressures simply by allowing ions to cross the plasma membrane freely.

The assumption of electroneutrality gives the two equations

$$[\text{Na}^+]_e + [\text{K}^+]_e - [\text{Cl}^-]_e = 0, \tag{2.158}$$

$$[\text{Na}^+]_i + [\text{K}^+]_i - [\text{Cl}^-]_i + z_x \frac{X}{w} = 0. \tag{2.159}$$

It is convenient to assume that the cell is in an infinite bath, so that ionic currents do not change the external concentrations, which are thus assumed to be fixed and known, and to satisfy (2.158).

The differential equations (2.150), (2.151), (2.152), and (2.153) together with the requirement of intracellular electroneutrality (2.159) describe the changes of cell volume and membrane potential as functions of time. Note that we have 4 differential equations and one algebraic equation for the five unknowns ($[Na^+]_i$, $[K^+]_i$, $[Cl^-]_i$, w and V).

Even though we formulated this model as a system of differential equations, we are interested, for the moment, only in their steady-state solution. Time-dependent currents and potentials become important in Chapter 5 for the discussion of excitability.

To understand these equations, we introduce the nondimensional variables $v = \frac{FV}{RT}$, $P = \frac{pFq}{RTg_{Na}}$, $\mu = \frac{w}{X}[Cl^-]_e$ and set $y = e^{-v}$. Then, the equation of intracellular electroneutrality becomes

$$\alpha y - \frac{1}{y} + \frac{z_x}{\mu} = 0, \tag{2.160}$$

and the equation of osmotic pressure balance becomes

$$\alpha y + \frac{1}{y} + \frac{1}{\mu} - 2 = 0, \tag{2.161}$$

where $\alpha = \frac{[Na^+]_e e^{-3P} + [K^+]_e e^{2P\gamma}}{[Na^+]_e + [K^+]_e}$ and $\gamma = g_{Na}/g_K$. In terms of these nondimensional variables, the ion concentrations are

$$\frac{[Na^+]_i}{[Na^+]_e} = e^{-3P}y, \tag{2.162}$$

$$\frac{[K^+]_i}{[K^+]_e} = e^{2P\gamma}y, \tag{2.163}$$

$$\frac{[Cl^-]_i}{[Cl^-]_e} = \frac{1}{y}. \tag{2.164}$$

Solving (2.160) for its unique positive root, we obtain

$$y = \frac{-z_x + \sqrt{z_x^2 + 4\alpha\mu^2}}{2\alpha\mu}, \tag{2.165}$$

and when we substitute for y back into (2.161), we find the quadratic equation for μ:

$$4(1 - \alpha)\mu^2 - 4\mu + 1 - z_x^2 = 0. \tag{2.166}$$

For $z_x \leq -1$, this quadratic equation has one positive root if and only if $\alpha < 1$. Expressed in terms of concentrations, the condition $\alpha < 1$ is

$$\rho(P) = \frac{[Na^+]_e e^{-3P} + [K^+]_e e^{2P\gamma}}{[Na^+]_e + [K^+]_e} < 1. \tag{2.167}$$

One can easily see that $\rho(0) = 1$ and that for large P, $\rho(P)$ is exponentially large and positive. Thus, the only hope for $\rho(P)$ to be less than one is if $\rho'(0) < 0$. This occurs if and only if

$$\frac{3[Na^+]_e}{g_{Na}} > \frac{2[K^+]_e}{g_K}, \tag{2.168}$$

in which case there is a range of values of P for which a finite, positive cell volume is possible and for which there is a corresponding nontrivial membrane potential.

To decide if this condition is ever satisfied we must determine "typical" values for g_{Na} and g_K. This is difficult to do, because, as is described in Chapter 5, excitability of nerve tissue depends strongly on the fact that conductances are voltage-dependent and can vary rapidly over a large range of values. However, at rest, in squid axon, reasonable values are $g_K = 0.367$ mS/cm^2 and $g_{Na} = 0.01$ mS/cm^2. For these values, and at the extracellular concentrations of 437 and 20 mM for Na$^+$ and K$^+$, respectively, the condition (2.168) is readily met.

One important property of the model is that the resting value of V is equal to the Nernst potential of Cl$^-$, as can be seen from (2.152) or (2.164). Thus, the membrane potential is set by the activity of the Na$^+$–K$^+$ ATPase, and the intracellular Cl$^-$ concentration is set by the membrane potential.

In Figs. 2.15 and 2.16 the nondimensional volume μ and the potential V (assuming $RT/F = 25.8$ mV) are plotted as functions of the pump rate P. In addition, in Fig. 2.16 are shown the Na$^+$ and K$^+$ equilibrium potentials. For these plots, γ was chosen to be 0.11, and $z_x = -1$. Then, at $P = 1.6$, the Na$^+$ and K$^+$ equilibrium potentials and the membrane potentials are close to their observed values for squid axon, of 56, −77 and −68 mV, respectively.

From these plots we can see the effect of changing pump rate on cell volume and membrane potential. At zero pump rate, the membrane potential is zero and the cell volume is infinite (dead cells swell). As the pump rate increases from zero, the cell volume and membrane potential rapidly decrease to their minimal values and then gradually increase until at some upper limit for pump rate, the volume and potential become infinite. The K$^+$ equilibrium potential is seen to decrease rapidly as a function of pump rate until it reaches a plateau at a minimum value. The Na$^+$ equilibrium potential increases monotonically.

Physically realistic values of the membrane potential are achieved fairly close to the local minimum. Clearly, there is little advantage for a higher pump rate, and since

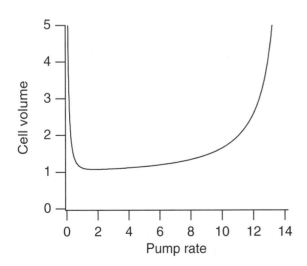

Figure 2.15 Cell volume as a function of the pump rate.

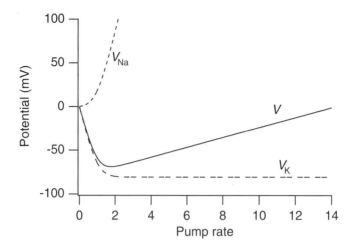

Figure 2.16 Membrane potential, Na^+ equilibrium potential, and K^+ equilibrium potential as functions of the pump rate.

Table 2.4 Resting potentials in some typical excitable cells.

Cell Type	Resting Potential (mV)
Neuron	−70
Skeletal muscle (mammalian)	−80
Skeletal muscle (frog)	−90
Cardiac muscle (atrial and ventricular)	−80
Cardiac Purkinje fiber	−90
Atrioventricular nodal cell	−65
Sinoatrial nodal cell	−55
Smooth muscle cell	−55

the pump rate is proportional to energy expenditure, it would seem that the pump rate is chosen approximately to minimize cell volume, membrane potential, and energy expenditure. However, no mechanism for the regulation of energy expenditure is suggested.

Generalizations

While the above model of volume control and membrane potential is useful and gives some insight into the control mechanisms, as with most models there are important features that have been ignored but that might lead to substantially different behavior.

There are (at least) two significant simplifications in the model presented here. First, the conductances g_{Na} and g_K are treated as constants. In Chapter 5 we show that the ability of cells to generate an electrical signal results from voltage and time dependence of the conductances. In fact, the discovery that ion channels have differing properties of voltage sensitivity was of fundamental importance to the understanding of neurons. The second simplification relates to the operation of the ion exchange pump. Figure 2.16 suggests that the minimal membrane potential is achieved at a particular pump rate and suggests the need for a tight control of pump rate that maintains the potential near this minimum. If indeed, such a tight control is required, it is natural to ask what that control mechanism might be. There is also the difficulty that in this simple model there is nothing preventing the complete depletion of Na^+ ions.

A different model of the pump activity might be beneficial. Recall from (2.98) that with each cycle of the ion exchange pump, three intracellular Na^+ ions are exchanged for two extracellular K^+ ions. Our previous analysis of the Na^+–K^+ ATPase (see (2.100)) suggests that at low internal Na^+ concentrations, the pump rate can be represented in nondimensional variables as

$$P = \rho u^3, \tag{2.169}$$

where $u = [Na^+]_i/[Na^+]_e$. This representation is appropriate at high pump rates, where effects of saturation are of no concern. Notice that P is proportional to the rate of ATP hydrolysis, and hence to energy consumption. Thus, as u decreases, so also does the rate of energy consumption. With this change, the equation for the Na^+ concentration becomes

$$u \exp(3\rho u^3) = y, \tag{2.170}$$

and this must be solved together with the quadratic polynomials (2.160) and (2.161), with (2.170) replacing (2.162).

In Fig. 2.17 are shown the membrane potential, and the Na^+ and K^+ equilibrium potentials, plotted as functions of the nondimensional reaction rate ρ. Here we

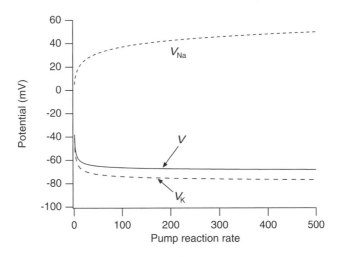

Figure 2.17 Membrane potential, Na^+ equilibrium potential, and K^+ equilibrium potential as functions of the pump rate, for the modified pump rate (2.169).

see something qualitatively different from what is depicted in Fig. 2.16. There the membrane potential had a noticeable local minimum and was sensitive to changes in pump rate. In this modified model, the membrane potential is insensitive to changes in the pump rate. The reason for this difference is clear. Since the effectiveness of the pump depends on the internal Na^+ concentration, increasing the speed of the pumping rate has little effect when the internal Na^+ is depleted, because of the diminished number of Na^+ ions available to be pumped.

While the pump rate is certainly ATP dependent, there are a number of drugs and hormones that are known to affect the pump rate. Catecholamines rapidly increase the activity of the pump in skeletal muscle, thereby preserving proper K^+ during strenuous exercise. Within minutes, insulin stimulates pump activity in the liver, muscle, and fat tissues, whereas over a period of hours, aldosterone and corticosterones increase activity in the intestine.

On the other hand, digitalis (clinically known as digoxin) is known to suppress pump activity. Digitalis is an important drug used in the treatment of congestive heart failure and during the 1980s was the fourth most widely prescribed drug in the United States. At therapeutic concentrations, digitalis inhibits a moderate fraction (say, 30–40%) of the Na^+–K^+ ATPase, by binding with the Na^+ binding site on the extracellular side. This causes an increase in internal Na^+, which has an inhibitory effect on the Na^+–Ca^{2+} exchanger, slowing the rate by which Ca^{2+} exits the cells. Increased levels of Ca^{2+} result in increased myocardial contractility, a positive and useful effect. However, it is also clear that at higher levels, the effect of digitalis is toxic.

2.8.2 Volume Regulation and Ionic Transport

Many cells have a more difficult problem to solve, that of maintaining their cell volume in widely varying conditions, while transporting large quantities of ions through the cell. Here we present a simplified model of transport and volume regulation in a Na^+-transporting epithelial cell.

As are virtually all models of transporting epithelia, the model is based on that of Koefoed-Johnsen and Ussing (1958), the so-called KJU model. In the KJU model, an epithelial cell is modeled as a single cell layer separating a mucosal solution from the serosal solution (Fig. 2.18). (The mucosal side of an epithelial cell is that side on which mucus is secreted and from which various chemicals are withdrawn, for example, from the stomach. The serosal side is the side of the epithelial cell facing the interstitium, wherein lie capillaries, etc.) Na^+ transport is achieved by separating the Na^+ pumping machinery from the channels that allow Na^+ entry into the cell. Thus, the mucosal membrane contains Na^+ channels that allow Na^+ to diffuse down its concentration gradient into the cell, while the serosal membrane contains the Na^+–K^+ ATPases which remove Na^+ from the cell. The overall result is the transport of Na^+ from the mucosal side of the cell to the serosal side. The important question is whether the cell can maintain a steady volume under widely varying concentrations of Na^+ on the mucosal side.

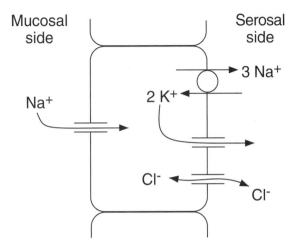

Figure 2.18 Schematic diagram of the model of a Na^+-transporting epithelial cell, based on the model of Koefoed-Johnsen and Ussing (1958).

We begin by letting N, K, and C denote Na^+, K^+, and Cl^- concentrations respectively, and letting subscripts m, i, and s denote mucosal, intracellular and serosal concentrations. Thus, for example, N_i is the intracellular Na^+ concentration, and N_m is the mucosal Na^+ concentration. We now write the conservation equations for Na^+, K^+, and Cl^- at steady state. The conservation equations are the same as those of the pump–leak model with some minor exceptions. First, instead of the linear I–V curve used in the pump–leak model, we use the GHK formulation to represent the ionic currents. This makes little qualitative change to the results but is more convenient because it simplifies the analysis that follows. Second, we assume that the rate of the Na^+–K^+ ATPase is proportional to the intracellular Na^+ concentration, N_i, rather than N_i^3, as was assumed in the generalized version of the pump–leak model. Thus,

$$P_{Na}v\frac{N_i - N_m e^{-v}}{1 - e^{-v}} + 3qpN_i = 0, \qquad (2.171)$$

$$P_K v\frac{K_i - K_s e^{-v}}{1 - e^{-v}} - 2qpN_i = 0, \qquad (2.172)$$

$$P_{Cl}v\frac{C_i - C_s e^{v}}{1 - e^{v}} = 0. \qquad (2.173)$$

Note that the voltage, v, is nondimensional, having been scaled by $\frac{F}{RT}$, and that the rate of the Na^+–K^+ ATPase is pN_i. Also note that the inward Na^+ current is assumed to enter from the mucosal side, and thus N_m appears in the GHK current expression, but that no other ions enter from the mucosa. Here the membrane potential is assumed to be the same across the lumenal membrane and across the basal membrane. This is not quite correct, as the potential across the lumenal membrane is typically −67 mV while across the basal membrane it is about −70 mV.

There are two further equations to describe the electroneutrality of the intracellular space and the osmotic balance. In steady state, these are, respectively,

$$w(N_i + K_i - C_i) + z_x X = 0, \tag{2.174}$$

$$N_i + K_i + C_i + \frac{X}{w} = N_s + K_s + C_s, \tag{2.175}$$

where X is the number of moles of protein, each with a charge of $z_x \leq -1$, that are trapped inside the cell, and w is the cell volume. Finally, the serosal solution is assumed to be electrically neutral, and so in specifying N_s, K_s, and C_s we must ensure that

$$N_s + K_s = C_s. \tag{2.176}$$

Since the mucosal and serosal concentrations are assumed to be known, we now have a system of 5 equations to solve for the 5 unknowns, N_i, K_i, C_i, v, and $\mu = w/X$. First, notice that (2.171), (2.172), and (2.173) can be solved for N_i, K_i, and C_i, respectively, to get

$$N_i(v) = \frac{v N_m e^{-v}}{v + 3\rho_n(1 - e^{-v})}, \tag{2.177}$$

$$K_i(v) = 2\rho_k N_i(v) \frac{1 - e^{-v}}{v} + K_s e^{-v}, \tag{2.178}$$

$$C_i(v) = C_s e^v, \tag{2.179}$$

where $\rho_n = pq/P_{\mathrm{Na}}$ and $\rho_k = pq/P_{\mathrm{K}}$.

Next, eliminating $N_i + K_i$ between (2.174) and (2.175), we find that

$$2\mu(C_i - C_s) = z_x - 1. \tag{2.180}$$

We now use (2.179) to find that

$$z_x - 1 = 2\mu C_s(e^v - 1), \tag{2.181}$$

and thus, using (2.181) to eliminate μ from (2.174), we get

$$N_i(v) + K_i(v) = \frac{C_s}{1 - z_x}[-2z_x + e^v(1 + z_x)] \equiv \phi(v). \tag{2.182}$$

Since $z_x - 1 < 0$, it must be (from (2.181)) that $v < 0$, and as $v \to 0$, the cell volume $w = \mu X$ becomes infinite. Thus, we wish to find a negative solution of (2.182), with $N_i(v)$ and $K_i(v)$ specified by (2.177) and (2.178).

It is instructive to consider when solutions for v (with $v < 0$) exist. First, notice that $\phi(0) = C_s$. Further, since $z_x \leq -1$, ϕ is a decreasing function of v, bounded above, with decreasing slope (i.e., concave down), as sketched in Fig. 2.19. Next, from (2.177) and (2.178) we determine that $N_i(v) + K_i(v)$ is a decreasing function of v that approaches ∞ as $v \to -\infty$ and approaches zero as $v \to \infty$. It follows that a negative solution for v exists if $N_i(0) + K_i(0) < C_s$, i.e., if

$$\frac{N_m}{1 + 3\rho_n} + \frac{2\rho_k N_m}{1 + 3\rho_n} + K_s < C_s. \tag{2.183}$$

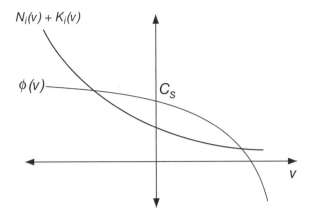

$N_i(v) + K_i(v)$

$\phi(v)$

C_s

v

Figure 2.19 Sketch (not to scale) of the function $\phi(v)$, defined as the right-hand side of (2.182), and of $N_i(v) + K_i(v)$, where N_i and K_i are defined by (2.177) and (2.178). $\phi(v)$ is sketched for $z_x < -1$.

Since $K_s + N_s = C_s$, this becomes

$$\frac{N_m}{N_s} < \frac{1 + 3\rho_n}{1 + 2\rho_k}. \tag{2.184}$$

This condition is sufficient for the existence of a solution, but not necessary. That is, if this condition is satisfied, we are assured that a solution exists, but if this condition fails to hold, it is not certain that a solution fails to exist. The problem, of course, is that negative solutions are not necessarily unique, nor is it guaranteed that increasing N_m through $N_s \frac{1+3\rho_n}{1+2\rho_k}$ causes a negative solution to disappear. It is apparent from (2.177) and (2.178) that $N_i(v)$ and $K_i(v)$ are monotone increasing functions of the parameter N_m, so that no negative solutions exist for N_m sufficiently large. However, for $N_m = N_s \frac{1+3\rho_n}{1+2\rho_k}$ to be the value at which the cell bursts by *increasing* N_m, it must also be true that

$$N_i'(0) + K_i'(0) < \phi'(0), \tag{2.185}$$

or that

$$4(1 + 3\rho_n)C_s + N_s(1 - z_x)\frac{3\rho_n - 2\rho_k}{1 + 2\rho_k} > 0. \tag{2.186}$$

For the remainder of this discussion we assume that this condition holds, so that the failure of (2.184) also implies that the cell bursts.

According to (2.184), a transporting epithelial cell can maintain its cell volume, provided the ratio of mucosal to serosal concentrations is not too large. When N_m/N_s becomes too large, μ becomes unbounded, and the cell bursts. Typical solutions for the cell volume and membrane potential, as functions of the mucosal Na$^+$ concentration, are shown in Fig. 2.20.

Obviously, this state of affairs is unsatisfactory. In fact, some epithelial cells, such as those in the loop of Henle in the nephron (Chapter 17), must work in environments with extremely high mucosal Na$^+$ concentrations. To do so, these Na$^+$-transporting epithelial cells have mechanisms to allow operation over a much wider range of mucosal Na$^+$ concentrations than suggested by this simple model.

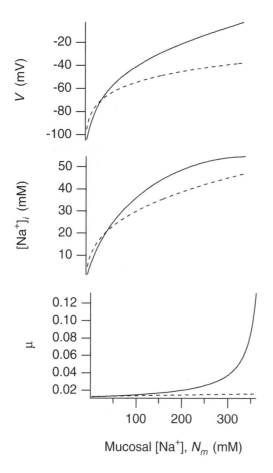

Figure 2.20 Numerical solutions of the model of epithelial cell volume regulation and Na^+ transport. The membrane potential, V, the scaled cell volume, μ, and the intracellular Na^+ concentration, $[Na^+]_i$, are plotted as functions of the mucosal Na^+ concentration. The solid lines are the solutions of the simpler version of the model, where P_{Na} and P_K are assumed to be constant. The dashed lines are the solutions of the model when P_{Na} is assumed to be a decreasing function of N_i, and P_K is assumed to be an increasing function of w, as described in the text. Parameter values are $K_s = 2.5$, $N_s = 120$, $C_s = 122.5$, $P = 2$, $\gamma = 0.3$, $z_x = -2$. All concentrations are in mM.

From (2.184) we can suggest some mechanisms by which a cell might avoid bursting at high mucosal concentrations. For example, the possibility of bursting is decreased if ρ_n is increased or if ρ_k is decreased. The reasons for this are apparent from (2.177) and (2.178), since $N_i(v) + K_i(v)$ is a decreasing function of ρ_n and an increasing function of ρ_k. From a physical perspective, increasing N_m causes an increase in N_i, which increases the osmotic pressure, inducing swelling. Decreasing the conductance of Na^+ ions from the mucosal side helps to control this swelling. Similarly, increasing the conductance of K^+ ions allows more K^+ ions to flow out of the cell, thereby decreasing the osmotic pressure from K^+ ions and counteracting the tendency to swell.

It has been conjectured for some time that epithelial cells use both of these mechanisms to control their volume (Schultz, 1981; Dawson and Richards, 1990; Beck et al., 1994). There is evidence that as N_i increases, epithelial cells decrease the Na^+ conductance on the mucosal side of the cell, thus restricting Na^+ entry. There is also evidence that as the cell swells, the K^+ conductance is increased, possibly by means of stretch-activated K^+ channels (Ussing, 1982. This assumption was used in the modeling work of Strieter et al., 1990).

To investigate the effects of these mechanisms in our simple model, we replace P_{Na} by $P_{Na}20/N_i$ (20 is a scale factor, so that when $N_i = 20$ mM, P_{Na} has the same value as in the original version of the model) and replace P_K by $P_K w/w_0$, where w_0 is the volume of the cell when $N_m = 100$ mM. As before, we can solve for v and μ as functions of N_m, and the results are shown in Fig. 2.20. Clearly the incorporation of these mechanisms decreases the variation of cell volume and allows the cell to survive over a much wider range of mucosal Na^+ concentrations.

The model of control of ion conductance used here is extremely simplistic, as for example, there is no parametric control of sensitivity, and the model is heuristic, not mechanistic. More realistic and mechanistic models have been constructed and analyzed in detail (Lew et al., 1979; Civan and Bookman, 1982; Strieter et al., 1990; Weinstein, 1992, 1994, 1996; Tang and Stephenson, 1996).

2.9 Appendix: Stochastic Processes

Although all of the models that have been presented so far in this text have been deterministic, the reality is that biological processes are fundamentally noisy. Furthermore, many of the assumptions underlying deterministic models are questionable, largely because of significant stochastic effects.

The purpose of this appendix is to outline some of the basic ideas of stochastic processes that play an important role in mathematical modeling of biological phenomena. Furthermore, there are a number of sections in the remainder of this book where stochastic models are crucial, and we hope that this appendix provides the necessary background for these. Of course, this is not a detailed treatment of these topics; for that one needs to consult a text on stochastic processes such as Gardiner (2004) or van Kampen (2007).

2.9.1 Markov Processes

A *Markov process* is any stochastic process that has no memory. More precisely, if the value of the state variable x is known at two times, say $t_1 < t_2$, to be x_1 and x_2, respectively, then

$$P(x, t \mid x_1, t_1, x_2, t_2) = P(x, t \mid x_2, t_2), \qquad (2.187)$$

where $P(x \mid y)$ denotes the conditional probability of x given y. In words, the conditional probability for the value of the state variable depends only on the most recent condition and not on any previous conditions.

A simple example of a Markov process is radioactive decay. For example, an atom of carbon-14 may lose an electron and decay to nitrogen-14. This is a Markov process, because the probability of decay does not depend on how old the carbon-14 atom is. A newly formed carbon-14 atom has exactly the same probability of decay in the next second as a very old carbon-14 atom.

To model this process, we suppose that the state variable has two possible values, say C (for carbon) and N (for nitrogen), and we denote the probability that the molecule is in state S at time t by $P(S, t)$. Now suppose that the probability of radioactive decay in a small interval of time dt is $\lambda\, dt$. (Note that λ need not be independent of time for this to be a Markov process.) Then,

$$P(N, t + \Delta t) = P(N, t) + P(C, t)\lambda \Delta t. \tag{2.188}$$

For this problem we assume that $P(C, t) + P(N, t) = 1$. It follows that in the limit $\Delta t \to 0$,

$$\frac{dP(N, t)}{dt} = \lambda(1 - P(N, t)). \tag{2.189}$$

By similar arguments, many of the differential equations in this chapter describing concentrations or fractions of molecules in a given state can be reinterpreted as equations for the probability that a single molecule is in a particular state. For example, for a transporter molecule that has two states,

$$C_i \underset{k}{\overset{k}{\rightleftarrows}} C_e, \tag{2.190}$$

under the assumption that transitions between states do not depend on how long the molecule has been in a given state, the probability P of being in state C_i at time t, $P(C_i, t)$, is given by

$$\frac{dP(C_i, t)}{dt} = k(1 - P(C_e, t)) - kP(C_i, t). \tag{2.191}$$

Similarly, for the chemical reaction

$$A \underset{k_-}{\overset{k_+}{\rightleftarrows}} B, \tag{2.192}$$

the probability $P(A, t)$ of a molecule being in state A at time t is determined by

$$\frac{dP(A, t)}{dt} = k_- P(B, t) - k_+ P(A, t). \tag{2.193}$$

Of course, this is the same as the equation found using the law of mass action for the conversion of A to B, namely,

$$\frac{da}{dt} = k_- b - k_+ a. \tag{2.194}$$

Even though these equations are identical, their interpretations are quite different. For example, at steady state, $\frac{a}{b} = \frac{k_-}{k_+}$ implies that the ratio of the number of molecules in state A to the number of those in state B is $\frac{k_-}{k_+}$. However, this cannot be the probabilistic interpretation, since there is no fractional state. Instead, the probabilistic interpretation is that the ratio of the time a single molecule spends in state A to the time it spends in state B is $\frac{k_-}{k_+}$.

2.9.2 Discrete-State Markov Processes

Often, the state space of a Markov process is discrete; for example, the model of the Na^+–K^+ ATPase (Fig. 2.11) assumes that the ATPase can be in only a small number of different states. The simple chemical reaction in the previous section is also a model of this type, since the molecule can exist in only two states, A or B. Since time is a continuous variable, such models are called *discrete-space continuous-time Markov processes*. Such Markov processes play an enormously important role in modeling; indeed, a large fraction of the models of chemical reactions, exchangers, pumps, and ion channels we discuss in this book are discrete-space continuous-time Markov processes. As is described in Chapter 3, one of the most important applications of such models is the analysis of data from single ion channels.

To describe the stochastic behavior of a discrete-space continuous-time Markov process with n possible states (open, closed, inactivated, etc.), we introduce a discrete random variable $S(t) \in 1, 2, \ldots, n$ defined so that $S(t) = i$ if the model is in state i at time t. Further, we suppose that the probability that the model changes from state i to state j in the time interval $(t, t + dt)$ is $k_{ij}dt$. In more condensed notation,

$$P(S(t + dt) = j \mid S(t) = i) = k_{ij}dt. \tag{2.195}$$

Note that (2.195) is valid only in the limit of small dt, since for sufficiently large dt and k_{ij} nonzero, this probability exceeds 1. Also, the probability that the model does not change state in the time interval $(t, t + dt)$ is given by

$$P(S(t + dt) = i \mid S(t) = i) = 1 - K_i dt, \tag{2.196}$$

where $K_i = \sum_{j \neq i} k_{ij}$.

Now we let $\Phi_j(t) = P(S(t) = j)$, i.e., $\Phi_j(t)$ is the probability that the model is in state j at time t. In words, the probability that the model is in state j at time $t + dt$ is the probability that it was in state j at time t and did not leave state j between times t and $t + dt$ plus the probability that at time t it was in another state and switched into state j in the time interval t to $t + dt$. In mathematical language,

$$\Phi_j(t + dt) = \Phi_j(t)P(S(t + dt) = i \mid S(t) = i) + \sum_{l \neq j} P(S(t + dt) = j \mid S(t) = l)\Phi_l(t)$$

$$= \Phi_j(t)\left(1 - K_j dt\right) + \sum_{l \neq j} k_{lj}\Phi_l(t)dt, \tag{2.197}$$

and thus, taking the limit $dt \to 0$,

$$\frac{d\Phi_j}{dt} = -K_j\Phi_j(t) + \sum_{l \neq j} k_{lj}\Phi_l(t). \tag{2.198}$$

In vector notation, with Φ the vector having elements Φ_j, we have

$$\frac{d\Phi}{dt} = A^T \Phi(t), \tag{2.199}$$

where A is the matrix

$$A = \begin{pmatrix} -K_1 & k_{12} & \cdots & k_{1n} \\ k_{21} & -K_2 & \cdots & k_{2n} \\ \vdots & \vdots & \vdots & \vdots \end{pmatrix}. \tag{2.200}$$

Of course, this is exactly the system of differential equations we would have written down for a system of first-order chemical reactions with reaction rates between species k_{ij}. Here, however, the variables Φ_j are probabilities and so must sum to one. In the terminology of stochastic processes, (2.199) is called the *master equation* for this Markov process.

The Waiting Time

One important question is how long a Markov model stays in a particular state before switching. This is called the waiting-time problem. To solve this we let T_i be the random time at which the model switches from state i to some other state, and we let $P_i(t) = P(T_i < t)$, that is, $P_i(t)$ is the probability that by time t the switch has occurred. In words, the probability that the switch has occurred by time $t + dt$ is the probability that it has occurred by time t plus the probability that the switch has not occurred by time t and occurs in the time interval between t and $t + dt$. In mathematical language,

$$P_i(t + dt) = P_i(t) + (1 - P_i)K_i dt. \tag{2.201}$$

Taking the limit $dt \to 0$ we obtain the differential equation for the waiting-time probability

$$\frac{dP_i}{dt} = K_i(1 - P_i), \tag{2.202}$$

which, since $P_i(0) = 0$ (assuming that the switch has not occurred at time $t = 0$), and if K_i is independent of time, yields

$$P_i(t) = 1 - \exp(-K_i t). \tag{2.203}$$

The function $P_i(t)$ is a cumulative probability distribution function, with probability density function

$$p_i(t) = \frac{dP_i}{dt} = K_i \exp(-K_i t). \tag{2.204}$$

The probability that the switch occurs between two specified times, t_1 and t_2, is

$$P(t_1 < T_i < t_2) = \int_{t_1}^{t_2} p_i(s)\, ds = P_i(t_2) - P_i(t_1), \tag{2.205}$$

and the expected switching time is

$$E(T_i) = \int_0^{\infty} t p_i(t)\, dt. \tag{2.206}$$

If K_i is time-independent,

$$E(T_i) = \frac{1}{K_i}.$$ (2.207)

The Transition Time

The waiting-time problem can be generalized to calculate the probability density function, ϕ_{ij}, for the time it takes to move from state i to state j.

To do this we make state j an absorbing state and then solve the master equations starting in state i. That is, we set $k_{jl} = 0$ for every l, and then solve for $\Phi_j(t)$, with the initial condition $\Phi_i(0) = 1$, $\Phi_l(0) = 0$ for $l \neq i$. Then, $\Phi_j(t)$ is the probability that the process is in state j at time t, given that it started in state i at time 0. Hence, $\Phi_j(t)$ is the cumulative probability that the transition to state j occurred at some time previous to t. Of course, this relies on the assumption that once state j is reached, it cannot be left. The probability density function for the transition time is the derivative of the cumulative probability. Hence,

$$\phi_{ij}(t) = \frac{d\Phi_j}{dt} = \sum_{l \neq j} k_{lj} \Phi_l.$$ (2.208)

Note that

$$\int_0^\infty \phi_{ij}\, dt = \Phi_j\big|_0^\infty = 1,$$ (2.209)

and thus ϕ_{ij} is indeed a probability density as claimed.

2.9.3 Numerical Simulation of Discrete Markov Processes

It is becoming increasingly important and valuable to do numerical simulations of discrete stochastic processes. The definition (2.195) provides a natural numerical algorithm for such a simulation. For a fixed small time step of size dt, divide the unit interval into $n-1$ regions of length $k_{ij}dt$ and one remaining region of length $1 - K_i dt$. Then, at each time step, pick a random number that is uniformly distributed on the unit interval and determine the next state by the subinterval in which the random number falls.

While this method of numerical simulation is simple and direct, it is not particularly efficient. Furthermore, it converges (i.e., gives the correct statistics) only in the limit that $dt \to 0$.

A method that is much more efficient is known as Gillespie's method (Gillespie, 1977). The idea of this method is to calculate the sequence of random switching times. For example, suppose that at time $t = 0$, the state variable is $S(0) = i$. We know from (2.204) that the probability that the first transition out of state i occurs in the interval $(t, t+dt)$ is $\int_t^{t+dt} K_i \exp(-K_i s)\, ds$. Hence, we can calculate the time to the next transition by selecting a random number from the exponential waiting-time distribution, $p_i(t)$. We do this by selecting a random number x uniformly distributed on the unit interval, and transforming that number via some transformation $t = f(x)$ to get the time at which the transition occurs. The transformation we use should preserve probabilities,

$$p_i(t)dt = q(x)dx, \qquad\qquad (2.210)$$

where $q(x) = 1$ is the uniform distribution. Integrating gives $x = \int_0^t p_i(s)\,ds = 1 - \exp(-K_i y) = P_i(t)$, and solving for t, we get

$$t = -\frac{1}{K_i}\ln(1 - x). \qquad\qquad (2.211)$$

However, since x is uniformly distributed on the unit interval, it is equivalent to replace $1 - x$ by x to get

$$t = -\frac{1}{K_i}\ln(x). \qquad\qquad (2.212)$$

Therefore, to calculate the next switching time, pick a random number that is uniformly distributed on the unit interval, say ξ, and then pick the time interval, T, to the next switch to be

$$T = -\frac{1}{K_i}\ln(\xi). \qquad\qquad (2.213)$$

To determine the state into which to switch, divide the unit interval into segments of length $\frac{k_{ij}}{K_i}$, and select the next interval to be the subinterval in which another uniformly distributed random number η resides. The reasoning for this is that if a switch is to occur, then the probability that the switch is into state j is $\frac{k_{ij}}{K_i}$.

There are numerous advantages to Gillespie's method. First, it is maximally efficient and it is exact. That is, since there is no time discretization step dt, accuracy does not require taking a limit $dt \to 0$. However, other nice features of the method are that it is easier to collect statistics such as closed time and open time distributions, because these quantities are directly, not indirectly, calculated.

A word of caution, however, is that while this method works well for time-independent processes (which we assumed in this discussion), for time-dependent processes, or processes in which the transition rates depend on other time-varying variables, determination of the next transition time requires a more sophisticated calculation (Alfonsi et al., 2005).

Further difficulties with stochastic simulations occur when there are reactions with vastly different time scales. Then, it is often the case that the rapid reactions achieve quasi-equilibrium, but most of the computational time for a stochastic simulation is taken in calculating the many fast transitions of the fast reactions. It is beyond the scope of this text to describe what to do in these situations.

Gillespie's method is the method of choice in many situations. It has been implemented by Adalsteinsson et al. (2004) in a useful software package that is readily available.

2.9.4 Diffusion

The diffusion equation (2.7) was derived to describe the evolution of a chemical concentration, under the assumption that the concentration is a continuous variable, even though the number of molecules involved is necessarily an integer. Einstein recognized that the solution of the diffusion equation could also be interpreted as the probability distribution function for the location of a single particle undergoing some kind of a random walk. That is, if $p(x,t)$ is the solution of the diffusion equation

$$\frac{\partial p}{\partial t} = D\nabla^2 p, \tag{2.214}$$

then $\int_\Omega p(x,t)\,dx$ could be identified as the probability that a particle is in the region Ω at time t. More specifically, if $p(x,t \mid x_0,t_0)$ is the probability distribution function for the particle to be at position x at time t, given that it was at position x_0 at time t_0, then

$$p(x,t_0 \mid x_0,t_0) = \delta(x - x_0), \tag{2.215}$$

and solving the diffusion equation (in one spatial dimension) gives

$$p(x,t \mid x_0,t_0) = \frac{1}{2\sqrt{\pi D(t - t_0)}} \exp\left(-\frac{(x - x_0)^2}{4D(t - t_0)}\right), \tag{2.216}$$

provided $t > t_0$. It follows immediately that the mean and variance of this distribution are

$$\langle x \rangle = \int_{-\infty}^{\infty} x p(x,t \mid 0,0)\,dx = 0 \tag{2.217}$$

and

$$\langle x^2 \rangle = \int_{-\infty}^{\infty} x^2 p(x,t \mid 0,0)\,dx = 2Dt. \tag{2.218}$$

The conditional probability $p(x,t \mid x_0,t_0)$ is the Green's function of the diffusion equation on an infinite domain (where we require, for physical reasons, that the solution and all its derivatives vanish at infinity). As a side issue (but a particularly interesting one), note that, from the transitive nature of Green's functions, it follows that

$$p(x_1,t_1 \mid x_3,t_3) = \int_{x_2} p(x_1,t_1 \mid x_2,t_2)p(x_2,t_2 \mid x_3,t_3)\,dx_2, \quad t_3 < t_2 < t_1, \tag{2.219}$$

which is known as the Chapman–Kolmogorov equation. From the point of view of conditional probabilities, the Chapman–Kolmogorov equation makes intuitive sense; for Markov processes, the probability of x_1 given x_3 is the sum of the probabilities of each path by which one can get from x_3 to x_1.

Now suppose we let $X(t)$ represent the position as a function of time of a sample path. We can readily calculate that $X(t)$ is continuous, since for any $\epsilon > 0$, the probability of escaping from a region of size ϵ in time Δt is

$$\int_{|x-z|>\epsilon} p(x,t+\Delta t \mid z,t)\,dx = 2\int_\epsilon^\infty \frac{1}{2\sqrt{\pi D\Delta t}}\exp\left(-\frac{x^2}{4D\Delta t}\right)dx$$

$$= \int_{\frac{\epsilon}{2\sqrt{D\Delta t}}}^\infty \exp\left(-x^2\right)dx,$$

which approaches zero in the limit $\Delta t \to 0$. On the other hand, the velocity of the particle is likely to be extremely large, since

$$\text{Prob}\left(\frac{1}{\Delta t}(X(t+\Delta t)-X(t)) > k\right) = \int_{k\Delta t}^\infty \frac{1}{2\sqrt{\pi D\Delta t}}\exp\left(-\frac{x^2}{4D\Delta t}\right)dx$$

$$= \int_{\frac{k}{2}\sqrt{\frac{\Delta t}{\pi D}}}^\infty \exp\left(-x^2\right)dx \to \frac{1}{2}, \qquad (2.220)$$

in the limit that $\Delta t \to 0$. In other words, with probability 1, the absolute value of the velocity is larger than any number k, hence infinite.

If $D = 1$, the stochastic process $X(t)$ is known as a *Wiener process*, is usually denoted by $W(t)$, and is a model of Brownian motion.

Diffusion as a Markov Process

A popular derivation of the diffusion equation is based on a Markovian random walk on a grid, as follows. We suppose that a particle moves along a one-dimensional line in discrete steps of length Δx at discrete times with time step Δt. At each step, however, the direction of motion is random, with probability $\frac{1}{2}$ of going to the left and probability $\frac{1}{2}$ of going to the right. If $p(x,t)$ is the probability of being at position x at time t, then

$$p(x,t+\Delta t) = \frac{1}{2}p(x+\Delta x,t) + \frac{1}{2}p(x-\Delta x,t). \qquad (2.221)$$

Now we make the assumption that $p(x,t)$ is a smooth function of both x and t and obtain the Taylor series expansion of (2.221),

$$\Delta t\frac{\partial p}{\partial t} + O(\Delta t^2) = \frac{\Delta x^2}{2}\frac{\partial^2 p}{\partial x^2} + O(\Delta x^4). \qquad (2.222)$$

In the limit that Δt and Δx both approach zero, keeping $\frac{\Delta x^2}{\Delta t} = 1$, we obtain the diffusion equation with diffusion coefficient $\frac{1}{2}$.

2.9.5 Sample Paths; the Langevin Equation

The diffusion equation describes the probability distribution that a particle is at a particular place at some time, but does not describe how the particle actually moves. The challenge, of course, is to write (and solve) an equation for motion that is random and continuous, but nowhere differentiable. Obviously, one cannot use a standard differential equation to describe the motion of such a particle. So, instead of writing

$\frac{dx}{dt}$ = something (which does not make sense, since the velocity $\frac{dx}{dt}$ of a Brownian particle is not finite), it is typical to write

$$dx = \sqrt{2D}\,dW. \tag{2.223}$$

To make careful mathematical sense of this expression requires a discussion of the Ito or Stratonovich calculus, topics that are beyond the scope of this text. However, a reasonable verbal description of what this means in practical terms is as follows. The term dW is intended to represent the fact that the displacement of a particle after a very short time interval, say dt, is a random variable having three properties, namely, it is uncorrelated with previous displacements (it has no memory and is therefore Markovian), it has zero mean, and it has variance dt, in the limit $dt \to 0$. This is also referred to as uncorrelated *Gaussian white noise*. In fact, this definition is rigged so that the probability distribution for this particle is described by the diffusion equation.

For this text, it is important to know how to numerically calculate representative sample paths, and to this end we write

$$dx = \sqrt{2D\,dt}N(0,1), \tag{2.224}$$

where $N(0,1)$ represents the Gaussian (normal) distribution with zero mean and variance 1. The interpretation of this is that at any given time one randomly chooses a number n from a normal distribution, takes a step of size $dx = \sqrt{2Ddt}n$, and then increments time by dt. It can be shown that in the limit that $dt \to 0$, this converges to the Wiener process (2.223).

Equation (2.223) is an example of a stochastic differential equation, also called a *Langevin equation*. More generally, Langevin equations are of the form

$$dx = a(x,t)\,dt + \sqrt{2b(x,t)}\,dW, \tag{2.225}$$

or, in a form that suggests a numerical algorithm,

$$dx = a(x,t)\,dt + \sqrt{2b(x,t)\,dt}N(0,1). \tag{2.226}$$

Here $a(x,t)$ represents the deterministic part of the velocity, since if there were no noise ($b(x,t) = 0$), this would be the same as the deterministic equation

$$\frac{dx}{dt} = a(x,t). \tag{2.227}$$

Thus, the displacement dx is a random variable, with mean value $a(x,t)dt$ and variance $2b(x,t)dt$, in the limit $dt \to 0$.

The special case $a(x,t) = -x$, $b(x,t) = 1$, called an *Ornstein–Uhlenbeck* process, is important in the study of molecular motors, described in Chapter 15.

2.9.6   The Fokker–Planck Equation and the Mean First Exit Time

The diffusion equation is the simplest example of an equation describing the evolution of the probability distribution function for the position of a particle. More generally, if we suppose that the position of the particle is continuous in time (no finite jumps are possible), that the Chapman–Kolmogorov equation (2.219) holds, and that the displacement of the particle in time dt has mean $a(x,t)dt$ and variance $2b(x,t)dt$, then one can derive that the probability distribution $p(x,t)$ for the position, x, of the particle at time t is governed by

$$\frac{\partial p}{\partial t} = -\frac{\partial}{\partial x}(a(x,t)p) + \frac{\partial^2}{\partial x^2}(b(x,t)p), \tag{2.228}$$

called the *Fokker–Planck* equation. Note that, since this equation models the motion of a particle which must be at a single position y at the starting time t_0, the initial condition must be $p(x,t_0 \mid y,t_0) = \delta(x - y)$. Thus, the probability distribution for the position of the particle is the Green's function of (2.228).

More generally, it is possible to start with the Chapman–Kolmogorov equation (2.219) and derive a general version of the Fokker–Planck equation that includes discrete jump processes. This is the point of view usually taken in the stochastic processes literature, which treats the Chapman–Kolmogorov equation as a fundamental requirement of a Markov process.

An extremely important problem is the so-called *mean first exit time problem*, in which we wish to determine how long a particle stays in a particular region of space. Before we can solve this problem we must first determine the equation for the conditional probability, $p(x,t \mid y,\tau)$, as a function of y and $\tau < t$, with x and t fixed. That is, we want to know the probability distribution function for a particle with known position x at time t to have been at the location y at time $\tau < t$.

The equation governing this conditional probability is most easily derived by using the fact that $p(x,t \mid y,\tau)$ is the Green's function of the Fokker–Planck equation. It follows from the properties of Green's functions (Keener, 1998) that

$$p(x,t \mid y,\tau) = p^*(y,\tau \mid x,t), \tag{2.229}$$

where p^* is the Green's function of the adjoint equation. The adjoint equation is easily calculated using integration by parts, and assuming that p and all its derivatives vanish at infinity. It follows that p, considered as a function of y and τ, satisfies the adjoint equation

$$\frac{\partial p}{\partial \tau} = -a(y,\tau)\frac{\partial p}{\partial y} - b(y,\tau)\frac{\partial^2 p}{\partial y^2}, \tag{2.230}$$

subject to the condition $p(x,t \mid y,t) = \delta(x - y)$. This equation for the backward conditional probability is called the *backward Fokker–Planck equation*. Notice that this is a *backward* diffusion equation, which in forward time is ill posed. However, it is well posed when solved for backward times $\tau < t$.

Armed with the backward Fokker–Planck equation, we now turn our attention to the mean first exit time problem. Suppose a particle is initially at position y, inside a one-dimensional region $\alpha < y < \beta$, and that the wall at $y = \alpha$ is impermeable, but the particle can leave the region freely at $y = \beta$. If we let $\tau(y)$ represent the time at which the particle first leaves the region having started from position y, then

$$P(\tau(y) > t) = G(y,t) = \int_\alpha^\beta p(x,t \mid y,0)\,dx, \tag{2.231}$$

which is the probability that the particle is inside the region at time t. Notice that since $p(x,0 \mid y,0) = \delta(x - y)$, $G(y,0) = 1$.

Since

$$P(\tau(y) > t) = G(y,t) = -\int_t^\infty G_t(y,s)\,ds, \tag{2.232}$$

it follows that $G_t(y,t)$ is the probability density function for the random variable $\tau(y)$. Thus, the expected value of $\tau(y)$ is

$$T(y) = E(\tau(y)) = -\int_0^\infty t G_t(y,t)\,dt = \int_0^\infty G(y,t)\,dt, \tag{2.233}$$

where we have integrated by parts to get the final expression. Note that $T(y)$ is the mean time at which a particle leaves the domain, given that it starts at y at time $t = 0$.

For a time-independent process (i.e., $a(y,t) = a(y)$, $b(y,t) = b(y)$) we have $p(x,t \mid y,0) = p(x,0 \mid y,-t)$. Hence, substituting $-t$ for t in (2.230), and writing $q(y,t \mid x) = p(x,t \mid y,0)$, it follows that q satisfies the negative backward Fokker–Planck equation

$$\frac{\partial q}{\partial t} = a(y)\frac{\partial q}{\partial y} + b(y)\frac{\partial^2 q}{\partial y^2}, \tag{2.234}$$

with $q(y,0 \mid x) = \delta(x - y)$. Integrating with respect to x from $x = \alpha$ to $x = \beta$ (since $G(y,t) = \int_\alpha^\beta p(x,t \mid y,0)\,dx$) then gives

$$\frac{\partial G}{\partial t} = a(y)\frac{\partial G}{\partial y} + b(y)\frac{\partial^2 G}{\partial y^2}, \tag{2.235}$$

with $G(y,0) = 1$. Finally, we can determine the equation for the expected value of $\tau(y)$ by integrating (2.235) in time to find

$$-1 = a(y)\frac{\partial T}{\partial y} + b(y)\frac{\partial^2 T}{\partial y^2}. \tag{2.236}$$

To completely specify the problem, we must specify boundary conditions. At impermeable boundaries, we require $\frac{\partial T}{\partial y} = 0$, while at absorbing boundaries (boundaries through which exit is allowed but reentry is not permitted) we require $T = 0$.

As an example, consider a pure diffusion process on a domain of length L with a reflecting boundary at $x = 0$ and an absorbing boundary at $x = L$. The mean first exit time satisfies the differential equation

$$DT_{xx} = -1, \tag{2.237}$$

subject to boundary conditions $T_x(0) = 0$, $T(L) = 0$. This has solution

$$T(x) = \frac{-x^2 + L^2}{2D}. \tag{2.238}$$

We readily calculate that $T(0) = \frac{L^2}{2D}$, as might be expected from (2.218). In addition, as x increases, $T(x)$ decreases, which again makes intuitive sense. The closer the particle starts to the absorbing boundary, the shorter is the mean first exit time.

2.9.7   Diffusion and Fick's Law

The Fokker–Planck equation describes the evolution of the probability distribution function for single particle diffusion. However, it also applies to the concentration of a dilute chemical species under the assumption that the chemical particles have no self-interaction. If the diffusion coefficient is homogeneous in space, then the Fokker–Planck equation and the diffusion equation are the same. However, if diffusion is not homogeneous in space, then the diffusion equation, derived using Fick's law, and the Fokker–Planck equation are not the same. With Fick's law, the diffusion equation is

$$\frac{\partial c}{\partial t} = \frac{\partial}{\partial x}\left(D\frac{\partial c}{\partial x}\right) \tag{2.239}$$

and the Fokker–Planck equation is

$$\frac{\partial c}{\partial t} = \frac{\partial^2 (Dc)}{\partial x^2}. \tag{2.240}$$

Which of these is correct?

There is a simple observation that can help answer this question. If Fick's law is correct, then at steady state the flux in a closed container is zero, so that

$$D\frac{\partial c}{\partial x} = 0, \tag{2.241}$$

implying that $c(x)$ is a uniform constant. On the other hand, if the Fokker–Planck equation is correct, then at steady state

$$D(x)c(x) = \text{constant}. \tag{2.242}$$

Notice that in this solution, the concentration varies inversely with D. That is, if D is low then c should be high, and vice versa. Further, if c is initially uniform, Fick's law predicts no change in the solution as a function of time, whereas the Fokker–Planck equation predicts transient behavior leading to a nonuniform distribution of c.

Van Milligen et al. (2006) reported a simple experimental test of this observation. They added green food coloring to water and then added differing amounts of gelatin to small quantities of the colored water. They then created gel bilayers consisting of two layers of the colored gelatin with differing gelatin concentrations under the assumption that diffusion of food coloring varies inversely with the amount of gelatin. They recorded the color intensity at several later times and compared these recordings with the solution of the partial differential equations.

The first observation is that the initially uniform concentration of food coloring did not remain uniform, but increased in regions where the gelatin density was highest. Even more striking, they were able to find very good fits of the data to the numerical solution of the Fokker–Planck equation. These observations lead to the conclusion that Fick's law is not the correct description of chemical diffusion in media where the diffusion coefficient is not constant. The more appropriate description, coming from the Fokker–Planck equation, is that

$$J = -\nabla(Dc). \tag{2.243}$$

2.10 EXERCISES

1. A rule of thumb (derived by Einstein) is that the diffusion coefficient for a globular molecule satisfies $D \sim M^{-1/3}$, where M is the molecular weight. Determine how well this relationship holds for the substances listed in Table 2.2 by plotting D and M on a log-log plot.

2. A fluorescent dye with a diffusion coefficient $D = 10^{-7}$ cm^2/s and binding equilibrium $K_{eq} = 30$ mM is used to track the spread of hydrogen ($D_h = 4.4 \times 10^{-5}$cm^2/s). Under these conditions the measured diffusion coefficient is 8×10^{-6}cm^2/s. How much dye is present? (Assume that the dye is a fast buffer of hydrogen and that the amount of hydrogen is much less that K_{eq}.)

3. Segel, Chet and Henis (1977) used (2.18) to estimate the diffusion coefficient for bacteria. With the external concentration C_0 at 7×10^7 ml^{-1}, at times $t = 2, 5, 10, 12.5, 15$, and 20 minutes, they counted N of 1,800, 3,700, 4,800, 5,500, 6,700, and 8,000 bacteria, respectively, in a capillary of length 32 mm with 1 μl total capacity. In addition, with external concentrations C_0 of 2.5, 4.6, 5.0, and 12.0 $\times 10^7$ bacteria per milliliter, counts of 1,350, 2,300, 3,400, and 6,200 were found at $t = 10$ minutes. Estimate D.

4. Calculate the effective diffusion coefficient of oxygen in a solution containing 1.2×10^{-5} M/cm^3 myoglobin. Assume that the rate constants for the uptake of oxygen by myoglobin are $k_+ = 1.4 \times 10^{10}$cm^3 M^{-1}s^{-1} and $k_- = 11$ s^{-1}.

5. Find the maximal enhancement for diffusive transport of carbon dioxide via binding with myoglobin using $D_s = 1.92 \times 10^{-5}$ cm^2/s, $k_+ = 2 \times 10^8$ cm^3/M $\cdot$ s, $k_- = 1.7 \times 10^{-2}$/s. Compare the amount of facilitation of carbon dioxide transport with that of oxygen at similar concentration levels.

6. Devise a model to determine the rate of production of product for a "one-dimensional" enzyme capsule of length L in a bath of substrate at concentration S_0. Assume that the enzyme is confined to the domain $0 \leq x \leq L$ and there is no flux through the boundary at $x = 0$. Assume that the enzyme cannot diffuse within the capsule but that the substrate and product can freely diffuse into, within, and out of the capsule. Show that the steady-state production per unit volume of enzyme is less than the production rate of a reactor of the same size in which substrate is homogeneously mixed (infinite diffusion).

7. Devise a model to determine the rate of production of product for a spherical enzyme capsule of radius R_0 in a bath of substrate at concentration S_0. Assume that the enzyme cannot diffuse within the capsule but that the substrate and product can freely diffuse into, within, and out of the capsule. Show that spheres of small radius have a larger rate of production than spheres of large radius.

Hint: Reduce the problem to the nondimensional boundary value problem

$$\frac{1}{y^2}(y^2\sigma')' - \alpha^2\frac{\sigma}{\sigma+1} = 0, \tag{2.244}$$

$$\sigma'(0) = 0, \tag{2.245}$$

$$\sigma(1) = \sigma_0, \tag{2.246}$$

and solve numerically as a function of α. How does the radius of the sphere enter the parameter α?

8. Red blood cells have a passive exchanger that exchanges a single Cl^- ion for a bicarbonate (HCO_3^-) ion. Develop a model of this exchanger and find the flux.

9. Almost immediately upon entering a cell, glucose is phosphorylated in the first reaction step of glycolysis. How does this rapid and nearly unidirectional reaction affect the trans-membrane flux of glucose as represented by (2.54)? How is this reaction affected by the concentration of ATP?

10. In the model of the glucose transporter (Fig. 2.6) the reaction diagram was simplified by assuming that each conformation of the transporter is equally likely, and that the affinity of the glucose binding site is unaffected by a change in conformation.

   (a) Construct a more detailed model in which these assumptions are relaxed, and calculate the flux through the model.

   (b) What is the total change in chemical potential after one cycle of the exchanger? What is the equilibrium condition?

   (c) Apply detailed balance to obtain a relationship between the rate constants.

11. Consider the model of a nonelectrogenic, 3 for 1, Na^+–Ca^{2+} exchanger. At equilibrium, the concentrations on either side of the membrane are related by the equation

$$\frac{n_e^3 c_i}{n_i^3 c_e} = 1. \tag{2.247}$$

Assume that the membrane separates two equal volumes. For a given set of initial concentrations and assuming there are no other exchange processes, what three additional conservation equations must be used to determine the equilibrium concentrations? Prove that there is a unique equilibrium solution. Hint: Give a graphical proof.

12. Simplify the model of the Na^+–Ca^{2+} exchanger (Fig. 2.9) by assuming that the binding and unbinding of Na^+ and Ca^{2+} are fast compared to the exchange processes between the inside and the outside of the cell. Write the new model equations and calculate the steady-state flux. Hint: The assumption of fast equilibrium gives

$$k_1 c_i x_1 = k_{-1} n_i^3 x_2, \tag{2.248}$$

$$k_3 n_e^3 y_2 = k_{-3} c_e y_1. \tag{2.249}$$

Then introduce the new variables $X = X_1 + X_2$ and $Y = Y_1 + Y_2$ and derive the equations for X and Y.

13. Simplify the model of Fig. 2.11 by assuming fast binding of Na^+ and K^+, and draw the reaction diagram of the simplified model. Calculate the expression for the steady-state flux.

Hint: Combine the states X_1, X_2, and X_3 into a single state, X, and similarly for Y_1, Y_2, and Y_3. Then use the equilibrium conditions

$$x_1 = K_1 x_2 \kappa_i^2, \tag{2.250}$$

$$n_i^3 x_2 = K_2 x_3, \tag{2.251}$$

$$y_3 = K_5 n_e^3 y_2, \tag{2.252}$$

$$\kappa_e^2 y_2 = K_6 y_1, \tag{2.253}$$

where $K_i = k_{-i}/k_i$, n denotes $[Na^+]$, and κ denotes $[K^+]$, to derive the differential equations for X and Y.

14. Calculate the flux of the Ran-GTP nuclear transporter. Use the information given in the text to estimate the concentrating ability of this transporter, assuming there is no difference in potential across the nuclear membrane.

15. Suppose that two compartments, each of one liter in volume, are connected by a membrane that is permeable to both K^+ and Cl^-, but not to water or protein (X). Suppose further that, as illustrated in Fig. 2.21, the compartment on the left initially contains 300 mM K^+ and 300 mM Cl^-, while the compartment on the right initially contains 200 mM protein, with valence -2, and 400 mM K^+.

    (a) Is the starting configuration electrically and osmotically balanced?

    (b) Find the concentrations at equilibrium.

    (c) Why is $[K^+]_i$ at equilibrium greater than its starting value, even though $[K^+]_i > [K^+]_e$ initially? Why does K^+ not diffuse from right to left to equalize the concentrations?

    (d) What is the equilibrium potential difference?

    (e) What would happen if the connecting membrane were suddenly made permeable to water when the system is at equilibrium? How large would the osmotic pressure be?

16. The derivation of the Gibbs–Donnan equilibrium for the case when $[S]_e$ is not fixed requires an additional constraint. Show that it is equivalent to use either $v_i[S]_i + v_e[S]_e = [S]_{tot}$ or $v_i[S']_i + v_e[S']_e = [S']_{tot}$. How must $[S]_{tot}$ and $[S']_{tot}$ be related so that the answers for the two are the same?

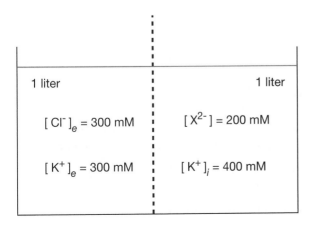

1 liter

$[Cl^-]_e = 300$ mM

$[K^+]_e = 300$ mM

1 liter

$[X^{2-}] = 200$ mM

$[K^+]_i = 400$ mM

Figure 2.21 The initial configuration for Exercise 15.

17. Suppose the intracellular macromolecule X can bind b molecules of the ion S via X + bS $\rightleftharpoons$ XS$_b$. What is the effect of this buffering on the Gibbs–Donnan equilibrium potential?

18. A 1.5 oz bag of potato chips (a typical single serving) contains about 200 mg of Na$^+$. When eaten and absorbed into the body, how many osmoles does this bag of potato chips represent?

19. (a) Confirm that π_S in (2.138) has units of pressure.

    (b) Confirm the statement that a pressure of 25 atm corresponds to a column of water over 250 meters high.

    (c) Consider a vertical tube with a cross-sectional area of 1 cm^2. The bottom of the tube is closed with a semipermeable membrane, and 1 gram of sugar is placed in the tube. The membrane-closed end of the tube is then put into an inexhaustible supply of pure water at $T = 300$ K. What will be the height of the water in the tube at equilibrium? (The weight of a sugar molecule is 3×10^{-22} gm, and the density of water is 1 gm/cm^3).

    (d) Two columns with cross-sectional area 1 cm^2 are initially filled to a height of one meter with water at $T = 300$ K. Suppose 0.001 gm of sugar is dissolved in one of the two columns. How high will the sugary column be when equilibrium is reached?

    (e) Suppose that, in the previous question, 1 gm of sugar is dissolved in one of the two columns. What is the equilibrium height of the two columns?

20. Suppose an otherwise normal cell is placed in a bath of high extracellular K$^+$. What happens to the cell volume and resting potentials?

21. Based on what you know about glycolysis from Chapter 1, how would you expect anoxia (insufficient oxygen) to affect the volume of the cell? How might you incorporate this into a model of cell volume? Hint: Lactic acid does not diffuse out of a cell as does carbon dioxide.

22. Suppose 90% of the Na$^+$ in the bath of a squid axon is replaced by inert choline, preserving electroneutrality. What happens to the equilibrium potentials and membrane potentials?

23. Determine the effect of temperature (through the Nernst equation) on cell volume and membrane potential.

24. Simulate the time-dependent differential equations governing cell volume and ionic concentrations. What happens if the extracellular ionic concentrations are suddenly increased or decreased?

25. Ouabain is known to compete with K$^+$ for external K$^+$ binding sites of the Na$^+$–K$^+$ ATPase. Many animal cells swell and burst when treated with the drug ouabain. Why? Hint: How would you include this effect in a model of cell volume control?

26. Since the Na$^+$–K$^+$ ATPase is electrogenic, the pump rate P in the pump-leak model must also include effects from the membrane potential. What effect does membrane potential have on the expression (2.169) and how does this modification affect the solution?

27. Use (2.224) to simulate a diffusion process, and verify that the mean and variance of the process are 0 and $2Dt$, respectively, as expected.

28. Find the steady-state probability distribution for the Ornstein–Uhlenbeck process.

29. A small particle (with diffusion constant D and viscosity v) experiences a constant load F directed to the left but is not permitted to move into the region with $x < 0$. Suppose that the particle is initially at $x = 0$. What is the expected time to first reach location $x = \delta$?

30. Suppose that a molecule enters a spherical cell of radius 5 μm at its boundary. How long will it take for the molecule to move by diffusion to find a binding target of radius 0.5 nm located at the center of the cell? Use a diffusion coefficient of 10^{-6} cm^2/s. (Hint: In higher dimensions the differential equation for the mean first exit time is $\nabla^2 T = -1$.)

CHAPTER 3

Membrane Ion Channels

Every cell membrane contains ion channels, macromolecular pores that allow specific ions to travel through the channels by a passive process, driven by their concentration gradient and the membrane potential. One of the most extensively studied problems in physiology is the regulation of such ionic currents. Indeed, in practically every chapter of this book there are examples of how the control of ionic current is vital for cellular function. Already we have seen how the cell membrane uses ion channels and pumps to maintain an intracellular environment that is different from the extracellular environment, and we have seen how such ionic separation results in a membrane potential. In subsequent chapters we will see that modulation of the membrane potential is one of the most important ways in which cells control their behavior or communicate with other cells. However, to understand the role played by ion channels in the control of membrane potential, it is first necessary to understand how membrane ionic currents depend on the voltage and ionic concentrations.

There is a vast literature, both theoretical and experimental, on the properties of ion channels. One of the best books on the subject is that of Hille (2001), to which the reader is referred for a more detailed presentation than that given here. The bibliography provided there also serves as a starting point for more detailed studies.

3.1 Current–Voltage Relations

Before we discuss specific models of ion channels, we emphasize an important fact that can be a source of confusion. Although the Nernst equation (2.104) for the equilibrium voltage generated by ionic separation can be derived from thermodynamic considerations and is thus universally applicable, there is no universal expression for the ionic current. An expression for, say, the Na^+ current cannot be derived from thermodynamic first principles and depends on the particular model used to describe the Na^+ channels. Already we have seen two different models of ionic currents. In the previous chapter we described two common models of Na^+ current as a function of the membrane potential and the internal and external Na^+ concentrations. In the simpler model, the Na^+ current across the cell membrane was assumed to be a linear function of the membrane potential, with a driving force given by the Na^+ Nernst potential. Thus,

$$I_{Na} = g_{Na}(V - V_{Na}), \tag{3.1}$$

where $V_{Na} = (RT/F) \ln([Na^+]_e/[Na^+]_i)$ is the Nernst potential of Na^+, and where $V = V_i - V_e$. (As usual, a subscript e denotes the external concentration, while a subscript i denotes the internal concentration.) Note that the Na^+ current is zero when V is the Nernst potential, as it must be. However, we also discussed an alternative model, where integration of the Nernst–Planck equation (2.114), assuming a constant electric field, gave the Goldman–Hodgkin–Katz (GHK), or constant-field, current equation:

$$I_{Na} = P_{Na} \frac{F^2}{RT} V \left[\frac{[Na^+]_i - [Na^+]_e \exp\left(\frac{-VF}{RT}\right)}{1 - \exp\left(\frac{-VF}{RT}\right)} \right]. \tag{3.2}$$

As before, the Na^+ current is zero when V equals the Nernst potential, but here the current is a nonlinear function of the voltage and linear in the ionic concentrations. In Fig. 3.1A we compare the linear and GHK I–V curves when there is only a single ion present.

There is no one "correct" expression for the Na^+ current, or any other ionic current for that matter. Different cells have different types of ion channels, each of which may have different current–voltage relations. The challenge is to determine the current–voltage, or I–V, curve for a given ion channel and relate it to underlying biophysical mechanisms.

Our choice of these two models as examples was not coincidental, as they are the two most commonly used in theoretical models of cellular electrical activity. Not only are they relatively simple (at least compared to some of the other models discussed later in this chapter), they also provide good quantitative descriptions of many ion channels. For example, the I–V curves of open Na^+ and K^+ channels in the squid giant axon are approximately linear, and thus the linear model was used by Hodgkin and Huxley in their classic model of the squid giant axon (discussed in detail in Chapter 5). However, the I–V curves of open Na^+ and K^+ channels in vertebrate axons are better described by the GHK equation, and so nonlinear I–V curves are often used for vertebrate models (Frankenhaeuser, 1960a,b, 1963; Campbell and Hille, 1976).

Because of the importance of these two models, we illustrate another way in which they differ. This also serves to illustrate the fact that although the Nernst potential is universal when there is only one ion present, the situation is more complicated when two or more species of ion can pass through the membrane. If both Na^+ and K^+ ions are present and both obey the GHK current equation, we showed in (2.127) that the reversal potential V_r at which there is no net current flow is

$$V_r = \frac{RT}{F} \ln\left(\frac{P_{Na}[Na^+]_e + P_K[K^+]_e}{P_{Na}[Na^+]_i + P_K[K^+]_i}\right). \qquad (3.3)$$

However, if instead that the I–V curves for Na^+ and K^+ are assumed to be linear, then the reversal potential is

$$V_r = \frac{g_{Na}V_{Na} + g_K V_K}{g_{Na} + g_K}, \qquad (3.4)$$

where V_K is the Nernst potential of K^+. Clearly, the reversal potential is model-dependent. This is due to the fact that at the reversal potential the net current flow is zero, but the individual Na^+ and K^+ currents are not. Thus, the equilibrium arguments used to derive the Nernst equation do not apply, and a universal form for the reversal potential does not exist. As an illustration of this, in Fig. 3.1B we plot the reversal potentials V_r from (3.3) and (3.4) as functions of $[K^+]_e$. Although the linear and GHK I–V curves predict different reversal potentials, the overall qualitative behavior is similar, making it difficult to distinguish between a linear and a GHK I–V curve on the basis of reversal potential measurements alone.

3.1.1 Steady-State and Instantaneous Current–Voltage Relations

Measurement of I–V curves is complicated by the fact that ion channels can open or close in response to changes in the membrane potential. Suppose that in a population of ion channels, I increases as V increases. This increase could be the result of two different factors. One possibility is that more channels open as V increases while the current through an individual channel remains unchanged. It is also possible that the same number of channels remain open but the current through each one increases. To understand how each channel operates, it is necessary to separate these two factors to determine the I–V curve of a single open channel. This has motivated the definition of *steady-state* and *instantaneous I–V* curves.

If channels open or close in response to a change in voltage, but this response is slower than the change in current in a channel that is already open, it should be possible to measure the I–V curve of a single open channel by changing the voltage quickly and measuring the channel current soon after the change. Presumably, if the measurement is performed fast enough, no channels in the population have time to open or close in response to the voltage change, and thus the observed current change reflects the current change through the open channels. Of course, this relies on the assumption that the current through each open channel changes instantaneously. The I–V curve measured in this way (at least in principle) is called the instantaneous I–V curve and reflects properties of the individual open channels. If the current measurement is performed after channels have had time to open or close, then the current change reflects

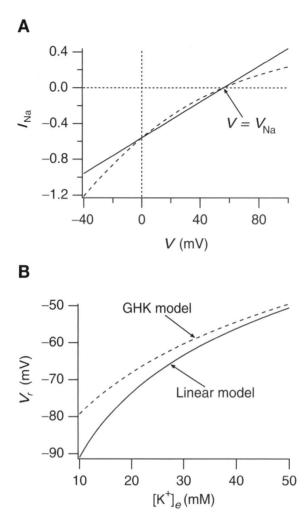

Figure 3.1 A: I–V curves of the linear and GHK models for Na^+ flux through a membrane. Both curves have the same reversal potential as expected, but the GHK model (dashed curve) gives a nonlinear I–V curve. Typical concentrations and conductances of the squid axon were used: $[Na^+]_i = 50$ mM, $[Na^+]_e = 437$ mM, and $g_{Na} = 0.01$ mS/cm^2. P_{Na} was chosen so that the GHK I–V curve intersects the linear I–V curve at $V = 0$. B: Reversal potentials of the linear and GHK models as functions of $[K^+]_e$. The membrane is permeable to both Na^+ and K^+. The same parameters as A, with $[K^+]_i = 397$ mM and $g_K = 0.367$ mS/cm^2. P_K was chosen so that the GHK I–V curve for K^+, with $[K^+]_e = 20$ mM, intersects the linear I–V curve for K^+ at $V = 0$.

the I–V curve of a single channel as well as the proportion of open channels. In this way one obtains a steady-state I–V curve.

There are two basic types of model that are used to describe ion flow through open channels, and we discuss simple versions of each. In the first type of model, the channel is described as a continuous medium, and the ionic current is determined by the Nernst–Planck electrodiffusion equation, coupled to the electric field by means of the Poisson equation. In more complex models of this type, channel geometry and the effects of induced charge on the channel wall are incorporated. In the second type of model the channel is modeled as a sequence of binding sites, separated by barriers that impede the ion's progress: the passage of an ion through the channel is described as a process of "hopping" over barriers from one binding site to another. The height of each barrier is determined by the properties of the channel, as well as by the membrane potential. Thus, the rate at which an ion traverses the channel is a function both of the membrane potential and of the channel type. An excellent summary of the advantages and disadvantages of the two model types is given by Dani and Levitt (1990).

We also discuss simple models of the kinetics of channel gating, and the stochastic behavior of a single channel. These models are of fundamental importance in Chapter 5, where we use an early model of the voltage-dependent gating of ion channels proposed by Hodgkin and Huxley as part of their model of the action potential in the squid giant axon. More detailed recent models of channel gating are not discussed at any length. The interested reader is referred to Hille (2001), Armstrong (1981), Armstrong and Bezanilla (1973, 1974, 1977), Aldrich et al. (1983), and Finkelstein and Peskin (1984) for a selection of models of how channels can open and close in response to changes in membrane potential. An important question that we do not consider here is how channels can discriminate between different ions. Detailed discussions of this and related issues are in Hille (2001) and the references therein.

3.2 Independence, Saturation, and the Ussing Flux Ratio

One of the most fundamental questions to be answered about an ion channel is whether the passage of an ion through the channel is independent of other ions. If so, the channel is said to obey the *independence principle*.

Suppose a membrane separates two solutions containing an ion species S with external concentration c_e and internal concentration c_i. If the independence principle is satisfied, the flow of S is proportional to its local concentration, independent of the concentration on the opposite side of the membrane, and thus the flux from outside to inside, J_{in}, is

$$J_{in} = k_e c_e, \tag{3.5}$$

for some constant k_e. Similarly, the outward flux is given by

$$J_{out} = k_i c_i, \tag{3.6}$$

where in general, $k_e \neq k_i$. We let V_S denote the Nernst potential of the ion S, and let V denote the potential difference across the membrane. Now we let c_e^* be the external concentration for which V is the Nernst potential. Thus,

$$\frac{c_e}{c_i} = \exp\left(\frac{zV_S F}{RT}\right), \tag{3.7}$$

and

$$\frac{c_e^*}{c_i} = \exp\left(\frac{zVF}{RT}\right). \tag{3.8}$$

If the external concentration were c_e^* with internal concentration c_i, then there would be no net flux across the membrane; i.e., the outward flux equals the inward flux, and so

$$k_e c_e^* = k_i c_i. \tag{3.9}$$

It follows that the flux ratio is given by

$$\frac{J_{\text{in}}}{J_{\text{out}}} = \frac{k_e c_e}{k_i c_i} = \frac{k_e c_e}{k_e c_e^*} = \frac{c_e}{c_e^*}$$

$$= \frac{\exp\left(\frac{zV_SF}{RT}\right)}{\exp\left(\frac{zVF}{RT}\right)}$$

$$= \exp\left[\frac{z(V_S - V)F}{RT}\right]. \tag{3.10}$$

This expression for the ratio of the inward to the outward flux is usually called the *Ussing flux ratio*. It was first derived by Ussing (1949), although the derivation given here is due to Hodgkin and Huxley (1952a). Alternatively, the Ussing flux ratio can be written as

$$\frac{J_{\text{in}}}{J_{\text{out}}} = \frac{c_e}{c_i} \exp\left(\frac{-zVF}{RT}\right). \tag{3.11}$$

Note that when $V = 0$, the ratio of the fluxes is equal to the ratio of the concentrations, as might be expected intuitively.

Although many ion channels follow the independence principle approximately over a range of ionic concentrations, most show deviations from independence when the ionic concentrations are sufficiently large. This has motivated the development of models that show saturation at high ionic concentrations. For example, one could assume that ion flow through the channel can be described by a barrier-type model, in which the ion jumps from one binding site to another as it moves through the channel. If there are only a limited number of binding sites available for ion passage through the channel, and each binding site can bind only one ion, then as the ionic concentration increases there are fewer binding sites available, and so the flux is not proportional to the concentration. Equivalently, one could say that each channel has a single binding site for ion transfer, but there are only a limited number of channels. However, in many of these models the Ussing flux ratio is obeyed, even though independence is not. Hence, although any ion channel obeying the independence principle must also satisfy the Ussing flux ratio, the converse is not true. We discuss saturating models later in this chapter.

Another way in which channels show deviations from independence is in flux-coupling. If ions can interact within a channel so that, for example, a group of ions must move through the channel together, then the Ussing flux ratio is not satisfied. The most common type of model used to describe such behavior is the so-called *multi-ion model*, in which it is assumed that there are a number of binding sites within a single channel and that the channel can bind multiple ions at the same time. The consequent interactions between the ions in the channel can result in deviations from the Ussing flux ratio. A more detailed consideration of multi-ion models is given later in this chapter. However, it is instructive to consider how the Ussing flux ratio is modified by a simple multi-ion channel mechanism in which the ions progress through the channel in single file (Hodgkin and Keynes, 1955).

Suppose a membrane separates two solutions, the external one (on the right) containing an ion S at concentration c_e, and the internal one (on the left) at concentration c_i. To keep track of where each S ion has come from, all the S ions on the left are labeled A, while those on the right are labeled B. Suppose also that the membrane contains n binding sites and that S ions traverse the membrane by binding sequentially to the binding sites and moving across in single file. For simplicity we assume that there are no vacancies in the chain of binding sites. It follows that the possible configurations of the chain of binding sites are $[A_r, B_{n-r}]$, for $r = 0, \ldots, n$, where $[A_r, B_{n-r}]$ denotes the configuration such that the r leftmost sites are occupied by A ions, while the rightmost $n - r$ sites are occupied by B ions. Notice that the only configuration that can result in the transfer of an A ion to the right-hand side is $[A_n B_0]$, i.e., if the chain of binding sites is completely filled with A ions.

Now we let α denote the total rate at which S ions are transferred from left to right. Since α denotes the total rate, irrespective of labeling, it does not take into account whether an A ion or a B ion is moved out of the channel from left to right. For this reason, α is not the same as the flux of labeled ions. Similarly, let β denote the total flux of S ions, irrespective of labeling, from right to left. It follows that the rate at which $[A_r B_{n-r}]$ is converted to $[A_{r+1} B_{n-r-1}]$ is $\alpha [A_r B_{n-r}]$, and the rate of the reverse conversion is $\beta [A_{r+1} B_{n-r-1}]$. According to Hodgkin and Keynes, it is reasonable to assume that if there is a potential difference V across the membrane, then the total flux ratio obeys the Ussing flux ratio,

$$\frac{\alpha}{\beta} = \frac{c_e}{c_i} \exp\left(\frac{-VF}{RT}\right). \tag{3.12}$$

This assumption is justified by the fact that a flux of one ion involves the movement of a single charge (assuming $z = 1$) through the membrane (as in the independent case treated above) and thus should have the same voltage dependence. We emphasize that α/β is not the flux ratio of labeled ions, but the total flux ratio.

To obtain the flux ratio of labeled ions, notice that the rate at which A ions are transferred to the right-hand side is $\alpha [A_n B_0]$, and the rate at which B ions are transferred to the left-hand side is $\beta [A_0 B_n]$. Thus, the flux ratio of labeled ions is

$$\frac{J_{\text{in}}}{J_{\text{out}}} = \frac{\alpha}{\beta} \frac{[A_n B_0]}{[A_0 B_n]}. \tag{3.13}$$

At steady state there can be no net change in the distribution of configurations, so that

$$\frac{[A_{r+1}B_{n-r-1}]}{[A_r B_{n-r}]} = \frac{\alpha}{\beta}. \tag{3.14}$$

Thus,

$$\frac{J_{\text{in}}}{J_{\text{out}}} = \frac{\alpha}{\beta} \frac{[A_n B_0]}{[A_0 B_n]} = \left(\frac{\alpha}{\beta}\right)^2 \frac{[A_{n-1}B_1]}{[A_0 B_n]} = \cdots = \left(\frac{\alpha}{\beta}\right)^{n+1}, \tag{3.15}$$

so that

$$\frac{J_{\text{in}}}{J_{\text{out}}} = \left[\frac{c_e}{c_i} \exp\left(\frac{-VF}{RT}\right)\right]^{n+1}. \tag{3.16}$$

A similar argument, taking into account the fact that occasional vacancies in the chain arise when ions at the two ends dissociate and that these vacancies propagate through the chain, gives

$$\frac{J_{\text{in}}}{J_{\text{out}}} = \left[\frac{c_e}{c_i} \exp\left(\frac{-VF}{RT}\right)\right]^{n}. \tag{3.17}$$

Experimental data confirm this theoretical prediction (although historically, the theory was motivated by the experimental result, as is often the case). Hodgkin and Keynes (1955) showed that flux ratios in the K^+ channel of the *Sepia* giant axon could be described by the Ussing flux ratio raised to the power 2.5. Their result, as presented in modified form by Hille (2001), is shown in Fig. 3.2. Unidirectional K^+ fluxes were measured with radioactive K^+, and the ratio of the outward to the inward flux was plotted as a function of $V - V_K$. The best-fit line on a semilogarithmic plot has a slope of 2.5, which suggests that at least 2 K^+ ions traverse the K^+ channel simultaneously.

3.3   Electrodiffusion Models

Most early work on ion channels was based on the theory of electrodiffusion. Recall from Chapter 2 that the movement of ions in response to a concentration gradient and an electric field is described by the Nernst–Planck equation,

$$J = -D\left(\frac{dc}{dx} + \frac{zF}{RT}c\frac{d\phi}{dx}\right), \tag{3.18}$$

where J denotes the flux density, c is the concentration of the ion under considera-tion, and ϕ is the electrical potential. If we make the simplifying assumption that the field $d\phi/dx$ is constant through the membrane, then (3.18) can be solved to give the Goldman–Hodgkin–Katz current and voltage equations (2.123) and (2.126). However, in general there is no reason to believe that the potential has a constant gradient in the membrane. Ions moving through the channel affect the local electric field, and this local

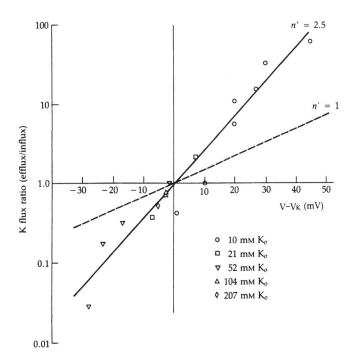

Figure 3.2 K^+ flux ratios as measured by Hodgkin and Keynes (1955), Fig. 7. Slightly modified into modern conventions by Hille (2001), page 487. K_o is the external K^+ concentration, and n' is the flux-ratio exponent, denoted by n in (3.17). (Hille, 2001, Fig. 15.7, p. 487.)

field in turn affects ionic fluxes. Thus, to determine the electric field and consequent ionic fluxes, one must solve a coupled problem.

3.3.1 Multi-Ion Flux: The Poisson–Nernst–Planck Equations

Suppose there are two types of ions, S_1 and S_2, with concentrations c_1 and c_2, passing through an ion channel, as shown schematically in Fig. 3.3.

For convenience, we assume that the valence of the first ion is $z > 0$ and that of the second is $-z$. Then, the potential in the channel $\phi(x)$ must satisfy Poisson's equation,

$$\frac{d^2\phi}{dx^2} = -\frac{zq}{\epsilon}N_a(c_1 - c_2), \tag{3.19}$$

where q is the unit electric charge, ϵ is the dielectric constant of the channel medium (usually assumed to be an aqueous solution), and N_a is Avogadro's number, necessary to convert units of concentration in moles per liter into number of molecules per liter. The flux densities J_1 and J_2 of S_1 and S_2 satisfy the Nernst–Planck equation, and at steady state dJ_1/dx and dJ_2/dx must both be zero to prevent charge buildup within the channel. Hence, the steady-state flux through the channel is described by (3.19)

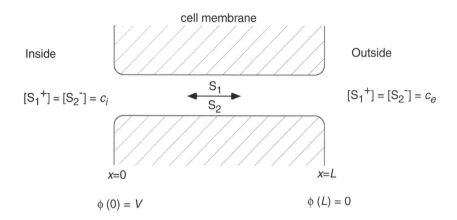

Figure 3.3 Schematic diagram of the electrodiffusion model of current through an ionic channel. Each side of the channel is electrically neutral, and both ion types can diffuse through the channel.

coupled with

$$J_1 = -D_1 \left(\frac{dc_1}{dx} + \frac{zF}{RT} c_1 \frac{d\phi}{dx} \right), \tag{3.20}$$

$$J_2 = -D_2 \left(\frac{dc_2}{dx} - \frac{zF}{RT} c_2 \frac{d\phi}{dx} \right), \tag{3.21}$$

where J_1 and J_2 are constants. To complete the specification of the problem, it is necessary to specify boundary conditions for c_1, c_2, and ϕ. We assume that the channel has length L, and that $x = 0$ denotes the left border, or inside, of the membrane. Then,

$$
\begin{aligned}
c_1(0) &= c_i, & c_1(L) &= c_e, \\
c_2(0) &= c_i, & c_2(L) &= c_e, \\
\phi(0) &= V, & \phi(L) &= 0.
\end{aligned}
\tag{3.22}
$$

Note that the solutions on both sides of the membrane are electrically neutral. V is the potential difference across the membrane, defined, as usual, as the internal potential minus the external potential. While at first glance it might appear that there are too many boundary conditions for the differential equations, this is in fact not so, as the constants J_1 and J_2 are additional unknowns to be determined.

In general, it is not possible to obtain an exact solution to the Poisson–Nernst–Planck (PNP) equations (3.19)–(3.22). However, some simplified cases can be solved approximately. A great deal of work on the PNP equations has been done by Eisenberg and his colleagues (Chen et al., 1992; Barcilon, 1992; Barcilon et al., 1992; Chen and Eisenberg, 1993). Here we present simplified versions of their models, ignoring, for example, the charge induced on the channel wall by the presence of ions in the channel, and considering only the movement of two ion types through the channel. Similar models have also been discussed by Peskin (1991).

It is convenient first to nondimensionalize the PNP equations. We let $y = x/L$, $\psi = \phi z F / RT$, $v = VF/RT$, $u_k = c_k/\tilde{c}$, for $k = 1, 2, i$ and e, where $\tilde{c} = c_e + c_i$. Substituting into (3.19)–(3.21), we find

$$-j_1 = \frac{du_1}{dy} + u_1 \frac{d\psi}{dy}, \tag{3.23}$$

$$-j_2 = \frac{du_2}{dy} - u_2 \frac{d\psi}{dy}, \tag{3.24}$$

$$\frac{d^2\psi}{dy^2} = -\lambda^2 (u_1 - u_2), \tag{3.25}$$

where $\lambda^2 = L^2 q F N_a \tilde{c}/(\epsilon RT)$, $j_1 = J_1 L/(\tilde{c} D_1)$, and $j_2 = J_2 L/(\tilde{c} D_1)$. The boundary conditions are

$$
\begin{aligned}
u_1(0) &= u_i, & u_1(1) &= u_e, \\
u_2(0) &= u_i, & u_2(1) &= u_e, \\
\psi(0) &= v, & \psi(1) &= 0.
\end{aligned}
\tag{3.26}
$$

The Short-Channel or Low Concentration Limit

If the channel is short or the ionic concentrations on either side of the membrane are small, so that $\lambda \ll 1$, we can find an approximate solution to the PNP equations by setting $\lambda = 0$. This gives

$$\frac{d^2\psi}{dy^2} = 0, \tag{3.27}$$

and thus

$$\frac{d\psi}{dy} = -v. \tag{3.28}$$

Hence, $\lambda \approx 0$ implies that the electric potential has a constant gradient in the membrane, which is exactly the constant field assumption that was made in the derivation of the GHK equations (Chapter 2). The equation for u_1 is then

$$\frac{du_1}{dy} - vu_1 = -j_1, \tag{3.29}$$

and thus

$$u_1 = \frac{j_1}{v} + K_1 e^{vy}. \tag{3.30}$$

From the boundary conditions $u_1(0) = u_i$, $u_1(1) = u_e$ it follows that

$$j_1 = v \frac{u_i - u_e e^{-v}}{1 - e^{-v}}. \tag{3.31}$$

In dimensional form, this is

$$I_1 = F J_1 = \frac{D_1}{L} \frac{F^2}{RT} V \left(\frac{c_i - c_e \exp(\frac{-zVF}{RT})}{1 - \exp(\frac{-zVF}{RT})} \right), \tag{3.32}$$

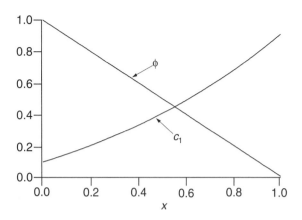

Figure 3.4 The concentration and potential profiles for the short-channel limit of the Poisson–Nernst–Planck equations. Dimensionless parameters were set arbitrarily at $u_i = 50/550 = 0.091$, $u_e = 500/550 = 0.909$, $v = 1$. In this limit the electric field is constant through the channel (the potential has a constant slope), the concentration profile is nonlinear, and the GHK I–V curve is obtained.

which is, as expected, the GHK current equation. Graphs of the concentration and voltage profiles through the membrane are shown in Fig. 3.4. It is reassuring that the widely used GHK equation for the ionic flux can be derived as a limiting case of a more general model.

The Long-Channel Limit

Another interesting limit is obtained by letting the length of the channel go to infinity. If we let $\eta = 1/\lambda$ denote a small parameter, the model equations are

$$-j_1 = \frac{du_1}{dy} + u_1 \frac{d\psi}{dy}, \tag{3.33}$$

$$-j_2 = \frac{du_2}{dy} - u_2 \frac{d\psi}{dy}, \tag{3.34}$$

$$-\eta^2 \frac{d^2\psi}{dy^2} = (u_1 - u_2). \tag{3.35}$$

Since there is a small parameter multiplying the highest derivative, this is a singular perturbation problem. The solution obtained by setting $\eta = 0$ does not, in general, satisfy all the boundary conditions, as the degree of the differential equation has been reduced, resulting in an overdetermined system. In the present case, however, this reduction of order is not a problem.

Setting $\eta = 0$ in (3.35) gives $u_1 = u_2$, which happens to satisfy both the left and right boundary conditions. Thus, u_1 and u_2 are identical throughout the channel. From (3.33) and (3.34) it follows that

$$\frac{d}{dy}(u_1 + u_2) = -j_1 - j_2. \tag{3.36}$$

Since both j_1 and j_2 are constants, it follows that du_1/dy is a constant, and hence, from the boundary conditions,

$$u_1 = u_2 = u_i + (u_e - u_i)y. \tag{3.37}$$

We are now able to solve for ψ. Subtracting (3.35) from (3.34) gives

$$2u_1 \frac{d\psi}{dy} = 2j, \tag{3.38}$$

where $2j = j_2 - j_1$, and hence

$$\psi = \frac{j}{u_e - u_i} \ln[u_i + (u_e - u_i)y] + K, \tag{3.39}$$

for some other constant K. Applying the boundary conditions $\psi(0) = v$, $\psi(1) = 0$ we determine j and K, with the result that

$$\psi = -\frac{v}{v_1} \ln \left[\frac{u_i}{u_e} + \left(1 - \frac{u_i}{u_e}\right) y \right], \tag{3.40}$$

where $v_1 = \ln(u_e/u_i)$ is the dimensionless Nernst potential of ion S_1. The flux density of one of the ions, say S_1, is obtained by substituting the expressions for u_1 and ψ into (3.33) to get

$$j_1 = \frac{u_e - u_i}{v_1}(v - v_1), \tag{3.41}$$

or in dimensional form,

$$J_1 = \frac{D_1}{L} \frac{zF}{RT} \frac{c_e - c_i}{\ln \frac{c_e}{c_i}} (V - \frac{RT}{zF} \ln \frac{c_e}{c_i}), \tag{3.42}$$

which is the linear I–V curve that we met previously. Graphs of the corresponding concentration and voltage profiles through the channel are shown in Fig. 3.5.

In summary, by taking two different limits of the PNP equations we obtain either the GHK I–V curve or a linear I–V curve. In the short-channel limit, ψ has a constant gradient through the membrane, but the concentration does not. In the long-channel limit the reverse is true, with a constant gradient for the concentration through the channel, but not for the potential. It is left as an exercise to prove that although the GHK equation obeys the independence principle and the Ussing flux ratio, the linear

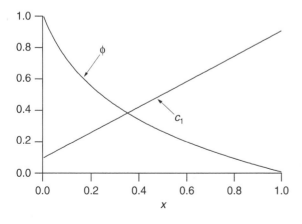

Figure 3.5 The concentration and potential profiles for the long-channel limit of the Poisson–Nernst–Planck equations. Dimensionless parameters were set arbitrarily at $u_i = 50/550 = 0.091$, $u_e = 500/550 = 0.909$, $v = 1$. In this limit the concentration profile has a constant slope, the potential profile is nonlinear, and the linear I–V curve is obtained.

I–V curve obeys neither. Given the above derivation of the linear I–V curve, this is not surprising. A linear I–V curve is obtained when either the channel is very long or the ionic concentrations on either side of the channel are very high. In either case, one should not expect the movement of an ion through the channel to be independent of other ions, and so that the independence principle is likely to fail. Conversely, the GHK equation is obtained in the limit of low ionic concentrations or short channels, in which case the independent movement of ions is not surprising.

3.4   Barrier Models

The second type of model that has been widely used to describe ion channels is based on the assumption that the movement of an ion through the channel can be modeled as the jumping of an ion over a discrete number of free-energy barriers (Eyring et al., 1949; Woodbury, 1971; Läuger, 1973). It is assumed that the potential energy of an ion passing through a channel is described by a potential energy profile of the general form shown in Fig. 3.6. The peaks of the potential energy profile correspond to barriers that impede the ion flow, while the local minima correspond to binding sites within the channel.

To traverse the channel the ion must hop from one binding site to another. According to the theory of chemical reaction rates, the rate at which an ion jumps from one binding site to the next is an exponential function of the height of the potential energy

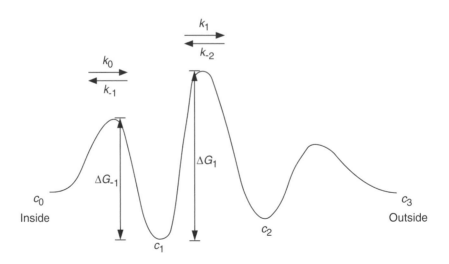

Figure 3.6   General potential energy profile for barrier models. The local minima correspond to binding sites within the channel, and the local maxima are barriers that impede the ion flow. An ion progresses through the channel by hopping over the barriers from one binding site to another.

barrier that it must cross. Thus, in the notation of the diagram,

$$k_j = \kappa \exp\left(\frac{-\Delta G_j}{RT}\right),$$
(3.43)

for some factor κ with units of 1/time. One of the most difficult questions in the use of this expression is deciding the precise form of the factor κ. According to Eyring rate theory (as used in this context by Hille (2001), for example), $\kappa = kT/h$, where k is Boltzmann's constant, T is the temperature, and h is Planck's constant. The derivation of this expression for κ relies on the quantization of the energy levels of the ion in some transition state as it binds to the channel binding sites. However, it is not clear that at biologically relevant temperatures energy quantization has an important effect on ionic flows. Using methods from nonequilibrium statistical thermodynamics, an alternative form of the factor has been derived by Kramers (1940), and discussions of this, and other, alternatives may be found in McQuarrie (1967) and Laidler (1969). In the appendix to this chapter, we give a derivation of Kramers' formula, but we do not enter into the debate of which answer is best. Instead, in what follows, we assume that κ is known, and independent of ΔG_j, even though, for Kramers' formula, such is not the case.

For simplicity, we assume that each local maximum occurs halfway between the local minima on each side. Barriers with this property are called *symmetrical*. An electric field in the channel also affects the rate constants. If the potential difference across the cell membrane is positive (so that the inside is more positive than the outside), it is easier for positive ions to cross the barriers in the outward direction but more difficult for positive ions to enter the cell. Thus, the heights of the barriers in the outward direction are reduced, while the heights in the inward direction are increased. If there is a potential difference of ΔV_j over the jth barrier, then

$$k_j = \kappa \exp\left[\frac{1}{RT}(-\Delta G_j + zF\Delta V_{j+1}/2)\right],$$
(3.44)

$$k_{-j} = \kappa \exp\left[\frac{1}{RT}(-\Delta G_{-j} - zF\Delta V_j/2)\right].$$
(3.45)

The factor 2 appears because the barriers are assumed to be symmetrical, so that the maxima are lowered by $zF\Delta V_j/2$. A simple illustration of this is given in Fig. 3.7A and B and is discussed in detail in the next section.

In addition to symmetry, the barriers are assumed to have another important property, namely, that in the absence of an electric field the ends of the energy profile are at the same height, and thus

$$\sum_{j=0}^{n-1} \Delta G_j - \sum_{j=1}^{n} \Delta G_{-j} = 0.$$
(3.46)

If this were not so, then in the absence of an electric field and with equal concentrations on either side of the membrane, there would be a nonzero flux through the membrane, a situation that is clearly unphysiological.

A number of different models have been constructed along these general lines. First, we consider the simplest type of barrier model, in which the ionic concentration in the channel can become arbitrarily large, i.e., the channel does not saturate. This is similar to the continuous models discussed above and can be thought of as a discrete approximation to the constant field model. Because of this, nonsaturating models give the GHK I–V curve in the limit of a homogeneous membrane. We then discuss saturating barrier models and multi-ion models. Before we do so, however, it is important to note that although barrier models can provide good quantitative descriptions of some experimental data, they are phenomenological. In other words, apart from the agreement between theory and experiment, there is often no reason to suppose that the potential energy barrier used to describe the channel corresponds in any way to physical properties of the channel. Thus, although their relative simplicity has led to their widespread use, mechanistic interpretations of the models should be made only with considerable caution. Of course, this does not imply that barrier models are inferior to continuous models such as the constant field model or the Poisson–Nernst–Planck equations, which suffer from their own disadvantages (Dani and Levitt, 1990).

3.4.1 Nonsaturating Barrier Models

In the simplest barrier model (Eyring et al., 1949; Woodbury, 1971), the potential energy barrier has the general form shown in Fig. 3.7A, and it is assumed that the movement of an ion S over a barrier is independent of the ionic concentrations at the neighboring barriers. This is equivalent to assuming that the concentration of S at any particular binding site can be arbitrarily large.

The internal concentration of S is denoted by c_0, while the external concentration is denoted by c_n. There are $n-1$ binding sites (and thus n barriers) in the membrane, and the concentration of S at the jth binding site is denoted by c_j. Note the slight change in notation from above. Instead of using c_e and c_i to denote the external and internal concentrations of S, we use c_n and c_0. This allows the labeling of the concentrations on either side of the membrane to be consistent with the labeling of the concentrations at the binding sites. There is an equal voltage drop across each barrier, and thus the electrical distance between each binding site, denoted by λ, is the same. For convenience, we assume the stronger condition, that the physical distance between the binding sites is the same also, which is equivalent to assuming a constant electric field in the membrane. In the absence of an electric field, we assume that the heights of the energy barriers decrease linearly through the membrane, as in Fig. 3.7, with

$$\Delta G_j = \Delta G_0 - j\delta G, \tag{3.47}$$

for some constant increment δG. Finally, it is assumed that the flux from left to right, say, across the jth barrier, is proportional to c_{j-1}, and similarly for the flux in the opposite direction. Thus, the flux over the jth barrier, J, is given by

$$J = \lambda(k_{j-1}c_{j-1} - k_{-j}c_j). \tag{3.48}$$

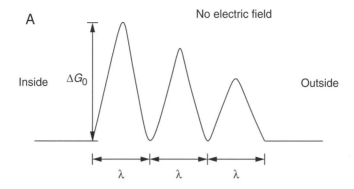

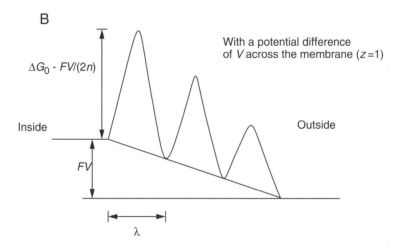

Figure 3.7 The potential energy diagram used in the nonsaturating model of Woodbury (1971). There is an equal distance between the binding sites, and the barriers are symmetrical. A. In the absence of an electric field the barrier height decreases linearly through the membrane. B. The presence of a constant electric field skews the energy profile, bringing the outside down relative to the inside. This increases the rate at which positive ions traverse the channel from inside to out and decreases their rate of entry.

Note that the units of J are concentration × distance/time, or moles per unit area per time, so J is a flux density. As usual, a flux from inside to outside (i.e., left to right) is defined as a positive flux.

At steady state the flux over each barrier must be the same, in which case we obtain a system of linear equations,

$$k_0 c_0 - k_{-1} c_1 = k_1 c_1 - k_{-2} c_2 = \cdots = k_{n-1} c_{n-1} - k_{-n} c_n = M, \qquad (3.49)$$

where $M = J/\lambda$ is a constant. Hence

$$k_0 c_0 = (k_1 + k_{-1})c_1 - k_{-2}c_2, \tag{3.50}$$

$$k_1 c_1 = (k_2 + k_{-2})c_2 - k_{-3}c_3, \tag{3.51}$$

$$k_2 c_2 = (k_3 + k_{-3})c_3 - k_{-4}c_4, \tag{3.52}$$

$$\vdots$$

We need to determine J in terms of the concentrations on either side of the membrane, c_0 and c_n. Solving (3.51) for c_1 and substituting into (3.50) gives

$$k_0 c_0 = c_2 k_2 \phi_2 - c_3 k_{-3} \phi_1, \tag{3.53}$$

where

$$\phi_j = \sum_{i=0}^{j} \pi_i, \tag{3.54}$$

$$\pi_j = \frac{k_{-1} \cdots k_{-j}}{k_1 \cdots k_j}, \qquad \pi_0 = 1. \tag{3.55}$$

Then solving (3.52) for c_2 and substituting into (3.53) gives

$$k_0 c_0 = c_3 k_3 \phi_3 - c_4 k_{-4} \phi_2. \tag{3.56}$$

Repeating this process of sequential substitutions, we find that

$$k_0 c_0 = k_{n-1} c_{n-1} \phi_{n-1} - c_n k_{-n} \phi_{n-2}. \tag{3.57}$$

Since

$$c_{n-1} = \frac{M + k_{-n}c_n}{k_{n-1}}, \tag{3.58}$$

it follows that

$$k_0 c_0 = \phi_{n-1}(M + k_{-n}c_n) - c_n k_{-n} \phi_{n-2}, \tag{3.59}$$

and hence

$$J = \lambda M = \frac{\lambda k_0 \left(c_0 - c_n \pi_n \frac{k_n}{k_0}\right)}{\phi_{n-1}}. \tag{3.60}$$

It remains to express the rate constants in terms of the membrane potential. If there is a potential difference V across the membrane (as shown in Fig. 3.7B), the constant electric field adds $FzV/(2n)$ to the barrier when moving from right to left, and $-FzV/(2n)$ when moving in the opposite direction. Hence

$$\Delta G_j = \Delta G_0 - j\delta G - \frac{FzV}{2n}, \tag{3.61}$$

$$\Delta G_{-j} = \Delta G_0 - (j-1)\delta G + \frac{FzV}{2n}. \tag{3.62}$$

Now we use (3.43) to get

$$\frac{k_{-j}}{k_{j-1}} = \exp(-v/n), \qquad \frac{k_{-j}}{k_j} = \exp(-g - v/n), \qquad (3.63)$$

where $g = \delta G/(RT)$ and $v = FzV/(RT)$. Hence

$$\pi_j = \exp(-j(g + v/n)), \qquad (3.64)$$

and

$$\phi_{n-1} = \sum_{j=0}^{n-1} \exp(-j(g + v/n)) = \frac{e^{-n(g+v/n)} - 1}{e^{-(g+v/n)} - 1}, \qquad (3.65)$$

so that

$$J = k_0 \lambda (c_0 - c_n e^{-v}) \frac{e^{-(g+v/n)} - 1}{e^{-n(g+v/n)} - 1}. \qquad (3.66)$$

As expected, (3.66) satisfies both the independence principle and the Ussing flux ratio. Also, the flux is zero when v is the Nernst potential of the ion.

The Homogeneous Membrane Simplification

One useful simplification of the nonsaturating barrier model is obtained if it is assumed that the membrane is homogeneous. We model a homogeneous membrane by setting $g = \delta G/(RT) = 0$ and letting $n \to \infty$. Thus, there is no increase in barrier height through the membrane, and the number of barriers approaches infinity. In this limit, keeping $n\lambda = L$ fixed,

$$J = \frac{k_{00} \lambda^2}{L} v \frac{c_0 - c_n e^{-v}}{1 - e^{-v}}, \qquad (3.67)$$

where k_{00} is the value of k_0 at $V = 0$, L is the width of the membrane, and $k_{00}\lambda^2$ is the diffusion coefficient of the ion over the first barrier in the absence of an electric field. Notice that for this to make sense it must be that k_{00} scales like λ^{-2} for small λ. In Section 3.7.3 we show that this is indeed the case.

It follows that in the homogeneous membrane case,

$$\begin{aligned} J &= \frac{D_S}{L} v \frac{c_0 - c_n e^{-v}}{1 - e^{-v}}, \\ &= P_S v \frac{c_0 - c_n e^{-v}}{1 - e^{-v}}, \end{aligned} \qquad (3.68)$$

which is exactly the GHK current equation (2.122) derived previously.

3.4.2   Saturating Barrier Models: One-Ion Pores

If an ion channel satisfies the independence principle, the flux of S is proportional to [S], even when [S] gets large. However, this is not usually found to be true experimentally.

It is more common for the flux to saturate as [S] increases, reaching some maximum value as [S] gets large. This has motivated the development of models in which the flux is not proportional to [S] but is a nonlinear, saturating, function of [S]. As one might expect, equations for such models are similar to those of enzyme kinetics.

The basic assumptions behind saturating barrier models are that to pass through the channel, ions must bind to binding sites in the channel, but that each binding site can hold only a single ion (Läuger, 1973; Hille, 2001). Hence, if all the binding sites are full, an increase in ionic concentration does not increase the ionic flux—the channel is saturated. Saturating barrier models can be further subdivided into one-ion pore models, in which each channel can bind only a single ion at any one time, and multi-ion pore models, in which each channel can bind multiple ions simultaneously. The theory of one-ion pores is considerably simpler than that of multi-ion pores, and so we discuss those models first.

The Simplest One-Ion Saturating Model

We begin by considering the simplest one-ion pore model, with a single binding site. If we let S_e denote the ion outside, S_i the ion inside, and X the binding site, the passage of an ion through the channel can be described by the kinetic scheme

$$X + S_i \underset{k_{-1}}{\overset{k_0}{\rightleftharpoons}} XS \underset{k_{-2}}{\overset{k_1}{\rightleftharpoons}} X + S_e. \tag{3.69}$$

Essentially, the binding site acts like an enzyme that transfers the ion from one side of the membrane to the other, such as was encountered in Chapter 2 for the transport of glucose across a membrane. Following the notation of the previous section, we let c_0 denote [S_i] and c_2 denote [S_e]. However, instead of using c_1 to denote the concentration of S at the binding site, it is more convenient to let c_1 denote the probability that the binding site is occupied. (In a population of channels, c_1 denotes the proportion of channels that have an occupied binding site.) Then, at steady state,

$$k_0 c_0 x - k_{-1} c_1 = k_1 c_1 - k_{-2} c_2 x, \tag{3.70}$$

where x denotes the probability that the binding site is empty. Note that (3.70) is similar to the corresponding equation for the nonsaturating pore, (3.49), with the only difference that x appears in the saturating model. In addition, we have a conservation equation for x,

$$x + c_1 = 1. \tag{3.71}$$

Solution of (3.70) and (3.71) gives the flux J as

$$J = k_0 c_0 x - k_{-1} c_1 = \frac{k_0 k_1 c_0 - k_{-1} k_{-2} c_2}{k_0 c_0 + k_{-2} c_2 + k_{-1} + k_1}. \tag{3.72}$$

It is important to note that J, as defined by (3.72), does not have the same units (concentration × distance/time) as in the previous model, but here has units of number of ions crossing the membrane per unit time. The corresponding transmembrane current,

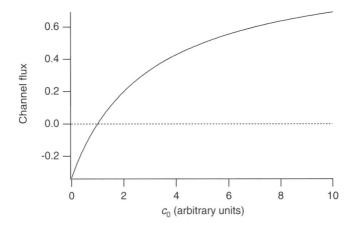

Figure 3.8 Plot of J against c_0 for the simplest saturating model with one binding site. When c_0 is small, the flux is approximately a linear function of c_0, but as c_0 increases, the flux saturates to a maximum value.

I, is given by $I = zqJ$, where q is the unit charge, and has the usual units of number of charges crossing the membrane per unit time. A plot of J as a function of c_0 is shown in Fig. 3.8. When c_0 is small, J is approximately a linear function of c_0, but as c_0 increases, J saturates at the maximum value k_1.

We now use (3.43) to express the rate constants in terms of the membrane potential. As before, we assume that the local maxima of the energy profile occur midway between the local minima; i.e., we assume that the barriers are symmetrical. However, we no longer assume that the barriers are equally spaced through the channel. If the local minimum occurs at an electrical distance δ from the left-hand side, it follows that

$$k_0 = \kappa \exp\left[\frac{1}{RT}(-\Delta G_0 + \delta zFV/2)\right], \tag{3.73}$$

$$k_1 = \kappa \exp\left[\frac{1}{RT}(-\Delta G_1 + (1-\delta)zFV/2)\right], \tag{3.74}$$

$$k_{-1} = \kappa \exp\left[\frac{1}{RT}(-\Delta G_{-1} - \delta zFV/2)\right], \tag{3.75}$$

$$k_{-2} = \kappa \exp\left[\frac{1}{RT}(-\Delta G_{-2} - (1-\delta)zFV/2)\right]. \tag{3.76}$$

Because δ denotes an electrical, not a physical, distance, it is not necessary to assume that the electric field in the membrane is constant, only that there is a drop of δV over the first barrier and $(1 - \delta)V$ over the second. In general, the energy profile of any particular channel is unknown. However, the number and positions of the binding sites and the values of the local maxima and minima can, in principle at least, be determined by fitting to experimental data. We consider an example of this procedure (for a slightly more complicated model) below.

The Ussing Flux Ratio

Earlier in this chapter we stated that it is possible for a model to obey the Ussing flux ratio but not the independence principle. Single-ion saturating models provide a simple example of this. First, note that they cannot obey the independence principle, since the flux is not linearly proportional to the ionic concentration. This nonlinear saturation effect is illustrated in Fig. 3.8.

To see that the model obeys the Ussing flux ratio, it is necessary to set up the model in a slightly different form. Suppose we have two isotopes, S and $\bar{S}$, similar enough so that they have identical energy profiles in the channel. Then, we suppose that a channel has only S on the left-hand side and only $\bar{S}$ on the right. We let c denote [S] and $\bar{c}$ denote [$\bar{S}$]. Since S and $\bar{S}$ have identical energy profiles in the channel, the rate constants for the passage of $\bar{S}$ through the channel are the same as those for S. From the kinetic schemes for S and $\bar{S}$ we obtain

$$k_0 c_0 x - k_{-1} c_1 = k_1 c_1 - k_{-2} c_2 x = J_S, \tag{3.77}$$

$$k_0 \bar{c}_0 x - k_{-1} \bar{c}_1 = k_1 \bar{c}_1 - k_{-2} \bar{c}_2 x = J_{\bar{S}}, \tag{3.78}$$

but here the conservation equation for x is

$$x + \bar{c}_1 + c_1 = 1. \tag{3.79}$$

To calculate the individual fluxes of S and $\bar{S}$ it is necessary to eliminate x from (3.77) and (3.78) using the conservation equation (3.79). However, to calculate the flux ratio this is not necessary. Solving (3.77) for J_S in terms of x, c_0, and c_2, we find

$$J_S = x \left(\frac{k_0 c_0 - \dfrac{k_{-1} k_{-2}}{k_1} c_2}{1 + k_{-1}/k_1} \right), \tag{3.80}$$

and similarly,

$$J_{\bar{S}} = x \left(\frac{k_0 \bar{c}_0 - \dfrac{k_{-1} k_{-2}}{k_1} \bar{c}_2}{1 + k_{-1}/k_1} \right). \tag{3.81}$$

If S is present only on the left-hand side and $\bar{S}$ only on the right, we then have $c_2 = 0$ and $\bar{c}_0 = 0$, in which case

$$\frac{J_S}{J_{\bar{S}}} = -\frac{k_0 k_1}{k_{-1} k_{-2}} \frac{c_0}{\bar{c}_2}. \tag{3.82}$$

The minus sign on the right-hand side appears because the fluxes are in different directions. Now we substitute for the rate constants, (3.73) to (3.76), and use the fact that the ends of the energy profile are at the same height (and thus $\Delta G_0 + \Delta G_1 - \Delta G_{-1} - \Delta G_{-2} = 0$) to find

$$\left| \frac{J_S}{J_{\bar{S}}} \right| = \exp\left(\frac{zVF}{RT} \right) \frac{c_0}{\bar{c}_2}, \tag{3.83}$$

which is the Ussing flux ratio, as proposed.

Multiple Binding Sites

When there are multiple binding sites within the channel, the analysis is essentially the same as the simpler case discussed above, but the details are more complicated. When there are n barriers in the membrane (and thus $n - 1$ binding sites), the steady-state equations are

$$k_0 c_0 x - k_{-1} c_1 = k_1 c_1 - k_{-2} c_2 = \cdots = k_{n-1} c_{n-1} - k_{-n} c_n x = J, \tag{3.84}$$

where x is the probability that all of the binding sites are empty and c_j is the probability that the ion is bound to the jth binding site. Because the channel must be in either state x or one of the states $c_1, \ldots, c_{n-1}$ (since there is only one ion in the channel at a time), it follows that

$$x = 1 - \sum_{i=1}^{n-1} c_i. \tag{3.85}$$

It is left as an exercise to show that

$$J = \frac{k_0 c_0 - k_{-n} c_n \pi_{n-1}}{\phi_{n-1} + \beta k_0 c_0 + k_{-n} c_n (\alpha \phi_{n-1} - \beta \phi_{n-2})}, \tag{3.86}$$

where

$$\alpha = \sum_{j=1}^{n-1} \frac{\phi_{n-2} - \phi_{j-1}}{k_j \pi_j}, \tag{3.87}$$

$$\beta = \sum_{j=1}^{n-1} \frac{\phi_{n-1} - \phi_{j-1}}{k_j \pi_j}, \tag{3.88}$$

where ϕ_j and π_j are defined in (3.54) and (3.55).

Equation (3.86) does not satisfy the independence principle, but it does satisfy the Ussing flux ratio. However, the details are left as an exercise (Exercise 5).

3.4.3 Saturating Barrier Models: Multi-Ion Pores

We showed above that single-ion models obey the Ussing flux ratio, even though they do not obey the independence principle. This means that to model the type of channel described in Fig. 3.2 it is necessary to use models that show flux coupling as predicted by Hodgkin and Keynes (1955). Such flux coupling arises in models in which more than one ion can be in the channel at any one time. Although the equations for such multi-ion models are essentially the same as the equations for the single-ion models described in the previous section, the analysis is complicated considerably by the fact that there are many more possible channel states. Hence, numerical techniques are the most efficient for studying such models. A great deal has been written about multi-ion models (e.g., Hille and Schwartz, 1978; Begenisich and Cahalan, 1980; Schumaker and MacKinnon, 1990; Urban and Hladky, 1979; Kohler and Heckmann, 1979). Space does not allow for a detailed discussion of the properties of these models, but so we present only a brief discussion of the simplest model. Hille and Schwartz (1978) and Hille (2001) give more detailed discussions.

Multi-ion models are based on assumptions similar to one-ion models. It is assumed that the passage of an ion through the channel can be described as the jumping of an ion over energy barriers, from one binding site to another. In one-ion models each binding site can either have an ion bound or not, and thus a channel with n binding sites can be in one of n independent states (i.e., the ion can be bound to any one of the binding sites). Hence, the steady-state ion distribution is found by solving a system of n linear equations, treating the concentrations on either side of the membrane as known. If more than one ion can be present simultaneously in the channel, the situation is more complicated. Each binding site can be in one of two states: binding an ion or empty. Therefore, a channel with n binding sites can be in any of 2^n states (at least; more states are possible if there is more than one ion type passing through the channel), and the steady-state probability distribution must be found by solving a large system of linear equations.

The simplest possible multi-ion model has three barriers and two binding sites, and so the channel can be in one of 4 possible states (Fig. 3.9). Arbitrary movements from one state to another are not possible. For example, the state OO (where both binding

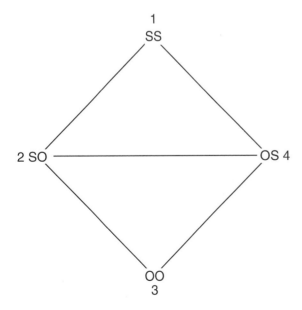

Figure 3.9 State diagram for a multi-ion barrier model with two binding sites and a single ion.

sites are empty) can change to OS or SO but cannot change to SS in a single step, as this would require two ions entering the channel simultaneously. We number the states as in Fig. 3.9 and let k_{ij} denote the rate of conversion of state i to state j. Also, let P_j denote the probability that the channel is in the jth state, and let c_e and c_i denote the external and internal ion concentrations, respectively. Then, the equations for the probabilities follow from the law of mass action; they are

$$\frac{dP_1}{dt} = -(k_{12} + k_{14})P_1 + k_{21}c_eP_2 + k_{41}c_iP_4, \tag{3.89}$$

$$\frac{dP_2}{dt} = -(k_{21}c_e + k_{23} + k_{24})P_2 + k_{12}P_1 + c_ik_{32}P_3 + k_{42}P_4, \tag{3.90}$$

$$\frac{dP_3}{dt} = -(c_ik_{32} + c_ek_{34})P_3 + k_{43}P_4 + k_{23}P_2, \tag{3.91}$$

$$\frac{dP_4}{dt} = -(k_{41}c_i + k_{42} + k_{43})P_4 + k_{14}P_1 + k_{24}P_2 + c_ek_{34}P_3. \tag{3.92}$$

The probabilities must also satisfy the conservation equation

$$\sum_{i=1}^{4} P_i = 1. \tag{3.93}$$

Using the conservation equation in place of the equation for P_4, the steady-state probability distribution is given by the linear system

$$\begin{pmatrix} -k_{12} - k_{14} & k_{21} & 0 & k_{41} \\ k_{12} & -k_{21} - k_{23} - k_{24} & c_ek_{32} & k_{42} \\ 0 & k_{23} & -c_ek_{32} - c_ik_{34} & k_{43} \\ 1 & 1 & 1 & 1 \end{pmatrix} \begin{pmatrix} P_1 \\ P_2 \\ P_3 \\ P_4 \end{pmatrix} = \begin{pmatrix} 0 \\ 0 \\ 0 \\ 1 \end{pmatrix}. \tag{3.94}$$

Since each rate constant is determined as a function of the voltage in the same way as one-ion models (as in, for example, (3.73)–(3.76)), solution of (3.94) gives each P_i as a function of voltage and the ionic concentrations on each side of the membrane. Finally, the membrane fluxes are calculated as the net rate of ions crossing any one barrier, and so, choosing the middle barrier arbitrarily, we have

$$J = P_2k_{24} - P_4k_{42}. \tag{3.95}$$

Although it is possible to solve such linear systems exactly (particularly with the help of symbolic manipulators such as Maple or Mathematica), it is often as useful to solve the equations numerically for a given energy profile. It is left as an exercise to show that the Ussing flux ratio is not obeyed by a multi-ion model with two binding sites and to compare the I–V curves of multi-ion and single-ion models.

3.4.4 Electrogenic Pumps and Exchangers

Recall from Chapter 2 that detailed balance required that rate constants in models of electrogenic exchangers and pumps be dependent on the membrane potential. See, for example, (2.78) or (2.96). However, although the arguments from chemical equilibrium show that voltage-dependency must exist, they do not specify exactly which rate constants depend on the voltage, or what the functional dependency is. As we have come to expect, it is much more difficult to answer these questions. Just as there are many ways to model ionic current flow, so there are many ways to model how rate constants depend on the membrane potential. In addition, depending on the exact assumptions, any of the steps in the model could depend on membrane potential. In other words, not only are there a number of ways to model the voltage-dependence when it occurs, there are also many places where it could occur. It is, in general, a very difficult task to determine the precise place and nature of the voltage-dependence.

One simple approach is to assume that the conformational change of the carrier protein is the step that moves the charge across the membrane, and thus requires the crossing of a free energy barrier. Consider the diagram shown in Fig. 2.9. If we assume that the transition from state X_2 to Y_2 involves the movement of 2 positive ions across an energy barrier and a potential difference V, then we can model the rate constants as

$$k_2 = \kappa \exp\left[\frac{1}{RT}(-\Delta G_+ + 2FV/2)\right] \tag{3.96}$$

$$= \bar{k}_2 \exp\left(\frac{FV}{RT}\right), \tag{3.97}$$

$$k_{-2} = \kappa \exp\left[\frac{1}{RT}(-\Delta G_- - 2FV/2)\right] \tag{3.98}$$

$$= \bar{k}_{-2} \exp\left(\frac{-FV}{RT}\right), \tag{3.99}$$

where $\bar{k}_2 = \kappa \exp[-\Delta G_+/(RT)]$ and similarly for $\bar{k}_{-2}$. In (3.96) and (3.98) $2FV$ is divided by 2 as we assume, for simplicity, that the energy barrier occurs halfway through the membrane.

If we make similar assumptions for k_4 and k_{-4}, i.e., that these transitions involve the reverse movement of 3 positive charges across an energy barrier and a potential difference, we obtain similar equations for those rate constants. Then

$$\frac{k_{-2}}{k_2}\frac{k_{-4}}{k_4} = \frac{\bar{k}_{-2}}{\bar{k}_2}\frac{\bar{k}_{-4}}{\bar{k}_4} \exp\left(\frac{-2FV}{RT}\right)\exp\left(\frac{3FV}{RT}\right)$$

$$= \frac{\bar{k}_{-2}}{\bar{k}_2}\frac{\bar{k}_{-4}}{\bar{k}_4} \exp\left(\frac{FV}{RT}\right), \tag{3.100}$$

from which it follows that $K_1 K_2 K_3 K_4 = \exp\left(\frac{FV}{RT}\right)$ (cf. (2.78)), which is the necessary equilibrium condition.

3.5 Channel Gating

So far in this chapter we have discussed how the current through a single open channel depends on the membrane potential and the ionic concentrations on either side of the membrane. However, it is of equal interest to determine how ionic channels open and close in response to voltage. As described in Chapter 5, the opening and closing of ionic channels in response to changes in the membrane potential is the basis for electrical excitability and is thus of fundamental significance in neurophysiology.

Recall that there is an important difference between the instantaneous and steady-state I–V curves. In general, the current through a population of channels is the product of three terms,

$$I = Ng(V,t)\phi(V),\qquad\qquad(3.101)$$

where $\phi(V)$ is the I–V curve of a single open channel, and $g(V,t)$ is the proportion of open channels in the population of N channels. In the previous sections we discussed electrodiffusion and barrier models of $\phi(V)$; in this section we discuss models of the dependence of g on voltage and time.

Consider, for example, the curves in Fig. 3.10, which show typical responses of populations of Na^+ and K^+ channels. When the voltage is stepped from -65 mV to -9 mV, and held fixed at the new level, the K^+ conductance (g_K) slowly increases to a new level, while the Na^+ conductance (g_{Na}) first increases and then decreases.

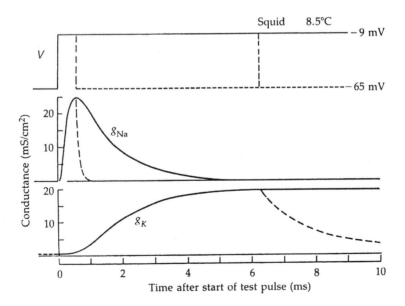

Figure 3.10 Na^+ and K^+ conductances as a function of time after a step change in voltage from -65 mV to -9 mV. The dashed line shows that after repolarization g_{Na} recovers quickly, and g_K recovers more slowly. (Hille, 2001, Fig. 2.11, p. 41.)

From these data we can draw several conclusions. First, as the voltage increases, the proportion of open K^+ channels increases. Second, although the proportion of open Na^+ channels initially increases, a second process is significant at longer times, as the Na^+ channel is inactivated. Thus, Na^+ channels first activate and then inactivate.

3.5.1 A Two-State K^+ Channel

The simplest model of the K^+ channel assumes that the channel can exist in either a closed state, C, or an open state, O, and that the rate of conversion from one state to another is dependent on the voltage. Thus,

$$C \underset{\beta(V)}{\overset{\alpha(V)}{\rightleftharpoons}} O. \tag{3.102}$$

If g denotes the proportion of channels in the open state (so $1 - g$ is the proportion of closed channels), the differential equation for the rate of change of g is

$$\frac{dg}{dt} = \alpha(V)(1 - g) - \beta(V)g. \tag{3.103}$$

Under voltage-clamp conditions (i.e., where the voltage is piecewise constant, as in Fig. 3.10), α and β are constants, and thus one can readily solve for g as a function of time. Equation (3.103) is often written as

$$\tau_g(V)\frac{dg}{dt} = g_\infty(V) - g, \tag{3.104}$$

where $g_\infty(V) = \alpha/(\alpha + \beta)$ is the steady-state value of g, and $\tau_g(V) = 1/(\alpha + \beta)$ is the time constant of approach to the steady state. From experimental data, such as those shown in Fig. 3.10, one can obtain values for g_∞ and τ_g, and thus α and β can be unambiguously determined.

The form of $g_\infty(V)$ can be determined from free energy arguments. The reason for voltage dependence must be that the subunit is charged and that to change from one conformation to another charges must move in the potential field. This movement of charge is a current, called the gating current. Now the difference in free energy between the two conformations is of the form

$$\Delta G = \Delta G^0 + aFV, \tag{3.105}$$

where ΔG^0 is the free energy difference between the two states in the absence of a potential, and a is a constant related to the number of charges that move and the relative distance they move during a change of conformation. It follows that the equilibrium constant for the subunit must be of the form

$$\frac{\beta}{\alpha} = k_0 \exp\left(\frac{aFV}{RT}\right), \tag{3.106}$$

in which case

$$g_\infty(V) = \frac{\alpha}{\alpha + \beta} = \frac{1}{1 + k_0 \exp(\frac{aFV}{RT})}, \tag{3.107}$$

which can also be expressed in the form

$$g_\infty(V) = \frac{1}{2} + \frac{1}{2}\tanh(b(V - V_0)).$$ (3.108)

3.5.2 Multiple Subunits

An important generalization of the two-state model occurs when the channel is assumed to consist of multiple identical subunits, each of which can be in either the closed or open state. For example, suppose that the channel consists of two identical subunits, each of which can be closed or open. Then, the channel can take any of four possible states, S_{00}, S_{10}, S_{01}, or S_{11}, where the subscripts denote the different subunits, with 1 and 0 denoting open and closed subunits, respectively. A general model of this channel involves three differential equations (although there is a differential equation for each of the four variables, one equation is superfluous because of the conservation equation $S_{00} + S_{10} + S_{01} + S_{11} = 1$), but we can simplify the model by grouping the channel states with the same number of closed and open subunits. Because the subunits are identical, there should be no difference between S_{10} and S_{01}, and thus they are amalgamated into a single variable.

So, we let S_i denote the group of channels with exactly i open subunits. Then, conversions between channel groups are governed by the reaction scheme

$$S_0 \underset{\beta}{\overset{2\alpha}{\rightleftarrows}} S_1 \underset{2\beta}{\overset{\alpha}{\rightleftarrows}} S_2.$$ (3.109)

The corresponding differential equations are

$$\frac{dx_0}{dt} = \beta x_1 - 2\alpha x_0,$$ (3.110)

$$\frac{dx_2}{dt} = \alpha x_1 - 2\beta x_2,$$ (3.111)

where x_i denotes the proportion of channels in state S_i, and $x_0 + x_1 + x_2 = 1$. We now make the change of variables $x_2 = n^2$, where n satisfies the differential equation

$$\frac{dn}{dt} = \alpha(1 - n) - \beta n.$$ (3.112)

A simple substitution shows that (3.110) and (3.111) are satisfied by $x_0 = (1 - n)^2$ and $x_1 = 2n(1 - n)$.

In fact, we can derive a stronger result. We let

$$x_0 = (1 - n)^2 + y_0,$$ (3.113)

$$x_2 = n^2 + y_2,$$ (3.114)

so that, of necessity, $x_1 = 2n(1-n) - y_0 - y_2$. It follows that

$$\frac{dy_0}{dt} = -2\alpha y_0 - \beta(y_0 + y_2), \tag{3.115}$$

$$\frac{dy_2}{dt} = -\alpha(y_0 + y_2) - 2\beta y_2. \tag{3.116}$$

This is a linear system of equations with eigenvalues $-(\alpha + \beta)$, $-2(\alpha + \beta)$, and so y_0, y_2 go exponentially to zero. This means that $x_0 = (1-n)^2$, $x_2 = n^2$ is an invariant stable manifold for the original system of equations; the solutions cannot leave this manifold, and with arbitrary initial data, the flow approaches this manifold exponentially. Notice that this is a stable invariant manifold even if α and β are functions of time (so they can depend on voltage or other concentrations).

This argument generalizes to the case of k identical independent binding sites where the invariant manifold for the flow is the binomial distribution with probability n satisfying (3.112) (see Exercise 14). Thus, the channel conductance is proportional to n^k, where n satisfies the simple equation (3.112). This multiple subunit model of channel gating provides the basis for the model of excitability that is examined in Chapter 5.

3.5.3 The Sodium Channel

A more complex model is needed to explain the behavior of the Na^+ channel, which both activates and inactivates. The simplest approach is to extend the above analysis to the case of multiple subunits of two different types, m and h, say, where each subunit can be either closed or open, and the channel is open, or conducting, only when all subunits are open. To illustrate, we assume that the channel has one h subunit and two m subunits. The reaction diagram of such a channel is shown in Fig. 3.11. We let S_{ij} denote the channel with i open m subunits and j open h subunits, and we let x_{ij} denote the fraction of channels in state S_{ij}. The dynamics of x_{ij} are described by a system of six differential equations. However, as above, direct substitution shows that this system of equations has an invariant manifold $x_{00} = (1-m)^2(1-h)$, $x_{10} = 2m(1-m)(1-h)$,

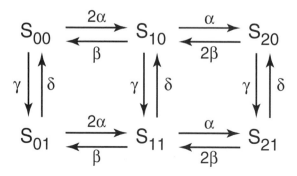

Figure 3.11 Diagram of the possible states in a model of the Na^+ channel.

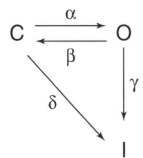

Figure 3.12 A: Schematic diagram of the states of the Na$^+$ channel. C, O, and I denote the closed, open, and inactivated states, respectively.

$x_{20} = m^2(1 - h)$, $x_{01} = (1 - m)^2 h$, $x_{11} = 2m(1 - m)h$, and $x_{21} = m^2 h$, provided

$$\frac{dm}{dt} = \alpha(1 - m) - \beta m, \tag{3.117}$$

$$\frac{dh}{dt} = \gamma(1 - h) - \delta h. \tag{3.118}$$

Furthermore, the invariant manifold is stable. A model of this type was used by Hodgkin and Huxley in their model of the nerve axon, which is discussed in detail in Chapter 5.

In an alternate model of the Na$^+$ channel (Aldrich et al., 1983; Peskin, 1991), it is assumed that the Na$^+$ channel can exist in three states, closed (C), open (O), or inactivated (I), and that once the channel is inactivated, it cannot return to either the closed or the open state (Fig. 3.12). Thus, the state I is absorbing. While this is clearly not true in general, it is a reasonable approximation at high voltages.

As before, we let g denote the proportion of open channels and let c denote the proportion of closed channels. Then,

$$\frac{dc}{dt} = -(\alpha + \delta)c + \beta g, \tag{3.119}$$

$$\frac{dg}{dt} = \alpha c - (\beta + \gamma)g, \tag{3.120}$$

where the proportion of channels in the inactivated state is $i = 1 - c - g$. Initial conditions are $c(0) = 1$, $g(0) = 0$, i.e., all the channels are initially in the closed state. This system of first-order differential equations can be solved directly to give

$$g(t) = a(e^{\lambda_1 t} - e^{\lambda_2 t}), \tag{3.121}$$

where $\lambda_2 < \lambda_1 < 0$ are the roots of the characteristic polynomial

$$\lambda^2 + (\alpha + \beta + \gamma + \delta)\lambda + (\alpha + \delta)(\beta + \gamma) - \alpha\beta = 0, \tag{3.122}$$

and where

$$g'(0) = \alpha = a(\lambda_1 - \lambda_2) > 0. \tag{3.123}$$

As in the simple two-state model, the function g can be fit to data to determine the parameters a, λ_1, and λ_2. However, unlike the two-state model, the rate constants cannot be determined uniquely from these parameters. For, since λ_1 and λ_2 are the roots of (3.122), it follows that

$$\alpha + \beta + \gamma + \delta = -\lambda_1 - \lambda_2, \tag{3.124}$$

$$(\alpha + \delta)(\beta + \gamma) - \alpha\beta = \lambda_1\lambda_2. \tag{3.125}$$

Thus, there are only three equations for the four unknowns, α, β, γ, and δ, so the system is underdetermined (see Exercise 16). This problem cannot be resolved using the macroscopic data that have been discussed so far, but requires data collected from a single channel, as described in Section 3.6.

3.5.4   Agonist-Controlled Ion Channels

Many ion channels are controlled by agonists, rather than by voltage. For example, the opening of ion channels in the postsynaptic membrane of the neuromuscular junction (Chapter 8) is controlled by the neurotransmitter acetylcholine, while in the central nervous system a host of neurotransmitters such as glutamate, dopamine, γ-aminobutyric acid (GABA), and serotonin have a similar role. The inositol trisphosphate receptor and ryanodine receptor are other important agonist-controlled ion channels (Chapter 7).

Early theories of agonist-controlled ion channels (Clark, 1933) assumed that the channel was opened simply by the binding of the agonist. Thus,

$$A + T \underset{k_{-1}}{\overset{k_1}{\rightleftharpoons}} AT, \tag{3.126}$$

where the state AT is open. However, this simple theory is unable to account for a number of experimental observations. For example, it can happen that only a fraction of channels are open at any given time, even at high agonist concentrations, a result that cannot be explained by this simple model.

In 1957, del Castillo and Katz proposed a model that explicitly separated the agonist-binding step from the gating step:

$$\begin{array}{ccccccc}
A & + & T & \underset{k_{-1}}{\overset{k_1}{\rightleftharpoons}} & AT & \underset{\alpha}{\overset{\beta}{\rightleftharpoons}} & AR. \\
\text{agonist} & & \text{state 3} & & \text{state 2} & & \text{state 1} \\
& & \text{closed} & & \text{closed} & & \text{open} \\
& & \text{unoccupied} & & \text{occupied} & & \text{occupied}
\end{array} \tag{3.127}$$

Note that the only open state is AR (state 1; the slightly unusual numbering of the states follows Colquhoun and Hawkes, 1981). Thus, in this model, binding of the agonist places the channel into an occupied state that allows, but does not require, opening. The agonist-binding step is controlled by the *affinity* of the channel for the agonist, while the gating is determined by the *efficacy* of the agonist. This separation of affinity and efficacy has proven to be an extremely powerful way of understanding agonist-controlled channels, and is at the heart of practically all modern approaches (Colquhoun, 2006).

The conductance of a population of agonist-controlled channels is determined as the solution of the system of differential equations

$$\frac{d\Phi_1}{dt} = \beta\Phi_2 - \alpha\Phi_1, \tag{3.128}$$

$$\frac{d\Phi_2}{dt} = \alpha\Phi_1 + k_1 a(1 - \Phi_2 - \Phi_1) - (\beta + k_{-1})\Phi_2, \tag{3.129}$$

where Φ_1, Φ_2, and $1 - \Phi_1 - \Phi_2$ represent the percentage of channels in states 1, 2, and 3, respectively, and a is the concentration of agonist A. The solution of this system of differential equations is easy to determine, provided a is constant. However, the practical usefulness of this exact solution is extremely limited, since in any realistic situation a is changing in time.

The steady-state solution is also readily found to be

$$\Phi_1 = \frac{1}{1 + \frac{\beta}{\alpha} + \frac{\beta k_{-1}}{\alpha k_1 a}}, \tag{3.130}$$

and this can be fit to data to find the equilibrium constants $\frac{\beta}{\alpha}$ and $\frac{k_{-1}}{k_1}$. However, complete determination of the four kinetic parameters is much more challenging. One could imagine a "concentration clamp" experiment, in which the concentration of a is suddenly switched from one level to another and the conductance of the channels monitored. From these data one could then determine the two eigenvalues of the system (3.128)–(3.129). However, usually such experiments are very difficult to perform. In Section 3.6 we show that there is more information contained in single-channel recordings and how this additional information can be used to determine the kinetic parameters of channel models.

3.5.5 Drugs and Toxins

Many drugs act by blocking a specific ion channel. There are numerous specific channel blockers, such as Na^+ channel blockers, K^+ channel blockers, Ca^{2+} channel blockers, and so on. In fact, the discovery of site-specific and channel-specific blockers has been of tremendous benefit to the experimental study of ion channels. Examples of important channel blockers include verapamil (Ca^{2+}-channel blocker), quinidine, sotolol, nicotine, DDT, various barbiturates (K^+-channel blockers), tetrodotoxin (TTX, the primary ingredient of puffer fish toxin), and scorpion toxins (Na^+-channel blockers).

To include the effects of a drug or toxin like TTX in a model of a Na^+ channel is a relatively simple matter. We assume that a population P of Na^+ channels is available for ionic conduction and that a population B is blocked because they are bound by the toxin. Thus,

$$P + D \underset{k_-}{\overset{k_+}{\rightleftharpoons}} B, \tag{3.131}$$

where D represents the concentration of the drug. Clearly, $P + B = P_0$, so that

$$\frac{dP}{dt} = k_-(P_0 - P) - k_+ DP, \qquad (3.132)$$

and the original channel conductance must be modified by multiplying by the percentage of unbound channels, P/P_0.

In steady state, we have

$$\frac{P}{P_0} = \frac{K_d}{K_d + D}. \qquad (3.133)$$

The remarkable potency of TTX is reflected by its small equilibrium constant K_d, as $K_d \approx 1$–5 nM for Na^+ channels in nerve cells, and $K_d \approx 1$–10 μM for Na^+ channels in cardiac cells. By contrast, verapamil has $K_d \approx 140$–940 μM.

Other important drugs, such as lidocaine, flecainide, and encainide are so-called *use-dependent* Na^+-channel blockers, in that they interfere with the Na^+ channel only when it is open. Thus, the more the channel is used, the more likely that it is blocked. Lidocaine is an important drug used in the treatment of cardiac arrhythmias. The folklore explanation of why it is useful is that because it is use-dependent, it helps prevent high-frequency firing of cardiac cells, which is commonly associated with cardiac arrhythmias. In fact, lidocaine, flecainide, and encainide are officially classified as antiarrhythmic drugs, even though it is now known that flecainide and encainide are proarrhythmic in certain postinfarction (after a heart attack) patients. A full explanation of this behavior is not known.

To keep track of the effect of a use-dependent drug on a two-state channel, we suppose that there are four classes of channels, those that are closed but unbound by the drug (C), those that are open and unbound by the drug (O), those that are closed and bound by the drug (CB), and those that are open and bound by the drug (OB) (but unable to pass a current). For this four-state model a reasonable reaction mechanism is

$$C \underset{\beta}{\overset{\alpha}{\rightleftharpoons}} O,$$

$$CB \underset{\beta}{\overset{\alpha}{\rightleftharpoons}} OB,$$

$$CB \overset{k_+}{\longrightarrow} C + D,$$

$$O + D \underset{k_-}{\overset{k_+}{\rightleftharpoons}} OB.$$

Notice that we have assumed that the drug does not interfere with the process of opening and closing, only with the flow of ionic current, and that the drug can bind the channel only when it is open. It is now a straightforward matter to find the differential equations governing these four states, and we leave this as an exercise.

This is not the only way that drugs might interfere with a channel. For example, for a channel with multiple subunits, the drug may bind only when certain of the subunits are in specific states. Indeed, the binding of drugs with channels can occur in many ways, and there are numerous unresolved questions concerning this complicated process.

3.6 Single-Channel Analysis

Since the late 1970s, the development of patch-clamp recording techniques has allowed the measurement of ionic current through a small piece of cell membrane, containing only a few, or even a single, ionic channel (Hamill et al., 1981; Sakmann and Neher, 1995; Neher and Sakmann received the 1991 Nobel Prize in Physiology or Medicine for their development of the patch-clamp technique).

Much of the mathematical theory of how to analyze single-channel recordings was worked out in a series of papers by Colquhoun and Hawkes (1977, 1981, 1982). As is true of most things written by Colquhoun and Hawkes, these are eminently readable. However, newcomers to the field should first read the two chapters in the *Plymouth Workshop Handbook on Microelectrode Techniques* (Colquhoun, 1994; Colquhoun and Hawkes, 1994), since these are an excellent introduction. The chapter by Colquhoun and Hawkes in the book by Sakmann and Neher (1995) is also a valuable reference.

An example of an experimental record for Na^+ channels is given in Fig. 3.13. The current through an individual channel is stochastic (panel A) and cannot be described by a deterministic process. Nevertheless, the ensemble average over many experiments (panel B) is deterministic and reproduces the same properties that are seen in the macroscopic measurements of Fig. 3.10. However, the single-channel recordings contain more information than does the ensemble average.

What information is available from single-channel recordings that is not available from ensemble averages? First of all, one can measure how long a channel is open on average, or more generally, the distribution of open times. Similarly, one can measure the distribution of times the channel is in the closed state. If there are additional dynamical processes underlying the opening and closing of channels, as there are with Na^+ channels, one can measure (for example) how many times a channel opens (and closes) before it is inactivated (permanently closed) or how many channels fail to open even once.

The most common models of ion channels are discrete-space continuous-time Markov processes, the basic theory of which was described in Section 2.9.2. Since it is this theory that lies at the heart of the analysis of single-channel data, the reader is encouraged to review the relevant sections of Chapter 2 before continuing.

3.6.1 Single-Channel Analysis of a Sodium Channel

Consider the Na^+ channel model shown in Fig. 3.12. For this model there are two obvious waiting-time problems. The probability that the amount of time spent in the closed state before opening is less than t is governed by the differential equation

$$\frac{dP}{dt} = \alpha(1 - P),\tag{3.134}$$

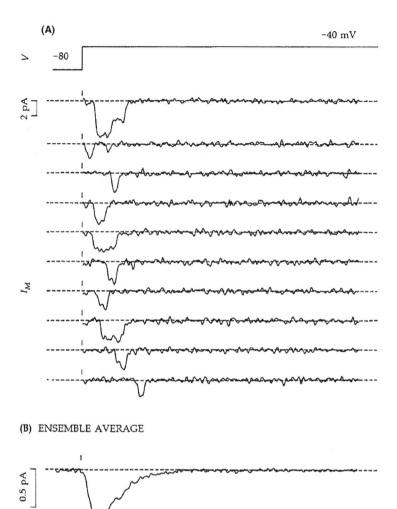

Figure 3.13 A: Na$^+$ currents from a single channel (or possibly two in the first trace) following a voltage step from −80 mV to −40 mV. B: Average open probability of the Na$^+$ channel, obtained by averaging over many traces of the type shown in A. (Hille, 2001, Fig. 3.16, p. 90.)

with $P(0) = 1$, and therefore is

$$P(t) = 1 - \exp(-\alpha t), \tag{3.135}$$

so that the closed time distribution is $\alpha \exp(-\alpha t)$.

Similarly, the probability that the amount of time spent in the open state is less than t is $1 - \exp(-(\beta + \gamma)t)$, so that the open time distribution is $(\beta + \gamma) \exp(-(\beta + \gamma)t)$.

According to this model, the channel can inactivate from the closed state at rate δ or from the open state at rate γ. At the beginning of the experiment the channel is in the closed state, and from there it can either open or inactivate. The probability that the first transition is to the open state is $A = \frac{\alpha}{\alpha+\delta}$, and the probability that the first transition is to the inactivated state is $1 - A$. Thus, $1 - A$ can be estimated by the proportion of experimental records in which no current is observed, even after the depolarizing stimulus was maintained for a long time.

A channel may open and close several times before it finally inactivates. To understand this, we let N be the number of times the channel opens before it finally inactivates and calculate the probability distribution for N. Clearly, $P[N = 0] = 1 - A$. Furthermore,

$$
\begin{aligned}
P[N = k] &= P[N = k \text{ and channel enters I from O}] \\
&\quad + P[N = k \text{ and channel enters I from C}] \\
&= A^k B^{k-1}(1 - B) + A^k B^k (1 - A) \\
&= (AB)^k \left(\frac{1 - AB}{B} \right),
\end{aligned}
\tag{3.136}
$$

where $B = \frac{\beta}{\beta+\gamma}$.

We now have enough information to estimate the four channel rate constants. Since A can be determined from the proportion of channels that never open, B can be determined from a plot of the experimental data for $P[N = k]$ vs. k. Then, $\beta + \gamma$ can be determined from the open time distribution of the channel and α can be determined from the closed time distribution.

Since the work of Hodgkin and Huxley (described in Chapter 5), the traditional view of a Na^+ channel has been that it activates quickly and inactivates slowly. According to this view, the decreasing portion of the g_{Na} curve in Fig. 3.10 is due entirely to inactivation of the channel. However, single-channel analysis has shown that this interpretation of macroscopic data is not always correct. It turns out that the rate of inactivation of some mammalian Na^+ channels is faster than the rate of activation. For example, Aldrich et al. (1983) found $\alpha = 1/ms$, $\beta = 0.4/ms$, $\gamma = 1.6/ms$, and $\delta = 1/ms$ at $V = 0$ for channels in a neuroblastoma cell line and a pituitary cell line. Although this reversal of activation and inactivation rates is not correct for all Na^+ channels in all species, the result does overturn some traditional ideas of how Na^+ channels work.

More modern models of the Na^+ channel are based on a wide range of experimental data, including single-channel recordings and macroscopic ionic and gating currents. It is a very difficult matter to decide, on the basis of these data, which is the best model of the channel. One of the most rigorous attempts is that of Vandenberg and Bezanilla (1991), who concluded that a sequential Markov model with three closed states, one open state, and one inactivated state was the best at reproducing the widest array of data. However, because of the ill-posed nature of this inverse problem, it is impossible to rule out the existence of multiple other states.

3.6.2   Single-Channel Analysis of an Agonist-Controlled Ion Channel

The single-channel analysis of the agonist-controlled ion channel (3.127) is more subtle than that of the Na$^+$ channel. This is because there are states that cannot be directly observed, but can only be inferred from the data. If the state AR is the only open state, then a typical single-channel recording might look (at least, in an ideal situation) like that shown in Fig. 3.14. The openings occur in bursts as the channel flicks between the AR and AT states, with longer interburst periods occurring when the channel escapes from AT into the closed state, T, because of agonist unbinding. However, because the binding and unbinding transitions are not directly observable, this process is called a *hidden Markov process*.

There are two distributions that are readily determined from the data. These are the open time and closed time distributions. Since the open state, AR, can close only by a single pathway to AT, the open time distribution is the exponential distribution $\alpha e^{-\alpha t}$, with mean $1/\alpha$.

The Closed Time Distribution

Every period during which the channel is closed must begin with the channel in state AT and end with the channel in state AR. However, during this time the channel can be in either state AT or state T. Thus, the closed time distribution is the transition time from state AT to state AR (see Section 2.9.2).

To calculate the closed time probability, we set state AR to be an absorbing state (i.e., set $\alpha = 0$), and impose the initial condition that the receptor starts in state AT. Hence,

$$\frac{d\Phi_1}{dt} = \beta\Phi_2, \tag{3.137}$$

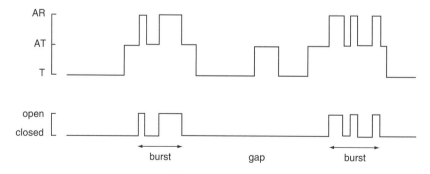

Figure 3.14   Schematic diagram of a possible single-channel recording in the model described by (3.127). The openings occur in bursts as the channel flickers between states AR and AT. However, the transitions between states T and AT cannot be observed. (Adapted from Colquhoun and Hawkes (1981), Fig. 1.)

$$\frac{d\Phi_2}{dt} = -(\beta + k_{-1})\Phi_2 + ak_1\Phi_3, \tag{3.138}$$

$$\frac{d\Phi_3}{dt} = k_{-1}\Phi_2 - ak_1\Phi_3, \tag{3.139}$$

with initial data $\Phi_1(0) = 0$, $\Phi_2(0) = 1$, $\Phi_3(0) = 0$.

We can readily solve this system of differential equations to determine that the transition time from state 2 to state 1 has the probability density

$$\phi_{21}(t) = \frac{d\Phi_1}{dt} = \beta x_2 = \frac{\beta}{\lambda_1 - \lambda_2}[(\lambda_1 + ak_1)e^{\lambda_1 t} - (\lambda_2 + ak_1)e^{\lambda_2 t}], \tag{3.140}$$

where the eigenvalues λ_1 and λ_2 are the roots (both negative) of $\lambda^2 + \lambda(\beta + k_{-1} + k_1 a) + ak_1\beta = 0$.

Since the closed time distribution is the sum of two exponentials, the open and closed time distributions along with the steady-state open probability (3.130) theoretically provide enough information to determine uniquely the four kinetic parameters of the model.

Other Distributions

There are other distributions that can be calculated, but obtaining the data for these is somewhat subjective. These are the distribution of closed times during a burst, the number of openings in a burst, and the distribution of gap closed times.

The distribution of closed times during a burst is the easiest to calculate, being simply $\beta \exp(-\beta t)$.

Each time the channel is in state AT a choice is made to go to state AR (with probability $\beta/(\beta + k_{-1})$) or to go to state T (with probability $k_{-1}/(\beta + k_{-1})$). Thus, for there to be N openings in a burst, the channel must reopen by going from state AT to state AR $N - 1$ times, and then end the burst by going from state AT to state T. Hence,

$$P(N \text{ openings}) = \left(\frac{\beta}{\beta + k_{-1}}\right)^{N-1} \left(\frac{k_{-1}}{\beta + k_{-1}}\right), \tag{3.141}$$

where $N \geq 1$, which has mean $1 + \beta/k_{-1}$.

To determine the closed time distribution for gaps, we observe that a gap begins in state AT and then moves back and forth between states T and AT before finally exiting into state AR. The waiting-time distribution for leaving state AT into state T is

$$\phi_{23}(t) = k_{-1}e^{-k_{-1}t}. \tag{3.142}$$

The transition time from state T to state AR is $\phi_{31} = \frac{d\Phi_1}{dt}$, determined as the solution of the system (3.137)–(3.139), subject to initial conditions $\Phi_1(0) = 0$, $\Phi_2(0) = 0$, and $\Phi_3(0) = 1$. One readily determines that

$$\phi_{31} = \beta \frac{(\lambda_2 + ak_1)(\lambda_1 + ak_1)}{k_{-1}(\lambda_2 - \lambda_1)}(e^{\lambda_1 t} - e^{\lambda_2 t}). \tag{3.143}$$

Now, to calculate the gap time distribution, we observe that the time in a gap is the sum of two times, namely the time in state AT before going to state T and the time to go from state T to state AR. Here we invoke a standard result from probability theory regarding the distribution for the sum of two random variables. That is, if $p_1(t_1)$ and $p_2(t_2)$ are the probability densities for random variables t_1 and t_2, then the probability density for the sum of these is the convolution

$$p_{\text{sum}}(t) = \int_0^t p_1(s)p_2(t-s)\,ds. \tag{3.144}$$

Thus, in this problem, the probability density for the gap time is

$$\phi_{\text{gap}}(t) = \int_0^t \phi_{23}(s)\phi_{31}(t-s)\,ds. \tag{3.145}$$

It is again straightforward to determine (use Laplace transforms and the convolution theorem) that

$$\phi_{\text{gap}}(t) = \beta \frac{(\lambda_2 + ak_1)(\lambda_1 + ak_1)}{(\lambda_2 - \lambda_1)} \left(\frac{e^{\lambda_1 t}}{\lambda_1 + k_{-1}} - \frac{e^{\lambda_2 t}}{\lambda_2 + k_{-1}} + \frac{(\lambda_1 - \lambda_2)e^{-k_{-1}t}}{(\lambda_1 + k_{-1})(\lambda_2 + k_{-1})} \right),$$

a sum of three exponentials.

Making use of these distributions is tricky, because it is not clear how to distinguish between a short gap and a long closed interval during a burst. In fact, if a is large enough and k_{-1} is not too small, then the mean gap length is shorter than the mean burst closed time, so that errors of classification are likely.

3.6.3 Comparing to Experimental Data

Experimental data typically come in lists of open and closed times. They are then displayed in a histogram, where the area under each histogram bar corresponds to the number of events in that interval.

However, it can be very difficult to determine from a histogram how many exponential components are in the distributions; an exponential distribution with three exponentials can look very similar to one with two exponentials, even when the time constants are widely separated.

This problem is sometimes avoided by first taking the log of the times, and plotting a histogram of the log(time) distributions. Since the log function is monotone increasing, we know that

$$P[t < t_1] = P[\log(t) < \log(t_1)], \tag{3.146}$$

and thus the cumulative distributions are the same, whether functions of the log or the linear times. However, the functions that are of interest to us are the probability density functions, which are the derivatives of the cumulative distribution functions. Suppose $\Phi(t)$ is a cumulative distribution function with corresponding probability density function $\phi(t) = \frac{d\Phi}{dt}$. For any monotone increasing function $g(x)$, $\Phi(g(x))$ is also a cumulative

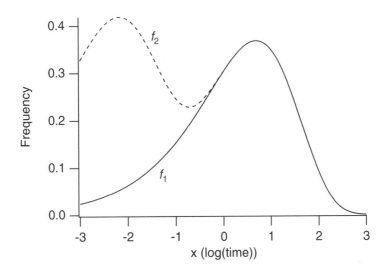

Figure 3.15 Two exponential distributions, transformed according to (3.147), and plotted against log(time). Solid line: $f_1(t) = 0.5e^{-0.5t}$. Dotted line: $f_2(t) = 0.5e^{-0.5t} + 10e^{-10t}$.

distribution function in x. However, the corresponding probability density function is

$$\frac{d\Phi(g(x))}{dx} = g'(x)\phi(g(x)). \tag{3.147}$$

Thus, to use a log(time) transformation, $x = \ln(t)$, for a given probability density function, $\phi(t)$, we plot $e^x \phi(e^x)$ and fit this to the histogram of the log(time).

There are significant advantages to this scaling of time, especially for exponential distributions, illustrated in Fig. 3.15. The solid line in Fig. 3.15 corresponds to the distribution $\phi_1(t) = 0.5e^{-0.5t}$, transformed according to (3.147). In other words, this is the plot of the function $0.5e^x e^{-0.5e^x}$ against x. Notice that the maximum of the curve occurs at the mean of the distribution, $t = 2$ ($x = \log 2 = 0.69$) (see Exercise 21). The dotted line is the distribution $f_2(t) = 0.5e^{-0.5t} + 10e^{-10t}$, again transformed according to (3.147). (This is not a true probability density function since the area under the curve is 2, rather than 1.) The two peaks occur at the means of the individual component exponential distributions, i.e., at $t = 2$ and $t = 0.1$ ($x = 0.69$ and $x = -2.3$ respectively).

Modern methods of fitting models to single-channel data are considerably more sophisticated than merely fitting histograms, as described above. Fitting directly to the set of open and closed times using the log likelihood is a common approach, but, more recently, methods to fit the model directly to the single-channel time course raw data (not simply to a list of open and closed times), using Markov chain Monte Carlo and Bayesian inference, have been developed (Fredkin and Rice, 1992; Ball et al., 1999; Hodgson and Green, 1999).

3.7   Appendix: Reaction Rates

In Section 3.4 we made extensive use of the formula (3.43), i.e.,

$$k_i = \kappa \exp\left(\frac{-\Delta G_i}{RT}\right),\qquad(3.148)$$

which states that the rate, k_i, at which a molecule leaves a binding site is proportional to the exponential of the height of the energy barrier ΔG_i that must be crossed to exit. This is called the *Arrhenius* equation, after Svante Arrhenius, who first discovered it experimentally in the late 1800s (Arrhenius received the 1903 Nobel Prize in Chemistry). Arrhenius determined, not the dependence of k_i on ΔG_i, but its dependence on temperature T. He showed experimentally that the rate of reaction is proportional to $\exp(-B/T)$, for some positive constant B. He then used the Boltzmann distribution to argue that $B = \Delta G_i/R$, as discussed below.

As was described in Section 1.2, the equilibrium constant, K_{eq}, for a reaction is related to the change in free energy, ΔG^0, by

$$K_{\text{eq}} = e^{\frac{\Delta G^0}{RT}}.\qquad(3.149)$$

Note that if κ is independent of ΔG_i, then (3.148) is consistent with (3.149). Given the potential energy profile in Fig. 3.16, it is clear that if $k_1 = \kappa \exp\left(\frac{-\Delta G_1}{RT}\right)$ and $k_{-1} = \kappa \exp\left(\frac{-\Delta G_{-1}}{RT}\right)$, then $K_{\text{eq}} = k_{-1}/k_1 = \exp\left(\frac{\Delta G_1 - \Delta G_{-1}}{RT}\right) = \exp\left(\frac{\Delta G^0}{RT}\right)$, where $\Delta G^0 = \Delta G_1 - \Delta G_{-1}$.

However, despite this consistency, (3.148) does not follow from (3.149); the equilibrium relationship tells us nothing about how each rate constant might depend on T or ΔG_i. Although the derivation of the equilibrium condition depends only on fundamental thermodynamic principles, derivation of a rate expression is much more difficult, and the exact rate expression depends, in general, on the choice of model. There is still enormous controversy over exactly how to derive rate equations, and which is most

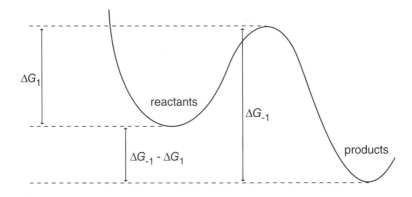

Figure 3.16 Schematic diagram of a potential energy profile of a hypothetical reaction.

suitable in any given situation. Here we give only a brief discussion of this important problem, enough to give a plausible derivation of the general form of (3.148). The exponential dependence occurs in every rate equation; it is the prefactor, κ, and its possible functional dependences, that is the source of so much discussion.

3.7.1 The Boltzmann Distribution

To show that $B = \Delta G/R$ (or $\Delta G/k$, depending on whether ΔG has units of per mole or per molecule; we have dropped the subscript i for this discussion), Arrhenius assumed that the rate of reaction was proportional to the fraction of molecules with energy greater than some minimum amount. Given this assumption, the Arrhenius equation follows from the Boltzmann distribution, which we now derive.

We begin with a brief digression. It is intuitively clear that were we to toss a fair coin 10^{20} times, the chance of obtaining any distribution of heads and tails significantly different from 50:50 is insignificant. We might, of course, get $\frac{10^{20}}{2} - 100$ heads and $\frac{10^{20}}{2} + 100$ tails, but the relative deviation from 50:50 is inconsequential.

To express this mathematically, suppose that we toss a fair coin n times to get a sequence of heads and tails. Of all the possible sequences, the total number that have h heads and $n - h$ tails (in any order) is $\frac{n!}{h!(n-h)!}$, and thus, since there are 2^n possible sequences, the probability of getting h heads and $n - h$ tails is given by

$$\text{Prob}[n \text{ heads}, n - h \text{ tails}] = \frac{n!}{2^n h!(n - h)!}. \tag{3.150}$$

As n gets very large, the graph of (3.150) becomes sharply peaked, with a maximum of 1 at $h = n/2$. (This can be shown easily using Stirling's formula, $\ln(n!) \approx n \ln(n) - n$ for n large). In other words, the probability of obtaining any sequence that does not contain an equal number of heads and tails is vanishingly small in the limit of large n.

An identical argument underlies the Boltzmann distribution. Suppose we have n particles each of which can be in one of k states, where state i has energy U_i. Let n_i denote the number of particles in state i. We assume that the total energy, U_{tot}, is fixed, so that

$$\sum_{i=1}^{k} n_i = n, \tag{3.151}$$

$$\sum_{i=1}^{k} U_i n_i = U_{\text{tot}}. \tag{3.152}$$

The number of ways, W, that these n particles can be partitioned into k states, with n_i particles in the state i, is given by the multinomial factor

$$W = \frac{n!}{\Pi_{i=1}^{k} n_i!}. \tag{3.153}$$

Now, like the function for the probability of heads and tails for a coin toss, the function W is sharply peaked when n is large, and the distribution corresponding to the peak is the one most likely to occur. Furthermore, the likelihood of distributions other than those near the peak is vanishingly small when n is large. Thus, to find this overwhelmingly most likely distribution we maximize W subject to the constraints (3.151) and (3.152). That is, we seek to maximize (using Lagrange multipliers)

$$F = \ln W - \lambda \left(\sum_{i=1}^{k} n_i - n \right) - \beta \left(\sum_{i=1}^{k} U_i n_i - U_{\text{tot}} \right).$$ (3.154)

(It is equivalent, and much more convenient, to use $\ln W$ rather than W.) According to (3.153),

$$\ln W = \ln(n!) - \sum_{i=1}^{k} \ln(n_i!)$$

$$\approx n \ln(n) - n + \sum_{i=1}^{k} n_i - \sum_{i=1}^{k} n_i \ln(n_i)$$

$$= n \ln(n) - \sum_{i=1}^{k} n_i \ln(n_i),$$ (3.155)

where we have used Stirling's formula, assuming all the n_i's are large. Thus,

$$\frac{\partial F}{\partial n_i} = -\ln(n_i) - 1 - \lambda - \beta U_i,$$ (3.156)

which is zero when

$$n_i = \alpha e^{-\beta U_i},$$ (3.157)

for positive constants α and β, which are independent of i. This most likely distribution of n_i is the *Boltzmann distribution*.

How does this relate to reaction rates? Suppose that we have a population of particles with two energy levels: a ground energy level U_0 and a reactive energy level $U_r > U_0$. If the particles are at statistical equilibrium, i.e., the Boltzmann distribution, then the proportion of particles in the reactive state is

$$e^{\beta(U_0 - U_r)} = e^{-\beta \Delta U}.$$ (3.158)

Since this is also assumed to be the rate at which the reaction takes place, we have that

$$k \propto e^{\beta(U_0 - U_r)} = e^{-\beta \Delta U}.$$ (3.159)

To obtain the Arrhenius rate equation it remains to show that $\beta = \frac{1}{RT}$. To do so rigorously is beyond the scope of this text, but a simple dimensional argument can at least demonstrate plausibility. Recall that it is known from experiment that

$$k \propto e^{-B/T},$$ (3.160)

for some constant $B > 0$. Thus, from (3.159), β must be proportional to $1/T$, and to get the correct units, $\beta \propto 1/(RT)$ or $1/(kT)$, depending on whether the units of U are per mole or per molecule.

3.7.2   A Fokker–Planck Equation Approach

The above derivation relies on the assumption that there is a large number of particles, each of which can be in one of a number of different states. It is less obvious how such a derivation can be applied to the behavior of a small number of molecules, or a single molecule. To do this, we need to use the methods developed in Appendix 2.9, and turn to a Fokker–Planck description of molecular motion.

Suppose that a molecule moves via Brownian motion, but also experiences a force generated by some potential, $U(x)$, and is subject to friction. If $x(t)$ denotes the position of the molecule, the Langevin equation for the molecular motion (Section 2.9.5) is

$$m\frac{d^2x}{dt^2} + \nu\frac{dx}{dt} + U'(x) = \sqrt{2\nu kT}W(t), \tag{3.161}$$

where W is a Wiener process. Here, ν is the friction coefficient, and is analogous to friction acting on a mass–spring system. If inertial effects can be neglected, which they can in most physiological situations, this simplifies to

$$\nu\frac{dx}{dt} = -U'(x) + \sqrt{2\nu kT}W(t). \tag{3.162}$$

Hence, the probability distribution that the particle is at position x at time t is given by $p(x,t)$, the solution of the Fokker–Planck equation

$$\nu\frac{\partial p}{\partial t} = \frac{\partial}{\partial x}(U'(x)p) + kT\frac{\partial^2 p}{\partial x^2}. \tag{3.163}$$

At steady state, i.e., when $\frac{\partial p}{\partial t} = 0$, (3.163) can be readily solved to give

$$p(x) = \frac{1}{A}\exp\left(-\frac{U(x)}{kT}\right), \tag{3.164}$$

where $A = \int_{-\infty}^{\infty}\exp(-\frac{U(x)}{kT})\,dx$ is chosen so that $\int_{-\infty}^{\infty}p(x)\,dx = 1$. We have thus regained a continuous version of the Boltzmann distribution; if $U(x)$ is a quadratic potential well ($U(x) = Ax^2$), then $p(x)$ is a Gaussian distribution.

If $U(x)$ is a double well potential with its maximum at $x = 0$ separating the two wells, then the ratio of the probability of finding the particle on the left to the probability of finding the particle on the right is

$$K_{eq} = \frac{\int_{-\infty}^{0}p(x)\,dx}{\int_{0}^{\infty}p(x)\,dx}. \tag{3.165}$$

Since it is difficult to calculate K_{eq} for general functions $U(x)$, it is useful to discretize the state space into a finite number of states $j = 1, 2, \ldots, n$ with energies U_j. For this, we know that the Boltzmann distribution is

$$p_j = \frac{1}{A} \exp\left(-\frac{U_j}{kT}\right), \tag{3.166}$$

where

$$A = \sum_j^n \exp\left(-\frac{U_j}{kT}\right). \tag{3.167}$$

We can make the association between the discrete case and the continuous case precise if we determine U_j by requiring

$$\exp\left(-\frac{U_j}{kT}\right) = \int_{x_{j-1}}^{x_j} \exp\left(-\frac{U(x)}{kT}\right) dx, \tag{3.168}$$

where x_j separates the $j-1$st from the jth potential well. Furthermore, if there are only two energy wells, the ratio of the probability of finding the particle in state one to the probability of finding the particle in state two is

$$K_{eq} = \frac{p_1}{p_2} = \exp\left(\frac{\Delta U}{kT}\right) = \exp\left(\frac{\Delta G^0}{RT}\right), \tag{3.169}$$

where $\Delta U = U_2 - U_1$ is the change in energy per molecule, so that ΔG^0 is the change in energy per mole. Here we have recovered (3.149) for the equilibrium distribution of a reaction in terms of the difference of standard free energy. However, one should note that with this identification, U_j is approximately, but not exactly, the value of U at the bottom of the jth potential well.

3.7.3 Reaction Rates and Kramers' Result

As noted in Chapter 1, equilibrium relationships give information only about the ratio of rate constants, not their individual values. To derive an expression for a rate constant, one must construct a model of how the reaction occurs. The consequent expression for the rate constant is only as good as the assumptions underlying the model.

One common model of a reaction rate is based on the mean first exit time of the time-dependent Fokker–Planck equation (3.163). (Mean first exit times are discussed in Section 2.9.6). This model assumes that a reactant particle can be modeled as a damped oscillator driven by a stochastic force, and that the reaction occurs once the particle reaches the peak of the energy profile between the reactant and product states. Although this model is based on a number of crude assumptions, it gives reasonably good results for a range of potential energy profiles, particularly those for which the energy wells are deep.

The mean first exit time is found from the solution of the ordinary differential equation

$$-U'(x)\frac{d\tau}{dx} + kT\frac{d^2\tau}{dx^2} = -v, \tag{3.170}$$

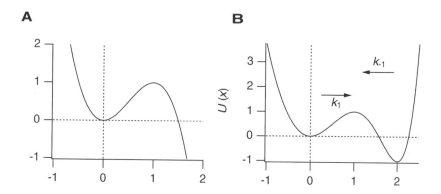

Figure 3.17   Potential energy profiles. A: a cubic profile, $U(x) = \Delta G(2x^2(3/2 - x))$, plotted for $\Delta G = 1$. B: a double well potential, $U(x) = \Delta G^0(\frac{19}{144}x^6 - \frac{1}{24}x^5 - \frac{77}{144}x^4 - \frac{4}{3}x^3 + \frac{25}{9}x^2)$, plotted for $\Delta G^0 = 1$.

subject to $\tau(x_0) = 0$ at any boundary point x_0 where escape is allowed, or $\tau'(x_1) = 0$ at any boundary point x_1 where escape is not allowed, but instead there is reflection. The off-rate, or unbinding rate, is defined as the inverse of the mean first exit time from the bottom of the potential well.

To be specific, consider a potential $U(x)$ such as shown in Fig. 3.17A. Here $U(x)$ is a cubic polynomial, with a minimum at $x = 0$ and a maximum at $x = 1$, with $U(1) = \Delta G$. We expect the particle to spend most of its time near $x = 0$. However, if the particle gets to $x = 1$ it can escape to $x = \infty$, and is assumed to have reacted. Thus, the time to react (the inverse of the reaction rate) is approximated by the mean first passage time from $x = 0$ to $x = 1$.

More generally, suppose $U(x) = \Delta Gu(\frac{x}{L})$, where $u'(0) = u'(1) = u(0) = 0$ and $u(1) = 1$, so that $x = 0$ is a local minimum and $x = L$ is a local maximum, and the height of the energy barrier is ΔG. The mean first passage time is the solution of (3.170) together with the boundary conditions $\tau(-\infty) = 0$ and $\tau(L) = 0$.

To find the solution it is useful to nondimensionalize (3.170). We set $y = \frac{x}{L}$ and $\sigma = \alpha\tau$ and obtain

$$-au'(y)\frac{d\sigma}{dy} + \frac{d^2\sigma}{dy^2} = -1, \tag{3.171}$$

where $a = \frac{\Delta G}{kT}$ and $\alpha = \frac{vL^2}{kT}$. Using an integrating factor, it is easily shown that

$$\sigma(y) = \int_x^1 e^{au(s')} \left(\int_{-\infty}^{s'} e^{-au(s)} ds \right) ds', \tag{3.172}$$

and thus the time to react is $\tau(0) = \frac{vL^2}{kT}\sigma(0)$, where

$$\sigma(0) = \int_0^1 e^{au(s')} \left(\int_{-\infty}^{s'} e^{-au(s)} ds \right) ds'. \tag{3.173}$$

As we demonstrate below, this formula does not agree with the Arrhenius rate law for all parameter values. However, when the potential well at $x = 0$ is deep (i.e., when $a = \Delta G/(kT) \gg 1$), the two are in agreement. Here we provide a demonstration of this agreement.

Notice first that

$$\sigma(0) = \int_0^1 e^{au(s')} \left(\int_{-\infty}^1 e^{-au(s)}\, ds - \int_{s'}^1 e^{-au(s)}\, ds \right) ds'. \tag{3.174}$$

Clearly,

$$\int_0^1 e^{au(s')} \left(\int_{s'}^1 e^{-au(s)}\, ds \right) ds' = \int_0^1 \left(\int_{s'}^1 e^{a(u(s')-u(s))}\, ds \right) ds'$$

$$< \int_0^1 \left(\int_{s'}^1 ds \right) ds' = \frac{1}{2}. \tag{3.175}$$

In fact, with a bit of work one can show that this integral approaches zero as $a \to \infty$. Thus,

$$\sigma(0) \approx \left(\int_0^1 e^{au(s')}\, ds' \right) \left(\int_{-\infty}^1 e^{-au(s)}\, ds \right). \tag{3.176}$$

We now use the fact that $y = 0$ and $y = 1$ are extremal values of $u(y)$ to approximate these integrals. When a is large, the integrands are well approximated by Gaussians, which decay to zero rapidly. Thus, near $y = 0$, $u(y) \approx \frac{1}{2}u''(0)y^2$, so that

$$\int_{-\infty}^1 e^{-au(s)}\, ds \approx \int_{-\infty}^1 e^{-\frac{1}{2}au''(0)s^2}\, ds$$

$$\approx \int_{-\infty}^{\infty} e^{-\frac{1}{2}au''(0)s^2}\, ds$$

$$= \sqrt{\frac{2\pi}{au''(0)}}. \tag{3.177}$$

Similarly, near $y = 1$, $u(y) \approx 1 - \frac{1}{2}|u''(1)|(y-1)^2$, so that

$$\int_0^1 e^{au(s)}\, ds \approx e^a \int_0^1 e^{-\frac{1}{2}|u''(1)|(s-1)^2}\, ds$$

$$\approx e^a \int_{-\infty}^0 e^{-\frac{1}{2}|u''(1)|s^2}\, ds$$

$$= \frac{1}{2}e^a \sqrt{\frac{2\pi}{a|u''(1)|}}. \tag{3.178}$$

Combining (3.176), (3.177), and (3.178) gives

$$\tau(0) \approx \frac{\pi v L^2}{\Delta G \sqrt{u''(0)|u''(1)|}} e^{\frac{\Delta G}{kT}}. \tag{3.179}$$

Since the reaction rate is the inverse of the mean first passage time, this gives the Arrhenius rate expression with

$$\kappa = \frac{\Delta G \sqrt{u''(0)|u''(1)|}}{\pi v L^2}, \tag{3.180}$$

which is independent of T, but not ΔG. This formula was first derived by Kramers (1940).

A Double Well Potential Profile

Now suppose that $U(x)$ is a double well potential, such as that shown in Fig. 3.17B. In particular, suppose that $U(x) = \Delta G^0 u(\frac{x}{L})$, where $u(x)$ has two local minima at $x = 0$ and $x = b > 1$, with a local maximum at $x = 1$. For the example in Fig. 3.17B, $\Delta G^0 = L = 1$ and $b = 2$. Note also that the potential profile is such that $\Delta G_{-1} = 2\Delta G^0$, $\Delta G_1 = \Delta G^0$.

According to Kramers' rate theory,

$$k_1 \approx \frac{\Delta G_1 \sqrt{u''(0)|u''(1)|}}{\pi v L^2} e^{-\frac{\Delta G_1}{kT}}, \tag{3.181}$$

$$k_{-1} \approx \frac{\Delta G_{-1} \sqrt{u''(b)|u''(1)|}}{\pi v L^2 (b-1)^2} e^{-\frac{\Delta G_{-1}}{kT}}. \tag{3.182}$$

To compare these with the exact solutions, in Fig. 3.18A we plot k_{-1} and k_1 for the double well potential shown Fig. 3.17B, calculated by numerical integration of (3.173), and using the approximations (3.181) and (3.182). Note that the reaction rates (both exact and approximate) are not exactly exponential functions of ΔG_i, and thus the curves in Fig. 3.18A are not straight lines (on a log scale). For the approximate rate constants this is because the prefactor is proportional to ΔG_i. Interestingly, the approximate

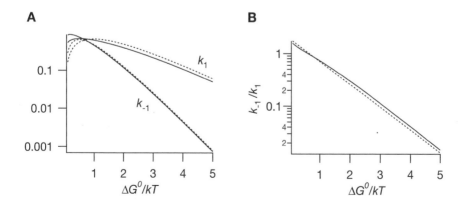

Figure 3.18 Reaction rates for the potential profile shown in Fig. 3.17B. A: exact (solid lines) and approximate (dashed lines) solutions for k_1 and k_{-1}, plotted as functions of $\Delta G^0/kT$. The exact solutions are calculated from (3.173), while the approximations are calculated from (3.179). For simplicity, we set $v L^2/kT = 1$. B: exact and approximate calculations of k_{-1}/k_1. As in A, the exact solution is plotted as a solid line.

solutions agree exactly with the Arrhenius rate law, when viewed as functions of T, while the exact solutions do not.

Next, we observe that, using Kramers' formula, the equilibrium constant is

$$K_{\text{eq}} = \frac{k_{-1}}{k_1} = \frac{1}{(b-1)^2} \frac{u(1) - u(b)}{u(1) - u(0)} \sqrt{\frac{u''(b)}{u''(0)}} e^{-\frac{\Delta G^0}{kT}}. \tag{3.183}$$

In Fig. 3.18B we plot $\frac{k_{-1}}{k_1}$ for the double well potential shown in Fig. 3.17B, with the exact ratio shown as a solid curve and the approximate ratio from (3.183) shown as a dashed curve. As before, the exact ratio k_{-1}/k_1 (solid line, Fig. 3.17B) is not an exact exponential function of ΔG^0, and thus does not give the correct equilibrium behavior. This results from the fact that, for small ΔG^0, the mean first exit time of the Fokker–Planck equation is not a good model of the reaction rate.

However, the ratio of the approximate expressions for the rate constants (3.183) is a true exponential function of ΔG^0, since the dependence of ΔG^0 in the prefactors cancels out in the ratio. Hence, the dashed line in Fig. 3.18B is straight. Thus, paradoxically, the approximate solution gives better agreement to the correct equilibrium behavior than does the exact solution. However, one must be somewhat cautious with this statement, since there is a factor multiplying the exponential that is not equal to one (as it should be for correct equilibrium behavior), but depends on the details of the shape of the energy function. Thus, if the shape of the potential energy function is modified by, for example, an external voltage potential, this factor is modified as well, in a voltage-dependent way.

3.8 EXERCISES

1. Show that the GHK equation (3.2) satisfies both the independence principle and the Ussing flux ratio, but that the linear I–V curve (3.1) satisfies neither.

2. Using concentrations typical of Na^+, determine whether the long channel limit or the short channel limit for (3.25) is the most appropriate approximation for Na^+ channels. (Estimate λ where $\lambda^2 = L^2 q F N_a \tilde{c}/(\epsilon RT)$, for Na^+ ions.)

3. In Section 3.3.1 the PNP equations were used to derive I–V curves when two ions with opposite valence are allowed to move through a channel. Extend this analysis by assuming that two types of ions with positive valence and one type of ion with negative valence are allowed to move through the channel. Show that in the high concentration limit, although the negative ion obeys a linear I–V curve, the two positive ions do not. Details can be found in Chen, Barcilon, and Eisenberg (1992), equations (43)–(45).

4. (a) Show that (3.66) satisfies the independence principle and the Ussing flux ratio.

   (b) Show that (3.66) can be made approximately linear by choosing g such that

   $$ng = \ln\left(\frac{c_n}{c_0}\right). \tag{3.184}$$

   Although a linear I–V curve does not satisfy the independence principle, why does this result not contradict part (a)?

5. Show that (3.86) does not satisfy the independence principle, but does obey the Ussing flux ratio.

6. Derive (3.86) by solving the steady-state equations (3.84) and (3.85). First show that

$$J = x\frac{k_0 c_0 - k_{-n}c_n\pi_{n-1}}{\phi_{n-1}}. \tag{3.185}$$

Then show that

$$k_0 c_0 x = k_{n-1}c_{n-1}\phi_{n-1} - xk_{-n}c_n\phi_{n-2}, \tag{3.186}$$

$$k_j c_j = \frac{k_{n-1}c_{n-1}}{\pi_j}(\phi_{n-1} - \phi_{j-1}) - \frac{k_{-n}c_n x}{\pi_j}(\phi_{n-2} - \phi_{j-1}), \tag{3.187}$$

for $j = 1, \dots, n-1$. Substitute these expressions into the conservation equation and solve for x.

7. Draw the state diagrams showing the channel states and the allowed transitions for a multi-ion model with two binding sites when the membrane is bathed with a solution containing:

   (a) Only ion S on the left and only ion S$'$ on the right.

   (b) Ion S on both sides and ion S$'$ only on the right.

   (c) Ions S and S$'$ on both the left and right.

   In each case write the corresponding system of linear equations that determine the steady-state ionic concentrations at the channel binding sites.

8. Using an arbitrary symmetric energy profile with two binding sites, show numerically that the Ussing flux ratio is not obeyed by a multi-ion model with two binding sites. (Note that since unidirectional fluxes must be calculated, it is necessary to treat the ions on each side of the membrane differently. Thus, an eight-state channel diagram must be used.) Hodgkin and Keynes predicted that the flux ratio is the Ussing ratio raised to the $(n+1)$st power (cf. (3.16)). How does n depend on the ionic concentrations on either side of the membrane, and on the energy profile?

9. Choose an arbitrary symmetric energy profile with two binding sites, and compare the I–V curves of the one-ion and multi-ion models. Assume that the same ionic species is present on both sides of the membrane, so that only a four-state multi-ion model is needed.

10. Suppose the Na^+ Nernst potential of a cell is 56 mV, its resting potential is -70 mV, and the extracellular Ca^{2+} concentration is 1 mM. At what intracellular Ca^{2+} concentration is the flux of a three-for-one Na^+–Ca^{2+} exchanger zero? (Use that $RT/F = 25.8$ mV at $27°$ C.)

11. Modify the pump–leak model of Chapter 2 to include a Ca^{2+} current and the 3-for-1 Na^+–Ca^{2+} exchanger. What effect does this modification have on the relationship between pump rate and membrane potential?

12. Because there is a net current, the Na^+–K^+ pump current must be voltage-dependent. Determine this dependence by including voltage dependence in the rates of conformational change in expression (2.100). How does voltage dependence affect the pump–leak model of Chapter 2?

13. Intestinal epithelial cells have a glucose–Na^+ symport that transports one Na^+ ion and one glucose molecule from the intestine into the cell. Model this transport process. Is the transport of glucose aided or hindered by the cell's negative membrane potential?

14. Suppose that a channel consists of k identical, independent subunits, each of which can be open or closed, and that a current can pass through the channel only if all units are open.

    (a) Let S_j denote the state in which j subunits are open. Show that the conversions between states are governed by the reaction scheme

    $$S_0 \underset{\beta}{\overset{k\alpha}{\rightleftarrows}} S_1 \ldots S_{k-1} \underset{k\beta}{\overset{\alpha}{\rightleftarrows}} S_k. \tag{3.188}$$

    (b) Derive the differential equation for x_j, the proportion of channels in state j.

    (c) By direct substitution, show that $x_j = \binom{k}{j} n^j (1-n)^{k-j}$, where $\binom{k}{j} = \frac{k!}{j!(k-j)!}$ is the *binomial coefficient*, is an invariant manifold for the system of differential equations, provided that

    $$\frac{dn}{dt} = \alpha(1-n) - \beta n. \tag{3.189}$$

15. Consider the model of the Na^+ channel shown in Fig. 3.11. Show that if α and β are large compared to γ and δ, then x_{21} is given (approximately) by

    $$x_{21} = \left(\frac{\alpha}{\alpha+\beta}\right)^2 h, \tag{3.190}$$

    $$\frac{dh}{dt} = \gamma(1-h) - \delta h, \tag{3.191}$$

    while conversely, if γ and δ are large compared to α and β, then (approximately)

    $$x_{21} = m^2 \left(\frac{\gamma}{\gamma+\delta}\right), \tag{3.192}$$

    $$\frac{dm}{dt} = \alpha(1-m) - \beta m. \tag{3.193}$$

16. Show that (3.122) has two negative real roots. Show that when $\beta = 0$ and $a \leq \frac{-\lambda_1}{\lambda_1 - \lambda_2}$, then (3.123)–(3.125) have two possible solutions, one with $\alpha + \delta = -\lambda_1$, $\gamma = -\lambda_2$, the other with $\alpha + \delta = -\lambda_2$, $\gamma = -\lambda_1$. In the first solution inactivation is faster than activation, while the reverse is true for the second solution.

17. Write a computer program to simulate the behavior of the stochastic three-state Na^+ channel shown in Fig. 3.12, assuming it starts in the closed state. Use $\alpha = 1/ms$, $\beta = 0.4/ms$, $\gamma = 1.6/ms$ and $\delta = 1/ms$. Take the ensemble average of many runs to reproduce its macroscopic behavior. Using the data from simulations, reconstruct the open time distribution, the latency distribution, and the distribution of N, the number of times the channel opens. From these distributions estimate the rate constants of the simulation and compare with the known values.

18. Consider the Markov model of a Na^+ channel (Patlak, 1991) shown in Fig. 3.19. Write a computer program to simulate the behavior of this stochastic channel assuming it starts in state C_1. Take the ensemble average of many runs to reproduce its macroscopic behavior. Using the data generated by these simulations, determine the open time distribution, the latency distribution, and the distribution of N, the number of times the channel opens. Compare these with the analytically calculated distributions.

19. Construct a set of differential equations that models the interaction of a two-state channel with a use-dependent blocker.

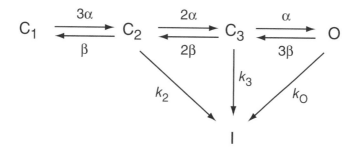

Figure 3.19 The Na$^+$ channel model of Exercise 18. Parameter values are $k_2 = 0.24$ ms^{-1}, $k_3 = 0.4$ ms^{-1}, $k_O = 1.5$ ms^{-1}, $\alpha = 1$ ms^{-1}, and $\beta = 0.4$ ms^{-1}.

20. Write a computer program to simulate the behavior of the stochastic three-state agonist-binding channel of (3.127) Use $\alpha = 1$/ms, $\beta = 0.4$/ms, $k_{-1} = 0.5$/ms, and $ak_1 = 0.2$/ms. Using the data from simulations, plot the open time distribution and the closed time distribution and estimate the parameters of the model. Is the closed time distribution obviously a double exponential distribution? Repeat this experiment for several different values of ak_1.

21. Show that for an exponential distribution $\phi(t) = \alpha \exp(-\alpha t)$ the plot of the corresponding distribution function on the $\ln(t)$ scale has a maximum at the expected value of the distribution, $t = -\frac{1}{a}$.

22. Find the distribution for the length of a burst in the model of (3.127).
    Hint: The apparent length of the burst is the time taken to get from AR to T, minus the length of one sojourn in AT. Use the Laplace transform and the convolution theorem.

23. Find the mean first exit time from the piecewise-linear potential

$$U(x) = \begin{cases} -\frac{\Delta G x}{L}, & -L < x < 0, \\ \frac{\Delta G x}{L}, & 0 < x < L, \end{cases} \tag{3.194}$$

with a reflecting boundary at $x = -L$ and absorbing boundary at $x = L$.

24. Find the most likely event for the binomial distribution

$$P(h) = \frac{n!}{h!(n-h)!} p^h (1-p)^{(n-h)} \tag{3.195}$$

when n is large. Show that the probability of this event approaches 1 in the limit $n \to \infty$.

CHAPTER 4

Passive Electrical Flow in Neurons

Neurons are among the most important and interesting cells in the body. They are the fundamental building blocks of the central nervous system and hence responsible for motor control, cognition, perception, and memory, among other things. Although our understanding of how networks of neurons interact to form an intelligent system is extremely limited, one prerequisite for an understanding of the nervous system is an understanding of how individual nerve cells behave.

A typical neuron consists of three principal parts: the *dendrites*; the cell body, or *soma*; and the *axon*. The structure of some typical neurons is shown in Fig. 4.1. Dendrites are the input stage of a neuron and receive synaptic input from other neurons. The soma contains the necessary cellular machinery such as a nucleus and mitochondria, and the axon is the output stage. At the end of the axon (which may also be branched, as are the dendrites) are synapses, which are cellular junctions specialized for the transmission of an electrical signal (Chapter 8). Thus, a single neuron may receive input along its dendrites from a large number of other neurons, which is called *convergence*, and may similarly transmit a signal along its axon to many other neurons, called *divergence*.

The behaviors of the dendrites, axon, and synapses are all quite different. The spread of electrical current in a dendritic network is (mostly) a passive process that can be well described by the diffusion of electricity along a leaky cable. The axon, on the other hand, has an excitable membrane of the type described in Chapter 5, and thus can propagate an electrical signal actively. At the synapse (Chapter 8), the membrane is specialized for the release or reception of chemical neurotransmitters.

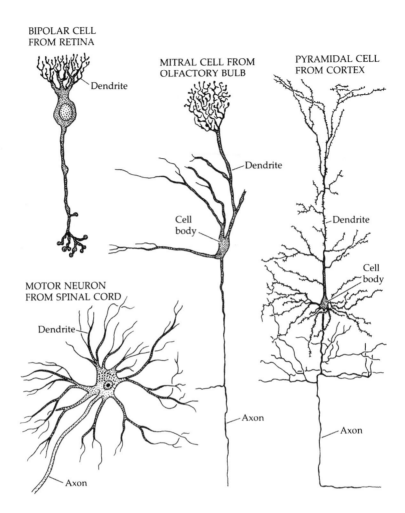

Figure 4.1 Structure of typical neurons. The motor neuron is from a mammalian spinal cord and was drawn by Dieters in 1869. The other cells were drawn by Ramón y Cajal. The pyramidal cell is from mouse cortex, and the mitral cell from the olfactory bulb of a cat. (Kuffler et al., 1984, Fig. 1, p. 10.)

In this chapter we discuss how to model the behavior of a cable, and then focus on the passive spread of current in a dendritic network; in the following chapter we show how an excitable membrane can actively propagate an electrical impulse, or action potential.

Although we discuss neurons in a number of chapters throughout this book, we cover them only in relatively little depth. For more comprehensive treatments of neurons and theoretical neuroscience the reader is referred to the excellent books by Jack et al. (1975), Koch and Segev (1998), Koch (1999), Dayan and Abbott (2001) and de Schutter (2000).

4.1   The Cable Equation

One of the first things to realize from the pictures in Fig. 4.1 is that it is unlikely that the membrane potential is the same at each point. In some cases spatial uniformity can be achieved experimentally (for example, by threading a silver wire along the axon, as did Hodgkin and Huxley), but *in vivo*, the intricate branched structure of the neuron can create spatial gradients in the membrane potential. Although this seems clear to us now, it was not until the pioneering work of Wilfrid Rall in the 1950s and 1960s that the importance of spatial effects gained widespread acceptance.

To understand something of how spatial distribution affects the behavior of a cable, we derive the *cable equation*. The theory of the flow of electricity in a leaky cable dates back to the work of Lord Kelvin in 1855, who derived the equations to study the transatlantic telegraph cable then under construction. However, the application of the cable equation to neuronal behavior is mainly due to Hodgkin and Rushton (1946), and then a series of classic papers by Rall (1957, 1959, 1960, 1969; an excellent summary of much of Rall's work on electrical flow in neurons is given in Segev et al., 1995.)

We view the cell as a long cylindrical piece of membrane surrounding an interior of cytoplasm (called a cable). We suppose that everywhere along its length, the potential depends only on the length variable and not on radial or angular variables, so that the cable can be viewed as one-dimensional. This assumption is called the *core conductor assumption* (Rall, 1977). We now divide the cable into a number of short pieces of isopotential membrane each of length dx. In any cable section, all currents must balance, and there are only two types of current, namely, transmembrane current and axial current (Fig. 4.2). The axial current has intracellular and extracellular components, both

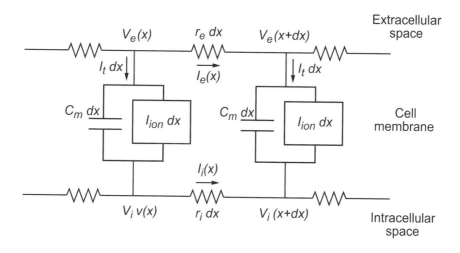

Figure 4.2   Schematic diagram of a discretized cable, with isopotential circuit elements of length dx.

of which are assumed to be ohmic, i.e., linear functions of the voltage. Hence,

$$V_i(x + dx) - V_i(x) = -I_i(x)r_i dx, \tag{4.1}$$

$$V_e(x + dx) - V_e(x) = -I_e(x)r_e dx, \tag{4.2}$$

where I_i and I_e are the intracellular and extracellular axial currents respectively. The minus sign on the right-hand side appears because of the convention that positive current is a flow of positive charges from left to right (i.e., in the direction of increasing x). If $V_i(x + dx) > V_i(x)$, then positive charges flow in the direction of decreasing x, giving a negative current. In the limit $dx \to 0$,

$$I_i = -\frac{1}{r_i}\frac{\partial V_i}{\partial x}, \tag{4.3}$$

$$I_e = -\frac{1}{r_e}\frac{\partial V_e}{\partial x}. \tag{4.4}$$

The numbers r_i and r_e are the resistances per unit length of the intracellular and extracellular media, respectively. In general,

$$r_i = \frac{R_c}{A_i}, \tag{4.5}$$

where R_c is the *cytoplasmic resistivity*, measured in units of Ohms-length, and A_i is the cross-sectional area of the cylindrical cable. A similar expression holds for the extracellular space, so if the cable is in a bath with large (effectively infinite) cross-sectional area, the extracellular resistance r_e is nearly zero.

Next, from Kirchhoff's laws, any change in extracellular or intracellular axial current must be due to a transmembrane current, and thus

$$I_i(x) - I_i(x + dx) = I_t dx = I_e(x + dx) - I_e(x), \tag{4.6}$$

where I_t is the total transmembrane current (positive outward) per unit length of membrane. In the limit as $dx \to 0$, this becomes

$$I_t = -\frac{\partial I_i}{\partial x} = \frac{\partial I_e}{\partial x}. \tag{4.7}$$

In a cable with no additional current sources, the total axial current is $I_T = I_i + I_e$, so using that $V = V_i - V_e$, we find

$$-I_T = \frac{r_i + r_e}{r_i r_e}\frac{\partial V_i}{\partial x} - \frac{1}{r_e}\frac{\partial V}{\partial x}, \tag{4.8}$$

from which it follows that

$$\frac{1}{r_i}\frac{\partial V_i}{\partial x} = \frac{1}{r_i + r_e}\frac{\partial V}{\partial x} - \frac{r_e}{r_i + r_e}I_T. \tag{4.9}$$

On substituting (4.9) into (4.7), we obtain

$$I_t = \frac{\partial}{\partial x}\left(\frac{1}{r_i + r_e}\frac{\partial V}{\partial x}\right), \tag{4.10}$$

where we have used (4.3) and the fact that I_T is constant. Finally, recall that the transmembrane current I_t is a sum of the capacitive and ionic currents, and thus

$$I_t = p \left(C_m \frac{\partial V}{\partial t} + I_{\text{ion}} \right) = \frac{\partial}{\partial x} \left(\frac{1}{r_i + r_e} \frac{\partial V}{\partial x} \right), \tag{4.11}$$

where p is the perimeter of the axon. Equation (4.11) is usually referred to as the cable equation. Note that C_m has units of capacitance per unit area of membrane, and I_{ion} has units of current per unit area of membrane. If a current I_{applied}, with units of current per unit area, is applied across the membrane (as before, taken positive in the outward direction), then the cable equation becomes

$$I_t = p \left(C_m \frac{\partial V}{\partial t} + I_{\text{ion}} + I_{\text{applied}} \right) = \frac{\partial}{\partial x} \left(\frac{1}{r_i + r_e} \frac{\partial V}{\partial x} \right). \tag{4.12}$$

It is useful to nondimensionalize the cable equation. To do so we define the *membrane resistivity* R_m as the resistance of a unit square area of membrane, having units of $\Omega\,\text{cm}^2$. For any fixed V_0, R_m is determined by measuring the change in membrane current when V is perturbed slightly from V_0. In mathematical terms,

$$\frac{1}{R_m} = \frac{dI_{\text{ion}}}{dV} \bigg|_{V=V_0}. \tag{4.13}$$

Although the value of R_m depends on the chosen value of V_0, it is typical to take V_0 to be the resting membrane potential to define R_m. Note that if the membrane is an ohmic resistor, then $I_{\text{ion}} = V/R_m$, in which case R_m is independent of V_0.

Assuming that r_i and r_e are constant, the cable equation (4.11) can now be written in the form

$$\tau_m \frac{\partial V}{\partial t} + R_m I_{\text{ion}} = \lambda_m^2 \frac{\partial^2 V}{\partial x^2}, \tag{4.14}$$

where

$$\lambda_m = \sqrt{\frac{R_m}{p(r_i + r_e)}} \tag{4.15}$$

has units of distance and is called the cable *space constant*, and where

$$\tau_m = R_m C_m \tag{4.16}$$

has units of time and is called the membrane *time constant*. If we ignore the extracellular resistance, then

$$\lambda_m = \sqrt{\frac{R_m d}{4 R_c}}, \tag{4.17}$$

where d is the diameter of the axon (assuming circular cross-section). Finally, we rescale the ionic current by defining $I_{\text{ion}} = -f(V,t)/R_m$ for some f, which, in general, is a function of both voltage and time and has units of voltage, and nondimensionalize

Table 4.1 Typical parameter values for a variety of excitable cells.

parameter	d	R_c	R_m	C_m	τ_m	λ_m
units	10^{-4} cm	Ω cm	$10^3\ \Omega$ cm^2	μF/cm^2	ms	cm
squid giant axon	500	30	1	1	1	0.65
lobster giant axon	75	60	2	1	2	0.25
crab giant axon	30	90	7	1	7	0.24
earthworm giant axon	105	200	12	0.3	3.6	0.4
marine worm giant axon	560	57	1.2	0.75	0.9	0.54
mammalian cardiac cell	20	150	7	1.2	8.4	0.15
barnacle muscle fiber	400	30	0.23	20	4.6	0.28

space and time by defining new variables $X = x/\lambda_m$ and $T = t/\tau_m$. In the new variables the cable equation is

$$\frac{\partial V}{\partial T} = \frac{\partial^2 V}{\partial X^2} + f(V, T). \tag{4.18}$$

Although f is written as a function of voltage and time, in many of the simpler versions of the cable equation, f is a function of V only (for example, (4.19) below). Typical parameter values for a variety of cells are shown in Table 4.1.

4.2 Dendritic Conduction

To complete the description of a spatially distributed cable, we must specify how the ionic current depends on voltage and time. In the squid giant axon, $f(V, t)$ is a function of m, n, h, and V as described in Chapter 5. This choice for f allows waves that propagate along the axon at constant speed and with a fixed profile. They require the input of energy from the axon, which must expend energy to maintain the necessary ionic concentrations, and thus they are often called *active waves*.

Any electrical activity for which the approximation $f = -V$ is valid (i.e., if the membrane is an Ohmic resistor) is said to be *passive* activity. There are some cables, primarily in neuronal dendritic networks, for which this is a good approximation in the range of normal activity. For other cells, activity is passive only if the membrane potential is sufficiently small. For simplicity in a passive cable, we shift V so that the resting potential is at $V = 0$. Thus,

$$\frac{\partial V}{\partial T} = \frac{\partial^2 V}{\partial X^2} - V, \tag{4.19}$$

which is called the *linear cable equation*. In the linear cable equation, current flows along the cable in a passive manner, leaking to the outside at a linear rate.

There is a vast literature on the application of the linear cable equation to dendritic networks. In particular, the books by Jack et al. (1975) and Tuckwell (1988) are largely devoted to this problem, and provide detailed discussions of the theory. Koch and Segev (1998) and Koch (1999) also provide excellent introductions.

4.2.1 Boundary Conditions

To determine the behavior of a single dendrite, we must first specify initial and boundary conditions. Usually, it is assumed that at time $T = 0$, the dendritic cable is in its resting state, $V = 0$, and so

$$V(X, 0) = 0. \tag{4.20}$$

Boundary conditions can be specified in a number of ways. Suppose that $X = X_b$ is a boundary point.

1. Voltage-clamp boundary conditions: If the voltage is fixed (i.e., clamped) at $X = X_b$, then the boundary condition is of Dirichlet type,

$$V(X_b, T) = V_b, \tag{4.21}$$

   where V_b is the specified voltage level.
2. Short circuit: If the ends of the cable are short-circuited, so that the extracellular and intracellular potentials are the same at $X = X_b$, then

$$V(X_b, T) = 0. \tag{4.22}$$

   This is a special case of the voltage clamp condition in which $V_b = 0$.
3. Current injection: Suppose a current $I(T)$ is injected at one end of the cable. Since

$$I_i = -\frac{1}{r_i}\frac{\partial V_i}{\partial x} = -\frac{1}{r_i \lambda_m}\frac{\partial V_i}{\partial X}, \tag{4.23}$$

   the boundary condition (ignoring extracellular resistance, so that the extracellular potential is uniform) is

$$\frac{\partial V(X_b, T)}{\partial X} = -r_i \lambda_m I(T). \tag{4.24}$$

   If X_b is at the left end, this corresponds to an inward current, while if it is on the right end, this is an outward current, if $I(T)$ is positive.
4. Sealed ends: If the end at $X = X_b$ is sealed to ensure that there is no current across the endpoint, then the boundary condition is the homogeneous Neumann condition,

$$\frac{\partial V(X_b, T)}{\partial X} = 0, \tag{4.25}$$

   a special case of an injected current for which $I(T) = 0$.

4.2.2 Input Resistance

One of the most important simple solutions of the cable equation corresponds to the situation in which a steady current is injected at one end of a semi-infinite cable. This is a common experimental protocol (although never with a truly semi-infinite cable) that can be used to determine the cable parameters R_m and R_c. Suppose the cable extends from $X = 0$ to $X = \infty$ and that a steady current I_0 is injected at $X = 0$. Then, the boundary condition at $X = 0$ is

$$\frac{dV(0)}{dX} = -r_i \lambda_m I_0. \tag{4.26}$$

Setting $\partial V/\partial T = 0$ and solving (4.19) subject to the boundary condition (4.19) gives

$$V(X) = \lambda_m r_i I_0 e^{-X} = V(0)e^{-X} = V(0)e^{-x/\lambda_m}. \tag{4.27}$$

Clearly, by measuring the rate at which the voltage decays along the cable, λ_m can be determined from experimental data. The *input resistance* R_{in} of the cable is defined to be the ratio $V(0)/I_0 = \lambda_m r_i$. Recall that when the extracellular resistance is ignored,

$$\lambda_m = \sqrt{\frac{R_m d}{4R_c}}. \tag{4.28}$$

Combining this with (4.5) gives

$$R_{\text{in}} = \lambda_m r_i = \sqrt{\frac{4R_m R_c}{\pi^2}} \frac{1}{d^{\frac{3}{2}}}. \tag{4.29}$$

Hence, the input resistance of the cable varies with the $-3/2$ power of the cable diameter, a fact that is of importance for the behavior of the cable equation in a branching structure. Since both the input resistance and the space constant of the cable can be measured experimentally, R_m and R_c can be calculated from experimental data.

Some solutions to the cable equation for various types of cable and boundary conditions are discussed in the exercises. Tuckwell (1988) gives a detailed discussion of the various types of solutions and how they are obtained.

4.2.3 Branching Structures

The property of neurons that is most obvious from Fig. 4.1 is that they are extensively branched. While the procedure to find solutions on a branched cable network is straightforward in concept, it can be quite messy in application. Thus, in what follows, we emphasize the procedure for obtaining the solution on branching structures, without calculating specific formulas.

The Steady-State Solution

It is useful first to consider the simplest branched cable, depicted in Fig. 4.3. The cable has a single branch point, or *node*, at $X = L_1$, and the two offspring branches extend

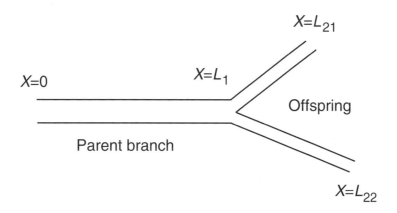

Figure 4.3 Diagram of the simplest possible branched cable.

to L_{21} and L_{22}, respectively. For convenience we express all lengths in nondimensional form, with the reminder that nondimensional length does not correspond to physical length, as the distance variable x along each branch of the cable is scaled by the length constant λ_m appropriate for that branch, and each branch may have a different length constant.

We construct the solution in three parts: V_1 on cylinder 1, and V_{21} and V_{22} on the two offspring cylinders. At steady state each V satisfies the differential equation $V'' = V$, and so we can immediately write the general solution as

$$V_1 = A_1 e^{-X} + B_1 e^X, \tag{4.30}$$

$$V_{21} = A_{21} e^{-X} + B_{21} e^X, \tag{4.31}$$

$$V_{22} = A_{22} e^{-X} + B_{22} e^X, \tag{4.32}$$

where the A's and B's are unknown constants. To determine the 6 unknown constants, we need 6 constraints, which come from the boundary and nodal conditions. For boundary conditions, we assume that a current I_0 is injected at $X = 0$ and that the terminal ends (at $X = L_{21}$ and $X = L_{22}$) are held fixed at $V = 0$. Thus,

$$\frac{dV_1(0)}{dX} = -r_i \lambda_m I_0, \tag{4.33}$$

$$V_{21}(L_{21}) = V_{22}(L_{22}) = 0. \tag{4.34}$$

The remaining three constraints come from conditions at the node. We require that V be a continuous function and that current be conserved at the node. It follows that

$$V_1(L_1) = V_{21}(L_1) = V_{22}(L_1), \tag{4.35}$$

and

$$d_1^{3/2} \sqrt{\frac{\pi^2}{4R_m R_c}} \frac{dV_1(L_1)}{dX} = d_{21}^{3/2} \sqrt{\frac{\pi^2}{4R_m R_c}} \frac{dV_{21}(L_1)}{dX} + d_{22}^{3/2} \sqrt{\frac{\pi^2}{4R_m R_c}} \frac{dV_{22}(L_1)}{dX}. \tag{4.36}$$

If we make the natural assumption that each branch of the cable has the same physical properties (and thus have the same R_m and R_c), although possibly differing in diameter, the final condition for conservation of current at the node becomes

$$d_1^{3/2}\frac{dV_1(L_1)}{dX} = d_{21}^{3/2}\frac{dV_{21}(L_1)}{dX} + d_{22}^{3/2}\frac{dV_{22}(L_1)}{dX}. \tag{4.37}$$

We thus have six linear equations for the six unknown constants; explicit solution of this linear system is left for Exercise 6.

More General Branching Structures

For this method to work for more general branching networks, there must be enough constraints to solve for the unknown constants. The following argument shows that this is the case. First, we know that each branch of the tree contributes two unknown constants, and thus, if there are N nodes, there are $1 + 2N$ individual cables with a total of $2 + 4N$ unknown constants. Each node contributes three constraints, and there are $2 + N$ terminal ends (including that at $X = 0$), each of which contributes one constraint, thus giving a grand total of $2 + 4N$ constraints. Thus, the resulting linear system is well-posed. Of course, a unique solution is guaranteed only if this system is invertible, which is not known a priori.

Equivalent Cylinders

One of the most important results in the theory of dendritic trees is due to Rall (1959), who showed that under certain conditions, the equations for passive electrical flow over a branching structure reduce to a single equation for electrical flow in a single cylinder, the so-called *equivalent cylinder*.

To see this reduction in a simple setting, consider again the branching structure of Fig. 4.3. To reduce this to an equivalent cylinder we need some additional assumptions. We assume, first, that the two offspring branches have the same dimensionless lengths, $L_{21} = L_{22}$, and that their terminals have the same boundary conditions. Since V_{21} and V_{22} obey the same differential equation on the same domain, obey the same boundary conditions at the terminals, and are equal at the node, it follows that they must be equal. That is,

$$\frac{dV_{21}(L_1)}{dX} = \frac{dV_{22}(L_1)}{dX}. \tag{4.38}$$

Substituting (4.38) into (4.37) we then get

$$d_1^{3/2}\frac{dV_1(L_1)}{dX} = (d_{21}^{3/2} + d_{22}^{3/2})\frac{dV_{21}(L_1)}{dX}. \tag{4.39}$$

Finally (and this is the crucial assumption), if we assume that

$$d_{21}^{3/2} + d_{22}^{3/2} = d_1^{3/2}, \tag{4.40}$$

then

$$\frac{dV_1(L_1)}{dX} = \frac{dV_{21}(L_1)}{dX}.$$

(4.41)

Thus V_1 and V_{21} have the same value and derivative at L_1 and obey the same differential equation. It follows that the composite function

$$V = \begin{cases} V_1(X), & 0 \le X \le L_1, \\ V_{21}(X), & L_1 \le X \le L_{21}, \end{cases}$$

(4.42)

is continuous with a continuous derivative on $0 < X < L_{21}$ and obeys the cable equation on that same interval. Thus, the simple branching structure is equivalent to a cable of length L_{21} and diameter d_1.

More generally, if the branching structure satisfies the following conditions:

1. R_m and R_c are the same for each branch of the cable;
2. At every node the cable diameters satisfy an equation analogous to (4.40). That is, if d_0 is the diameter of the parent branch, and $d_1, d_2, \ldots$ are the diameters of the offspring, then

$$d_0^{3/2} = d_1^{3/2} + d_2^{3/2} + \cdots;$$

(4.43)

3. The boundary conditions at the terminal ends are all the same;
4. Each terminal is the same dimensionless distance L from the origin of the tree (at $X = 0$);

then the entire tree is equivalent to a cylinder of length L and diameter d_1, where d_1 is the diameter of the cable at $X = 0$. Using an inductive argument, it is not difficult to show that this is so (although a rigorous proof is complicated by the notation). Working from the terminal ends, one can condense the outermost branches into equivalent cylinders, then work progressively inwards, condensing the equivalent cylinders into other equivalent cylinders, and so on, until only a single cylinder remains. It is left as an exercise (Exercise 7) to show that during this process the requirements for condensing branches into an equivalent cylinder are never violated.

4.2.4   A Dendrite with Synaptic Input

Suppose we have a dendrite with a time-dependent synaptic input at some point along the dendrite. Then the potential along the cable satisfies the equation

$$\frac{\partial V}{\partial T} = \frac{\partial^2 V}{\partial X^2} - V + g(T)\delta(X - X_s)(V_e - V),$$

(4.44)

with $V_x = 0$ at both ends of the cable $x = 0, L$ (assuming the ends of the cable are sealed). (For a derivation of the form of the synaptic input, see Chapter 8.)

There are two questions one might ask. First, one might want to know the voltage at the end of the cable with a given input function $g(t)$. However, it is more likely that the voltage at the ends of the cable can be measured so it is the input function $g(t)$ and its location X_s that one would like to determine. This latter is the question we address here.

We suppose that $V(0,t) = V_0(t)$ and $V(L,t) = V_1(t)$ are known. Notice that we can integrate the governing equation with respect to time and find that

$$V_{0XX} - V_0 = 0, \qquad (4.45)$$

provided $X \neq X_s$, where $V_0(x) = \int_{-\infty}^{\infty} V(X,T)\,dX$, and V_0 must satisfy boundary conditions $V_0(0) = V_0^0$, $V_0(L) = V_0^1$, where $V_0^j = \int_{-\infty}^{\infty} V_j(T)\,dT$, $j = 0,1$. It follows that

$$V_0(X) = \begin{cases} V_0^0 \cosh(x), & X < X_s, \\ V_0^1 \cosh(L - x), & X > X_s. \end{cases} \qquad (4.46)$$

Since $V_0(X)$ must be continuous at $X = X_s$, it must be that

$$\frac{V_0^0}{V_0^1} = \frac{\cosh(L - X_s)}{\cosh(X_s)} = F(X_s). \qquad (4.47)$$

The function $F(X_s)$ is a monotone decreasing function of X_s, so there is at most one value of X_s which satisfies (4.47).

Next, notice that integrating (4.44) across $X = X_s$ gives the jump condition

$$V_X|_{X_s^-}^{X_s^+} = g(T)(V(X_s) - V_e), \qquad (4.48)$$

so that $g(T)$ is determined from

$$g(T) = \frac{V_X(X_s^+) - V_X(X_s^-)}{V(X_s) - V_e}. \qquad (4.49)$$

Now, the Fourier transform of V is

$$\hat{V}(x, \omega) = \int_{-\infty}^{\infty} V(X,T)e^{-i\omega T}\,dT, \qquad (4.50)$$

and the Fourier transformed equation is

$$\hat{V}_{XX} - (1 - i\omega)\hat{V} = 0, \qquad (4.51)$$

for $X \neq X_s$. It follows that

$$\hat{V}(X) = \begin{cases} \hat{V}_0(\omega) \cosh(\mu(\omega)x), & X < X_s, \\ \hat{V}_0(\omega) \cosh(\mu(\omega)(L - x)), & X > X_s, \end{cases} \qquad (4.52)$$

where $\mu^2(\omega) = 1 - i\omega$. We now calculate $V(x)$ using the inverse Fourier transform and find

$$V(X_s) = \frac{1}{2\pi} \int_{-\infty}^{\infty} \hat{V}_0(\omega) \cosh(\mu(\omega)X_s)e^{i\omega T}\,d\omega, \qquad (4.53)$$

$$V_X(X_s^-) = \frac{1}{2\pi} \int_{-\infty}^{\infty} \hat{V}_0(\omega)\mu(\omega) \sinh(\mu(\omega)X_s)e^{i\omega T}\,d\omega, \qquad (4.54)$$

and

$$V(X_s^+) = \frac{1}{2\pi} \int_{-\infty}^{\infty} \hat{V}_0(\omega)\mu(\omega) \sinh(\mu(\omega)(X_s - L))e^{i\omega T}\, d\omega. \tag{4.55}$$

These combined with (4.49), uniquely determine $g(T)$.

This calculation and its extension to multiple synaptic inputs is due to Cox (2004).

4.3   The Rall Model of a Neuron

When studying a model of a neuron, the item of greatest interest is often the voltage at the cell body, or soma. This is primarily because the voltage at the cell body can be measured experimentally with greater ease than can the voltage in the dendritic network, and further, it is the voltage at the soma that determines whether or not the neuron fires an action potential. Therefore, it is important to determine the solution of the cable equation on a dendritic network when one end of the network is connected to a soma. The most common approach to incorporating a soma into the model is due to Rall (1960), and is called the *Rall lumped-soma model*.

The three basic assumptions of the Rall model are, first, that the soma is isopotential (i.e., that the soma membrane potential is the same at all points), second, that the soma acts like a resistance (R_s) and a capacitance (C_s) in parallel, and, third, that the dendritic network can be collapsed into a single equivalent cylinder. This is illustrated in Fig. 4.4.

The potential V satisfies the cable equation on the equivalent cylinder. The boundary condition must account for current flow within the soma and into the cable. Thus, if I_0 denotes an applied current at $X = 0$, then the boundary condition is

$$I_0 = -\frac{1}{r_i} \frac{\partial V(0,t)}{\partial x} + C_s \frac{\partial V(0,t)}{\partial t} + \frac{V(0,t)}{R_s}, \tag{4.56}$$

so that

$$R_s I_0 = -\gamma \frac{\partial V(0,T)}{\partial X} + \sigma \frac{\partial V(0,T)}{\partial T} + V(0,T), \tag{4.57}$$

where $\sigma = C_s R_s/\tau_m = \tau_s/\tau_m$ and $\gamma = R_s/(r_i\lambda_m)$. For convenience we assume that the time constant of the soma is the same as the membrane time constant, so that $\sigma = 1$.

4.3.1   A Semi-Infinite Neuron with a Soma

We first calculate the steady response of a semi-infinite neuron to a current I_0 injected at $X = 0$, as in Section 4.2.2. As before, we set the time derivative to zero to get

$$\frac{d^2V}{dX^2} = V, \tag{4.58}$$

$$V(0) - \gamma \frac{dV(0)}{dX} = R_s I_0, \tag{4.59}$$

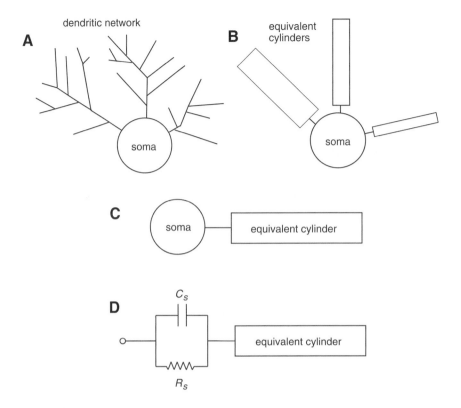

Figure 4.4 Schematic diagram of the Rall lumped-soma model of the neuron. First, it is assumed that the dendritic network pictured in A is equivalent to the equivalent cylinders shown in B, and that these cylinders are themselves equivalent to a single cylinder as in C. The soma is assumed to be isopotential and to behave like a resistance and capacitance in parallel, as in D.

which can easily be solved to give

$$V(X) = \frac{R_s}{r_i\lambda_m + R_s} r_i\lambda_m I_0 e^{-X}. \tag{4.60}$$

This solution is nearly the same as the steady response of the equivalent cylinder without a soma to an injected current, except that V is decreased by the constant factor $R_s/(R_s + r_i\lambda_m) < 1$. As $R_s \to \infty$, in which limit the soma carries no current, the solution to the lumped-soma model approaches the solution to the simple cable.

The input resistance R_{in} of the lumped-soma model is

$$R_{\text{in}} = \frac{V(0)}{I_0} = \frac{r_i\lambda_m R_s}{r_i\lambda_m + R_s}, \tag{4.61}$$

and thus

$$\frac{1}{R_{\text{in}}} = \frac{1}{r_i\lambda_m} + \frac{1}{R_s}. \tag{4.62}$$

Since $r_i \lambda_m$ is the input resistance of the cylinder, the input conductance of the lumped-soma model is the sum of the input conductance of the soma and the input conductance of the cylinder. This is as expected, since the equivalent cylinder and the soma are in parallel.

4.3.2   A Finite Neuron and Soma

We now calculate the time-dependent response of a finite cable and lumped soma to a delta function current input at the soma, as this is readily observed experimentally.

   We assume that the equivalent cylinder has finite length L. Then the potential satisfies

$$\frac{\partial V}{\partial T} = \frac{\partial^2 V}{\partial X^2} - V, \qquad 0 < X < L, T > 0, \tag{4.63}$$

$$[6bp]V(X,0) = 0, \tag{4.64}$$

with boundary conditions

$$\frac{\partial V(L,T)}{\partial X} = 0, \tag{4.65}$$

$$[6bp]\frac{\partial V(0,T)}{\partial T} + V(0,T) - \gamma \frac{\partial V(0,T)}{\partial X} = R_s \delta(T). \tag{4.66}$$

Note that the boundary condition (4.66) is equivalent to

$$\frac{\partial V(0,T)}{\partial T} + V(0,T) - \gamma \frac{\partial V(0,T)}{\partial X} = 0, \qquad T > 0, \tag{4.67}$$

together with the initial condition

$$V(0,0) = R_s. \tag{4.68}$$

   We begin by seeking a generalized Fourier series expansion of the solution. Using separation of variables, we find solutions of the form

$$V(X,T) = \phi(X)e^{-\mu^2 T}, \tag{4.69}$$

where ϕ satisfies the differential equation

$$\phi'' - (1 - \mu^2)\phi = 0, \tag{4.70}$$

with boundary conditions (when $T > 0$)

$$\phi'(L) = 0, \tag{4.71}$$

$$\phi'(0) = \phi(0)\frac{1 - \mu^2}{\gamma}. \tag{4.72}$$

The solution of (4.70) is

$$\phi = A\cos(\lambda X) + B\sin(\lambda X), \tag{4.73}$$

for some constants A and B and $\lambda^2 = \mu^2 - 1$, and applying the boundary conditions, we find

$$B = \frac{-\lambda A}{\gamma} \qquad (4.74)$$

and

$$\tan(\lambda L) = -\frac{\lambda}{\gamma}. \qquad (4.75)$$

The roots of (4.75) determine the eigenvalues. Although the eigenvalues cannot be found analytically, they can be determined numerically. A graph of the left- and right-hand sides of (4.75), showing the location of the eigenvalues as intersections of these curves, is given in Fig. 4.5. There is an infinite number of discrete eigenvalues, labeled λ_n, with $\lambda_0 = 0$. Expecting the full solution to be a linear combination of the eigenfunctions, we write

$$V(X,T) = \sum_{n=0}^{\infty} A_n \phi_n(X) \exp(-(1 + \lambda_n^2)T), \qquad (4.76)$$

where

$$\phi_n(X) = \cos(\lambda_n X) - \frac{\lambda_n}{\gamma} \sin(\lambda_n X). \qquad (4.77)$$

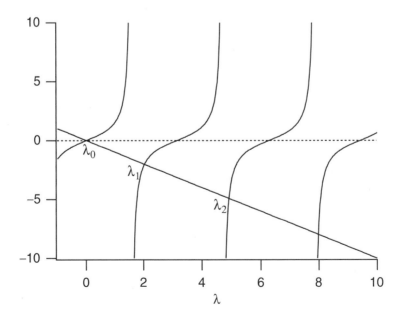

Figure 4.5 The eigenvalues of (4.63)–(4.66) are determined by the intersections of the curves $\tan(\lambda L)$ and $-\lambda/\gamma$. In this figure, $L = \gamma = 1$.

Note that if λ_n is an eigenvalue, so also is $-\lambda_n$, but the eigenfunction $\phi_n(X) = \cos(\lambda_n X) - \frac{\lambda_n}{\gamma}\sin(\lambda_n X)$ is an even function of λ_n, so it suffices to include only positive eigenvalues in the expansion.

We now strike a problem. The standard procedure is to expand the initial condition in terms of the eigenfunctions and thereby determine the coefficients A_n. However, the eigenfunctions are not mutually orthogonal in the usual way, so the standard approach fails. One way around this is to construct a nonorthogonal expansion of the initial condition, an approach used by Durand (1984). A different approach (Bluman and Tuckwell, 1987; Tuckwell, 1988) is to calculate the Laplace transform of the solution, and then, by matching the two forms of the solution, obtain expressions for the unknown coefficients.

Here we present a different approach. The standard Fourier series approach can be rescued by introducing a slightly different operator and inner product (Keener, 1998). Suppose we consider a Hilbert space of vectors of the form

$$U = \begin{pmatrix} u(x) \\ \alpha \end{pmatrix}, \tag{4.78}$$

where α is a real scalar, with the inner product

$$\langle U, V \rangle = \int_0^L u(x)v(x)\,dx + \gamma\alpha\beta, \tag{4.79}$$

where $V = \begin{pmatrix} v(x) \\ \beta \end{pmatrix}$. The differential operator L on this space is defined by

$$LU = \begin{pmatrix} u''(x) \\ u'(0) \end{pmatrix}, \tag{4.80}$$

with the boundary conditions $u'(L) = 0$ and $u(0) = \gamma\alpha$. Now, the usual calculation shows that the operator L is a self-adjoint operator with the inner product defined by (4.79). Furthermore, the eigenfunctions of L, $L\Phi = \kappa\Phi$ satisfy the two equations $\phi''(x) = \kappa\phi(x)$ and $\gamma\phi'(0) = \kappa\phi(0)$, which are exactly the equations (4.70) and (4.72) with $\kappa = 1 - \mu^2$. It follows immediately that the eigenfunctions of the operator L are orthogonal and complete on this Hilbert space and are given by

$$\Phi_n(X) = \begin{pmatrix} \phi_n(X) \\ \frac{\phi_n(0)}{\gamma} \end{pmatrix}. \tag{4.81}$$

Furthermore, the solution of the full problem can be written in this Hilbert space as

$$\begin{pmatrix} V(X,T) \\ \frac{1}{\gamma}V(0,T) \end{pmatrix} = \sum_{n=0}^{\infty} A_n\Phi_n(X)\exp(-(1+\lambda_n^2)T), \tag{4.82}$$

and the coefficients are found by requiring

$$\begin{pmatrix} V(X,0) \\ \frac{1}{\gamma}V(0,0) \end{pmatrix} = \sum_{n=0}^{\infty} A_n \Phi_n(X) = \begin{pmatrix} 0 \\ \frac{R_s}{\gamma} \end{pmatrix}. \tag{4.83}$$

The coefficients A_n are now found by taking the inner product of (4.83) with $\Phi_n(X)$ with the result

$$A_n = \frac{R_s}{\gamma \langle \Phi_n, \Phi_n \rangle}, \tag{4.84}$$

where

$$\gamma \langle \Phi_n, \Phi_n \rangle = \gamma \int_0^L \left(\cos(\lambda_n X) - \frac{\lambda_n}{\gamma} \sin(\lambda_n X) \right)^2 dX + 1$$
$$= \frac{1}{2} + \frac{L}{2\gamma}(\gamma^2 + \mu_n^2). \tag{4.85}$$

4.3.3 Other Compartmental Models

The methods presented above give some idea of the difficulty of calculating analytical solutions to the cable equation on branching structures, with or without a soma termination. Since modern experimental techniques can determine the detailed structure of a neuron (for example, by staining with horseradish peroxidase), it is clear that more experimental information can be obtained than can be incorporated into an analytical model (as is nearly always the case). Thus, one common approach is to construct a large computational model of a neuron and then determine the solution by a numerical method.

In a numerical approach, a neuron is divided into a large number of small pieces, or compartments, each of which is assumed to be isopotential. Within each compartment the properties of the neuronal membrane are specified, and thus some compartments may have excitable kinetics, while others are purely passive. The compartments are then connected by an axial resistance, resulting in a large system of coupled ordinary differential equations, with the voltage specified at discrete places along the neuron.

Compartmental models, numerical methods for their solution, and software packages used for these kinds of models are discussed in detail in Koch and Segev (1998) and de Schutter (2000), to which the interested reader is referred.

4.4 Appendix: Transform Methods

To follow all of the calculations and complete all the exercises in this chapter, you will need to know about Fourier and Laplace transforms, generalized functions and the delta function, Green's functions, as well as some aspects of complex variable theory, including contour integration and the residue theorem. If you have made it this far into this book, then you are probably familiar with these classic techniques. However, should you need a reference for these techniques, there are many books with the generic title "Advanced Engineering Mathematics," from which to choose (see, for example, Kreyszig (1994), O'Neill (1983), or Kaplan (1981)). At an intermediate level one might consider Strang (1986) or Boyce and DiPrima (1997). Keener (1998) provides a more advanced coverage of this material.

4.5 EXERCISES

1. Calculate the input resistance of a cable with a sealed end at $X = L$. Determine how the length of the cable, and the boundary condition at $X = L$, affects the input resistance, and compare to the result for a semi-infinite cable.

2. (a) Find the fundamental solution K of the linear cable equation satisfying

$$-\frac{d^2K}{dX^2} + K = \delta(X - \xi), \quad -\infty < X < \infty, \tag{4.86}$$

   where $\delta(X - \xi)$ denotes an *inward* flow of positive current at the point $X = \xi$.

   (b) Use the fundamental solution to construct a solution of the cable equation with inhomogeneous current input

$$-\frac{d^2V}{dX^2} + V = I(X). \tag{4.87}$$

3. Solve

$$-\frac{d^2G(X)}{dx^2} + G(X) = \delta(X - \xi), \quad 0 < X, \xi < L, \tag{4.88}$$

   subject to (i) sealed end, and (ii) short circuit, boundary conditions.

4. (a) Use Laplace transforms to find the solution of the semi-infinite (time-dependent) cable equation with clamped voltage

$$V(X, 0) = 0 \tag{4.89}$$

   and current input

$$\frac{\partial V(0, T)}{\partial X} = -r_i \lambda_m I_0 H(T), \tag{4.90}$$

   where H is the Heaviside function.
   Hint: Use the identity

$$\frac{2}{s\sqrt{s+1}} = \frac{1}{s+1-\sqrt{s+1}} - \frac{1}{s+1+\sqrt{s+1}}, \tag{4.91}$$

   and then use

$$\mathcal{L}^{-1}\left\{\frac{e^{-a\sqrt{s}}}{s+b\sqrt{s}}\right\} = e^{b^2T+ab}\,\text{erfc}\left(\frac{a}{2\sqrt{T}} + b\sqrt{T}\right), \tag{4.92}$$

   where $\mathcal{L}^{-1}$ denotes the inverse Laplace transform.

   (b) Show that

$$V(X, T) \to r_i \lambda_m I_0 e^{-X} \tag{4.93}$$

   as $T \to \infty$.

5. Calculate the time-dependent Green's function for a finite cylinder of length L for (i) sealed end, and (ii) short circuit, boundary conditions. These may be calculated in two different ways, either using Fourier series or by constructing sums of fundamental solutions.

6. By solving for the unknown constants, calculate the solution of the cable equation on the simple branching structure of Fig. 4.3. Show explicitly that this solution is the same as the equivalent cylinder solution, provided the necessary conditions are satisfied.

7. Show that if the conditions in Section 4.2.3 are satisfied, a branching structure can be condensed into a single equivalent cylinder.

8. Show that

$$\frac{\partial V}{\partial T} = \frac{\partial^2 V}{\partial X^2} - V + \delta(X)\delta(T), \qquad T \geq 0, \tag{4.94}$$

with $V(X,T) = 0$ for $T < 0$, is equivalent to

$$\frac{\partial V}{\partial T} = \frac{\partial^2 V}{\partial X^2} - V, \qquad T > 0, \tag{4.95}$$

with initial condition

$$V(X,0) = \delta(X). \tag{4.96}$$

9. Find the numerical solution of (4.44) with $g(T) = Te^{-aT}$. From this determine $V_0(T)$ and $V_1(T)$ and use (4.47) to determine X_s. How does the computed value of X_s compare with the original value?

10. Show that as $n \to \infty$, the eigenvalues λ_n of (4.75) are approximately $(2n-1)\pi/(2L)$.

11. Using the method of Section 4.3.2, find the Green's function for the finite cylinder and lumped soma; i.e., solve

$$\frac{\partial V}{\partial T} = \frac{\partial^2 V}{\partial X^2} - V + \delta(X - \xi)\delta(T), \tag{4.97}$$

$$V(X,0) = 0, \tag{4.98}$$

with boundary conditions

$$\frac{\partial V(L,T)}{\partial X} = 0, \tag{4.99}$$

$$\frac{\partial V(0,T)}{\partial T} + V(0,T) - \gamma\frac{\partial V(0,T)}{\partial X} = 0. \tag{4.100}$$

Show that as $\xi \to 0$ the solution approaches that found in Section 4.3.2, scaled by the factor γ/R_s.

CHAPTER 5

Excitability

We have seen in previous chapters that the control of cell volume results in a potential difference across the cell membrane, and that this potential difference causes ionic currents to flow through channels in the cell membrane. Regulation of this membrane potential by control of the ionic channels is one of the most important cellular functions. Many cells, such as neurons and muscle cells, use the membrane potential as a signal, and thus the operation of the nervous system and muscle contraction (to name but two examples) are both dependent on the generation and propagation of electrical signals.

To understand electrical signaling in cells, it is helpful (and not too inaccurate) to divide all cells into two groups: excitable cells and nonexcitable cells. Many cells maintain a stable equilibrium potential. For some, if currents are applied to the cell for a short period of time, the potential returns directly to its equilibrium value after the applied current is removed. Such cells are nonexcitable, typical examples of which are the epithelial cells that line the walls of the gut. Photoreceptors (Chapter 19) are also nonexcitable, although in their case, membrane potential plays an extremely important signaling role nonetheless.

However, there are cells for which, if the applied current is sufficiently strong, the membrane potential goes through a large excursion, called an *action potential*, before eventually returning to rest. Such cells are called *excitable*. Excitable cells include cardiac cells, smooth and skeletal muscle cells, some secretory cells, and most neurons. The most obvious advantage of excitability is that an excitable cell either responds in full to a stimulus or not at all, and thus a stimulus of sufficient amplitude may be reliably distinguished from background noise. In this way, noise is filtered out, and a signal is reliably transmitted.

There are many examples of excitability that occur in nature. A simple example of an excitable system is a household match. The chemical components of the match head are stable to small fluctuations in temperature, but a sufficiently large temperature fluctuation, caused, for example, by friction between the head and a rough surface, triggers the abrupt oxidation of these chemicals with a dramatic release of heat and light. The fuse of a stick of dynamite is a one-dimensional continuous version of an excitable medium, and a field of dry grass is its two-dimensional version. Both of these spatially extended systems admit the possibility of wave propagation (Chapter 6). The field of grass has one additional feature that the match and dynamite fuse fail to have, and that is recovery. While it is not very rapid by physiological standards, given a few months of growth, a burned-over field of grass will regrow enough fuel so that another fire may spread across it.

Although the generation and propagation of signals have been extensively studied by physiologists for at least the past 100 years, the most important landmark in these studies is the work of Alan Hodgkin and Andrew Huxley, who developed the first quantitative model of the propagation of an electrical signal along a squid giant axon (deemed "giant" because of the size of the axon, *not* the size of the squid). Their model was originally used to explain the action potential in the long giant axon of a squid nerve cell, but the ideas have since been extended and applied to a wide variety of excitable cells. Hodgkin–Huxley theory is remarkable, not only for its influence on electrophysiology, but also for its influence, after some filtering, on applied mathematics. FitzHugh (in particular) showed how the essentials of the excitable process could be distilled into a simpler model on which mathematical analysis could make some progress. Because this simplified model turned out to be of such great theoretical interest, it contributed enormously to the formation of a new field of applied mathematics, the study of excitable systems, a field that continues to stimulate a vast amount of research.

Because of the central importance of cellular electrical activity in physiology, because of the importance of the Hodgkin–Huxley equations in the study of electrical activity, and because it forms the basis for the study of excitability, it is no exaggeration to say that the Hodgkin–Huxley equations are the most important model in all of the physiological literature.

5.1   The Hodgkin–Huxley Model

In Chapter 2 we described how the cell membrane can be modeled as a capacitor in parallel with an ionic current, resulting in the equation

$$C_m \frac{dV}{dt} + I_{\text{ion}}(V, t) = 0, \tag{5.1}$$

where V, as usual, denotes the internal minus the external potential ($V = V_i - V_e$). In the squid giant axon, as in many neural cells, the principal ionic currents are the

Figure 5.1 The infamous giant squid (or even octopus, if you wish to be pedantic), having nothing to do with the work of Hodgkin and Huxley on squid giant axon. From *Dangerous Sea Creatures*, © 1976, 1977 Time-Life Films, Inc.

Na$^+$ current and the K$^+$ current. Although there are other ionic currents, primarily the Cl$^-$ current, in the Hodgkin–Huxley theory they are small and lumped together into one current called the *leakage current*. Since the instantaneous *I–V* curves of open Na$^+$ and K$^+$ channels in the squid giant axon are approximately linear, (5.1) becomes

$$C_m \frac{dV}{dt} = -g_{Na}(V - V_{Na}) - g_K(V - V_K) - g_L(V - V_L) + I_{app}, \tag{5.2}$$

where I_{app} is the applied current. During an action potential there is a measured influx of 3.7 pmoles/cm^2 of Na$^+$ and a subsequent efflux of 4.3 pmoles/cm^2 of K$^+$. These amounts are so small that it is realistic to assume that the ionic concentrations, and hence the equilibrium potentials, are constant and unaffected by an action potential. It is important to emphasize that the choice of linear I–V curves for the three different channel types is dictated largely by experimental data. Axons in other species (such as vertebrates) have ionic channels that are better described by other I–V curves, such as the GHK current equation (2.123). However, the qualitative nature of the results remains largely unaffected, and so the discussion in this chapter, which is mostly of a qualitative nature, remains correct for models that use more complex I–V curves to describe the ionic currents.

Equation (5.2) is a first-order ordinary differential equation and can be written in the form

$$C_m \frac{dV}{dt} = -g_{\text{eff}}(V - V_{\text{eq}}) + I_{\text{app}}, \tag{5.3}$$

where $g_{\text{eff}} = g_{\text{Na}} + g_{\text{K}} + g_{\text{L}}$ and $V_{\text{eq}} = (g_{\text{Na}}V_{\text{Na}} + g_{\text{K}}V_{\text{K}} + g_{\text{L}}V_{\text{L}})/g_{\text{eff}}$. V_{eq} is the membrane resting potential and is a balance between the reversal potentials for the three ionic currents. In fact, at rest, the Na^+ and leakage conductances are small compared to the K^+ conductance, so that the resting potential is close to the K^+ equilibrium potential.

The quantity $R_m = 1/g_{\text{eff}}$, the passive membrane resistance, is on the order of 1000 $\Omega \, \text{cm}^2$. The time constant for this equation is

$$\tau_m = C_m R_m, \tag{5.4}$$

on the order of 1 msec. It follows that, with a steady applied current, the membrane potential should equilibrate quickly to

$$V = V_{\text{eq}} + R_m I_{\text{app}}. \tag{5.5}$$

For sufficiently small applied currents this is indeed what happens. However, for larger applied currents the response is quite different. Assuming that the model (5.2) is correct, the only possible explanation for these differences is that the conductances are not constant but depend in some way on the voltage. Historically, the key step to determining the conductances was being able to measure the individual ionic currents and from this to deduce the changes in conductances. This was brilliantly accomplished by Hodgkin and Huxley in 1952.

5.1.1 History of the Hodgkin–Huxley Equations

(This section is adapted from Rinzel, 1990.) In a series of five articles that appeared in the *Journal of Physiology* in 1952, Alan Lloyd Hodgkin and Andrew Fielding Huxley, along with Bernard Katz, who was a coauthor of the lead paper and a collaborator in several related studies, unraveled the dynamic ionic conductances that generate the nerve action potential (Hodgkin et al., 1952; Hodgkin and Huxley, 1952a,b,c,d). They were awarded the 1963 Nobel Prize in Physiology or Medicine (shared with John C. Eccles, for his work on potentials and conductances at motorneuron synapses).

Before about 1939, the membrane potential was believed to play an important role in the membrane's state, but there was no way to measure it. It was known that a cell's membrane separated different ionic concentrations inside and outside the cell. Applying the Nernst equation, Bernstein (1902) was led to suggest that the resting membrane was semipermeable to K^+, implying that, at rest, V should be around -70 mV. He believed that during activity there was a breakdown in the membrane's resistance to all ionic fluxes, and potential differences would disappear, i.e., V would approach zero.

In 1940, Cole and Curtis, using careful electrode placement coupled with biophysical and mathematical analysis, obtained the first convincing evidence for a substantial transient increase in membrane conductivity during passage of the action potential. While they estimated a large conductance increase, it was not infinite, so without a direct measurement of membrane potential it was not possible to confirm or nullify Bernstein's hypothesis. During a postdoctoral year in the U.S. in 1937–1938, Hodgkin established connections with Cole's group at Columbia and worked with them at Woods Hole in the summer. He and Curtis nearly succeeded in measuring V directly by tunneling along the giant axon with a glass micropipette. When each succeeded later (separately, with other collaborators), they found, surprisingly, that V rose transiently toward zero, but with a substantial overshoot. This finding brought into serious question the hypothesis of Bernstein and provided much food for thought during World War II, when Hodgkin, Huxley, and many other scientists were involved in the war effort.

By the time postwar experimental work was resuming in England, Cole and Marmont had developed the *space-clamp technique*. This method allowed one to measure directly the total transmembrane current, uniform through a known area, rather than spatially nonuniform as generated by a capillary electrode. To achieve current control with space clamping, the axon was threaded with a metallic conductor (like a thin silver wire) to provide low axial resistance and thereby eliminate voltage gradients along the length of the axon. Under these conditions the membrane potential is no longer a function of distance along the axon, only of time. In addition, during the 1947 squid season, Cole and company made substantial progress toward controlling the membrane potential as well.

In 1948, Hodgkin went to visit Cole (then at Chicago) to learn directly of their methods. With some further developments of their own, Hodgkin, Huxley, and Katz applied the techniques with great success to record transient ionic fluxes over the physiological ranges of voltages. Working diligently, they collected most of the data for their papers in the summer of 1949. Next came the step of identifying the individual contributions of the different ion species. Explicit evidence that both Na^+ and K^+ were important came from the work of Hodgkin and Katz (1949). This also explained the earlier puzzling observations that V overshoots zero during an action potential, opposing the suggestion of Bernstein. Instead of supposing that there was a transient increase in permeability identical for all ions, Hodgkin and Katz realized that different changes in permeabilities for different ions could account for the V time course, as V would approach the Nernst potential for the ion to which the membrane was predominantly permeable, and this dominance could change with time. For example, at rest the membrane is most permeable to K^+, so that V is close to V_K. However, if g_K were to decrease and g_{Na} were to increase, then V would be pushed toward V_{Na}, which is positive, thus depolarizing the cell.

The question of how the changes in permeability were dynamically linked to V was not completely stated until the papers of 1952. In fact, the substantial delay from data collection in 1949 until final publication in 1952 can be attributed to the considerable time devoted to data analysis, model formulation, and testing. Computer downtime was also a factor, as some of the solutions of the Hodgkin–Huxley equations were computed on a desktop, hand-cranked calculator. As Hodgkin notes, "The propagated action potential took about three weeks to complete and must have been an enormous labour for Andrew [Huxley]" (Hodgkin, 1976, p. 19).

The final paper of the 1952 series is a masterpiece of the scientific art. Therein they present their elegant experimental data, a comprehensive theoretical hypothesis, a fit of the model to the experimental data (obtained for fixed values of the membrane potential), and then, presto, a prediction (from their numerical computations) of the time course of the propagated action potential. In biology, where quantitatively predictive theories are rare, this work stands out as one of the most successful combinations of experiment and theory.

5.1.2   Voltage and Time Dependence of Conductances

The key step to sorting out the dynamics of the conductances came from the development of the *voltage clamp*. A voltage clamp fixes the membrane potential, usually by a rapid step from one voltage to another, and then measures the current that must be supplied in order to hold the voltage constant. Since the supplied current must equal the transmembrane current, the voltage clamp provides a way to measure the transient transmembrane current that results. The crucial point is that the voltage can be stepped from one constant level to another, and so the ionic currents can be measured at a constant, known, voltage. Thus, even when the conductances are functions of the voltage (as is actually the case), a voltage clamp eliminates any voltage changes and permits measurement of the conductances as functions of time only.

Hodgkin and Huxley found that when the voltage was stepped up and held fixed at a higher level, the total ionic current was initially inward, but at later times an outward current developed (Fig. 5.2). For a number of reasons, not discussed here, they argued that the initial inward current is carried almost entirely by Na^+, while the outward current that develops later is carried largely by K^+. With these assumptions, Hodgkin and Huxley were able to use a clever trick to separate the total ionic current into its constituent ionic parts. They replaced 90% of the extracellular Na^+ in the normal seawater bath with choline (a viscous liquid vitamin B complex found in many animal and vegetable tissues), which rendered the axon nonexcitable but changed the resting potential only slightly. Since it is assumed that immediately after the voltage has been stepped up, the ionic current is all carried by Na^+, it is possible to measure

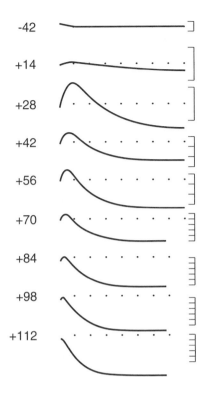

-42

+14

+28

+42

+56

+70

+84

+98

+112

Figure 5.2 Experimental results describing the total membrane current in response to a step depolarization. The numbers on the left give the final value of the membrane potential, in mV. The interval between dots on the horizontal scale is 1 ms, while one division on the vertical scale represents 0.5 mA/cm^2. (Hodgkin and Huxley, 1952a, Fig. 2a.)

the initial Na$^+$ currents in response to a voltage step. Note that although the Na$^+$ currents can be measured directly immediately after the voltage step, they cannot be measured directly over a longer time period, as the total ionic current begins to include a contribution from the K$^+$ current. If we denote the Na$^+$ currents for the two cases of normal extracellular Na$^+$ and zero extracellular Na$^+$ by I_{Na}^1 and I_{Na}^2 respectively, then the ratio of the two currents,

$$I_{Na}^1/I_{Na}^2 = K, \tag{5.6}$$

say, can be measured directly from the experimental data.

Next, Hodgkin and Huxley made two further assumptions. First, they assumed that the Na$^+$ current ratio K is independent of time and is thus constant over the course of each voltage clamp experiment. In other words, the amplitude and direction of the Na$^+$ current may be affected by the low extracellular Na$^+$ solution, but its time course is not. Second, they assumed that the K$^+$ channels are unaffected by the change in extracellular Na$^+$ concentration. There is considerable evidence that the Na$^+$ and K$^+$ channels are independent. Tetrodotoxin (TTX) is known to block Na$^+$ currents while leaving the K$^+$ currents almost unaffected, while tetraethylammonium (TEA) has the opposite effect of blocking the K$^+$ current but not the Na$^+$ current. To complete the argument, since $I_{ion} = I_{Na} + I_K$, and $I_K^1 = I_K^2$, it follows that $I_{ion}^1 - I_{Na}^1 = I_{ion}^2 - I_{Na}^2$, and thus

$$I_{\mathrm{Na}}^1 = \frac{K}{K-1}(I_{\mathrm{ion}}^1 - I_{\mathrm{ion}}^2), \tag{5.7}$$

$$I_{\mathrm{K}} = \frac{I_{\mathrm{ion}}^1 - KI_{\mathrm{ion}}^2}{1-K}. \tag{5.8}$$

Hence, given measurements of the total ionic currents in the two cases, and given the ratio K of the Na$^+$ currents, it is possible to determine the complete time courses of both the Na$^+$ and K$^+$ currents.

Finally, from knowledge of the individual currents, one obtains the conductances as

$$g_{\mathrm{Na}} = \frac{I_{\mathrm{Na}}}{V - V_{\mathrm{Na}}}, \qquad g_{\mathrm{K}} = \frac{I_{\mathrm{K}}}{V - V_{\mathrm{K}}}. \tag{5.9}$$

Note that this result relies on the specific (linear) model used to describe the *I–V* curve of the Na$^+$ and K$^+$ channels, but, as stated above, we assume throughout that the instantaneous *I–V* curves of the Na$^+$ and K$^+$ channels are linear.

Samples of Hodgkin and Huxley's data are shown in Fig. 5.3. The plots show ionic conductances as functions of time following a step increase or decrease in the membrane potential. The important observation is that with voltages fixed, the conductances are time-dependent. For example, when V is stepped up and held fixed at a higher level, g_{K} does not increase instantaneously, but instead increases over time to a final steady level. Both the time constant of the increase and the final value of g_{K} are dependent on the value to which the voltage is stepped. Further, g_{K} increases in a sigmoidal fashion, with a slope that first increases and then decreases (Fig. 5.3A and B). Following a step decrease in the voltage, g_{K} falls in a simple exponential fashion (Fig. 5.3A). This particular feature of g_{K}—a sigmoidal increase coupled with an exponential decrease—is important in what follows when we model g_{K}. The behavior of g_{Na} is more complex. Following a step increase in voltage, g_{Na} first increases, but then decreases again, *all at the same fixed voltage* (Fig. 5.3C). Hence, the time dependence of g_{Na} requires a more complex model than for that of g_{K}.

The Potassium Conductance

From the experimental data shown in Fig. 5.3A and B, it is reasonable to expect that g_{K} obeys some differential equation,

$$\frac{dg_{\mathrm{K}}}{dt} = f(v, t), \tag{5.10}$$

say, where $v = V - V_{\mathrm{eq}}$; i.e., v is the difference between the membrane potential and the resting potential. (Of course, since V_{eq} is a constant, $dv/dt = dV/dt$.) However, for g_{K} to have the required sigmoidal increase and exponential decrease, Hodgkin and Huxley realized that it would be easier to write g_{K} as some power of a different variable, n say,

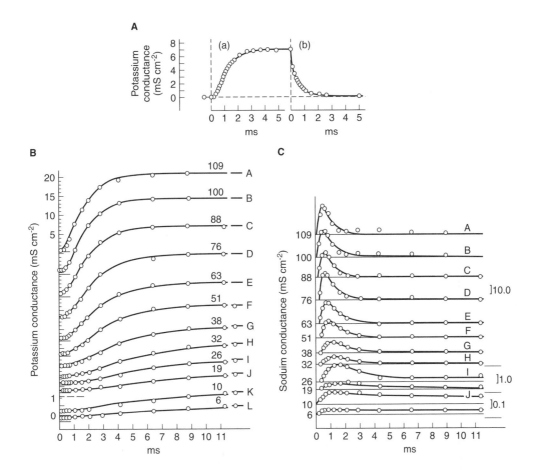

Figure 5.3 Conductance changes as a function of time at different voltage clamps. A: The response of g_K to a step increase in V and then a step decrease. B: Responses of g_K to step increases in V of varying magnitudes. The number on each curve gives the depolarization in mV, and the smooth curves are calculated from solution of (5.11) and (5.12), with the initial condition $g_K(t = 0) = 0.24$ mS/cm^2. The vertical scale is the same in curves A–J, but is increased by a factor of four in the lower two curves. For clarity, the baseline of each curve has been shifted up. C: Responses of g_{Na} to step increases in V of magnitudes given by the numbers on the left, in mV. The smooth curves are the model solutions. The vertical scales on the right are in units of mS/cm^2. (Hodgkin and Huxley, 1952d, Figs. 2, 3, and 6.)

where n satisfies a first-order differential equation. Thus, they wrote

$$g_K = \bar{g}_K n^4, \tag{5.11}$$

for some constant $\bar{g}_K$. The fourth power was chosen not for physiological reasons, but because it was the smallest exponent that gave acceptable agreement with the experimental data. The secondary variable n obeys the differential equation

$$\tau_n(v)\frac{dn}{dt} = n_\infty(v) - n, \tag{5.12}$$

for some functions $\tau_n(v)$ and $n_\infty(v)$ that must be determined from the experimental data in a manner that is described below. Equation (5.12) is often written in the form

$$\frac{dn}{dt} = \alpha_n(v)(1-n) - \beta_n(v)n, \tag{5.13}$$

where

$$n_\infty(v) = \frac{\alpha_n(v)}{\alpha_n(v) + \beta_n(v)}, \tag{5.14}$$

$$\tau_n(v) = \frac{1}{\alpha_n(v) + \beta_n(v)}. \tag{5.15}$$

At elevated potentials $n(t)$ increases monotonically and exponentially toward its resting value, thereby turning on, or *activating*, the K^+ current. Since the Nernst potential is below the resting potential, the K^+ current is an outward current at potentials greater than rest. The function $n(t)$ is called the K^+ *activation*.

It is instructive to consider in detail how such a formulation for g_K results in the required sigmoidal increase and exponential decrease. Suppose that at time $t = 0$, v is increased from 0 to v_0 and then held constant, and suppose further that n is at steady state when $t = 0$, i.e., $n(0) = n_\infty(0)$. For simplicity, we assume that $n_\infty(0) = 0$, although this assumption is not necessary for the argument. Solving (5.12) then gives

$$n(t) = n_\infty(v_0)\left[1 - \exp\left(\frac{-t}{\tau_n(v_0)}\right)\right], \tag{5.16}$$

which is an increasing curve (with monotonically decreasing slope) that approaches its maximum at $n_\infty(v_0)$. Raising n to the fourth power gives a sigmoidally increasing curve as required. Higher powers of n result in curves with a greater maximum slope at the point of inflection. However, in response to a step decrease in v, from v_0 to 0 say, the solution for n is

$$n(t) = n_\infty(v_0)\exp\left(\frac{-t}{\tau_n(0)}\right), \tag{5.17}$$

in which case n^4 is exponentially decreasing, with no inflection point.

It remains to describe how the functions n_∞ and τ_n are determined from the experimental data. For any given voltage step, the time constant τ_n, and the final value of n, namely n_∞, can be determined by fitting (5.16) to the experimental data. By this procedure one can determine τ_n and n_∞ at a discrete set of values for v, i.e., those values used experimentally. Typical data points for n_∞ are shown in Fig. 5.4 as symbols. To obtain a complete description of g_K, valid for all voltages and not only those used in the experiments, Hodgkin and Huxley fitted a smooth curve through the data points. The functional form of the smooth curve has no physiological significance, but is a convenient way of providing a continuous description of n_∞. A similar procedure is followed for τ_n. The continuous descriptions of n_∞ and τ_n (expressed in terms of α_n and β_n) are given in (5.28) and (5.29) below.

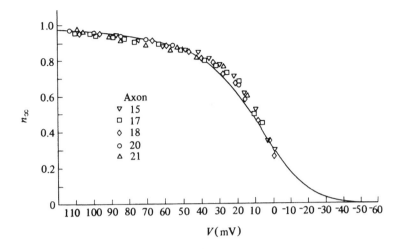

Figure 5.4 Data points (symbols) of n_∞, determined by fitting (5.16) to the experimental time courses. The smooth curve through the symbols provides a continuous description of n_∞, and its functional form has no physiological significance. In the original plot (Hodgkin and Huxley, 1952d, Fig. 5) V was calculated with a reverse sign, which has here been changed to agree with modern conventions. Thus, the horizontal axis appears reversed.

The Sodium Conductance

The time dependence for the Na^+ conductance is more difficult to unravel. From the experimental data it is suggested that there are two processes at work, one that turns on the Na^+ current and one that turns it off. Hodgkin and Huxley proposed that the Na^+ conductance is of the form

$$g_{Na}(v) = \bar{g}_{Na} m^3 h, \tag{5.18}$$

and they fitted the time-dependent behavior of m and h to exponentials with dynamics

$$\frac{dw}{dt} = \alpha_w (1 - w) - \beta_w w, \tag{5.19}$$

where $w = m$ or h. Because m is small at rest and first increases, it is called the *sodium activation* variable, and because h shuts down, or inactivates, the Na^+ current, it is called the *sodium inactivation* variable. When $h = 0$, the Na^+ current is completely inactivated. The overall procedure is similar to that used in the specification of g_K. For any fixed voltage step, the unknown functions α_w and β_w are determined by fitting to the experimental curves (Fig. 5.3C), and then smooth curves, with arbitrary functional forms, are fitted through the data points for α_w and β_w.

Summary of the Equations

In summary, the Hodgkin–Huxley equations for the space-clamped axon are

$$C_m \frac{dv}{dt} = -\bar{g}_K n^4 (v - v_K) - \bar{g}_{Na} m^3 h (v - v_{Na}) - \bar{g}_L (v - v_L) + I_{app}, \tag{5.20}$$

$$\frac{dm}{dt} = \alpha_m(1-m) - \beta_m m, \tag{5.21}$$

$$\frac{dn}{dt} = \alpha_n(1-n) - \beta_n n, \tag{5.22}$$

$$\frac{dh}{dt} = \alpha_h(1-h) - \beta_h h. \tag{5.23}$$

The specific functions α and β proposed by Hodgkin and Huxley are, in units of $(ms)^{-1}$,

$$\alpha_m = 0.1 \frac{25-v}{\exp\left(\frac{25-v}{10}\right) - 1}, \tag{5.24}$$

$$\beta_m = 4 \exp\left(\frac{-v}{18}\right), \tag{5.25}$$

$$\alpha_h = 0.07 \exp\left(\frac{-v}{20}\right), \tag{5.26}$$

$$\beta_h = \frac{1}{\exp\left(\frac{30-v}{10}\right) + 1}, \tag{5.27}$$

$$\alpha_n = 0.01 \frac{10-v}{\exp\left(\frac{10-v}{10}\right) - 1}, \tag{5.28}$$

$$\beta_n = 0.125 \exp\left(\frac{-v}{80}\right). \tag{5.29}$$

For these expressions, the potential v is the deviation from rest ($v = V - V_{eq}$), measured in units of mV, current density is in units of $\mu A/cm^2$, conductances are in units of mS/cm^2, and capacitance is in units of $\mu F/cm^2$. The remaining parameters are

$$\bar{g}_{Na} = 120, \qquad \bar{g}_K = 36, \qquad \bar{g}_L = 0.3, \qquad C_m = 1, \tag{5.30}$$

with (shifted) equilibrium potentials $v_{Na} = 115, v_K = -12$, and $v_L = 10.6$. (The astute reader will notice immediately that these values are not quite consistent with the values given in Table 2.1. Instead, these correspond to $V_{Na} = 50$ mV, $V_K = -77$ mV, $V_L = -54.4$ mV, with an equilibrium membrane potential of $V_{eq} = -65$ mV. These values are close enough to those of Table 2.1 to be of no concern.) In Fig. 5.5 are shown the steady-state functions and the time constants.

In Chapter 3 we discussed simple models of the gating of Na^+ and K^+ channels and showed how the rate constants in simple kinetic schemes could be determined from whole-cell or single-channel data. We also showed how models of the form (5.20)–(5.23) can be derived by modeling the ionic channels as consisting of multiple subunits, each of which obeys a simple two-state model. For example, the Hodgkin–Huxley Na^+ gating equations can be derived from the assumption that the Na^+ channel consists of three "m" gates and one "h" gate, each of which can be either closed or open. If the gates operate independently, then the fraction of open Na^+ channels is $m^3 h$, where m and h obey the equation of the two-state channel model. Similarly, if there are four "n" gates

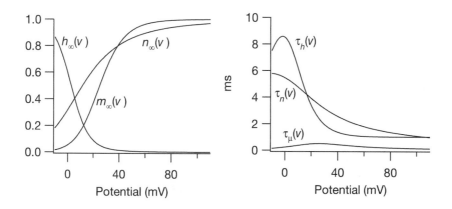

Figure 5.5 In the left panel are the steady-state functions, and in the right panel are the time constants of the Hodgkin–Huxley equations (5.20)–(5.23).

per K^+ channel, all of which must be open for K^+ to flow, then the fraction of open K^+ channels is n^4.

Now comes the most interesting challenge facing these equations. Having incorporated the measurements of conductance found from voltage-clamp experiments, one wonders whether these equations reproduce a realistic action potential, and if so, by what mechanism is the action potential produced? We can describe in qualitative terms how the Hodgkin–Huxley equations should work. If small currents are applied to a cell for a short period of time, the potential returns rapidly to its equilibrium $v = 0$ after the applied current is removed. The equilibrium potential is close to the K^+ Nernst potential $v_K = -12$, because at rest, the Na^+ and leakage conductances are small. There is always competition among the three ionic currents to drive the potential to the corresponding resting potential. For example, if the K^+ and leakage currents could be blocked or the Na^+ conductance dramatically increased, then the term $g_{Na}(V - V_{Na})$ should dominate (5.2), and as long as v is below v_{Na}, an inward Na^+ current would drive the potential toward v_{Na}. Similarly, while v is above v_K, the K^+ current is outward in an attempt to drive v toward v_K. Notice that since $v_K < v_L < v_{Na}$, v is necessarily restricted to lie in the range $v_K < v < v_{Na}$.

If g_{Na} and g_K were constant, that would be the end of the story. The equilibrium at $v = 0$ would be a stable equilibrium, and, following any stimulus, the potential would return exponentially to rest. But since g_{Na} and g_K can change, the different currents can exert their respective influences. The actual sequence of events is determined by the dynamics of m, n, and h. The most important observation for the moment is that $\tau_m(v)$ is much smaller than either $\tau_n(v)$ or $\tau_h(v)$, so that $m(t)$ responds much more quickly to changes in v than either n or h. We can now understand why the Hodgkin–Huxley system is an excitable system. As noted above, if the potential v is raised slightly by a small stimulating current, the system returns to its stable equilibrium. However, during the period of time that the potential v is elevated, the Na^+ activation m tracks $m_\infty(v)$. If the stimulating current is large enough to raise the potential and therefore $m_\infty(v)$ to a high enough level (above its threshold), then before the system can return to rest, m will increase sufficiently to change the sign of the net current, resulting in an autocatalytic inward Na^+ current. Now, as the potential rises, m continues to rise, and the inward Na^+ current is increased, further adding to the rise of the potential.

If nothing further were to happen, the potential would be driven to a new equilibrium at v_{Na}. However, here is where the difference in time constants plays an important role. When the potential is at rest, the Na^+ inactivation variable, h, is positive, about 0.6. As the potential increases, h_∞ decreases toward zero, and as h approaches zero, the Na^+ current is inactivated because g_{Na} approaches zero. However, because the time constant $\tau_h(v)$ is much larger than $\tau_m(v)$, there is a considerable delay between turning on the Na^+ current (as m increases) and turning off the Na^+ current (as h decreases). The net effect of the two different time scales of m and h is that the Na^+ current is at first turned on and later turned off, and this is seen as an initial increase of the potential, followed by a decrease toward rest.

At about the same time that the Na^+ current is inactivated, the outward K^+ current is activated. This is because of the similarity of the time constants $\tau_n(v)$ and $\tau_h(v)$. Activation of the K^+ current drives the potential below rest toward v_K. When v is negative, n declines, and the potential eventually returns to rest, and the whole process can start again. Fig. 5.6A shows a plot of the potential $v(t)$ during an action potential following a superthreshold stimulus. Fig. 5.6B shows $m(t)$, $n(t)$, and $h(t)$ during the same action potential.

There are four recognizable phases of an action potential: the *upstroke*, *excited*, *refractory*, and *recovery* phases. The refractory period is the period following the excited phase when additional stimuli evoke no substantial response, even though the potential is below or close to its resting value. There can be no response, since the Na^+ channels are inactivated because h is small. As h gradually returns to its resting value, further responses once again become possible.

Oscillations in the Hodgkin–Huxley Equations

There are two ways that the Hodgkin–Huxley system can be made into an autonomous oscillator. The first is to inject a steady current of sufficient strength, i.e., by increasing I_{app}. Such a current raises the resting potential above the threshold for an action potential, so that after the axon has recovered from an action potential, the potential rises to a superthreshold level at which another action potential is evoked.

In Fig. 5.7A we plot the steady state v (i.e., $V - V_{eq}$) as a function of the applied current, I_{app}. The stable steady state is plotted as a solid line, and an unstable steady state is plotted with a dashed line. As I_{app} increases, so does v, and the steady state is stable for $I_{app} < 9.78$, at which value it loses stability in a subcritical Hopf bifurcation. This bifurcation gives rises to a branch of unstable limit cycle oscillations which bends backwards initially. Unstable limit cycles are drawn with a dashed line, and stable ones with a solid line.

In Fig. 5.7A we also plot the minimum and maximum of the oscillations (i.e., osc min and osc max) as functions of I_{app}. The branch of unstable limit cycles terminates at

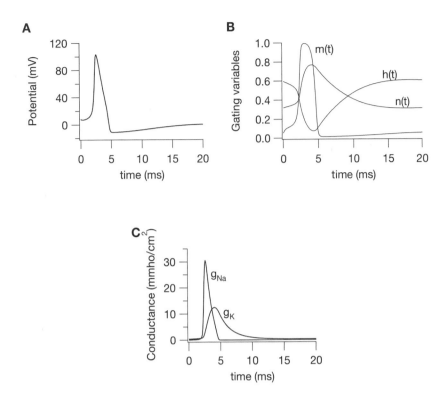

Figure 5.6 An action potential in the Hodgkin–Huxley equations. A: The action potential; B: the gating variables during an action potential, and C: the conductances during an action potential.

a limit point (a saddle-node of periodics, or SNP, bifurcation) where it coalesces with a branch of stable limit cycles. The stable periodic solutions are observed by direct numerical simulation of the differential equations. At larger values of I_{app}, the limit cycles disappear in another Hopf bifurcation, this time a supercritical one, leaving only a branch of stable steady-state solutions for higher values of I_{app}.

Hence, for intermediate values of I_{app}, stable oscillations exist. Two examples are shown in Fig. 5.7B. When I_{app} is too high, the model exhibits only a raised steady state. Furthermore, for a narrow range of values of I_{app}, slightly below the lower Hopf bifurcation, a stable steady state, an unstable periodic orbit, and a stable periodic orbit coexist.

Immersing the axon in a bath of high extracellular K^+ has the same effect through a slightly different mechanism. Increased extracellular K^+ has the effect of increasing the K^+ Nernst potential, raising the resting potential (since the resting potential is close to the K^+ Nernst potential). If this increase of the K^+ Nernst potential is sufficiently large, the resting potential becomes superthreshold, and autonomous oscillations result. This mechanism of creating an autonomous oscillator out of normally excitable but nonoscillatory cells is important for certain cardiac arrhythmias.

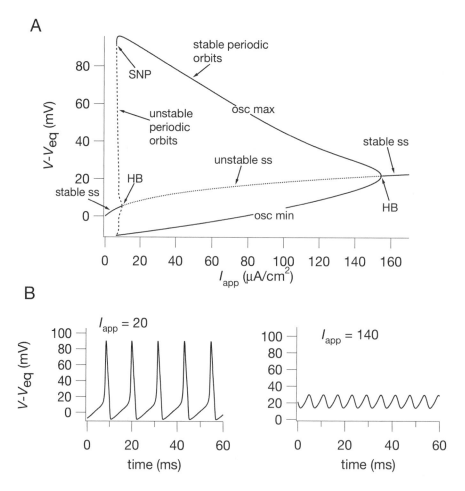

Figure 5.7 A: Bifurcation diagram of the Hodgkin–Huxley equations, with the applied current, I_{app} as the bifurcation parameter. HB denotes a Hopf bifurcation, SNP denotes a saddle-node of periodics bifurcation, osc max and osc min denote, respectively, the maximum and minimum of an oscillation, and ss denotes a steady state. Solid lines denote stable branches, dashed or dotted lines denote unstable branches. B: Sample oscillations at two different values of I_{app}.

5.1.3 Qualitative Analysis

FitzHugh (1960, 1961, 1969) provided a particularly elegant qualitative description of the Hodgkin–Huxley equations that allows a better understanding of the model's behavior. More detailed analyses have also been given by Rinzel (1978), Troy (1978), Cole et al. (1955), and Sabah and Spangler (1970). FitzHugh's approach is based on the fact that some of the model variables have fast kinetics, while others are much slower. In particular, m and v are fast variables (i.e., the Na$^+$ channel activates quickly, and the membrane potential changes quickly), while n and h are slow variables (i.e.,

Na$^+$ channels are inactivated slowly, and the K$^+$ channels are activated slowly). Thus, during the initial stages of the action potential, n and h remain essentially constant while m and v vary. This allows the full four-dimensional phase space to be simplified by fixing the slow variables and considering the behavior of the model as a function only of the two fast variables. Although this description is accurate only for the initial stages of the action potential, it provides a useful way to study the process of excitation.

The Fast Phase Plane

Thus motivated, we fix the slow variables n and h at their respective resting states, which we call n_0 and h_0, and consider how m and v behave in response to stimulation. The differential equations for the fast phase plane are

$$C_m \frac{dv}{dt} = -\bar{g}_K n_0^4 (v - v_K) - \bar{g}_{Na} m^3 h_0 (v - v_{Na}) - \bar{g}_L (v - v_L), \tag{5.31}$$

$$\frac{dm}{dt} = \alpha_m (1 - m) - \beta_m m, \tag{5.32}$$

or, equivalently,

$$\tau_m \frac{dm}{dt} = m_\infty - m. \tag{5.33}$$

This is a two-dimensional system and can be studied in the (m, v) phase plane, a plot of which is given in Fig. 5.8. The curves defined by $dv/dt = 0$ and $dm/dt = 0$ are the v and m nullclines, respectively. The m nullcline is the curve $m = m_\infty(v)$, which we have seen before (in Fig. 5.5), while the v nullcline is the curve

$$v = \frac{\bar{g}_{Na} m^3 h_0 v_{Na} + \bar{g}_K n_0^4 v_K + \bar{g}_L v_L}{\bar{g}_{Na} m^3 h_0 + \bar{g}_K n_0^4 + \bar{g}_L}. \tag{5.34}$$

For the parameters of the Hodgkin–Huxley equations, the m and v nullclines intersect in three places, corresponding to three steady states of the fast equations. Note that these three intersections are not steady states of the full model, only of the fast subsystem, and, to be precise, should be called pseudo-steady states. However, in the context of the fast phase plane we continue to call them steady states. We label the three steady states v_r, v_s, and v_e (for resting, saddle, and excited).

It is left as an exercise to show that v_r and v_e are stable steady states of the fast subsystem, while v_s is a saddle point. Since v_s is a saddle point, it has a one-dimensional stable manifold, shown as a dot-dash line in Fig. 5.8. This stable manifold divides the (m, v) plane into two regions: any trajectory starting to the left of the stable manifold is prevented from reaching v_e and must eventually return to the resting state, v_r. However, any trajectory starting to the right of the stable manifold is prevented from returning to the resting state and must eventually end up at the excited state, v_e. Hence, the stable manifold, in combination with the two stable steady states, gives rise to a threshold phenomenon. Any perturbation from the resting state that is not large enough to cross the stable manifold eventually dies away, but a perturbation that crosses the

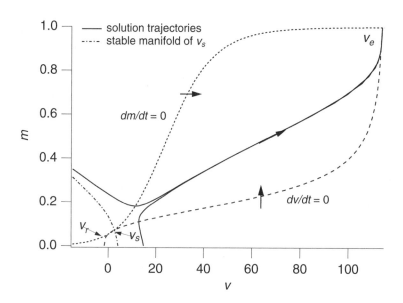

Figure 5.8 The Hodgkin–Huxley fast phase plane, showing the nullclines $dv/dt = 0$ and $dm/dt = 0$ (with $h_0 = 0.596$, $n_0 = 0.3176$), two sample trajectories and the stable manifold of the saddle point v_s.

stable manifold results in a large excursion in the voltage to the excited state. Sample trajectories are sketched in Fig. 5.8.

If m and v were the only variables in the model, then v would stay at v_e indefinitely. However, as pointed out before, v_e is not a steady state of the full model. Thus, to see what happens on a longer time scale, we must consider how slow variations in n and h affect the fast phase plane. First note that since $v_e > v_r$, it follows that $h_\infty(v_e) < h_\infty(v_r)$ and $n_\infty(v_e) > n_\infty(v_r)$. Hence, while v is at the excited state, h begins to decrease, thus inactivating the Na^+ conductance, and n starts to increase thus activating the K^+ conductance. Next note that although the m nullcline in the fast phase plane is independent of n and h, the v nullcline is not. In Fig. 5.8 the nullclines were drawn using the steady-state values for n and h: different values of n and h change the shape of the v nullcline. As n increases and h decreases, the v nullcline moves to the left and up, as illustrated in Fig. 5.9. As the v nullcline moves up and to the left, v_e and v_s move toward each other, while v_r moves to the left. During this phase the voltage is at v_e and thus decreases slowly. Eventually, v_e and v_s coalesce and disappear in a saddle-node bifurcation. When this happens v_r is the only remaining steady state, and so the solution must return to the resting state. Note that since the v nullcline has moved up and to the left, v_r is not a steady state of the full system. However, when v decreases to v_r, n and h both return to their steady states and as they do so, v_r slowly increases until the steady state of the full system is reached and the action potential is complete. A schematic diagram of a complete action potential is shown in Fig. 5.10.

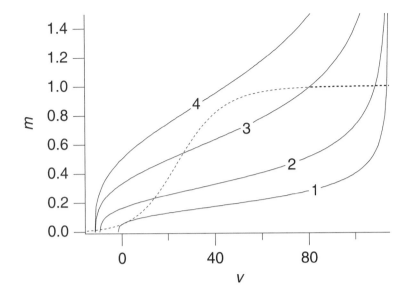

Figure 5.9 The Hodgkin–Huxley fast phase plane as a function of the slow variables, showing the m nullcline (dashed), the movement of the v nullcline (solid) and the disappearance of the steady states. For these curves, parameter values are (1) $h_0 = 0.596$, $n_0 = 0.3176$; (2) $h_0 = 0.4$, $n_0 = 0.5$; (3) $h_0 = 0.2$, $n_0 = 0.7$; and (4) $h_0 = 0.1$, $n_0 = 0.8$.

The Fast–Slow Phase Plane

In the above analysis, the four-dimensional phase space was simplified by taking a series of two-dimensional cross-sections, those with various fixed values of n and h. However, by taking a different cross-section other aspects of the action potential can be highlighted. In particular, by taking a cross-section involving one fast variable and one slow variable we obtain a description of the Hodgkin–Huxley equations that has proven to be extraordinarily useful.

We extract a single fast variable by assuming that m is always in instantaneous equilibrium, and thus $m = m_\infty(v)$. This corresponds to assuming that activation of the Na$^+$ conductance is on a time scale faster than that of the voltage. Next, FitzHugh noticed that during the course of an action potential, $h+n \approx 0.8$ (notice the approximate symmetry of $n(t)$ and $h(t)$ in Fig. 5.6), and thus h can be eliminated by setting $h = 0.8-n$. With these simplifications, the Hodgkin–Huxley equations contain one fast variable v and one slow variable n, and can be written as

$$-C_m \frac{dv}{dt} = \bar{g}_K n^4 (v - v_K) + \bar{g}_{Na} m_\infty^3(v)(0.8 - n)(v - v_{Na}) + \bar{g}_L(v - v_L), \qquad (5.35)$$

$$\frac{dn}{dt} = \alpha_n(1 - n) - \beta_n n. \qquad (5.36)$$

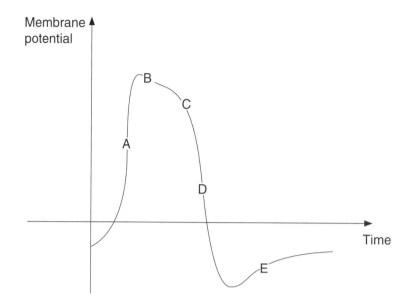

Figure 5.10 Schematic diagram of a complete action potential. A: Superthreshold stimulus causes a fast increase of v to the excited state. B: v is sitting at the excited state, v_e, decreasing slowly as n increases and h decreases, i.e., as v_e moves toward v_s. C: v_e and v_s disappear at a saddle-node bifurcation, and so, D: The solution must return to the resting state v_r. E: n and h slowly return to their resting states, and as they do so, v_r slowly increases until the steady state of the full four-dimensional system is reached.

For convenience we let $f(v, n)$ denote the right-hand side of (5.35), i.e.,

$$-f(v, n) = \bar{g}_K n^4 (v - v_K) + \bar{g}_{Na} m_\infty^3(v)(0.8 - n)(v - v_{Na}) + \bar{g}_L(v - v_L). \qquad (5.37)$$

A plot of the nullclines of the fast–slow subsystem is given in Fig. 5.11A. The v nullcline is defined by $f(v, n) = 0$ and has a cubic shape, while the n nullcline is $n_\infty(v)$ and is monotonically increasing. There is a single intersection (at least for the given parameter values) and thus a single steady state. Because v is a fast variable and n is a slow one, the solution trajectories are almost horizontal except where $f(v, n) \approx 0$. The curve $f(v, n) = 0$ is called the *slow manifold*. Along the slow manifold the solution moves slowly in the direction determined by the sign of dn/dt, but away from the slow manifold the solution moves quickly in a horizontal direction. From the sign of dv/dt it follows that the solution trajectories move away from the middle branch of the slow manifold and toward the left and right branches. Thus, the middle branch is termed the unstable branch of the slow manifold. This unstable branch acts as a threshold. If a perturbation from the steady state is small enough so that v does not cross the unstable manifold, then the trajectory moves horizontally toward the left and returns to the steady state. However, if the perturbation is large enough so that v crosses the unstable manifold, then the trajectory moves to the right until it reaches

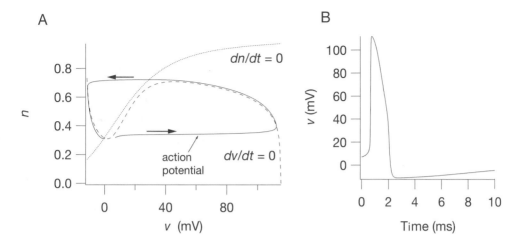

Figure 5.11 A: Fast–slow phase plane of the Hodgkin–Huxley equations (with $I_{app} = 0$), showing the nullclines and an action potential. B: Action potential of panel A, plotted as a function of time.

the right branch of the slow manifold, which corresponds to the excited state. On this right branch $dn/dt > 0$, and so the solution moves slowly up the slow manifold until the turning point is reached. At the turning point, n cannot increase any further, as the right branch of the slow manifold ceases to exist, and so the solution moves over to the left branch of the slow manifold. On this left branch $dn/dt < 0$, and so the solution moves down the left branch until the steady state is reached, completing the action potential (Fig. 5.11A). A plot of the potential as a function of time is shown in Fig. 5.11B.

The variables v and n are usually called the excitation and recovery variables, respectively: excitation because v governs the rise to the excited state, and recovery because n causes the return to the steady state. In the absence of n the solution would stay at the excited state indefinitely.

There is a close relationship between the fast phase plane and the fast–slow phase plane. Recall that in the fast phase plane, the v and m nullclines have three intersection points when $n = n_0$ and $h = h_0$. These three intersections correspond to the three branches of the curve $f(v, n_0) = 0$. In other words, when n is fixed at n_0, the equation $f(v, n_0) = 0$ has three possible solutions, corresponding to v_r, v_s and v_e in the fast phase plane. However, consideration of Fig. 5.11 shows that, as n increases, the two rightmost branches of the slow manifold (i.e., the dashed line) coalesce and disappear. This is analogous to the merging and disappearance of v_e and v_s seen in the fast phase plane (Fig. 5.9). The fast–slow phase plane is a convenient way of summarizing how v_r, v_s, and v_e depend on the slow variables.

This representation of the Hodgkin–Huxley equations in terms of two variables, one fast and one slow, is the basis of the FitzHugh–Nagumo model of excitability, and models of this generic type are discussed in some detail throughout this book.

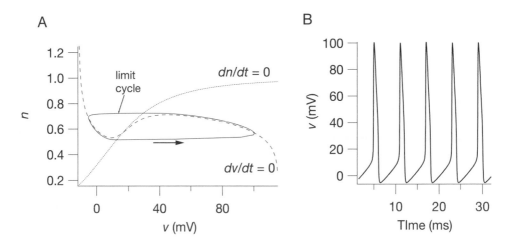

Figure 5.12 A: Fast–slow phase plane of the Hodgkin–Huxley equations, with $I_{app} = 50$, showing the nullclines and an oscillation. B: The oscillations of panel A, plotted as a function of time.

Oscillations in the Fast–Slow Phase Plane

As was true for the full Hodgkin–Huxley equations, the addition of an applied current to the fast–slow phase plane gives rise to oscillations. Why this is so can be seen in Fig. 5.12. As I_{app} increases, the cubic nullcline moves across and up, until the two null-clines intersect on the middle branch of the cubic. The trajectory can never approach this steady state, always falling off each of the branches of the cubic, and alternating periodically between the two stable branches in what is called a relaxation limit cycle.

5.2   The FitzHugh–Nagumo Equations

There is considerable value in studying systems of equations that are simpler than the Hodgkin–Huxley equations but that retain many of their qualitative features. This is the motivation for the FitzHugh–Nagumo equations and their variants. Basically, the FitzHugh–Nagumo equations extract the essential behavior of the Hodgkin–Huxley fast–slow phase plane and presents it in a simplified form. Thus, the FitzHugh–Nagumo equations have two variables, one fast (v) and one slow (w). The fast variable has a cubic nullcline and is called the excitation variable, while the slow variable is called the recovery variable and has a nullcline that is monotonically increasing. The nullclines have a single intersection point, which, without loss of generality, is assumed to be at the origin. A schematic diagram of the phase plane is given in Fig. 5.13, where we introduce some of the notation used later in this section.

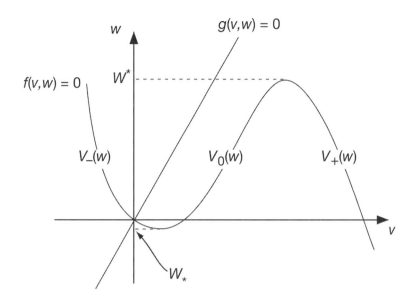

Figure 5.13 Schematic diagram of the generalized FitzHugh–Nagumo phase plane.

The traditional FitzHugh–Nagumo equations are obtained by assuming a cubic nullcline for v and a linear nullcline for w. Thus,

$$\epsilon \frac{dv}{dt} = f(v) - w + I_{\text{app}}, \tag{5.38}$$

$$\frac{dw}{dt} = v - \gamma w, \tag{5.39}$$

where

$$f(v) = v(1 - v)(v - \alpha), \qquad \text{for } 0 < \alpha < 1, \epsilon \ll 1. \tag{5.40}$$

I_{app} is the applied current. Typical values would be $\alpha = 0.1$, $\gamma = 0.5$ and $\epsilon = 0.01$.

Other choices for $f(v)$ include the McKean model (McKean, 1970), for which

$$f(v) = H(v - \alpha) - v, \tag{5.41}$$

where H is the Heaviside function. This choice recommends itself because then the model is piecewise linear, allowing explicit solutions of many interesting problems. Another piecewise-linear model (also proposed by McKean, 1970) has

$$f(v) = \begin{cases} -v, & \text{for } v < \alpha/2, \\ v - \alpha, & \text{for } \frac{\alpha}{2} < v < \frac{1+\alpha}{2}, \\ 1 - v, & \text{for } v > \frac{1+\alpha}{2}. \end{cases} \tag{5.42}$$

A third piecewise-linear model that has found widespread usage is the *Pushchino* model, so named because of its development in Pushchino (about 70 miles south of

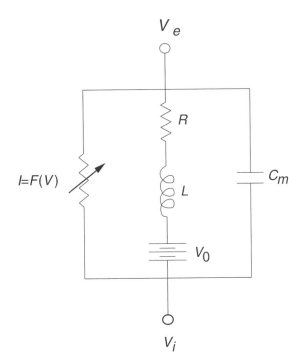

Figure 5.14 Circuit diagram for the FitzHugh–Nagumo equations.

Moscow), by Krinsky, Panfilov, Pertsov, Zykov, and their coworkers. The details of the Pushchino model are described in Exercise 13.

The FitzHugh–Nagumo equations can be derived from a simplified model of the cell membrane (Fig. 5.14). Here the cell (or membrane patch) consists of three components, a capacitor representing the membrane capacitance, a nonlinear current–voltage device for the fast current, and a resistor, inductor, and battery in series for the recovery current. In the 1960s Nagumo, a Japanese electrical engineer, built this circuit using a tunnel diode as the nonlinear element (Nagumo et al., 1964), thereby attaching his name to this system.

Using Kirchhoff's laws, we can write down equations for the behavior of this membrane circuit diagram. We find

$$C_m \frac{dV}{d\tau} + F(V) + i = -I_0, \tag{5.43}$$

$$L \frac{di}{d\tau} + Ri = V - V_0, \tag{5.44}$$

where I_0 is the applied external current, i is the current through the resistor–inductor, $V = V_i - V_e$ is the membrane potential, and V_0 is the potential gain across the battery. Here τ is used to represent dimensional time because we shortly introduce t as a dimensionless time variable. The function $F(V)$ is assumed to be of cubic shape, having three zeros, of which the smallest $V = 0$ and largest $V = V_1$ are stable solutions of the differential equation $dV/d\tau = -F(V)$. We take R_1 to be the passive resistance of the nonlinear element, $R_1 = 1/F'(0)$. Now we introduce the dimensionless variables $v = V/V_1, w = R_1 i/V_1, f(v) = -R_1 F(V_1 v)/V_1$, and $t = L\tau/R_1$. Then (5.43) and (5.44) become

$$\epsilon \frac{dv}{dt} = f(v) - w - w_0, \tag{5.45}$$

$$\frac{dw}{dt} = v - \gamma w - v_0, \tag{5.46}$$

where $\epsilon = R_1^2 C_m/L, w_0 = R_1 I_0/V_1, v_0 = V_0/V_1$, and $\gamma = R/R_1$.

An important variant of the FitzHugh–Nagumo equations is the *van der Pol oscillator*. An electrical engineer, van der Pol built the circuit using triodes because it exhibits stable oscillations. As there was little interest in oscillatory circuits at the time, he proposed his circuit as a model of an oscillatory cardiac pacemaker (van der Pol and van der Mark, 1928). Since then it has become a classic example of a system with limit cycle behavior and relaxation oscillations, included in almost every textbook on oscillations (see, for example, Stoker, 1950, or Minorsky, 1962).

If we eliminate the resistor R from the circuit in Fig. 5.14, differentiate (5.43), and eliminate the current i, we get the second-order differential equation

$$C_m \frac{d^2 V}{d\tau^2} + F'(V)\frac{dV}{d\tau} + \frac{V}{L} = \frac{V_0}{L}. \tag{5.47}$$

Following rescaling, and setting $F(v) = A(v^3/3 - v)$, we arrive at the *van der Pol equation*

$$v'' + a(v^2 - 1)v' + v = 0. \tag{5.48}$$

5.2.1 The Generalized FitzHugh-Nagumo Equations

From now on, by the *generalized FitzHugh–Nagumo equations* we mean the system of equations

$$\epsilon \frac{dv}{dt} = f(v, w), \tag{5.49}$$

$$\frac{dw}{dt} = g(v, w), \tag{5.50}$$

where the nullcline $f(v, w) = 0$ is of "cubic" shape. By this we mean that for a finite range of values of w, there are three solutions $v = v(w)$ of the equation $f(v, w) = 0$. These we denote by $v = V_-(w), v = V_0(w)$, and $v = V_+(w)$, and, where comparison is possible (since these functions need not all exist for the same range of w),

$$V_-(w) \le V_0(w) \le V_+(w). \tag{5.51}$$

We denote the minimal value of w for which $V_-(w)$ exists by W_*, and the maximal value of w for which $V_+(w)$ exists by W^*. For values of w above the nullcline $f(v, w) = 0, f(v, w) < 0$, and below the nullcline, $f(v, w) > 0$ (in other words, $f_w(v, w) < 0$).

The nullcline $g(v, w) = 0$ is assumed to have precisely one intersection with the curve $f(v, w) = 0$. Increasing v beyond the curve $g(v, w) = 0$ makes $g(v, w)$ positive (i.e., $g_v(v,w) > 0$), and decreasing w below the curve $g(v,w) = 0$ increases $g(v,w)$ (hence $g_w(v,w) < 0$). The nullclines f and g are illustrated in Fig. 5.13.

5.2.2 Phase-Plane Behavior

One attractive feature of the FitzHugh–Nagumo equations is that because they form a two-variable system, they can be studied using phase-plane techniques. (For an example of a different approach, see Troy, 1976.) There are two characteristic phase portraits possible (shown in Figs. 5.15 and 5.16). By assumption, there is only one steady state, at $v = v^*, w = w^*$, with $f(v^*, w^*) = g(v^*, w^*) = 0$. Without loss of generality, we assume that this steady state occurs at the origin, as this involves only a shift of the variables. Furthermore, it is typical that the parameter ϵ is a small number. For small ϵ, if the steady state lies on either the left or right solution branch of $f(v,w) = 0$, i.e., the curves $v = V_\pm(w)$, it is linearly stable. Somewhere on the middle solution branch $v = V_0(w)$, near the extremal values of the curve $f(v,w) = 0$, there is a Hopf bifurcation point. If parameters are varied so that the steady-state solution passes through this point, a periodic orbit arises as a continuous solution branch and bifurcates into a stable limit cycle oscillation.

When the steady state is on the leftmost branch, but close to the minimum (Fig. 5.15), the system is excitable. This is because even though the steady state is

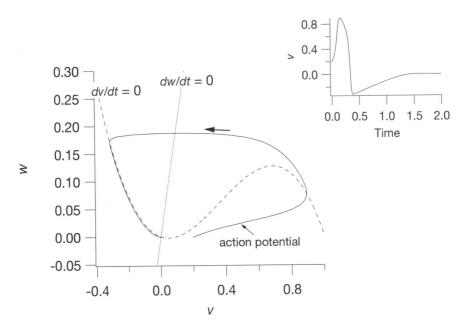

Figure 5.15 Phase portrait for the FitzHugh–Nagumo equations, (5.38)–(5.40), with $\alpha = 0.1$, $\gamma = 0.5$, $\epsilon = 0.01$ and zero applied current. For these parameter values the system has a unique globally stable rest point, but is excitable. The inset at top right shows the action potential as a function of time.

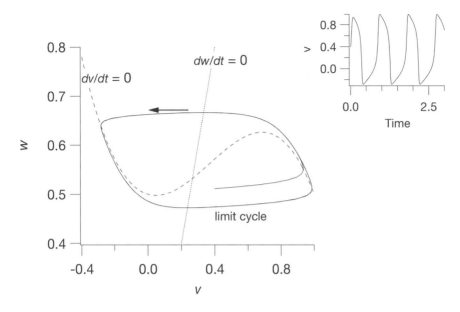

Figure 5.16 Phase portrait for the FitzHugh–Nagumo equations, (5.38)–(5.40), with $\alpha = 0.1$, $\gamma = 0.5$, $\epsilon = 0.01$ and $I_{app} = 0.5$. For these parameter values, the unique rest point is unstable and there is a globally stable periodic orbit. The inset at top right shows the periodic orbit plotted against time.

linearly stable, a sufficiently large perturbation from the steady state sends the state variable on a trajectory that runs away from the steady state before eventually returning to rest. Such a trajectory goes rapidly to the rightmost branch, which it hugs as it gradually creeps upward, where upon reaching the maximum, it goes rapidly to the leftmost branch and then gradually returns to rest, staying close to this branch as it does. Plots of the variables v and w are shown as functions of time in Fig. 5.15.

The mathematical description of these events follows from singular perturbation theory. With $\epsilon \ll 1$, the variable v is a fast variable and the variable w is a slow variable. This means that if possible, v is adjusted rapidly to maintain a pseudo-equilibrium at $f(v, w) = 0$. In other words, if possible, v clings to the stable branches of $f(v, w) = 0$, namely $v = V_{\pm}(w)$. Along these branches the dynamics of w are governed by the reduced dynamics

$$\frac{dw}{dt} = g(V_{\pm}(w), w) = G_{\pm}(w). \qquad (5.52)$$

When it is not possible for v to be in quasi-equilibrium, the motion is governed approximately by the differential equations,

$$\frac{dv}{d\tau} = f(v, w), \qquad \frac{dw}{d\tau} = 0, \qquad (5.53)$$

found by making the change of variables to the fast time scale $t = \epsilon\tau$, and then setting $\epsilon = 0$. On this time scale, w is constant, while v equilibrates to a stable solution of $f(v, w) = 0$.

The evolution of v and w starting from specified initial conditions v_0 and w_0 can now be described. Suppose v_0 is greater than the rest value v^*. If $v_0 < V_0(w)$, then v returns directly to the steady state. If $v_0 > V_0(w)$, then v goes rapidly to the upper branch $V_+(w)$ with w remaining nearly constant at w_0. The curve $v = V_0(w)$ is a *threshold curve*. While v remains on the upper branch, w increases according to

$$\frac{dw}{dt} = G_+(w), \tag{5.54}$$

as long as possible. However, in the finite time

$$T_e = \int_{w_0}^{W^*} \frac{dw}{G_+(w)}, \tag{5.55}$$

w reaches the "knee" of the nullcline $f(v, w) = 0$. This period of time constitutes the *excited phase* of the action potential.

When w reaches W^* it is no longer possible for v to stay on the excited branch, so it must return to the lower branch $V_-(w)$. Once on this branch, w decreases following the dynamics

$$\frac{dw}{dt} = G_-(w). \tag{5.56}$$

If the rest point lies on the lower branch, then $G_-(w^*) = 0$, and w gradually returns to rest on the lower branch.

Applied Current and Oscillations

When a current is applied to the generalized FitzHugh–Nagumo equations, they become

$$\epsilon\frac{dv}{dt} = f(v, w) + I_{\text{app}}, \tag{5.57}$$

$$\frac{dw}{dt} = g(v, w). \tag{5.58}$$

As with the fast–slow phase plane of the Hodgkin–Huxley equations, the cubic nullcline moves up as I_{app} increases. Thus, when I_{app} takes values in some intermediate range, the steady state lies on the middle branch, $V_0(w)$, and is unstable. Instead of returning to rest after one excursion on the excited branch, the trajectory alternates periodically between the upper and lower branches, with w varying between W_* and W^* (Fig. 5.16). This limit cycle behavior, where there are fast jumps between regions in which the solution moves more slowly, is called a *relaxation oscillation*. In this figure, the relaxation nature of the oscillations is not very pronounced; however, as ϵ decreases, the jumps

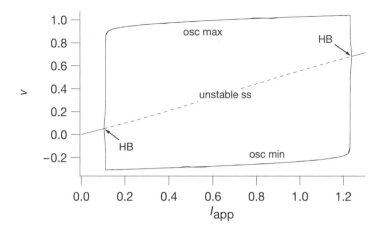

Figure 5.17 Bifurcation diagram of the FitzHugh–Nagumo equations, (5.38)–(5.40), with $\alpha = 0.1$, $\gamma = 0.5$, $\epsilon = 0.01$, with the applied current as the bifurcation parameter. The steady-state solution is labeled ss, while osc max and osc min denote, respectively, the maximum and minimum of v over an oscillation. HB denotes a Hopf bifurcation point.

become faster. For small ϵ, the period of the oscillation is approximately

$$T = \int_{W_*}^{W^*} \left(\frac{1}{G_+(w)} - \frac{1}{G_-(w)} \right) dw. \tag{5.59}$$

This number is finite because $G_+(w) > 0$, and $G_-(w) < 0$ for all appropriate w.

As with the Hodgkin–Huxley equations, the behavior of the periodic orbits as I_{app} varies can be summarized in a bifurcation diagram. For each value of I_{app} we plot the value of v at the steady state, and (where appropriate) the maximum and minimum values of v over a periodic orbit. As I_{app} increases, a branch of periodic orbits appears in a Hopf bifurcation at $I_{app} = 0.1$ and disappears again in another Hopf bifurcation at $I_{app} = 1.24$. Between these two points there is a branch of stable periodic orbits. The bifurcation diagram is shown in Fig. 5.17.

5.3 EXERCISES

1. Show that, if $k > 1$, then $(1 - e^{-x})^k$ has an inflection point, but $(e^{-x})^k$ does not.

2. Explain why replacing the extracellular Na$^+$ with choline has little effect on the resting potential of an axon. Calculate the new resting potential with 90% of the extracellular Na$^+$ removed. Why is the same not true if K$^+$ is replaced? (Assume the conductances are constant.)

3. Plot the nullclines of the Hodgkin–Huxley fast subsystem. Show that v_r and v_e in the Hodgkin–Huxley fast subsystem are stable steady states, while v_s is a saddle point. Compute the stable manifold of the saddle point and compute sample trajectories in the fast phase plane, demonstrating the threshold effect.

4. Show how the Hodgkin–Huxley fast subsystem depends on the slow variables; i.e., show how the v nullcline moves as n and h are changed, and demonstrate the saddle-node bifurcation in which v_e and v_s disappear.

5. Plot the nullclines of the fast–slow Hodgkin–Huxley phase plane and compute a complete action potential.

6. How does the phase plane of the fast–slow Hodgkin–Huxley equations change with applied current? How much applied current in the fast–slow Hodgkin–Huxley equations is needed to generate oscillations? Plot a typical oscillation in the phase plane. Plot the maximum of the oscillation against the applied current to construct a bifurcation diagram.

7. Suppose that in the Hodgkin–Huxley fast–slow phase plane, v is slowly decreased to $v^* < v_0$ (where v_0 is the steady state), held there for a considerable time, and then released. Describe what happens in qualitative terms, i.e., without actually computing the solution. This is called *anode break excitation* (Hodgkin and Huxley, 1952d. Also see Peskin, 1991). What happens if v is instantaneously decreased to v^* and then released immediately? Why do these two solutions differ?

8. In the text, the Hodgkin–Huxley equations are written in terms of $v = V - V_{eq}$. Show that in terms of V the equations are

$$C_m \frac{dV}{dt} = -\bar{g}_K n^4(V - V_K) - \bar{g}_{Na}m^3 h(V - V_{Na})$$
$$- \bar{g}_L(V - V_L) + I_{app}, \tag{5.60}$$

$$\frac{dm}{dt} = \alpha_m(1 - m) - \beta_m m, \tag{5.61}$$

$$\frac{dn}{dt} = \alpha_n(1 - n) - \beta_n n, \tag{5.62}$$

$$\frac{dh}{dt} = \alpha_h(1 - h) - \beta_h h, \tag{5.63}$$

where (in units of $(ms)^{-1}$),

$$\alpha_m = 0.1 \frac{-40 - V}{\exp\left(\frac{-40-V}{10}\right) - 1}, \tag{5.64}$$

$$\beta_m = 4 \exp\left(\frac{-V - 65}{18}\right), \tag{5.65}$$

$$\alpha_h = 0.07 \exp\left(\frac{-V - 65}{20}\right), \tag{5.66}$$

$$\beta_h = \frac{1}{\exp(\frac{-35-V}{10}) + 1}, \tag{5.67}$$

$$\alpha_n = 0.01 \frac{-55 - V}{\exp(\frac{-55-V}{10}) - 1}, \tag{5.68}$$

$$\beta_n = 0.125 \exp\left(\frac{-V - 65}{80}\right), \tag{5.69}$$

and

$$\bar{g}_{Na} = 120, \quad \bar{g}_K = 36, \quad \bar{g}_L = 0.3, \tag{5.70}$$
$$V_{Na} = 55, \quad V_K = -77, \quad V_L = -54.4, \quad V_{eq} = -65. \tag{5.71}$$

9. Solve the full Hodgkin–Huxley equations numerically with a variety of constant current inputs. For what range of inputs are there self-sustained oscillations? Construct the bifurcation diagram as in Exercise 6.

10. The Hodgkin–Huxley equations are for the squid axon at 6.3°C. Using that the absolute temperature enters the equations through the Nernst equation, determine how changes in temperature affect the behavior of the equations. In particular, simulate the equations at 0°C and 30°C to determine whether the equations become more or less excitable with an increase in temperature.

11. Show that a Hopf bifurcation occurs in the generalized FitzHugh–Nagumo equations when $f_v(v^*, w^*) = -\epsilon g_w(v^*, w^*)$, assuming that

$$f_v(v^*, w^*)g_w(v^*, w^*) - g_v(v^*, w^*)f_w(v^*, w^*) > 0.$$

On which side of the minimum of the v nullcline can this condition be satisfied?

12. Morris and Lecar (1981) proposed the following two-variable model of membrane potential for a barnacle muscle fiber:

$$C_m \frac{dV}{dT} + I_{\text{ion}}(V, W) = I_{\text{app}}, \tag{5.72}$$

$$\frac{dW}{dT} = \phi \Lambda(V)[W_\infty(V) - W], \tag{5.73}$$

where V = membrane potential, W = fraction of open K^+ channels, T = time, C_m = membrane capacitance, I_{app} = externally applied current, ϕ = maximum rate for closing K^+ channels, and

$$I_{\text{ion}}(V, W) = g_{\text{Ca}}M_\infty(V)(V - V_{\text{Ca}}) + g_K W(V - V_K) + g_L(V - V_L), \tag{5.74}$$

$$M_\infty(V) = \frac{1}{2}\left(1 + \tanh\left(\frac{V - V_1}{V_2}\right)\right), \tag{5.75}$$

$$W_\infty(V) = \frac{1}{2}\left(1 + \tanh\left(\frac{V - V_3}{V_4}\right)\right), \tag{5.76}$$

$$\Lambda(V) = \cosh\left(\frac{V - V_3}{2V_4}\right). \tag{5.77}$$

Typical rate constants in these equations are shown in Table 5.1.

Table 5.1 Typical parameter values for the Morris–Lecar model.

$C_m = 20 \ \mu\text{F/cm}^2$	$I_{\text{app}} = 0.06 \ \text{mA/cm}^2$
$g_{\text{Ca}} = 4.4 \ \text{mS/cm}^2$	$g_K = 8 \ \text{mS/cm}^2$
$g_L = 2 \ \text{mS/cm}^2$	$\phi = 0.04 \ (\text{ms})^{-1}$
$V_1 = -1.2 \ \text{mV}$	$V_2 = 18 \ \text{mV}$
$V_3 = 2$	$V_4 = 30 \ \text{mV}$
$V_{\text{Ca}} = 120 \ \text{mV}$	$V_K = -84 \ \text{mV}$
$V_L = -60 \ \text{mV}$	

(a) Make a phase portrait for the Morris–Lecar equations. Plot the nullclines and show some typical trajectories, demonstrating that the model is excitable.

(b) Does the Morris–Lecar model exhibit anode break excitation (see Exercise 7)? If not, why not?

13. The Pushchino model is a piecewise-linear model of FitzHugh–Nagumo type proposed as a model of the ventricular action potential. The model has

$$f(v, w) = F(v) - w, \tag{5.78}$$

$$g(v, w) = \frac{1}{\tau(v)}(v - w), \tag{5.79}$$

where

$$F(v) = \begin{cases} -30v, & \text{for } v < v_1, \\ \gamma v - 0.12, & \text{for } v_1 < v < v_2, \\ -30(v - 1), & \text{for } v > v_2, \end{cases} \tag{5.80}$$

$$\tau(v) = \begin{cases} 2, & \text{for } v < v_1, \\ 16.6, & \text{for } v > v_1, \end{cases} \tag{5.81}$$

with $v_1 = 0.12/(30 + \gamma)$ and $v_2 = 30.12/(30 + \gamma)$.

Simulate the action potential for this model. What is the effect on the action potential of changing $\tau(v)$?

14. Perhaps the most important example of a nonphysiological excitable system is the Belousov–Zhabotinsky reaction. This reaction denotes the oxidation of malonic acid by bromate in acidic solution in the presence of a transition metal ion catalyst. Kinetic equations describing this reaction are (Tyson and Fife, 1980)

$$\epsilon \frac{du}{dt} = -fv\frac{u - q}{u + q} + u - u^2, \tag{5.82}$$

$$\frac{dv}{dt} = u - v, \tag{5.83}$$

where u denotes the concentration of bromous acid and v denotes the concentration of the oxidized catalyst metal. Typical values for parameters are $\epsilon \approx 0.01, f = 1, q \approx 10^{-4}$. Describe the phase portrait for this system of equations.

15. It is not particularly difficult to build an electrical analogue of the FitzHugh–Nagumo equations with inexpensive and easily obtained electronic components. The parts list for one "cell" (shown in Fig. 5.20) includes two op-amps (operational amplifiers), two power supplies, a few resistors, and two capacitors, all readily available from any consumer electronics store (Keener, 1983).

The key component is an operational amplifier (Fig. 5.18). An op-amp is denoted in a circuit diagram by a triangle with two inputs on the left and a single output from the vertex on the right. Only three circuit connections are shown on a diagram, but two more

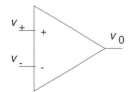

Figure 5.18 Diagram for an operational amplifier (op-amp).

are assumed, being necessary to connect with the power supply to operate the op-amp. Corresponding to the supply voltages V_{s-} and V_{s+}, there are voltages V_{r-} and V_{r+}, called the *rail voltages*, which determine the operational range for the output of an op-amp. The job of an op-amp is to compare the two input voltages v_+ and v_-, and if $v_+ > v_-$, to set (if possible) the output voltage v_0 to the high rail voltage V_{r+}, whereas if $v_+ < v_-$, then v_0 is set to V_{r-}. With reliable electronic components it is a good first approximation to assume that the input draws no current, while the output v_0 can supply whatever current is necessary to maintain the required voltage level.

The response of an op-amp to changes in input is not instantaneous, but is described reasonably well by the differential equation

$$\epsilon_s \frac{dv_0}{dt} = g(v_+ - v_-) - v_0. \tag{5.84}$$

The function $g(v)$ is continuous, but quite close to the piecewise-constant function

$$g(v) = V_{r+}H(v) + V_{r-}H(-v), \tag{5.85}$$

with $H(v)$ the Heaviside function. The number ϵ_s is small, and is the inverse of the *slew-rate*, which is typically on the order of 10^6–10^7 V/sec. For all of the following circuit analysis, take $\epsilon_s \to 0$.

(a) Show that the simple circuit shown in Fig. 5.19 is a linear amplifier, with

$$v_0 = \frac{R_1 + R_2}{R_2} v_+, \tag{5.86}$$

provided that v_0 is within the range of the rail voltages.

(b) Show that if $R_1 = 0, R_2 = \infty$, then the device in Fig. 5.19 becomes a *voltage follower* with $v_0 = v_+$.

(c) Find the governing equations for the circuit in Fig. 5.20, assuming that the rail voltages for op-amp 2 are well within the range of the rail voltages for op-amp 1. Show that

$$C_1 \frac{dv}{dt} + i_2 \left(1 - \frac{R_4}{R_5}\right) + \frac{F(v)}{R_3} + \frac{v - v_g}{R_5} = 0, \tag{5.87}$$

$$C_2 R_5 \frac{di_2}{dt} + R_4 i_2 = v - v_g, \tag{5.88}$$

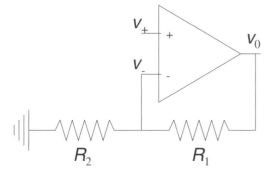

Figure 5.19 Linear amplifier using an op-amp.

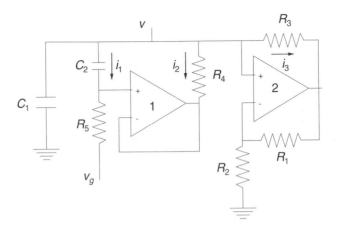

Figure 5.20 FitzHugh–Nagumo circuit using op-amps.

Table 5.2 Parts list for the FitzHugh–Nagumo analog circuit.

2 LM 741 op-amps (National Semiconductor)	
$R_1 = R_2 = 100\text{k}\Omega$	$R_3 = 2.4\Omega$
$R_4 = 1\text{k}\Omega$	$R_5 = 10\text{k}\Omega$
$C_1 = 0.01\mu\text{F}$	$C_2 = 0.5\mu\text{F}$
Power supplies:	
±15V for op-amp #1	±12V for op-amp #2

where $F(v)$ is the piecewise-linear function

$$F(v) = \begin{cases} v - V_{r_+}, & \text{for} \quad v > \alpha V_{r_+}, \\ -\frac{R_1}{R_2}v, & \text{for} \quad \alpha V_{r_-} \le v \le \alpha V_{r_+}, \\ v - V_{r_-}, & \text{for} \quad v < \alpha V_{r_-}, \end{cases} \tag{5.89}$$

and $\alpha = \frac{R_2}{R_1+R_2}$.

(d) Sketch the phase portrait for these circuit equations. Show that this is a piecewise-linear FitzHugh–Nagumo system.

(e) Use the singular perturbation approximation (5.59) to estimate the period of oscillation for the piecewise-linear analog FitzHugh–Nagumo circuit in Fig. 5.20.

CHAPTER 6

Wave Propagation in Excitable Systems

The problem of current flow in the axon of a nerve is much more complicated than that of flow in dendritic networks (Chapter 4). Recall from Chapter 5 that the voltage dependence of the ionic currents can lead to excitability and action potentials. In this chapter we show that when an excitable membrane is incorporated into a *nonlinear* cable equation, it can give rise to traveling waves of electrical excitation. Indeed, this property is one of the reasons that the Hodgkin–Huxley equations are so important. In addition to producing a realistic description of a space-clamped action potential, Hodgkin and Huxley showed that this action potential propagates along an axon with a fixed speed, which could be calculated.

However, the nerve axon is but one of many examples of a spatially extended excitable system in which there is propagated activity. For example, propagated waves of electrical or chemical activity are known to occur in skeletal and cardiac tissue, in the retina, in the cortex of the brain, and within single cells of a wide variety of types. In this chapter we describe this wave activity, beginning with a discussion of propagated electrical activity along one-dimensional cables, then concluding with a brief discussion of waves in higher-dimensional excitable systems.

6.1 Brief Overview of Wave Propagation

There is a vast literature on wave propagation in biological systems. In addition to the books by Murray (2002), Britton (1986), and Grindrod (1991), there are numerous articles in journals and books, many of which are cited in this chapter.

There are many different kinds of waves in biological systems. For example, there are waves in excitable systems that arise from the underlying excitability of the cell.

An excitable wave acts as a model of, among other things, the propagation of an action potential along the axon of a nerve or the propagation of a grass fire on a prairie. However, if the underlying kinetics are oscillatory but not excitable, and a large number of individual oscillatory units are coupled by diffusion, the resulting behavior is oscillatory waves and periodic wave trains. In this chapter we focus our attention on waves in excitable media, and defer consideration of the theory of coupled oscillators to Chapters 12 and 18.

We emphasize at the outset that by a *traveling wave* we mean a solution of a partial differential equation on an infinite domain (a fictional object, of course) that travels at constant velocity with fixed shape. It is also helpful to make a distinction between the two most important types of traveling waves in excitable systems. First, there is the wave that looks like a moving plateau, or transition between different levels. If v denotes the wave variable, then ahead of the wave, v is steady at some low value, and behind the wave, v is steady at a higher value (Fig. 6.1A). Such waves are called *traveling fronts*. The second type of wave begins and ends at the same value of v (Fig. 6.1B) and resembles a moving bump. This type of wave is called a *traveling pulse*.

These two wave types can be interpreted in the terminology of the Hodgkin–Huxley fast–slow phase plane discussed in Chapter 5. When the recovery variable is fixed at the steady state, the fast–slow phase plane has two stable steady states, $v = v_r$ and $v = v_e$ (i.e., it is bistable). Under appropriate conditions there exists a traveling front with $v = v_r$ ahead of the wave and $v = v_e$ behind the wave. Thus, the traveling front acts like a zipper, switching the domain from the resting to the excited state. However, if the recovery variable n is allowed to vary, the solution is eventually forced to return to the resting state and the traveling solution becomes a traveling pulse. The primary difference between the traveling front and the traveling pulse is that in the former case there is no recovery (or recovery is static), while in the latter case recovery plays an important dynamic role.

One of the simplest models of biological wave propagation is Fisher's equation. Although this equation is used extensively in population biology and ecology, it is much

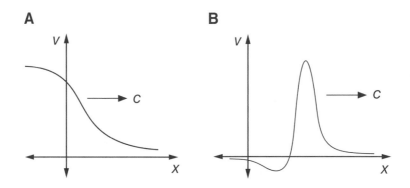

Figure 6.1   Schematic diagram of A: Traveling front, B: Traveling pulse.

less relevant in a physiological context, and so is not discussed here (see Exercise 14 and Fife, 1979).

The next level of complexity is the *bistable equation*. The bistable equation is so named because it has two stable rest points, and it is related to the FitzHugh–Nagumo equations without recovery. For the bistable equation, one expects to find traveling fronts but not traveling pulses. Inclusion of the recovery variable leads to a more complex model, the spatially distributed FitzHugh–Nagumo equations, for which one expects to find traveling pulses (among other types of waves). Wave propagation in the FitzHugh–Nagumo equations is still not completely understood, especially in higher-dimensional domains. At the highest level of complexity are the spatially distributed models of Hodgkin–Huxley type, systems of equations that are resistant to analytical approaches.

6.2 Traveling Fronts

6.2.1 The Bistable Equation

The bistable equation is a specific version of the cable equation (4.18), namely

$$\frac{\partial V}{\partial t} = \frac{\partial^2 V}{\partial x^2} + f(V), \qquad (6.1)$$

where $f(V)$ has three zeros at $0, \alpha$, and 1, where $0 < \alpha < 1$. The values $V = 0$ and $V = 1$ are stable steady solutions of the ordinary differential equation $dV/dt = f(V)$. Notice that the variable V has been scaled so that 0 and 1 are zeros of $f(V)$. In the standard nondimensional form, $f'(0) = -1$. (Recall from (4.13) that the passive cable resistance was defined so that the ionic current has slope 1 at rest.) However, this restriction on $f(V)$ is often ignored.

An example of such a function can be found in the Hodgkin–Huxley fast–slow phase plane. When the recovery variable n is held fixed at its steady state, the Hodgkin–Huxley fast–slow model is bistable. Two other examples of functions that are often used in this context are the cubic polynomial

$$f(V) = aV(V - 1)(\alpha - V), \qquad 0 < \alpha < 1, \qquad (6.2)$$

and the piecewise-linear function

$$f(V) = -V + H(V - \alpha), \qquad 0 < \alpha < 1. \qquad (6.3)$$

where $H(V)$ is the Heaviside function (Mckean, 1970). This piecewise-linear function is not continuous, nor does it have three zeros, yet it is useful in the study of traveling wave solutions of the bistable equation because it is an analytically tractable model that retains many important qualitative features.

By a traveling wave solution, we mean a translation-invariant solution of (6.1) that provides a transition between the two stable rest states (zeros of the nonlinear function

$f(V)$) and travels with constant speed. That is, we seek a solution of (6.1) of the form

$$V(x,t) = U(x + ct) = U(\xi) \tag{6.4}$$

for some (yet to be determined) value of c. The new variable ξ, called the traveling wave coordinate, has the property that fixed values move with fixed speed c. When written as a function of ξ, the wave appears stationary. Note that, because we use $\xi = x + ct$ as the traveling wave coordinate, a solution with c positive corresponds to a wave moving from right to left. We could equally well have used $x - ct$ as the traveling wave coordinate, to obtain waves moving from left to right (for positive c).

By substituting (6.4) into (6.1) it can be seen that any traveling wave solution must satisfy

$$U_{\xi\xi} - cU_\xi + f(U) = 0, \tag{6.5}$$

and this, being an ordinary differential equation, should be easier to analyze than the original partial differential equation. For $U(\xi)$ to provide a transition between rest points, it must be that $f(U(\xi)) \to 0$ as $\xi \to \pm\infty$.

It is convenient to write (6.5) as two first-order equations,

$$U_\xi = W, \tag{6.6}$$
$$W_\xi = cW - f(U). \tag{6.7}$$

To find traveling front solutions for the bistable equation, we look for a solution of (6.6) and (6.7) that connects the rest points $(U, W) = (0,0)$ and $(U, W) = (1,0)$ in the (U, W) phase plane. Such a trajectory, connecting two different steady states, is called a heteroclinic trajectory, and in this case is parameterized by ξ; the trajectory approaches $(0,0)$ as $\xi \to -\infty$ and approaches $(1,0)$ as $\xi \to +\infty$ (see the dashed line in Fig. 6.2A). The steady states at $U = 0$ and $U = 1$ are both saddle points, while for the steady state $U = \alpha$, the real part of both eigenvalues have the same sign, negative if c is positive and positive if c is negative, so that this is a node or a spiral point. Since the points at $U = 0$ and $U = 1$ are saddle points, the goal is to determine whether the parameter c can be chosen so that the trajectory that leaves $U = 0$ at $\xi = -\infty$ connects with the saddle point $U = 1$ at $\xi = +\infty$. This mathematical procedure is called *shooting*, and some sample trajectories are shown in Fig. 6.2A.

First, we can determine the sign of c. Supposing a monotone increasing ($U_\xi > 0$) connecting trajectory exists, we multiply (6.5) by U_ξ and integrate from $\xi = -\infty$ to $\xi = \infty$ with the result that

$$c \int_{-\infty}^{\infty} W^2 \, d\xi = \int_0^1 f(u) \, du. \tag{6.8}$$

In other words, if a traveling wave solution exists, then the sign of c is the same as the sign of the area under the curve $f(u)$ between $u = 0$ and $u = 1$. If this area is positive, then the traveling solutions move the state variable U from $U = 0$ to $U = 1$, and the state at $U = 1$ is said to be *dominant*. In both of the special cases (6.2) and (6.3), the state $U = 1$ is dominant if $\alpha < 1/2$.

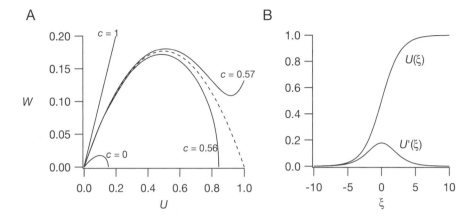

Figure 6.2 A: Trajectories in the (U, W) phase plane leaving the rest point $U = 0, W = 0$ for the equation $U_{\xi\xi} - cU_\xi + U(U - 0.1)(1 - U) = 0$, with $c = 0.0, 0.56, 0.57$, and 1.0. Dashed curve shows the connecting heteroclinic trajectory. B: Profile of the traveling wave solution of panel A. In the original coordinates, this front moves to the left with speed c.

Suppose $\int_0^1 f(u)\,du > 0$. We want to determine what happens to the unstable trajectory that leaves the saddle point $U = 0, U_\xi = 0$ for different values of c. With $c = 0$, an explicit expression for this trajectory is found by multiplying (6.5) by U_ξ and integrating to get

$$\frac{W^2}{2} + \int_0^U f(u)\,du = 0. \tag{6.9}$$

If this trajectory were to reach $U = 1$ for some value of W, then

$$\frac{W^2}{2} + \int_0^1 f(u)\,du = 0, \tag{6.10}$$

which can occur only if $\int_0^1 f(u)\,du \leq 0$. Since this contradicts our assumption that $\int_0^1 f(u)\,du \geq 0$, we conclude that this trajectory cannot reach $U = 1$. Neither can this trajectory remain in the first quadrant, as $W > 0$ implies that U is increasing. Thus, this trajectory must intersect the $W = 0$ axis at some value of $U < 1$ (Fig. 6.2A). It cannot be the connecting trajectory.

Next, suppose c is large. In the (U, W) phase plane, the slope of the unstable trajectory leaving the rest point at $U = 0$ is the positive root of $\lambda^2 - c\lambda + f'(0) = 0$, which is always larger than c (Exercise 1). Let K be the smallest positive number for which $f(u)/u \leq K$ for all u on the interval $0 < u \leq 1$ (Exercise: How do we know K exists?), and let σ be any fixed positive number. On the line $W = \sigma U$ the slope of trajectories satisfies

$$\frac{dW}{dU} = c - \frac{f(U)}{W} = c - \frac{f(U)}{\sigma U} \geq c - \frac{K}{\sigma}. \tag{6.11}$$

By picking c large enough, we are assured that $c - K/\sigma > \sigma$, so that once trajectories are above the line $W = \sigma U$, they stay above it. We know that for large enough c, the trajectory leaving the saddle point $U = 0$ starts out above this curve. Thus, this trajectory always stays above the line $W = \sigma U$, and therefore passes above the rest point at $(U, W) = (1, 0)$.

Now we have two trajectories, one with $c = 0$, which misses the rest point at $U = 1$ by crossing the $W = 0$ axis at some point $U < 1$, and one with c large, which misses this rest point by staying above it at $U = 1$. Since trajectories depend continuously on the parameters of the problem, there is a continuous family of trajectories depending on the parameter c between these two special trajectories, and therefore there is at least one trajectory that hits the point $U = 1, W = 0$ exactly.

The value of c for which this heteroclinic connection occurs is unique. To verify this, notice from (6.11) that the slope dW/dU of trajectories in the (U, W) plane is a monotone increasing function of the parameter c. Suppose at some value of $c = c_0$ there is known to be a connecting trajectory. For any value of c that is larger than c_0, the trajectory leaving the saddle point at $U = 0$ must lie above the connecting curve for c_0. For the same reason, with $c > c_0$, the trajectory approaching the saddle point at $U = 1$ as $\xi \to \infty$ must lie below the connecting curve with $c = c_0$. A single curve cannot simultaneously lie above and below another curve, so there cannot be a connecting trajectory for $c > c_0$. By a similar argument, there cannot be a connecting trajectory for a smaller value of c, so the value c_0, and hence the connecting trajectory, is unique.

For most functions $f(V)$, it is necessary to calculate the speed of propagation of the traveling front solution numerically. However, in the two special cases (6.2) and (6.3) the speed of propagation can be calculated explicitly. In the piecewise linear case (6.3) one calculates directly that

$$c = \frac{1 - 2\alpha}{\sqrt{\alpha - \alpha^2}} \tag{6.12}$$

(see Exercise 4).

Suppose $f(u)$ is the cubic polynomial

$$f(u) = -A^2(u - u_0)(u - u_1)(u - u_2), \tag{6.13}$$

where the zeros of the cubic are ordered $u_0 < u_1 < u_2$. We want to find a heteroclinic connection between the smallest zero, u_0, and the largest zero, u_2, so we guess that

$$W = -B(U - u_0)(U - u_2). \tag{6.14}$$

We substitute this guess into the governing equation (6.5), and find that we must have

$$B^2(2U - u_0 - u_2) - cB - A^2(U - u_1) = 0. \tag{6.15}$$

This is a linear function of U that can be made identically zero only if we choose $B = A/\sqrt{2}$ and

$$c = \frac{A}{\sqrt{2}}(u_2 - 2u_1 + u_0). \tag{6.16}$$

It follows from (6.14) that

$$U(\xi) = \frac{u_0 + u_2}{2} + \frac{u_2 - u_0}{2} \tanh\left(\frac{A}{\sqrt{2}}\frac{u_2 - u_0}{2}\xi\right),$$ (6.17)

which is independent of u_1. In the case that $u_0 = 0, u_1 = \alpha$, and $u_2 = 1$, the speed reduces to

$$c = \frac{A}{\sqrt{2}}(1 - 2\alpha),$$ (6.18)

showing that the speed is a decreasing function of α and the direction of propagation changes at $\alpha = 1/2$. The profile of the traveling wave in this case is

$$U(\xi) = \frac{1}{2}\left[1 + \tanh\left(\frac{A}{2\sqrt{2}}\xi\right)\right].$$ (6.19)

A plot of this traveling wave profile is shown in Fig. 6.2B.

Once the solution of the nondimensional cable equation (6.1) is known, it is a simple matter to express the solution in terms of physical parameters as

$$V(x,t) = U\left(\frac{x}{\lambda_m} + c\frac{t}{\tau_m}\right),$$ (6.20)

where λ_m and τ_m are, respectively, the space and time constants of the cable, as described in Chapter 4. The speed of the traveling wave is

$$s = \frac{c\lambda_m}{\tau_m} = \frac{c}{2C_m}\sqrt{\frac{d}{R_m R_c}},$$ (6.21)

which shows how the wave speed depends on capacitance, membrane resistance, cytoplasmic resistance, and axonal diameter. The dependence of the speed on ionic channel conductances is contained (but hidden) in c. According to empirical measurements, a good estimate of the speed of an action potential in an axon is

$$s = \sqrt{\frac{d}{10^{-6}\text{m}}}\ \text{m/sec}.$$ (6.22)

Using $d = 500\ \mu\text{m}$ for squid axon, this estimate gives $s = 22.4$ mm/ms, which compares favorably to the measured value of $s = 21.2$ mm/ms.

Scaling arguments can also be used to find the dependence of speed on certain other parameters. Suppose, for example, that a drug is applied to the membrane that blocks a percentage of all ion channels, irrespective of type. If ρ is the fraction of remaining operational channels, then the speed of propagation is reduced by the factor $\sqrt{\rho}$. This follows directly by noting that the bistable equation with a reduced number of ion channels,

$$V'' - sV' + \rho f(V) = 0,$$ (6.23)

can be related to the original bistable equation (6.5) by taking $V(\xi) = U(\sqrt{\rho}\xi), s = c\sqrt{\rho}$.

Table 6.1 Sodium channel densities in selected excitable tissues.

Tissue	Channel density (channels/μm^2)
Mammalian	
Vagus nerve (nonmyelinated)	110
Node of Ranvier	2100
Skeletal muscle	205–560
Other animals	
Squid giant axon	166–533
Frog sartorius muscle	280
Electriceel electroplax	550
Garfish olfactory nerve	35
Lobster walking leg nerve	90

Thresholds and Stability

There are many other features of the bistable equation, the details of which are beyond the scope of this book. Perhaps the most important of these features is that solutions of the bistable equation satisfy a comparison property: any two solutions of the bistable equation, say $u_1(x,t)$ and $u_2(x,t)$, that are ordered with $u_1(x,t_0) \leq u_2(x,t_0)$ at some time $t = t_0$, remain ordered for all subsequent times, i.e., $u_1(x,t) \leq u_2(x,t)$ for $t \geq t_0$.

With comparison arguments it is possible to prove a number of additional facts (Aronson and Weinberger, 1975). For example, the bistable equation exhibits threshold behavior. Specifically, if initial data are sufficiently small, then the solution of the bistable equation approaches zero in the limit $t \to \infty$. However, there are initial functions with compact support lying between 0 and 1 for which the solution approaches 1 in the limit $t \to \infty$. Because of the comparison theorem any larger initial function also initiates a solution that approaches 1 in the limit $t \to \infty$. Such initial data are said to be *superthreshold*.

Furthermore, the traveling wave solution of the bistable equation is stable in a very strong way (Fife, 1979; Fife and McLeod, 1977), as follows. Starting from any initial data that lie between 0 and α in the limit $x \to -\infty$ and between α and 1 in the limit $x \to \infty$, the solution approaches some phase shift of the traveling wave solution in the limit of large time.

6.2.2 Myelination

Most nerve fibers are coated with a lipid material called *myelin* with periodic gaps of exposure called *nodes of Ranvier* (Fig. 6.3). The myelin sheath consists of a single cell, called a *Schwann cell*, which is wrapped many times (roughly 100 times) around the axonal membrane. This wrapping of the axon increases the effective membrane resistance by a factor of about 100 and decreases the membrane capacitance by a factor of about 100. Indeed, rough data are that R_m is 10^3 Ω cm^2 for cell membrane

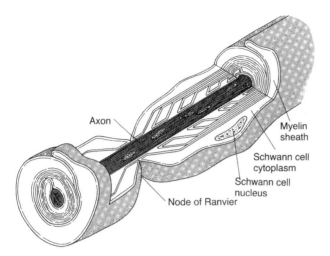

Axon

Myelin sheath

Schwann cell cytoplasm

Schwann cell nucleus

Node of Ranvier

Figure 6.3 Schematic diagram of the myelin sheath. (Guyton and Hall, 1996, Fig. 5–16, p. 69.)

and $10^5 \, \Omega \, \text{cm}^2$ for myelin sheath, and that C_m is $10^{-6} \, \mu\text{F}/\text{cm}^2$ for cell membrane and $10^{-8} \, \mu\text{F}/\text{cm}^2$ for a myelinated fiber. The length of myelin sheath is typically 1 to 2 mm (close to 100 d, where d is the fiber diameter), and the width of the node of Ranvier is about 1 μm.

Propagation along myelinated fiber is faster than along nonmyelinated fiber. This is presumably caused by the fact that there is little transmembrane ionic current and little capacitive current in the myelinated section, allowing the axon to act as a simple resistor. An action potential does not propagate along the myelinated fiber but rather jumps from node-to-node. This node-to-node propagation is said to be *saltatory* (from the Latin word *saltare*, to leap or dance).

The pathophysiological condition of nerve cells in which damage of the myelin sheath impairs nerve impulse transmission in the central nervous system is called *multiple sclerosis*. Multiple sclerosis is an autoimmune disease that usually affects young adults between the ages of 18 and 40, occurring slightly more often in females than males. With multiple sclerosis, there is an immune response to the white matter of the brain and spinal cord, causing demyelination of nerve fibers at various locations throughout the central nervous system, although the underlying nerve axons and cell bodies are not usually damaged. The loss of myelin slows or stops the transmission of action potentials, with the resultant symptoms of muscle fatigue and weakness or extreme "heaviness."

To model the electrical activity in a myelinated fiber we assume that the capacitive and transmembrane ionic currents are negligible, so that, along the myelin sheath, the axial currents

$$I_e = -\frac{1}{r_e}\frac{\partial V_e}{\partial x}, \qquad I_i = -\frac{1}{r_i}\frac{\partial V_i}{\partial x} \tag{6.24}$$

are constant (using the same notation as in Chapter 4). We also assume that V does not vary within each node of Ranvier (i.e., that the nodes are isopotential), and that

V_n is the voltage at the nth node. Then the axial currents between node n and node $n+1$ are

$$I_e = -\frac{1}{Lr_e}(V_{e,n+1} - V_{e,n}), \qquad I_i = -\frac{1}{Lr_i}(V_{i,n+1} - V_{i,n}), \tag{6.25}$$

where L is the length of the myelin sheath between nodes. The total transmembrane current at a node is given by

$$\mu p \left(C_m \frac{\partial V_n}{\partial t} + I_{\text{ion}} \right) = I_{i,n} - I_{i,n+1}$$

$$= \frac{1}{L(r_i + r_e)}(V_{n+1} - 2V_n + V_{n-1}), \tag{6.26}$$

where μ is the length of the node.

We can introduce dimensionless time $\tau = \frac{t}{C_m R_m} = t/\tau_m$ (but not dimensionless space), to rewrite (6.26) as

$$\frac{dV_n}{d\tau} = f(V_n) + D(V_{n+1} - 2V_n + V_{n-1}), \tag{6.27}$$

where $D = \frac{R_m}{\mu L p (r_i + r_e)}$ is the coupling coefficient. We call this equation the *discrete cable equation*.

6.2.3 The Discrete Bistable Equation

The discrete bistable equation is the system of equations (6.27) where $f(V)$ has typical bistable form, as, for example, (6.2) or (6.3). The study of the discrete bistable equation is substantially more difficult than that of the continuous version (6.1). While the discrete bistable equation looks like a finite difference approximation of the continuous bistable equation, solutions of the two have significantly different behavior.

It is a highly nontrivial matter to prove that traveling wave solutions of the discrete system exist (Zinner, 1992). However, a traveling wave solution, if it exists, satisfies the special relationship $V_{n+1}(\tau) = V_n(\tau - \tau_d)$. In other words, the $(n+1)$st node experiences exactly the same time course as the nth node, with time delay τ_d. Furthermore, if $V_n(\tau) = V(\tau)$, it follows from (6.27) that $V(\tau)$ must satisfy the delay differential equation

$$\frac{dV}{d\tau} = D(V(\tau + \tau_d) - 2V(\tau) + V(\tau - \tau_d)) + f(V(\tau)). \tag{6.28}$$

If the function $V(\tau)$ is sufficiently smooth and if τ_d is sufficiently small, then we can approximate $V(\tau + \tau_d)$ with its Taylor series $V(\tau + \tau_d) = \sum_{n=0} \frac{1}{n!} V^{(n)}(\tau)\tau_d^n$, so that (6.28) is approximated by the differential equation

$$D\left(\tau_d^2 V_{\tau\tau} + \frac{\tau_d^4}{12} V_{\tau\tau\tau\tau} \right) - V_\tau + f(V) = 0, \tag{6.29}$$

ignoring terms of order τ_d^6 and higher.

Now we suppose that τ_d is small. The leading-order equation is

$$D\tau_d^2 V_{\tau\tau} - V_\tau + f(V) = 0, \tag{6.30}$$

which has solution $V_0(\tau) = U(c\tau)$, provided that $D\tau_d^2 = 1/c^2$, where U is the traveling front solution of the bistable equation (6.5) and c is the dimensionless wave speed for the continuous equation. The wave speed s is the internodal distance $L + \mu$ divided by the time delay $\tau_m \tau_d$, so that

$$s = \frac{L + \mu}{\tau_m \tau_d} = (L + \mu)c \frac{\sqrt{D}}{\tau_m}. \tag{6.31}$$

For myelinated nerve fiber we know that $D = \frac{R_m}{\mu L p (r_i + r_e)}$. If we ignore extracellular resistance, we find a leading-order approximation for the velocity of

$$s = \frac{L + \mu}{\sqrt{\mu L}} \frac{c}{2C_m} \sqrt{\frac{d}{R_m R_c}}, \tag{6.32}$$

giving a change in velocity compared to nonmyelinated fiber by the factor $\frac{L + \mu}{\sqrt{\mu L}}$. With the estimates $L = 100\, d$ and $\mu = 1\ \mu$m, this increase in velocity is by a factor of $10\sqrt{\frac{d}{10^{-6}\text{m}}}$, which is substantial. Empirically it is known that the improvement of velocity for myelinated fiber compared to nonmyelinated fiber is by a factor of about $6\sqrt{\frac{d}{10^{-6}\text{m}}}$.

Higher-Order Approximation

We can find a higher-order approximation to the speed of propagation by using a standard regular perturbation argument. We set $\epsilon = 1/D$ and seek a solution of (6.29) of the form

$$V(\tau) = V_0(\tau) + \epsilon V_1(\tau) + \cdots, \tag{6.33}$$

$$\tau_d^2 = \frac{\epsilon}{c^2} + \epsilon^2 \tau_1 + \cdots. \tag{6.34}$$

We expand (6.29) into its powers of ϵ and set the coefficients of ϵ to zero. The first equation we obtain from this procedure is (6.30), and the second equation is

$$L[V_1] = \frac{1}{c^2}V_1'' - V_1' + f'(V_0)V_1 = -\frac{V_0''''}{12c^4} - \tau_1 V_0''. \tag{6.35}$$

Note that here we are using $L[\cdot]$ to denote a linear differential operator. The goal is to find solutions of (6.35) that are square integrable on the infinite domain, so that the solution is "close" to V_0. The linear operator $L[\cdot]$ is not an invertible operator in this space by virtue of the fact that $L[V_0'(\tau)] = 0$. (This follows by differentiating (6.30) once with respect to τ.) Thus, it follows from the Fredholm alternative theorem (Keener, 1998) that a solution of (6.35) exists if and only if the right-hand side of the equation is orthogonal to the null space of the adjoint operator L^*. Here the adjoint differential operator is

$$L^*[V] = \frac{1}{c^2}V'' + V' + f'(V_0)V, \tag{6.36}$$

and the one element of the null space (a solution of $L^*[V] = 0$) is

$$V^*(\tau) = \exp\left(-c^2\tau\right) V_0'(\tau). \tag{6.37}$$

This leads to the solvability condition

$$\tau_1 \int_{-\infty}^{\infty} \exp\left(-c^2\tau\right) V_0'(\tau) V_0''(\tau) \, d\tau = -\frac{1}{12c^4} \int_{-\infty}^{\infty} \exp\left(-c^2\tau\right) V_0'(\tau) V_0''''(\tau) \, d\tau. \tag{6.38}$$

As a result, τ_1 can be calculated (either analytically or numerically) by evaluating two integrals, and the speed of propagation is determined as

$$s = (L + \mu)\frac{c}{\tau_m}\sqrt{D}\left(1 - \frac{\tau_1 c^2}{2D} + O\left(\frac{c^2}{D}\right)^2\right). \tag{6.39}$$

This exercise is interesting from the point of view of numerical analysis, as it shows the effect of numerical discretization on the speed of propagation. This method can be applied to other numerical schemes for an equation with traveling wave solutions (Exercise 20).

Propagation Failure

The most significant difference between the discrete and continuous equations is that the discrete system has a coupling threshold for propagation, while the continuous model allows for propagation at all coupling strengths. It is readily seen from (6.21) that for the continuous cable equation, continuous changes in the physical parameters lead to continuous changes in the speed of propagation, and the speed cannot be driven to zero unless the diameter is zero or the resistances or capacitance are infinite. Such is not the case for the discrete system, and propagation may fail if the coupling coefficient is too small. This is easy to understand when we realize that if the coupling strength is very weak, so that the effective internodal resistance is large, the current flow from an excited node to an unexcited node may be so small that the threshold of the unexcited node is not exceeded, and propagation cannot continue.

To study propagation failure, we seek standing (time-independent, i.e., $dV_n/d\tau = 0$) solutions of the discrete equation (6.27). The motivation for this comes from the maximum principle and comparison arguments. One can show that if two sets of initial data for the discrete bistable equation are initially ordered, the corresponding solutions remain ordered for all time. It follows that if the discrete bistable equation has a monotone increasing stationary front solution, then there cannot be a traveling wave front solution.

A standing front solution of the discrete bistable equation is a sequence $\{V_n\}$ satisfying the finite difference equation

$$0 = D(V_{n+1} - 2V_n + V_{n-1}) + f(V_n) \tag{6.40}$$

for all integers n, for which $V_n \to 1$ as $n \to \infty$ and $V_n \to 0$ as $n \to -\infty$.

One can show (Keener, 1987) that for any bistable function f, there is a number $D^* > 0$ such that for $D \leq D^*$, the discrete bistable equation has a standing solution, that is, propagation fails. To get a simple understanding of the behavior of this coupling threshold, we solve (6.40) in the special case of piecewise-linear dynamics (6.3). Since the discrete equation with dynamics (6.3) is linear, the homogeneous solution can be expressed as a linear combination of powers of some number λ as

$$V_n = A\lambda^n + B\lambda^{-n}, \tag{6.41}$$

where λ is a solution of the characteristic polynomial equation

$$\lambda^2 - \left(2 + \frac{1}{D}\right)\lambda + 1 = 0. \tag{6.42}$$

Note that this implies that

$$D = \frac{\lambda}{(\lambda - 1)^2}. \tag{6.43}$$

The characteristic equation has two positive roots, one larger and one smaller than 1. Let λ be the root that is smaller than one. Then, taking the conditions at $\pm\infty$ into account, we write the solution as

$$V_n = \begin{cases} 1 + A\lambda^n, & \text{for } n \geq 0, \\ B\lambda^{-n}, & \text{for } n < 0. \end{cases} \tag{6.44}$$

This expression for V_n must also satisfy the piecewise-linear discrete bistable equation for $n = -1, 0$. Thus,

$$D(V_1 - 2V_0 + V_{-1}) = V_0 - 1, \tag{6.45}$$

$$D(V_0 - 2V_{-1} + V_{-2}) = V_{-1}, \tag{6.46}$$

where we have assumed that $V_n \geq \alpha$ for all $n \geq 0$, and $V_n < \alpha$ for all $n < 0$. Substituting in (6.43) for D, and solving for A and B, then gives $B = A + 1 = \frac{1}{1+\lambda}$.

Finally, this is a solution for all n, provided that $V_0 \geq \alpha$. Since $V_0 = B = \frac{1}{1+\lambda}$, we need $\frac{1}{1+\lambda} \geq \alpha$, or $\lambda \leq \frac{1-\alpha}{\alpha}$. However, when $\lambda < 1$, D is an increasing function of λ, and thus $\lambda \leq \frac{1-\alpha}{\alpha}$ whenever

$$D \leq D\left(\frac{1-\alpha}{\alpha}\right) = \frac{\alpha(1-\alpha)}{(2\alpha - 1)^2} = D^*. \tag{6.47}$$

In other words, there is a standing wave, precluding propagation, whenever the coupling is small, with $D \leq D^*$. Since α is a measure of the excitability of this medium, we see that when the medium is weakly excitable (α is near $1/2$), then D^* is large and very little resistance is needed to halt propagation. On the other hand, when α is small, so that the medium is highly excitable, the resistance threshold is large, and propagation is relatively difficult to stop.

6.3 Traveling Pulses

A traveling pulse (often called a *solitary pulse*) is a traveling wave solution that starts
and ends at the same steady state of the governing equations. Recall that a traveling
front solution corresponds to a heteroclinic trajectory in the (U, W) phase plane, i.e., a
trajectory, parameterized by ξ, that connects two different steady states of the system.
A traveling pulse solution is similar, corresponding to a trajectory that begins and ends
at the *same* steady state in the traveling wave coordinate system. Such trajectories are
called *homoclinic orbits*.

There are three main approaches to finding traveling pulses for excitable systems.
First, one can approximate the nonlinear functions with piecewise-linear functions,
and then find traveling pulse solutions as exact solutions of transcendental equations.
Second, one can use perturbation methods exploiting the different time scales to find
approximate analytical expressions. Finally, one can use numerical simulations to solve
the governing differential equations. We illustrate each of these techniques in turn.

6.3.1 The FitzHugh–Nagumo Equations

To understand the structure of a traveling pulse it is helpful first to study traveling pulse
solutions in the FitzHugh–Nagumo equations

$$\epsilon \frac{\partial v}{\partial t} = \epsilon^2 \frac{\partial^2 v}{\partial x^2} + f(v, w), \tag{6.48}$$

$$\frac{\partial w}{\partial t} = g(v, w), \tag{6.49}$$

where ϵ is assumed to be a small positive number. Without any loss of generality, space
has been scaled so that the diffusion coefficient is ϵ^2. It is important to realize that this
does not imply anything about the magnitude of the physical diffusion coefficient. This
is simply a scaling of the space variable so that in the new coordinate system, the wave
front appears steep, a procedure that facilitates the study of the wave as a whole. The
variable v is spatially coupled with diffusion, but the variable w is not, owing to the
fact that v represents the membrane potential, while w represents a slow ionic current
or gating variable.

To study traveling waves, we place the system of equations (6.48)–(6.49) in a traveling coordinate frame of reference. We define the traveling wave coordinate $\xi = x - ct$,
where $c > 0$ is the wave speed, yet to be determined. Note that the traveling wave variable, $\xi = x - ct$, is different from the one previously used in this chapter, $x + ct$. Hence,
$c > 0$ here corresponds to a wave moving from left to right.

The partial differential equations (6.48)–(6.49) become the ordinary differential
equations

$$\epsilon^2 v_{\xi\xi} + c\epsilon v_\xi + f(v, w) = 0, \tag{6.50}$$

$$c w_\xi + g(v, w) = 0. \tag{6.51}$$

A Piecewise-Linear Model

We begin by examining the simplest case, the piecewise-linear dynamics (Rinzel and Keller, 1973)

$$f(v, w) = H(v - \alpha) - v - w, \tag{6.52}$$

$$g(v, w) = v. \tag{6.53}$$

Because the dynamics are piecewise linear, the exact solution can be constructed in a piecewise manner. We look for solutions of the form sketched in Fig. 6.4. The position of the wave along the ξ axis is specified by fixing $v(0) = v(\xi_1) = \alpha$. As yet, ξ_1 is unknown, and is to be determined as part of the solution process. Note that the places where $v = \alpha$ are those where the dynamics change (since α is the point of discontinuity of f). Let I, II, and III denote, respectively, the regions $\xi < 0$, $0 < \xi < \xi_1$, and $\xi_1 < \xi$. In each region, the differential equation is linear and so can be solved exactly. The three regional solutions are then joined at $\xi = 0$ and $\xi = \xi_1$ by stipulating that v and w be continuous at the boundaries and that v have a continuous derivative there. These constraints are sufficient to determine the solution unambiguously.

   In regions I and III, $v < \alpha$, and so the differential equation is

$$\epsilon^2 v_{\xi\xi} + c\epsilon v_\xi - v - w = 0, \tag{6.54}$$

$$cw_\xi + v = 0. \tag{6.55}$$

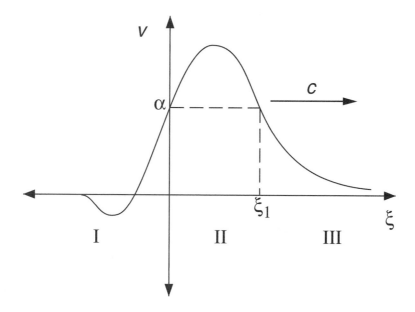

Figure 6.4 Schematic diagram of the traveling pulse of the piecewise-linear FitzHugh–Nagumo equations.

Looking for solutions of the form $v = A \exp(\lambda\xi)$, $w = B \exp(\lambda\xi)$, we find that A and B must satisfy

$$\begin{pmatrix} \lambda^2\epsilon^2 + c\epsilon\lambda - 1 & -1 \\ 1 & c\lambda \end{pmatrix} \begin{pmatrix} A \\ B \end{pmatrix} = \begin{pmatrix} 0 \\ 0 \end{pmatrix}, \tag{6.56}$$

which has a nontrivial solution if and only if

$$\begin{vmatrix} \lambda^2\epsilon^2 + c\epsilon\lambda - 1 & -1 \\ 1 & c\lambda \end{vmatrix} = 0. \tag{6.57}$$

Hence, λ must be a root of the characteristic polynomial

$$\epsilon^2 p(\lambda) = \epsilon^2\lambda^3 + \epsilon c\lambda^2 - \lambda + 1/c = 0. \tag{6.58}$$

There is exactly one negative root, call it λ_1, and the real parts of the other two roots, λ_2 and λ_3, are positive.

In region II, the differential equation is

$$\epsilon^2 v_{\xi\xi} + c\epsilon v_\xi + 1 - v - w = 0, \tag{6.59}$$

$$cw_\xi + v = 0. \tag{6.60}$$

The inhomogeneous solution is $w = 1, v = 0$, and the homogeneous solution is a sum of exponentials of the form $e^{\lambda_i\xi}$.

Since we want the solution to approach zero in the limit $\xi \to \pm\infty$, the traveling pulse can be represented as the exponential $e^{\lambda_1\xi}$ for large positive ξ, the sum of the two exponentials $e^{\lambda_2\xi}$ and $e^{\lambda_3\xi}$ for large negative ξ, and the sum of all three exponentials for the intermediate range of ξ for which $v(\xi) > \alpha$. We take

$$w(\xi) = \begin{cases} Ae^{\lambda_1\xi}, & \text{for} \quad \xi \geq \xi_1, \\ 1 + \sum\limits_{i=1}^{3} B_i e^{\lambda_i\xi}, & \text{for} \quad 0 \leq \xi \leq \xi_1, \\ \sum\limits_{i=2}^{3} C_i e^{\lambda_i\xi}, & \text{for} \quad \xi \leq 0, \end{cases} \tag{6.61}$$

with $v = -cw_\xi$. We also require $w(\xi), v(\xi)$, and $v_\xi(\xi)$ to be continuous at $\xi = 0, \xi_1$, and that $v(0) = v(\xi_1) = \alpha$.

There are six unknown constants and two unknown parameters c and ξ_1 that must be determined from the six continuity conditions and the two constraints. Following some calculation, we eliminate the coefficients A, B_i, and C_i, leaving the two constraints

$$e^{\lambda_1\xi_1} + \epsilon^2 p'(\lambda_1)\alpha - 1 = 0, \tag{6.62}$$

$$\frac{e^{-\lambda_2\xi_1}}{p'(\lambda_2)} + \frac{e^{-\lambda_3\xi_1}}{p'(\lambda_3)} + \frac{1}{p'(\lambda_1)} + \epsilon^2\alpha = 0. \tag{6.63}$$

There are now two unknowns, c and ξ_1, and two equations. In general, (6.62) could be solved for ξ_1, and (6.63) could then be used to determine c for each fixed α and ϵ.

However, it is more convenient to treat c as known and α as unknown, and then find α as a function of c. So, we set $s = e^{\lambda_1 \xi_1}$, in which case (6.63) becomes

$$h(s) = 2 - s + \frac{p'(\lambda_1)}{p'(\lambda_2)} e^{-\lambda_2 \ln(s)/\lambda_1} + \frac{p'(\lambda_1)}{p'(\lambda_3)} e^{-\lambda_3 \ln(s)/\lambda_1} = 0, \qquad (6.64)$$

where α has been eliminated using (6.62). We seek a solution of $h(s) = 0$ with $0 < s < 1$.

We begin by noting that $h(0) = 2, h(1) = 0, h'(1) = 0$, and $h''(1) = p'(\lambda_1)/\lambda_1^2 - 2$. The first of these relationships follows from the fact that the real parts of λ_2 and λ_3 are of different sign from λ_1, and therefore, in the limit as $s \to 0$, the exponential terms disappear as the real parts of the exponents approach $-\infty$. The second relationship, $h(1) = 0$, follows from the fact that $1/p'(\lambda_1) + 1/p'(\lambda_2) + 1/p'(\lambda_3) = 0$ (Exercise 9). The final two relationships are similar and are left as exercises (Exercises 9, 10).

If $h''(1) < 0$, then the value $s = 1$ is a local maximum of $h(s)$, so for s slightly less than 1, $h(s) < 0$. Since $h(0) > 0$, a root of $h(s) = 0$ in the interval $0 < s < 1$ is assured.

When λ_2 and λ_3 are real, $h(s)$ can have at most one inflection point in the interval $0 < s < 1$. This follows because the equation $h''(s) = 0$ can be written in the form $e^{(\lambda_2 - \lambda_3)\xi_1} = c$, which can have at most one root. Thus, if $h''(1) < 0$ there is precisely one root, while if $h''(1) > 0$ there can be no roots. If the roots λ_2 and λ_3 are complex, uniqueness is not assured, although the condition $h''(1) < 0$ guarantees that there is at least one root.

Differentiating the defining polynomial (6.58) with respect to λ, we observe that the condition $h''(1) < 0$ is equivalent to requiring $\epsilon^2 \lambda_1^2 + 2c\epsilon\lambda_1 - 1 < 0$. Furthermore, from the defining characteristic polynomial, we know that $\epsilon^2 \lambda_1^2 - 1 = -c\epsilon\lambda_1 + \epsilon^2/(\lambda_1 c)$, and thus it follows that $h''(1) < 0$ if $\lambda_1 < -\frac{1}{c\sqrt{\epsilon}}$. Since the polynomial $p(\lambda)$ is increasing at λ_1, we are assured that $\lambda_1 < -\frac{1}{c\sqrt{\epsilon}}$ if $p(-\frac{1}{c\sqrt{\epsilon}}) > 0$, i.e., if

$$c^2 > \epsilon. \qquad (6.65)$$

Thus, whenever $c > \sqrt{\epsilon}$, a root of $h(s) = 0$ with $0 < s < 1$ is guaranteed to exist.

Once s is known, α can be found from the relationship (6.62) whereby

$$\alpha = \frac{1-s}{\epsilon^2 p'(\lambda_1)}. \qquad (6.66)$$

In Fig. 6.5, we show the results of solving (6.64) numerically. Shown plotted is the speed c against α for three values of ϵ. The dashed curve is the asymptotic limit (6.12) for the curves in the limit $\epsilon \to 0$. The important feature to notice is that for each value of α and ϵ small enough there are two traveling pulses, while for large α there are no traveling pulses. In Fig. 6.6A is shown the fast traveling pulse, and in Fig. 6.6B is shown the slow traveling pulse, both for $\alpha = 0.1, \epsilon = 0.1$, and with $v(\xi)$ shown solid and $w(\xi)$ shown dashed.

Note that the amplitude of the slow pulse in Fig. 6.6B is substantially smaller than that of the fast pulse in Fig. 6.6A. Generally speaking, the fast pulse is stable (Jones, 1984; Yanagida, 1985), and the slow pulse is unstable (Maginu, 1985). Also note that there is nothing in the construction of these wave solutions requiring ϵ to be small.

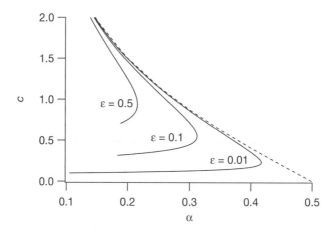

Figure 6.5 Speed c as a function of α for the traveling pulse solution of the piecewise-linear FitzHugh–Nagumo system, shown for $\epsilon = 0.5, 0.1, 0.01$. The dashed curve shows the asymptotic limit as $\epsilon \to 0$, found by singular perturbation arguments.

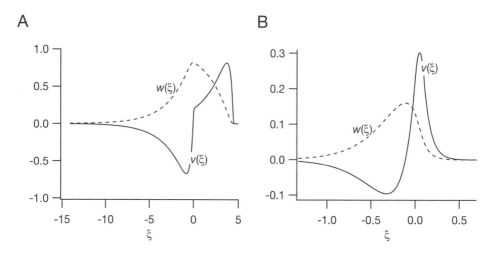

Figure 6.6 A: Plots of $v(\xi)$ and $w(\xi)$ for the fast traveling pulse ($c = 2.66$) for the piecewise-linear FitzHugh–Nagumo system with $\alpha = 0.1, \epsilon = 0.1$. B: Plots of $v(\xi)$ and $w(\xi)$ for the slow traveling pulse ($c = 0.34$) for the same piecewise-linear FitzHugh–Nagumo system as in panel A.

Singular Perturbation Theory

The next way to extract information about the traveling pulse solution of (6.48)–(6.49) is to exploit the smallness of the parameter ϵ (Keener, 1980a; for a different approach, see Rauch and Smoller, 1978). One reason we expect this to be fruitful is because of similarities with the phase portrait of the FitzHugh–Nagumo equations without diffusion, shown in Fig. 5.15. By analogy, we expect the solution to stay close to the nullcline $f(v, w) = 0$ wherever possible, with rapid transitions between the two outer branches.

The details of this behavior follow from singular perturbation analysis. (This analysis was first given for a simplified FitzHugh–Nagumo system by Casten et al., 1975.) The first observation follows simply from setting ϵ to zero in (6.48). Doing so, we obtain the *outer equations*

$$w_t = g(v, w), \qquad f(v, w) = 0. \tag{6.67}$$

Because the equation $f(v, w) = 0$ is assumed to have three solutions for v as a function of w, and only two of these solutions, the upper and lower solution branches, are stable (cf. Fig. 5.13 and the discussion in Section 5.2), the outer equations (6.67) reduce to

$$\frac{\partial w}{\partial t} = G_{\pm}(w). \tag{6.68}$$

A region of space in which $v = V_+(w)$ is called an *excited region*, and a region in which $v = V_-(w)$ is called a *recovering region*. The outer equation is valid whenever diffusion is not large. However, we anticipate that there are regions of space (*interfaces*) where diffusion is large and in which (6.68) cannot be correct.

To find out what happens when diffusion is large we rescale space and time. Letting $y(t)$ denote the position of the wave front, we set $\tau = t$ and $\xi = \frac{x - y(t)}{\epsilon}$, after which the original system of equations (6.48)–(6.49) becomes

$$v_{\xi\xi} + y'(\tau)v_{\xi} + f(v, w) = \epsilon \frac{\partial v}{\partial \tau}, \tag{6.69}$$

$$-y'(\tau)w_{\xi} = \epsilon \left(g(v, w) - \frac{\partial w}{\partial \tau} \right). \tag{6.70}$$

Upon setting $\epsilon = 0$, we find the reduced *inner equations*

$$v_{\xi\xi} + y'(\tau)v_{\xi} + f(v, w) = 0, \tag{6.71}$$

$$y'(\tau)w_{\xi} = 0. \tag{6.72}$$

Even though the inner equations (6.71)–(6.72) are partial differential equations, the variable τ occurs only as a parameter, and so (6.71)–(6.72) can be solved as if they were ordinary differential equations. This is because the traveling wave is stationary in the moving coordinate system ξ, τ. It follows that w is independent of ξ (but not necessarily τ). Finally, since the inner equation is supposed to provide a transition layer between regions where outer dynamics hold, we require the *matching condition* that $f(v, w) \to 0$ as $\xi \to \pm\infty$. Note that here we use $y(t)$ to locate the wave front, rather than ct as before, and then $y'(\tau)$ is the instantaneous wave velocity.

We recognize (6.71) as a bistable equation for which there are heteroclinic orbits. That is, for fixed w, if the equation $f(v, w) = 0$ has three roots, two of which are stable as solutions of the equation $dv/dt = f(v, w)$, then there is a number $c = c(w)$ for which the equation

$$v'' + c(w)v' + f(v,w) = 0 \tag{6.73}$$

has a heteroclinic orbit connecting the two stable roots of $f(v,w) = 0$. This heteroclinic orbit corresponds to a moving transition layer, traveling with speed c. It is important to note that since the roots of $f(v,w) = 0$ are functions of w, c is also a function of w. To be specific, we define $c(w)$ to be the unique parameter value for which (6.73) has a solution with $v \to V_-(w)$ as $\xi \to \infty$, and $v \to V_+(w)$ as $\xi \to -\infty$. In the case that $c(w) > 0$, we describe this transition as an "upjump" moving to the right. If $c(w) < 0$, then the transition is a "downjump" moving to the left.

We are now able to describe a general picture of wave propagation. In most of space, outer dynamics (6.68) are satisfied. At any transition between the two types of outer dynamics, continuity of w is maintained by a sharp transition in v that travels at the speed $y'(t) = c(w)$ if $v = V_-(w)$ on the right and $v = V_+(w)$ on the left, or at speed $y'(t) = -c(w)$ if $v = V_+(w)$ on the right and $v = V_-(w)$ on the left, where w is the value of the recovery variable in the interior of the transition layer. As a transition layer passes any particular point in space, there is a switch of outer dynamics from one to the other of the possible outer solution branches.

This singular perturbation description of wave propagation allows us to examine in more detail the specific case of a traveling pulse. The phase portrait for a solitary pulse is sketched in Fig. 6.7. A traveling pulse consists of a single excitation front followed by a single recovery back. We suppose that far to the right, the medium is at rest, and that a wave front of excitation has been initiated and is moving from left to right. Of course, for the medium to be at rest there must be a rest point of the dynamics on the lower branch, say $G_-(w_+) = 0$. Then, a wave that is moving from left to right has $v = V_-(w_+)$ on its right and $v = V_+(w_+)$ on its left, traveling at speed $y'(t) = c(w_+)$. Necessarily, it must be that $c(w_+) > 0$. Following the same procedure used to derive (6.8), one can show that

$$c(w) = \frac{\int_{V_-(w)}^{V_+(w)} f(v,w)\, dv}{\int_{-\infty}^{\infty} v_\xi^2\, d\xi}, \tag{6.74}$$

and thus $c(w_+) > 0$ if and only if

$$\int_{V_-(w_+)}^{V_+(w_+)} f(v,w_+)\, dv > 0. \tag{6.75}$$

If (6.75) fails to hold, then the medium is not sufficiently excitable to sustain a propagating pulse. It is important also to note that if $f(v,w)$ is of generalized FitzHugh–Nagumo form, then $c(w)$ has a unique zero in the interval (W_*, W^*), where W_* and W^* are defined in Section 5.2.

Immediately to the left of the excitation front, the medium is excited and satisfies the outer dynamics on the upper branch $v = V_+(w)$. Because (by assumption) $G_+(w) > 0$, this can hold for at most a finite amount of time before the outer dynamics force another transition layer to appear. This second transition layer provides a transition

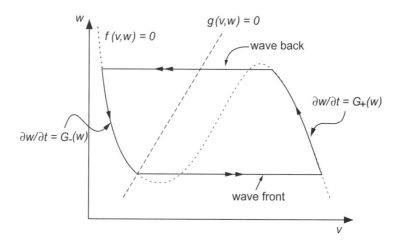

Figure 6.7 Sketch of the phase portrait of the fast traveling solitary pulse for FitzHugh–Nagumo dynamics in the singular limit $\epsilon \to 0$.

between the excited region on the right and a recovering region on the left and travels with speed $y'(t) = -c(w)$, where w is the value of the recovery variable in the transition layer. The minus sign here is because the second transition layer must be a downjump. For this to be a steadily propagating traveling pulse, the speeds of the upjump and downjump must be identical. Thus, the value of w at the downjump, say w_-, must be such that $c(w_-) = -c(w_+)$.

It may be that the equation $c(w_-) = -c(w_+)$ has no solution. In this case, the down-jump must occur at the knee, and then the wave is called a *phase wave* since the timing of the downjump is determined solely by the timing, or phase, of the outer dynamics, and not by any diffusive processes. That such a wave can travel at any speed greater than some minimal speed can be shown using standard arguments. The dynamics for phase waves are different from those for the bistable equation because the downjump must be a heteroclinic connection between a saddle point and a saddle-node. That is, at the knee, two of the three steady solutions of the bistable equation are merged into one. The demonstration of the existence of traveling waves in this situation is similar to the case of Fisher's equation, where the nonlinearity $f(v, w)$ in (6.73) has two simple zeros, rather than three as in the bistable case. In the phase wave problem, however, one of the zeros of $f(v, w)$ is not simple, but quadratic in nature, a canonical example of which is $f(v, w) = v^2(1 - v)$. We do not pursue this further except to say that such waves exist (see Exercise 16).

In summary, from singular perturbation theory we learn that the value of w ahead of the traveling pulse is given by the steady-state value w_+, and the speed of the rising wave front is then determined from the bistable equation (6.73) with $w = w_+$. The wave front switches the value of v from $v = V_-(w_+)$ (ahead of the wave) to $v = V_+(w_+)$ (behind the wave front). A wave back then occurs at $w = w_-$, where w_- is determined from $c(w_-) = -c(w_+)$. The wave back switches the value of v from $v = V_+(w_-)$ to $v = V_-(w_-)$. The duration of the excited phase of the traveling pulse is

$$T_e = \int_{w_+}^{w_-} \frac{dw}{G^+(w)}. \tag{6.76}$$

The duration of the absolute refractory period is

$$T_{\text{ar}} = \int_{w_-}^{w_0} \frac{dw}{G_-(w)}, \tag{6.77}$$

where w_0 is that value of w for which $c(w) = 0$ (Exercise 11). This approximate solution is said to be a singular solution, because derivatives of the solution become infinite (are singular) in the limit $\epsilon \to 0$.

6.3.2 The Hodgkin–Huxley Equations

The traveling pulse for the Hodgkin–Huxley equations must be computed numerically. The most direct way to do this is to simulate the partial differential equation on a long one-dimensional spatial domain. Alternately, one can use the technique of shooting. In fact, shooting was used by Hodgkin and Huxley in their 1952 paper to demonstrate that the Hodgkin–Huxley equations support a traveling wave solution. Shooting is also the method by which a rigorous proof of the existence of traveling waves has been given (Hastings, 1975; Carpenter, 1977).

The shooting argument is as follows. We write the Hodgkin–Huxley equations in the form

$$\tau_m \frac{\partial v}{\partial t} = \lambda_m^2 \frac{\partial^2 v}{\partial x^2} + f(v, m, n, h), \tag{6.78}$$

$$\frac{dw}{dt} = \alpha_w(v)(1 - w) - \beta_w(v)w, \text{ for } w = n, m, \text{ and } h. \tag{6.79}$$

Now we look for solutions in x, t that are functions of the translating variable $\xi = x/c + t$, and find the system of ordinary differential equations

$$\frac{\lambda_m^2}{c^2} \frac{d^2 v}{d\xi^2} + f(v, m, n, h) - \tau_m \frac{dv}{d\xi} = 0, \tag{6.80}$$

$$\frac{dw}{d\xi} = \alpha_w(v)(1 - w) - \beta_w(v)w, \text{ for } w = n, m, \text{ and } h. \tag{6.81}$$

Linearizing the system (6.80) and (6.81) about the resting solution at $v = 0$, one finds that there are four negative eigenvalues and one positive eigenvalue. A reasonable approximation to the unstable manifold is found by neglecting variations in g_K and g_{Na}, from which

$$v(t) = v_0 e^{\mu t}, \tag{6.82}$$

where

$$\frac{\lambda_m^2}{c^2}\mu^2 - \tau_m\mu - 1 = 0$$

or

$$\mu = \frac{1}{2}\left(\tau_m\frac{c^2}{\lambda_m^2} + \frac{c}{\lambda_m}\sqrt{\tau_m^2\frac{c^2}{\lambda_m^2} + 4}\right).$$

To implement shooting, one chooses a value of c, and initial data close to the rest point but on the unstable manifold (6.82), and then integrates numerically until (in all likelihood) the potential becomes very large. It could be that the potential becomes large positive or large negative. In fact, once values of c are found that do both, one uses bisection to home in on the homoclinic orbit that returns to the rest point in the limit $\xi \to \infty$.

For the Hodgkin–Huxley equations one finds a traveling pulse for $c = 3.24\,\lambda_m$ ms^{-1}. Using typical values for squid axon (from Table 4.1, $\lambda_m = 0.65$ cm), we find $c = 21$ mm/ms, which is close to the value of 21.2 mm/ms found experimentally by Hodgkin and Huxley. Hodgkin and Huxley estimated the space constant for squid axon as $\lambda_m = 0.58$ cm, from which they calculated that $c = 18.8$ mm/ms. Their calculated speeds agreed very well with experimental data and thus their model, which was based only on measurements of ionic conductance, was used to predict accurately macroscopic behavior of the axon. It is rare that quantitative models can be applied so successfully. Propagation velocities for several types of excitable tissue are listed in Table 6.2.

Table 6.2 Propagation velocities in nerve and muscle.

Excitable Tissue	Velocity (m/sec)
Myelinated nerve fibers	
Large diameter (16–20 μm)	100–120
Mid-diameter (10–12 μm)	60–70
Small diameter (4–6 μm)	30–50
Nonmyelinated nerve fibers	
Mid-diameter (3–5 μm)	15–20
Skeletal muscle fibers	6
Heart	
Purkinje fibers	1.0
Cardiac muscle	0.5
Smooth muscle	0.05

6.4 Periodic Wave Trains

Excitable systems are characterized by both excitability and refractoriness. That is, after the system has responded to a superthreshold stimulus with a large excursion from rest, there is a refractory period during which no subsequent responses can be evoked, followed by a recovery period during which excitability is gradually restored. Once excitability is restored, another wave of excitation can be evoked. However, the speed at which subsequent waves of excitation travel depends strongly on the time allowed for recovery of excitability following the last excitation wave. Generally (but not always), the longer the period of recovery, the faster the new wave of excitation can travel.

One might guess that a nerve axon supports, in addition to a traveling pulse, periodic wave trains of action potentials. With a periodic wave train, if recovery is a monotonic process, one expects propagation to be slower than for a traveling pulse, because subsequent action potentials occur before the medium is fully recovered, so that the Na^+ upstroke is slower than for a traveling pulse. The relationship between the speed and period is called the *dispersion curve*.

There are at least two ways to numerically calculate the dispersion curve for the Hodgkin–Huxley equations. The most direct method is to construct a ring, that is, a one-dimensional domain with periodic boundary conditions, initiate a pulse that travels in one direction on the ring, and solve the equations numerically until the solution becomes periodic in time. One can then use this waveform as initial data for a ring of slightly different length, and do the calculation again. While this method is relatively easy, its principal disadvantage is that it requires the periodic solution to be stable. Dispersion curves often have regions whose periodic solutions are unstable, and this method cannot find those. Of course, only the stable solutions are physically realizable, so this disadvantage may not be so serious to the realist.

The second method is to look for periodic solutions of the equations in their traveling wave coordinates (6.80)–(6.81), using a numerical continuation method (an automatic continuation program such as AUTO recommends itself here). With this method, periodic solutions are found without reference to their stability, so that the entire dispersion curve can be calculated.

Dispersion curves for excitable systems have a typical shape, depicted in Figs. 6.8 and 6.9. Here we see a dispersion curve having two branches, one denoting fast waves, the other slow. The two branches meet at a knee or corner at the *absolute refractory period*, and for shorter periods no periodic solutions exist. The solutions on the fast branch are typical of action potentials and are usually (but not always) stable. The solutions on the slow branch are small amplitude oscillations and are unstable.

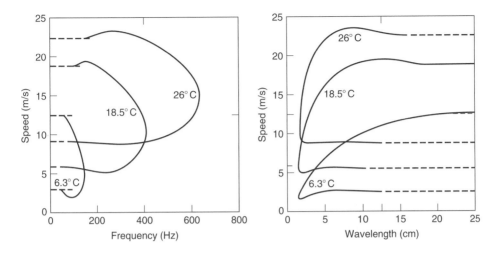

Figure 6.8 Numerically computed dispersion curve (speed vs. frequency and speed vs. wavelength for various temperatures) for the Hodgkin–Huxley equations. (Miller and Rinzel, 1981, Figs. 1 and 2.)

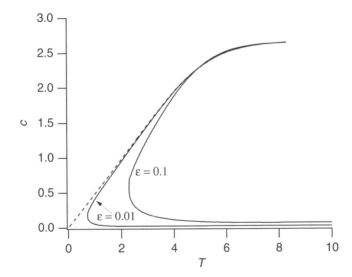

Figure 6.9 Dispersion curves for the piecewise-linear FitzHugh–Nagumo equations shown for $\epsilon = 0.1$ and 0.01. The dashed curve shows the singular perturbation approximation to the dispersion curve.

6.4.1 Piecewise-Linear FitzHugh–Nagumo Equations

The dispersion curve in Fig. 6.9 was found for the FitzHugh–Nagumo system (6.48)–(6.49) with piecewise-linear functions (6.52) and (6.53). The calculation is similar to that for the traveling pulse (Rinzel and Keller, 1973). Since this system is piecewise linear, we can express its solution as the sum of three exponentials,

$$w(\xi) = \sum_{i=1}^{3} A_i e^{\lambda_i \xi}, \tag{6.83}$$

on the interval $0 \le \xi < \xi_1$, and as

$$w(\xi) = 1 + \sum_{i=1}^{3} B_i e^{\lambda_i \xi} \tag{6.84}$$

on the interval $\xi_1 \le \xi \le \xi_2$, where $v = -cw_\xi$. We also assume that $v > \alpha$ on the interval $\xi_1 \le \xi \le \xi_2$. The numbers $\lambda_i, i = 1, 2, 3$, are roots of the characteristic polynomial (6.58).

We require that $w(\xi), v(\xi)$, and $v'(\xi)$ be continuous at $\xi = \xi_1$, and that $w(0) = w(\xi_2), v(0) = v(\xi_2)$, and $v'(0) = v'(\xi_2)$ for periodicity. Finally, we require that $v(0) = v(\xi_1) = \alpha$. This gives a total of eight equations in nine unknowns, $A_1, \ldots, A_3, B_1, \ldots, B_3$, ξ_1, ξ_2, and c. After some calculation we find two equations for the three unknowns ξ_1, ξ_2, and c given by

$$\frac{e^{\lambda_1(P-\xi_1)} - 1}{p'(\lambda_1)(e^{\lambda_1 P} - 1)} + \frac{e^{\lambda_2(P-\xi_1)} - 1}{p'(\lambda_2)(e^{\lambda_2 P} - 1)} + \frac{e^{\lambda_3(P-\xi_1)} - 1}{p'(\lambda_3)(e^{\lambda_3 P} - 1)} + \epsilon^2 \alpha = 0, \tag{6.85}$$

$$\frac{e^{\lambda_1 P} - e^{\lambda_1 \xi_1}}{p'(\lambda_1)(e^{\lambda_1 P} - 1)} + \frac{e^{\lambda_2 P} - e^{\lambda_2 \xi_1}}{p'(\lambda_2)(e^{\lambda_2 P} - 1)} + \frac{e^{\lambda_3 P} - e^{\lambda_3 \xi_1}}{p'(\lambda_3)(e^{\lambda_3 P} - 1)} + \epsilon^2 \alpha = 0, \tag{6.86}$$

where $P = \xi_2/c$. It is important to note that since there are only two equations for the three unknowns, (6.85) and (6.86) define a family of periodic waves, parameterized by either the period or the wave speed. The relationship between the period and the speed of this wave family is the dispersion curve. In Fig. 6.9 are shown examples of the dispersion curve for a sampling of values of ϵ with $\alpha = 0.1$. Changing α has little qualitative effect on this plot. The dashed curve shows the limiting behavior of the upper branch (the fast waves) in the limit $\epsilon \to 0$. Of significance in this plot is the fact that there are fast and slow waves, and in the limit of large wavelength, the periodic waves approach the solitary traveling pulses represented by Fig. 6.5 (Exercise 12). In fact, periodic solutions look much like evenly spaced periodic repeats of (truncated) solitary pulses.

The dispersion curve for the piecewise-linear FitzHugh–Nagumo system is typical of dispersion curves for excitable media, with a fast and slow branch meeting at a corner. In general, the location of the corner depends on the excitability of the medium (in this case, the parameter α) and on the ratio of time scales ϵ.

6.4.2 Singular Perturbation Theory

The fast branch of the dispersion curve can be found for a general FitzHugh–Nagumo system in the limit $\epsilon \to 0$ using singular perturbation theory. A periodic wave consists of an alternating series of upjumps and downjumps, separated by regions of outer

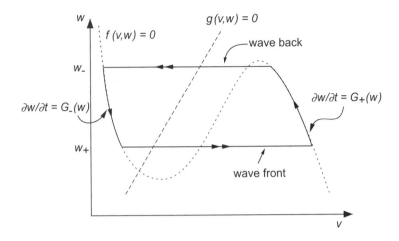

Figure 6.10 Sketch of the phase portrait for the fast traveling periodic wave train for FitzHugh–Nagumo dynamics in the singular limit $\epsilon \to 0$.

dynamics. The phase portrait for a periodic wave train is sketched in Fig. 6.10. To be periodic, if w_+ is the value of the recovery variable in the upjump, traveling with speed $c(w_+)$, then the value of w in the downjump must be w_-, where $c(w_+) = -c(w_-)$. The amount of time spent on the excited branch is

$$T_e = \int_{w_+}^{w_-} \frac{dw}{G_+(w)}, \tag{6.87}$$

and the amount of time spent on the recovery branch is

$$T_r = \int_{w_-}^{w_+} \frac{dw}{G_-(w)}. \tag{6.88}$$

The dispersion curve is then the relationship between speed $c(w_+)$ and period

$$T = T_e + T_r, \tag{6.89}$$

parameterized by w_+. This approximate dispersion curve (calculated numerically) is shown in Fig. 6.9 as a dashed curve.

The slow branch of the dispersion curve can also be found using perturbation methods, although in this case since the speed is small of order ϵ, a regular perturbation expansion is appropriate. The details of this expansion are beyond the scope of this book, although the interested reader is referred to Dockery and Keener (1989). In general, the slow periodic solutions are unstable (Maginu, 1985) and therefore are of less physical interest than the fast solutions. Again, stability theory for the traveling wave solutions is beyond the scope of this book.

6.4.3 Kinematics

Not all waves are periodic. There can be wave trains with action potentials that are irregularly spaced and that travel with different velocities. A *kinematic theory* of wave propagation is one that attempts to follow the progress of individual action potentials without tracking the details of the structure of the pulse (Rinzel and Maginu, 1984). The simplest kinematic theory is to interpret the dispersion curve in a local way. That is, suppose we know the speed as a function of period for the stable periodic wave trains, $c = C(T)$. We suppose that the wave train consists of action potentials, and that the nth action potential reaches position x at time $t_n(x)$. To keep track of time of arrival at position x we note that

$$\frac{dt_n}{dx} = \frac{1}{c}. \tag{6.90}$$

We complete the description by taking $c = C(t_n(x) - t_{n-1}(x))$, realizing that $t_n(x) - t_{n-1}(x)$ is the instantaneous period of the wave train that is felt by the medium at position x.

A more sophisticated kinematic theory can be derived from the singular solution of the FitzHugh–Nagumo equations. For this derivation we assume that recovery is always via a phase wave, occurring with recovery value W^*. Suppose that the front of the nth action potential has speed $c(w_n)$, corresponding to the recovery value w_n in the transition layer. Keeping track of the time until the next action potential, we find

$$t_{n+1}(x) - t_n(x) = T_e(x) + T_r(x) \tag{6.91}$$

$$= \int_{w_n}^{W^*} \frac{dw}{G_+(w)} + \int_{W^*}^{w_{n+1}} \frac{dw}{G_-(w)}. \tag{6.92}$$

Differentiating (6.92) with respect to x, we find the differential equation for $w_{n+1}(x)$:

$$\frac{1}{G_-(w_{n+1})} \frac{dw_{n+1}}{dx} = \frac{1}{G_+(w_n)} \frac{dw_n}{dx} + \frac{1}{c(w_{n+1})} - \frac{1}{c(w_n)}. \tag{6.93}$$

With this equation one can track the variable w_{n+1} as a function of x given $w_n(x)$ and from it reconstruct the speed and time of arrival of the $(n + 1)$st action potential wave front.

While this formulation is useful for the FitzHugh–Nagumo equations, it can be given a more general usefulness as follows. Since there is a one-to-one relationship between the speed of a front and the value of the recovery variable w in that front, we can represent these functions in terms of the speeds of the fronts as

$$t_{n+1}(x) - t_n(x) = A(c_n) + t_r(c_{n+1}), \tag{6.94}$$

where $A(c_n) = T_e$ is the *action potential duration* (APD) since the nth upstroke, and $t_r(c_{n+1}) = T_r$ is the recovery time preceding the $(n+1)$st upstroke. When we differentiate this conservation law with respect to x, we find a differential equation for the speed of the $(n + 1)$st wave front as a function of the speed of the nth wave front, given by

$$t'_r(c_{n+1})\frac{dc_{n+1}}{dx} = \frac{1}{c_{n+1}} - \frac{1}{c_n} - A'(c_n)\frac{dc_n}{dx}. \tag{6.95}$$

The advantage of this formulation is that the functions A and t_r may be known for other reasons, perhaps from experimental data. It is generally recognized that the action potential duration is functionally related to the speed of the previous action potential, and it is also reasonable that the speed of a subsequent action potential can be related functionally to the time of recovery since the end of the last action potential. Thus, the model (6.95) has applicability that goes beyond the FitzHugh–Nagumo context. For an example of how this idea has been used for the Beeler–Reuter dynamics, see Courtemanche et al. (1996).

6.5 Wave Propagation in Higher Dimensions

Not all excitable media can be viewed as one-dimensional cables, neither is all propagated activity one-dimensional. Tissues for which one-dimensional descriptions of cellular communication are inadequate include skeletal and cardiac tissue, the retina, and the cortex of the brain. To understand communication and signaling in these tissues requires more complicated mathematical analysis than for one-dimensional cables.

When beginning a study of two- or three-dimensional wave propagation, it is tempting to extend the cable equation to higher dimensions by replacing first derivatives in space with spatial gradients, and second spatial derivatives with the Laplacian operator. Indeed, all the models discussed here are of this type. However, this replacement is not always appropriate.

Some cells, such as *Xenopus* oocytes (frog eggs), are large enough so that waves of chemical activity can be sustained within a single cell. This is unusual, however, as most waves in normal physiological situations serve the purpose of communication between cells. For chemical waves in single cells, a reasonable first guess is that spatial coupling is by chemical diffusion. In that case, if the local chemical dynamics are described by the differential equation $\frac{\partial u}{\partial t} = kf$, then with spatial coupling the dynamics are represented by

$$\frac{\partial u}{\partial t} = \nabla \cdot (D\nabla u) + kf, \tag{6.96}$$

where D is the (scalar) diffusion coefficient of the chemical species and ∇ is the three-dimensional gradient operator. Here we have included the time constant k (with units time^{-1}), so that f has the same dimensional units as u.

For many cell types, however, intercellular communication is through gap junctions between immediate neighbors, so that diffusion is not spatially uniform. In this case, the first guess (or hope) is that the length constant of the phenomenon to be described is much larger than the typical cell size, and homogenization can be used to find an effective diffusion coefficient, D_e, as described in Chapter 8. Then (6.96) with the effective diffusion coefficient is a reasonable model. Note that there is no a priori reason to believe that cellular coupling is isotropic or that D_e is a scalar quantity.

If this assumption, that the space constant of the signaling phenomenon is larger than the size of the cell, and hence that homogenization gives a valid approximation, is not justified, then we are stuck with the unpleasant business of studying communication between discretely coupled cells. Descriptions of such studies are included in Chapter 7.

For electrically active cells, such as cardiac or muscle cells, the situation is further complicated by the fact that the signal is the transmembrane potential, and to determine this one needs to know both the intracellular and extracellular potentials. Consequently, spatial coupling cannot be represented by the Laplacian of the transmembrane potential. We address this problem in Chapter 12, where we discuss waves in myocardial tissue.

Even more complicated are neural networks, where axons may extend long distances and cells may be connected to many other cells, far more than their nearest neighbors, so that spatial coupling is nonlocal.

Suffice it to say, while (6.96) is an interesting place to start the study of waves in higher dimensions, it is by no means obvious that all the results of this study are applicable to cellular media.

6.5.1   Propagating Fronts

Plane Waves

The simplest wave to look for in a higher-dimensional medium is a plane wave. Suppose that the canonical problem

$$U'' + c_0 U' + f(U) = 0 \tag{6.97}$$

(with dimensionless independent variable) is bistable and has a wave front solution $U(\xi)$ for some unique value of c_0, the value of which depends on f.

To find plane wave solutions of (6.96), we suppose that u is a function of the single variable $\xi = \mathbf{n} \cdot \mathbf{x} - ct$, where $\mathbf{n}$ is a unit vector pointing in the forward direction of wave front propagation. In the traveling wave coordinate ξ, the time derivative $\frac{d}{dt}$ is replaced by $-c \frac{d}{d\xi}$, and the spatial gradient operator ∇ is replaced by $\mathbf{n} \frac{d}{d\xi}$, so that the governing equation reduces to the ordinary differential equation

$$(\mathbf{n} \cdot D\mathbf{n})u'' + cu' + kf(u) = 0. \tag{6.98}$$

We compare (6.98) with the canonical equation (6.97) and note that the solution of (6.98) can be found by a simple rescaling to be

$$u(\mathbf{x}, t) = U\left(\frac{\mathbf{n} \cdot \mathbf{x} - ct}{\Lambda(\mathbf{n})}\right), \tag{6.99}$$

where $c = c_0 k \Lambda(\mathbf{n})$ is the (directionally dependent) speed and $\Lambda(\mathbf{n}) = \sqrt{\frac{\mathbf{n} \cdot D\mathbf{n}}{k}}$ is the directionally dependent space constant.

Waves with Curvature

Wave fronts in two- or three-dimensional media are not expected to be plane waves. They are typically initiated at a specific location, and so might be circular in shape. Additionally, the medium may be structurally inhomogeneous or have a complicated geometry, all of which introduce curvature into the wave front.

It is known that curvature plays an important role in the propagation of a wave front in an excitable medium. A physical explanation makes this clear. Suppose that a circular wave front is moving inward, so that the circle is collapsing. Because different parts of the front are working to excite the same points, we expect the region directly in front of the wave front to be excited more quickly than if the wave were planar. Similarly, an expanding circular wave front should move more slowly than a plane wave because the efforts of the wave to excite its neighbors are more spread out, and excitation should be slower than for a plane wave.

While these curvature effects are well known in many contexts, we are interested here in a quantitative description of this effect. In this section we derive an equation for action-potential spread called the *eikonal-curvature equation*, the purpose of which is to show the contribution of curvature to wave-front velocity. Eikonal-curvature equations have been used in a number of biological contexts, including the study of wave-front propagation in the excitable Belousov–Zhabotinsky reagent (Foerster et al., 1989; Keener, 1986; Keener and Tyson, 1986; Tyson and Keener, 1988; Ohta et al., 1989), Ca^{2+} waves in *Xenopus* oocytes (Lechleiter et al., 1991b; Sneyd and Atri, 1993; Jafri and Keizer, 1995) and in studies of myocardial tissue (Keener, 1991a; Colli-Franzone et al., 1990, 1993; also see Chapter 12). Eikonal-curvature equations have a long history in other scientific fields as well, including crystal growth (Burton et al., 1951) and flame front propagation (Frankel and Sivashinsky, 1987, 1988).

The derivation of the eikonal-curvature equation uses standard mathematical arguments of singular perturbation theory, which we summarize here. The key observation is that hidden inside (6.96) is the bistable equation (6.97), and the idea to be explored is that in some moving coordinate system that is yet to be determined, (6.96) is well approximated by the bistable equation (6.97).

Our goal is to rewrite (6.96) in terms of a moving coordinate system chosen so that it takes the form of (6.97). In three dimensions, we must have three spatial coordinates, one of which is locally orthogonal to the wave front, while the other two are coordinates describing the wave-front surface. By assumption, the function u is approximately independent of the wave-front coordinates. We introduce a scaling of the variables such that the derivatives with respect to the first variable are of most importance, and all other derivatives are less so. From this computation, we learn how the coordinate system must move in order to maintain itself as a wave-front coordinate system, and this law of coordinate system motion is the eikonal-curvature equation.

To begin, we introduce a general (as yet unknown) moving coordinate system

$$\mathbf{x} = \mathbf{X}(\xi, \tau), \qquad t = \tau. \tag{6.100}$$

According to the chain rule,

$$\frac{\partial}{\partial \xi_i} = \frac{\partial X_j}{\partial \xi_i}\frac{\partial}{\partial x_j}, \qquad \frac{\partial}{\partial \tau} = \frac{\partial}{\partial t} + \frac{\partial X_j}{\partial \tau}\frac{\partial}{\partial x_j}. \tag{6.101}$$

Here and in what follows, the summation convention is followed (i.e., unless otherwise noted, repeated indices are summed from 1 to 3). It follows that

$$\frac{\partial}{\partial x_i} = \alpha_{ij}\frac{\partial}{\partial \xi_j}, \qquad \frac{\partial}{\partial t} = \frac{\partial}{\partial \tau} - \frac{\partial X_j}{\partial \tau}\alpha_{jk}\frac{\partial}{\partial \xi_k}, \tag{6.102}$$

where the matrix with entries α_{ij} is the inverse of the matrix with entries $\frac{\partial X_j}{\partial \xi_i}$ (the Jacobian of the coordinate transformation (6.100)).

We identify the variable ξ_1 as the coordinate normal to level surfaces of u, so ξ_2 and ξ_3 are the coordinates of the moving level surfaces. Then, we define the tangent vectors $\mathbf{r}_i = \frac{\partial X_j}{\partial \xi_i}, i = 1, 2, 3$, and the normal vectors $\mathbf{n}_i = \mathbf{r}_j \times \mathbf{r}_k$, where $i \neq j, k$, and $j < k$. Without loss of generality we take $\mathbf{r}_1 = \sigma(\mathbf{r}_2 \times \mathbf{r}_3)$, so that $\mathbf{r}_1$ is always normal to the level surfaces of u. Here σ is an arbitrary (unspecified) scale factor. The vectors $\mathbf{r}_2$ and $\mathbf{r}_3$ are tangent to the moving level surface, although they are not necessarily orthogonal. While one can force the vectors $\mathbf{r}_2$ and $\mathbf{r}_3$ to be orthogonal, the actual construction of such a coordinate description on a moving surface is generally quite difficult. Furthermore, it is preferred to have an equation of motion that does not have additional restrictions, since the motion should be independent of the coordinate system by which the surface is described.

We can calculate the entries α_{ij} explicitly. It follows from Cramer's rule (Exercise 23) that

$$\alpha_{ij} = \frac{(\mathbf{n}_j)_i}{\mathbf{r}_j \cdot \mathbf{n}_j}\text{(no summation)}, \tag{6.103}$$

where by $(\mathbf{n}_j)_i$ we mean the ith component of the jth normal vector $\mathbf{n}_j$.

Now we can write out the full change of variables. We calculate that (treating the coefficients α_{ij} as functions of $\mathbf{x}$)

$$\frac{\partial u}{\partial t} = \frac{\partial u}{\partial \tau} - \frac{\partial X_j}{\partial \tau}\alpha_{jk}\frac{\partial u}{\partial \xi_k}, \tag{6.104}$$

$$[6bp]\nabla^2 u = \alpha_{ip}\alpha_{ik}\frac{\partial^2 u}{\partial \xi_p \partial \xi_k} + \frac{\partial \alpha_{ip}}{\partial x_i}\frac{\partial u}{\partial \xi_p}, \tag{6.105}$$

and rewrite (6.96) in terms of these new variables, finding (in the case that D is a constant scalar) that

$$0 = D\alpha_{ip}\alpha_{iq}\frac{\partial^2 u}{\partial \xi_p \partial \xi_q} + D\frac{\partial \alpha_{ip}}{\partial x_i}\frac{\partial u}{\partial \xi_p} - \left(u_\tau - \frac{\partial X_j}{\partial \tau}\alpha_{jk}\frac{\partial u}{\partial \xi_k}\right) + kf(u). \tag{6.106}$$

There are two important assumptions that are now invoked, namely that the spatial scale of variation in ξ_1 is much shorter than the spatial scale for variations in the variables ξ_2 and ξ_3. We quantify this by supposing that there is a small parameter ϵ

and that $\alpha_{j1} = O(1)$, while $\alpha_{jk} = O(\epsilon)$ for all j and $k \neq 1$. In addition, we assume that to leading order in ϵ, u is independent of ξ_2, ξ_3, and τ. Consequently, not all of the terms in (6.106) are of equal importance. If we take into account the ϵ dependence of α_{ij}, then (6.106) simplifies to

$$D|\alpha|^2 \frac{\partial^2 u}{\partial \xi_1^2} + \left(D\nabla \cdot \alpha + \frac{\partial \mathbf{X}}{\partial \tau} \cdot \alpha\right) \frac{\partial u}{\partial \xi_1} + kf(u) = O(\epsilon), \tag{6.107}$$

where α is the vector with components α_{j1}, and hence, from (6.103), proportional to the normal vector $\mathbf{n}_1$. All of the terms on the left-hand side of (6.107) are large compared to ϵ.

Here we see an equation that resembles the bistable equation (6.97). If the coefficients of (6.107) are constant, we can identify (6.107) with (6.97) by setting

$$\frac{\partial \mathbf{X}}{\partial \tau} \cdot \alpha + D\nabla \cdot \alpha = kc_0, \tag{6.108}$$

while requiring $D|\alpha|^2 = k$.

Equation (6.108) tells us how the coordinate system should move, and since α is proportional to $\mathbf{n}_1$, setting $D|\alpha|^2 = k$ determines the scale of the coordinate normal to the wave front, i.e., the thickness of the wave front. In reality, the coefficients of (6.107) are not constants, and these two requirements overdetermine the full coordinate transformation $\mathbf{X}(\xi, \tau)$. To overcome this difficulty, we assume that since the wave front and the coordinate system are slowly varying in space, we interpret (6.108) as determining the motion of only the midline of the coordinate system, at the location of the largest gradient of the front, rather than the entire coordinate system.

Equation (6.108) is the equation we seek that describes the motion of an action potential front, called the *eikonal-curvature equation*. However, for numerical simulations it is essentially useless. Numerical algorithms to simulate this equation reliably are extremely hard to construct. Instead, it is useful to introduce a function $S(x, t)$ that acts as an indicator function for the fronts (think of S as determining the "shock" location). That is, if $S(x, t) > 0$, the medium is activated, while if $S(x, t) < 0$, the medium is in the resting state. Taking α to be in the direction of forward wave-front motion means that $\alpha = -\sqrt{\frac{k}{D}} \frac{\nabla S}{|\nabla S|}$, where we have used the fact that $|\alpha| = \sqrt{k/D}$. Since the zero level surface of $S(x, t)$ denotes the wave-front location, and thus S is constant along the wave front, it follows that $0 = \nabla S \cdot X_t + S_t$, so that $X_t \cdot \alpha = \sqrt{\frac{k}{D}} \frac{S_t}{|\nabla S|}$. Hence,

$$S_t = |\nabla S| c_0 \sqrt{Dk} + D|\nabla S| \nabla \cdot \left(\frac{\nabla S}{|\nabla S|}\right). \tag{6.109}$$

The use of an indicator function $S(x, t)$ to determine the motion of an interface is called the *level set method* (Osher and Sethian, 1988), and is both powerful and easy to implement.

Equation (6.109) is called the eikonal-curvature equation because of the physical interpretation of each of its terms. If we ignore the diffusive term, then we have the eikonal equation

$$\frac{\partial S}{\partial t} = |\nabla S| c_0 \sqrt{Dk}.$$ (6.110)

If $\mathbf{R}$ is a level surface of the function $S(x, t)$ and if $\mathbf{n}$ is the unit normal vector to that surface at some point, then (6.110) implies that the normal velocity of the surface $\mathbf{R}$, denoted by $\mathbf{R}_t \cdot \mathbf{n}$, satisfies

$$\mathbf{R}_t \cdot \mathbf{n} = c_0 \sqrt{Dk}.$$ (6.111)

In other words, the front moves in the normal direction $\mathbf{n}$ with speed $c = c_0 \sqrt{Dk}$.

Equation (6.111) is the basis of a geometrical "Huygens" construction for front propagation, but the numerical integration of either (6.110) or (6.111) is fraught with difficulties. In particular, cusp singularities develop, and the indicator function $S(x, t)$ becomes ill-defined in finite time (usually very quickly). The second term of the right-hand side of (6.109) is a curvature correction, appropriately named because the term $\nabla \cdot \left(\frac{\nabla S}{|\nabla S|} \right)$ is twice the mean curvature (in three-dimensional space) or the curvature (in two-dimensional space) of the level surfaces of S (see Exercise 24). In fact, the eikonal-curvature equation can be written as

$$\mathbf{R}_t \cdot \mathbf{n} = c_0 \sqrt{Dk} - D\kappa,$$ (6.112)

or

$$\tau \mathbf{R}_t \cdot \mathbf{n} = c_0 \Lambda - \Lambda^2 \kappa,$$ (6.113)

where κ is the curvature (in two dimensions) or twice the mean curvature (in three dimensions) of the front, $\Lambda = \sqrt{D/k}$ is the space constant, and $\tau = 1/k$ is the time constant. Even though it usually represents only a small correction to the normal velocity of fronts, the curvature correction is important for physical and stability reasons, to prevent singularity formation. The sign of the curvature correction is such that a front with ripples is gradually smoothed into a plane wave.

Experiments on the Belousov–Zhabotinsky reagent have verified this relationship between speed and curvature of propagating fronts. For example, Foerster et al. (1988) measured the speed and curvature at different positions of a rotating spiral wave and at intersections of two spiral waves (thus obtaining curvatures of different signs) and found that the relationship between normal velocity and curvature was well approximated by a straight line with slope that was the diffusion coefficient of the rapidly reacting species.

6.5.2 Spatial Patterns and Spiral Waves

Now that we have some idea of how wave fronts propagate in an excitable medium, we next wish to determine the spatial patterns that may result. The most common pattern is created by, and spreads outward from, a single source. If the medium is sufficiently large so that more than one wave front can exist at the same time, then these are referred to as *target patterns*. Target patterns require a periodic source and so cannot exist in a homogeneous nonoscillatory medium.

A second type of spatial pattern is a spiral wave. Spiral waves do not require a periodic source for their existence, as they are typically self-sustained. Because they are self-sustained, spirals usually occur only in pathophysiological situations. That is, it is usually not a good thing for a system that relies on faithful propagation of a signal to be taken over by a self-sustained pattern. Thus, spirals on the heart are fatal, spirals in the cortex may lead to epileptic seizures, and spirals on the retina or visual cortex may cause hallucinations. One particularly famous example of spiral waves in an excitable medium is in the Belousov–Zhabotinsky reaction (Winfree, 1972, 1974), which we do not discuss here.

The mathematical discussion of spiral waves centers on the nature of periodic solutions of a system of differential equations with excitable dynamics spatially coupled by diffusion. A specific example is the FitzHugh–Nagumo equations with diffusive coupling in two spatial dimensions,

$$\epsilon \frac{\partial v}{\partial t} = \epsilon^2 \nabla^2 v + f(v, w), \tag{6.114}$$

$$\frac{\partial w}{\partial t} = g(v, w). \tag{6.115}$$

The leading-order singular perturbation analysis (i.e., with $\epsilon = 0$) suggests that the domain be separated into two, in which outer dynamics

$$\frac{\partial w}{\partial t} = G_{\pm}(w) \tag{6.116}$$

hold (using the notation of Section 6.3.1). The region in which $\frac{\partial w}{\partial t} = G_+(w)$ is identified as the excited region, and the region in which $\frac{\partial w}{\partial t} = G_-(w)$ is called the recovering region. Separating these are moving interfaces in which v changes rapidly (with space and time constant ϵ), so that diffusion is important, while w remains essentially constant. At any point in space the solution should be periodic in time, so at large radii, where the wave fronts are nearly planar, the solution should lie on the dispersion curve.

The first guess as to how the interface should move is to assume that the interface is nearly planar and therefore has the same velocity as a plane wave, namely

$$\mathbf{R}_t = c(w)\mathbf{n}, \tag{6.117}$$

where $\mathbf{R}$ is the position vector for the interface, $\mathbf{n}$ is the unit normal vector of $\mathbf{R}$, and $c(w)$ is the plane-wave velocity as a function of w.

To see the implications of the eikonal equation, we suppose that the spiral interface is a curve $\mathbf{R}$ given by

$$X(r, t) = r\cos(\theta(r) - \omega t), \qquad Y(r, t) = r\sin(\theta(r) - \omega t). \tag{6.118}$$

Notice that the interface is a curve parameterized by r, and so the tangent is (X_r, Y_r). We then calculate that

$$\mathbf{R}_t = \begin{pmatrix} -\omega r \sin(\theta - \omega t) \\ \omega r \cos(\theta - \omega t) \end{pmatrix}, \tag{6.119}$$

and

$$\sqrt{1 + r^2 \theta'^2} \, \mathbf{n} = \begin{pmatrix} -\sin(\theta - \omega t) - r\theta' \cos(\theta - \omega t) \\ \cos(\theta - \omega t) - r\theta' \sin(\theta - \omega t) \end{pmatrix}, \tag{6.120}$$

so that the eikonal equation becomes

$$c(w)\sqrt{1 + r^2 \theta'^2} = \omega r. \tag{6.121}$$

An integration then gives

$$\theta(r) = \rho(r) - \tan(\rho(r)), \qquad \rho(r) = \sqrt{\frac{r^2}{r_0^2} - 1}, \tag{6.122}$$

where $r_0 = c/\omega$, so that the interface is given by

$$X = r_0 \cos(s) + r_0 \rho(r) \sin(s), \qquad Y = r_0 \sin(s) - r_0 \rho(r) \cos(s), \tag{6.123}$$

where $s = \rho(r) - \omega t$. This interface is the involute of a circle of radius r_0. (The involute of a circle is the locus of points at the end of a string that is unwrapped from a circle.)

There are significant difficulties with this as a spiral solution, the most significant of which is that it exists only for $r \geq r_0$. The parameter r_0 is arbitrary, but positive, so that this spiral is rotating about some hole of finite size. The frequency of rotation is determined by requiring consistency with the dispersion curve. Note that the spiral has wavelength $2\pi r_0$, and period $\frac{2\pi}{\omega}$, and so $c = r_0 \omega$. However, since the dispersion curve generally has a knee, and thus periodic waves do not exist for small enough wavelength, there is a lower bound on the radii for which this can be satisfied. Numerical studies of spirals suggest no such lower bound on the inner core radius and also suggest that there is a unique spiral frequency for a medium without a hole at the center. Unfortunately, the use of the eikonal equation gives no hint of the way a unique frequency is selected, so a different approach, using the eikonal-curvature equation, is required.

To apply the eikonal-curvature equation to find rotating spiral waves, we assume that the wave front is expressed in the form (6.118), so that the curvature is

$$\kappa = \frac{X'Y'' - Y'X''}{(X'^2 + Y'^2)^{3/2}} = \frac{\psi'}{(1 + \psi^2)^{3/2}} + \frac{\psi}{r(1 + \psi^2)^{1/2}}, \tag{6.124}$$

where $\psi = r\theta'(r)$ is called the shape function. Thus the eikonal-curvature equation (6.113) becomes

$$r\frac{d\psi}{dr} = (1 + \psi^2) \left[\frac{rc(w)}{\epsilon}(1 + \psi^2)^{1/2} - \frac{\omega r^2}{\epsilon} - \psi \right]. \tag{6.125}$$

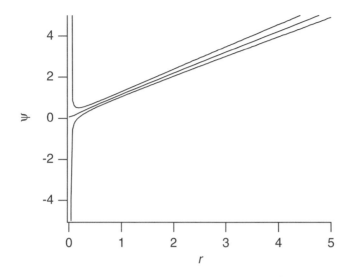

Figure 6.11 Trajectories of (6.125) with $\omega/\epsilon = 2.8, 9m^*$, and 3.2, and $c/\epsilon = 3.0$.

If we suppose that w is constant along the spiral front, then (6.125) can be solved numerically by "shooting" from $r = \infty$. A portrait of sample trajectories in the (r, ψ) plane is shown in Fig. 6.11. The trajectories of (6.125) are stiff, meaning that, for large r, the trajectory $c(1 + \psi^2)^{1/2} - \omega r = 0$ is a strong attractor. This stiffness can be readily observed when (6.125) is written in terms of the variable $\phi = \frac{r\psi}{\sqrt{1+\psi^2}}$ as

$$\frac{\epsilon}{r}\frac{d\phi}{dr} = c - \omega\sqrt{r^2 - \phi^2}, \tag{6.126}$$

since ϵ multiplies the derivative term in (6.126).

Integrating from $r = \infty$, trajectories of (6.125) approach the origin by either blowing up or down near the origin $r = 0$. Since the origin is a saddle point, if parameters are chosen exactly right, the trajectory approaches $\psi = 0$. Thus, there is a unique relationship between ω and c of the form $\omega/\epsilon = F(c/\epsilon)$ that yields trajectories that go all the way to the origin $r = 0, \psi = 0$.

Notice that the rescaling of variables $r \to \alpha r, c \to c/\alpha, \omega \to \omega/\alpha^2$ leaves (6.125) invariant, so that the relationship between ω and c for which a trajectory approaches the saddle point at the origin must be of the form

$$\frac{\omega}{\epsilon\alpha^2} = F\left(\frac{c}{\epsilon\alpha}\right). \tag{6.127}$$

It follows that $F(c/\alpha) = \frac{1}{\alpha^2}F(c)$ and thus $F(x) = m^* x^2$, for some constant m^*, so that

$$\omega = \frac{c^2 m^*}{\epsilon}. \tag{6.128}$$

Numerically, one determines that $m^* = 0.330958$.

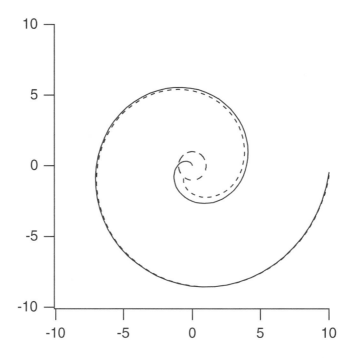

Figure 6.12 Spiral arm corresponding to the trajectory of (6.125) that approaches the origin (shown solid), compared with the involute spiral (shown dashed) with the same parameter values. For this plot, $c/\epsilon = 3.0$.

An example of this spiral front is shown in Fig. 6.12, compared with the comparable involute spiral (6.123), shown dashed. For this figure $c/\epsilon = 3.0$.

At this point we have a family of possible spiral trajectories that have correct asymptotic (large r) behavior and approach $\psi = 0$ as $r \to 0$. This family is parameterized by the speed c. To determine which particular member of this family is the correct spiral front, we also require that the spirals be periodic waves; that is, they must satisfy the dispersion relationship. These two requirements uniquely determine the spiral properties. To see that this is so, in Fig. 6.13 are plotted the critical curve (6.128) and the approximate dispersion curve (6.89). For this plot we used piecewise-linear dynamics, $f(v, w) = H(v - \alpha) - v - w$, $g(v, w) = v - \gamma w$ with $\alpha = 0.1$, $\gamma = 0$, $\epsilon = 0.05$.

More About Spirals

This discussion of higher-dimensional waves is merely the tip of the iceberg (or tip of the spiral), and there are many interesting unresolved questions.

While the spirals that are observed in physiological systems share certain qualitative similarities, their details are certainly different. The FitzHugh–Nagumo equations discussed here show only the qualitative behavior for generic excitable systems and so have little quantitative relevance. Other physiological systems are likely governed by other dynamics. For example, a model of spreading cortical depression in the cortex has

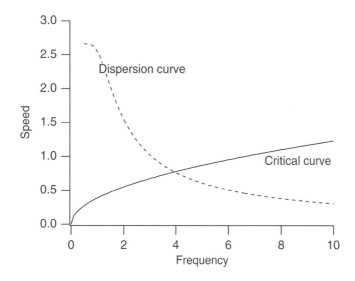

Figure 6.13 Critical curve (6.128) and the approximate dispersion curve (6.89) using piecewise-linear dynamics $f(v, w) = H(v - \alpha) - v - w$, $g(v, w) = v - \gamma w$ with $\alpha = 0.1$, $\gamma = 0$, $\epsilon = 0.05$.

been proposed by Tuckwell and Miura (1978; Miura, 1981), and numerical simulations have shown rotating spirals. However, a detailed mathematical study of these equations has not been given. Similarly, spiral waves of Ca^{2+} have been studied numerically in some detail (Chapter 7) but are not well understood anaytically.

This analytical calculation for the FitzHugh–Nagumo system is based on singular perturbation theory and therefore is not mathematically rigorous. In fact, as yet there is not a rigorous proof of the existence of spiral waves in an excitable medium. While the approximate solution presented here is known to be asymptotically valid, the structure of the core of the spiral is not correct. This problem has been addressed by Pelce and Sun (1991) and Keener (1992) for FitzHugh–Nagumo models with a single diffusing variable, and by Keener (1994) and Kessler and Kupferman (1996) for FitzHugh–Nagumo models with two diffusing variables, relevant for chemical reaction systems.

A second issue of concern is the stability of spirals. This also is a large topic, which is not addressed here. The interested reader should consult the work of Winfree (1991), Jahnke and Winfree (1991), Barkley (1994), Karma (1993, 1994), Panfilov and Hogeweg (1995), Kessler and Kupferman (1996).

Because the analytical study of excitable media is so difficult, simpler models have been sought, with the result that finite-state automata are quite popular. A finite-state automaton divides the state space into a few discrete values (for example $v = 0$ or 1), divides the spatial domain into discrete cells, and discretizes time into discrete steps. Then, rules are devised for how the states of the cells change in time. Finite-state automata are extremely easy to program and visualize. They give some useful insight into the behavior of excitable media, but they are also beguiling and can give "wrong" answers that are not easily detected. The literature on finite-state automata is vast (see, for instance, Moe et al., 1964; Smith and Cohen, 1984; and Gerhardt et al., 1990).

The obvious generalization of a spiral wave in a two-dimensional region to three dimensions is called a *scroll wave* (Winfree, 1973, 1991; Keener and Tyson, 1992). Scroll waves have been observed numerically (Jahnke et al., 1988; Lugosi and Winfree, 1988), in three-dimensional BZ reagent (Gomatam and Grindrod, 1987), and in cardiac tissue (Chen et al., 1988), although in experimental settings they are extremely difficult to visualize. In numerical simulations it is possible to initiate scroll waves with interesting topology, including closed scroll rings, knotted scrolls, or linked pairs of scroll rings.

The mathematical theory of scroll waves is also in its infancy. Attributes of the topology of closed scrolls were worked out by Winfree and Strogatz (1983a,b,c; 1984), and a general asymptotic theory for their evolution has been suggested (Keener, 1988) and tested against numerical experiments on circular scroll rings and helical scrolls. There is not sufficient space here to discuss the theory of scroll waves. However, scroll waves are mentioned again briefly in later chapters on cardiac waves and rhythmicity.

6.6 EXERCISES

1. Show that the (U, W) phase plane of the bistable equation, (6.6) and (6.7), has three steady states, two of which are saddle points. What is the nature of the third steady state? Show that the slope of the unstable manifold at the origin is given by the positive root of $\lambda^2 - c\lambda + f'(0) = 0$ and is always larger than c. What is the slope of the stable manifold at $(U, W) = (1, 0)$? Show that the slopes of both these manifolds are increasing functions of c.

2. Use cable theory to find the speed of propagation for each of the axons listed in Table 4.1, assuming that the ionic currents are identical and the speed of propagation for the squid giant axon is 21 mm/ms.

3. Find the space constant for a myelinated fiber.

4. Construct a traveling wave solution to the piecewise-linear bistable equation, (6.1) and (6.3). Show that the wave travels with speed $\frac{1-2\alpha}{\sqrt{\alpha-\alpha^2}}$.

5. Find the shape of the traveling wave profile for (6.1) in the case that the function $f(v)$ is

$$f(v) = \begin{cases} -v, & \text{for } v < \alpha/2, \\ v - \alpha, & \text{for } \alpha/2 < v < \frac{1+\alpha}{2}, \\ 1 - v, & \text{for } v > \frac{1+\alpha}{2}. \end{cases} \qquad (6.129)$$

6. (a) Solve the bistable equation

$$v_t = v_{xx} + v(0.1 - v)(v - 1) \qquad (6.130)$$

       numerically and plot a traveling wave.

   (b) Solve the FitzHugh–Nagumo equations

$$v_t = v_{xx} + v(0.1 - v)(v - 1) - w, \qquad (6.131)$$
$$w_t = 0.1(v - w) \qquad (6.132)$$

       numerically and plot a traveling wave.

7. Find the speed of traveling fronts for barnacle muscle fiber using the Morris–Lecar model (Chapter 5, (5.72)–(5.73)).

   Answer: About 6 cm/s.

8. Write a program to numerically calculate the speed of propagation for the bistable equation. Use the program to determine the effect of Na^+ channel density (Table 6.1) on the speed of propagation in various axons, assuming that all other currents are the same as for the Hodgkin–Huxley equations.

9. Show that $1/p'(\lambda_1) + 1/p'(\lambda_2) + 1/p'(\lambda_3) = 0$, where p is defined by (6.58). Hence, show that $h(1) = 0$, where h is defined by (6.64). Show also that $\lambda_1/p'(\lambda_1) + \lambda_2/p'(\lambda_2) + \lambda_3/p'(\lambda_3) = 0$ and thus $h'(1) = 0$. Finally, show that $h''(1) = p'(\lambda_1)/\lambda_1^2 - 2$.

10. The results of Exercise 9 can be generalized. Use contour integration in the complex plane to show that for an nth order polynomial $p(z) = z^n + \cdots$ with simple roots $z_k, k = 1, \ldots, n$,

$$\sum_{k=1}^{n} \frac{z_k^j}{p'(z_k)} = 0, \tag{6.133}$$

   provided that $n > j + 1$. In addition, show that

$$\sum_{k=1}^{n} \frac{z_k^{n-1}}{p'(z_k)} = 1. \tag{6.134}$$

11. Show that the duration of the absolute refractory period of the traveling pulse for the generalized FitzHugh–Nagumo equations is (approximately)

$$T_{\mathrm{ar}} = \int_{w_-}^{w_0} \frac{dw}{G_-(w)}, \tag{6.135}$$

   where w_0 is that value of w for which $c(w) = 0$.

12. Show that in the limit as the period approaches infinity, (6.85) and (6.86) reduce to the equations for a solitary pulse (6.62) and (6.63).

13. Suppose a nearly singular (ϵ small) FitzHugh–Nagumo system has a stable periodic oscillatory solution when there is no diffusive coupling. How do the phase portraits for the spatially independent solutions and the periodic traveling waves differ? How are these differences reflected in the temporal behavior of the solutions?

14. (Fisher's equation.) The goal of this exercise is to find traveling wave solutions of the equation

$$u_t = u_{xx} + f(u), \tag{6.136}$$

   where $f(u) = u(1 - u)$. This equation, sometimes known as Fisher's equation, is commonly used in mathematical ecology and population biology to model traveling waves. In this context u corresponds to a population density or a gene density.

   (a) Convert to the traveling wave coordinate $\xi = x + ct$, where $c \geq 0$ is the wave speed, and show that the PDE converts to the system of ODEs

$$u' = w, \tag{6.137}$$

$$w' = cw - u(1 - u). \tag{6.138}$$

   (b) What are the steady states of the ODE system? Determine the stability of the steady states and sketch the phase plane, including the directions of the eigenvectors at the saddle point.

(c)   Show that a biologically relevant traveling wave cannot exist when $c < 2$.

(d)   Starting at the saddle point $(1,0)$, follow the stable manifold backwards in time. Show that this stable manifold cannot leave the first quadrant across the u axis.

(e)   Show that if $c \geq 2$ there exists a line, $u = mw$, such that the stable manifold of the saddle point cannot be above this line.

(f)   Use this to draw a conclusion about the existence of traveling wave solutions to Fisher's equation.

15.   Here we examine Fisher's equation (6.136) again in a more general form. Suppose $f(0) = f(1) = 0, f'(0) > 0, f'(1) < 0$, and $f(u) > 0$ for $0 < u < 1$.

(a)   Show that when $c = 0$ there are no trajectories with $u \geq 0$ connecting the two rest points $u = 0$ and $u = 1$.

(b)   Show that if there is a value of c for which there is no heteroclinic connection with $u \geq 0$, then there is no such connecting trajectory for any smaller value of c.

(c)   Show that if there is a value of c for which there is a connecting trajectory with $u \geq 0$, then there is a connecting trajectory for every larger value of c.

(d)   Let μ be the smallest positive number for which $f(u) \leq \mu u$ for $0 \leq u \leq 1$. Show that a heteroclinic connection exists for all $c \geq 2\sqrt{\mu}$.

16.   (Phase waves.) Suppose $f(0) = f(1) = f'(0) = 0, f'(1) < 0$ and $f(v) > 0$ for $0 < v < 1$. Show that all the statements of Exercise 14 hold.

(a)   Show that there are values of c for which there are no trajectories connecting critical points for the equation $V'' + cV' + f(V) = 0$, with $f(0) = f'(0) = f(1) = 0$, and $f(v) > 0$ for $0 < v < 1$. Hint: What is the behavior of trajectories for $c = 0$?

(b)   Show that if a connecting trajectory exists for one value of $c < 0$, then it exists for all smaller (larger in absolute value) values of c. Hint: What happens to trajectories when c is decreased (increased in absolute value) slightly? How do trajectories for different values of c compare?

(c)   Suppose $f(v) = v^2(1-v)$. Show that there is a connecting trajectory for $c \leq -\frac{1}{\sqrt{2}}$. Hint: Find an exact solution of the form $v' = Av(1-v)$.

17.   Construct a traveling wave solution for the equation $v_t = v_{xx} + f(v)$ with $v(-\infty, t) = 0$ and $v(+\infty, t) = 1$ where $f(v) = 0$ for $0 < v < q, f(v) = 1 - v$ for $q < v < 1$. What is the speed of propagation? For what values of q does the wave exist? Plot the traveling waveform on the same graph as the traveling waveform for the piecewise-linear bistable equation for values of q and α for which the speeds are the same.

18.   Show that the equation $v_t = v_{xx} + f(v)$, where $f(0) = f(1) = 0$, has a monotone traveling wave solution connecting $v = 0$ to $v = 1$ only if $\int_0^1 f(v) dv > 0$. Interpret the results of Exercises 4 and 17 in light of this result.

19.   Given a dispersion curve $c = C(T)$, use the simple kinetic theory,

$$\frac{dt_n(x)}{dx} = \frac{1}{C(t_n(x) - t_{n-1}(x))}, \qquad (6.139)$$

to determine the stability of periodic waves on a ring of length L.

Hint: On a ring of length L, $t_n(x) = t_{n-1}(x - L)$. Suppose that $T^*C(T^*) = L$. Perform linear stability analysis for the solution $t_n(x) = \frac{x}{C(T^*)}$.

20. Use perturbation arguments to estimate the error in the calculated speed of propagation when Euler's method (forward differencing in time) with second-order centered differencing in space is used to approximate the solution of the bistable equation.

21. Generalize the kinematic theory (6.93)–(6.95) to the case in which wave backs are not phase waves, by tracking both fronts and backs and the corresponding recovery value (Cytrynbaum and Keener, 2002).

22. What is the eikonal-curvature equation for (6.96) when the medium is anisotropic and D is a symmetric matrix, slowly varying in space?

Answer: $S_t = \sqrt{\nabla S \cdot D \nabla S} \left(c_0 \sqrt{k} + \nabla \cdot \left(\frac{D \nabla S}{\sqrt{\nabla S \cdot D \nabla S}} \right) \right)$.

23. Verify (6.103).

24. Verify that in two spatial dimensions, $\nabla \cdot \left(\frac{\nabla S}{|\nabla S|} \right)$ is the curvature of the level surface of the function S.

25. The following are the rules for a simple finite automaton on a rectangular grid of points:

    (a) The state space consists of three states, 0, 1, and 2, 0 meaning at rest, 1 meaning excited, and 2 meaning refractory.

    (b) A point in state 1 goes to state 2 on the next time step. A point in state 2 goes to 0 on the next step.

    (c) A point in state 0 remains in state 0 unless at least one of its nearest neighbors is in state 1, in which case it goes to state 1 on the next step.

    Write a computer program that implements these rules. What initial data must be supplied to initiate a spiral? Can you initiate a double spiral by supplying two stimuli at different times and different points?

26. (a) Numerically simulate spiral waves for the Pushchino model of Chapter 5, Exercise 13.

    (b) Numerically simulate spiral waves for the Pushchino model with

$$ f(V) = \begin{cases} C_1 V, & \text{when} \quad V < V_1, \\ -C_2 V + a, & \text{when} \quad V_1 < V < V_2, \\ C_3(V - 1), & \text{when} \quad V > V_2, \end{cases} \tag{6.140} $$

    and

$$ \tau(V, w) = \begin{cases} \tau_1, & \text{when} \quad V_1 < V < V_2, \\ \tau_1, & \text{when} \quad V < V_1, w > w_1, \\ \tau_2, & \text{when} \quad V > V_2, \\ \tau_3, & \text{when} \quad V < V_1, w < w_1. \end{cases} $$

Use the parameters $V_1 = 0.0026$, $V_2 = 0.837$, $w_1 = 1.8$, $C_1 = 20$, $C_2 = 3$, $C_3 = 15$, $a = 0.06$, $\tau_1 = 75$, $\tau_2 = 1.0$, $\tau_3 = 2.75$, and $k = 3$. What is the difference between these spirals and those for the previous model?

Answer: There are no stable spirals for this model, but spirals continually form and break apart, giving a chaotic appearance.

CHAPTER 7

Calcium Dynamics

Calcium is critically important for a vast array of cellular functions, as can be seen by a quick look through any physiology book. For example, in this book we discuss the role that Ca^{2+} plays in muscle mechanics, cardiac electrophysiology, bursting oscillations and secretion, hair cells, and adaptation in photoreceptors, among other things. Clearly, the mechanisms by which a cell controls its Ca^{2+} concentration are of central interest in cell physiology.

There are a number of Ca^{2+} control mechanisms operating on different levels, all designed to ensure that Ca^{2+} is present in sufficient quantity to perform its necessary functions, but not in too great a quantity in the wrong places. Prolonged high cytoplasmic concentrations of Ca^{2+} are toxic. For example, cellular Ca^{2+} overload can trigger apoptotic cell death, a process in which the cell kills itself. In muscle cells, high intracellular Ca^{2+} is responsible for prolonged muscle tension and rigor mortis.

There are many reviews of Ca^{2+} physiology in the literature: in 2003 an entire issue of *Nature Reviews* was devoted to the subject and contains reviews of Ca^{2+} homeostasis (Berridge et al., 2003), extracellular Ca^{2+} sensing (Hofer and Brown, 2003), Ca^{2+} signaling during embryogenesis (Webb and Miller, 2003), the Ca^{2+}-apoptosis link (Orrenius et al., 2003), and the regulation of cardiac contractility by Ca^{2+} (MacLennan and Kranias, 2003). Other useful reviews are Berridge (1997) and Carafoli (2002).

In vertebrates, the majority of body Ca^{2+} is stored in the bones, from where it can be released by hormonal stimulation to maintain an extracellular Ca^{2+} concentration of around 1 mM, while active pumps and exchangers keep the cytoplasmic Ca^{2+} concentration at around 0.1 μM. Since the cytoplasmic concentration is low, there is a steep concentration gradient from the outside of a cell to the inside. This disparity has the advantage that cells are able to raise their Ca^{2+} concentration quickly, by opening Ca^{2+} channels and relying on passive flow down a steep concentration gradient, but it has the disadvantage that energy must be expended to keep the cytoplasmic Ca^{2+} concentration low. Thus, cells have finely tuned mechanisms to control how Ca^{2+} is allowed into, or removed from, the cytoplasm.

Calcium is removed from the cytoplasm in two principal ways: it is pumped out of the cell across the plasma membrane, and it is sequestered into internal membrane-bound compartments such as the mitochondria, the endoplasmic reticulum (ER) or sarcoplasmic reticulum (SR), and secretory granules. Since the Ca^{2+} concentration in the cytoplasm is much lower than either the extracellular concentration or the concentration inside the internal compartments, both methods of Ca^{2+} removal require expenditure of energy. Some of this is by a Ca^{2+} ATPase, similar to the Na^+–K^+ ATPase discussed in Chapter 2, that uses energy stored in ATP to pump Ca^{2+} out of the cell or into an internal compartment. There is also a Na^+–Ca^{2+} exchanger (NCX) in the cell membrane that uses the energy of the Na^+ electrochemical gradient to remove Ca^{2+} from the cell at the expense of Na^+ entry (also discussed in Chapters 2 and 3).

Calcium influx also occurs via two principal pathways: inflow from the extracellular medium through Ca^{2+} channels in the plasma membrane and release from internal stores. The plasma membrane Ca^{2+} channels are of several different types. For example, voltage-controlled channels open in response to depolarization of the cell membrane, receptor-operated channels open in response to the binding of an external ligand, second-messenger-operated channels open in response to the binding of a cellular second messenger, and mechanically operated channels open in response to mechanical stimulation. Voltage-controlled Ca^{2+} channels are of great importance in other chapters of this book (in particular, for bursting oscillations in Chapter 9 or cardiac cells in Chapter 12), but are not discussed here. We also omit consideration of the other plasma membrane channels, concentrating instead on the properties of Ca^{2+} release from internal stores.

Calcium release from internal stores such as the ER is the second major way in which Ca^{2+} enters the cytoplasm, and this is mediated principally by two types of Ca^{2+} channels that are also receptors: the ryanodine receptor and the inositol (1,4,5)-trisphosphate (IP$_3$) receptor. The ryanodine receptor, so called because of its sensitivity to the plant alkaloid ryanodine, plays an integral role in excitation–contraction coupling in skeletal and cardiac muscle cells, and is believed to underlie Ca^{2+}-induced Ca^{2+} release, whereby a small amount of Ca^{2+} entering the cardiac or skeletal muscle cell through voltage-gated Ca^{2+} channels initiates an explosive release of Ca^{2+} from the sarcoplasmic reticulum. Excitation–contraction coupling is discussed in detail in Chapter 15. Ryanodine receptors are also found in a variety of nonmuscle cells such as neurons, pituitary cells, and sea urchin eggs. The IP$_3$ receptor, although similar in structure to the ryanodine receptor, is found predominantly in nonmuscle cells, and is sensitive to the second messenger IP$_3$. The binding of an extracellular agonist such as a hormone or a neurotransmitter to a receptor in the plasma membrane can cause, via a G-protein link to phospholipase C (PLC), the cleavage of phosphotidylinositol (4,5)-bisphosphate (PIP$_2$) into diacylglycerol (DAG) and IP$_3$ (Fig. 7.1). The water-soluble IP$_3$ is free to diffuse through the cell cytoplasm and bind to IP$_3$ receptors situated on the ER

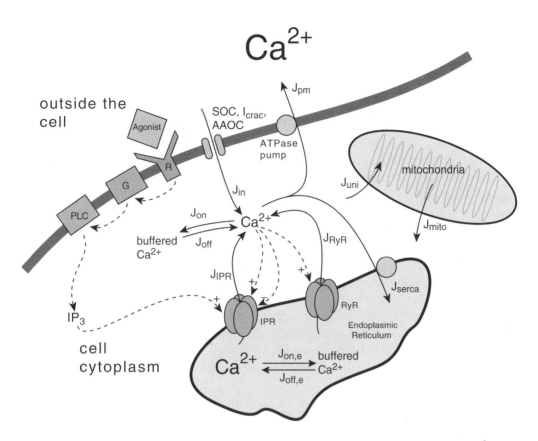

Figure 7.1 Diagram of the major fluxes involved in the control of cytoplasmic Ca^{2+} concentration. Binding of agonist to a cell membrane receptor (R) leads to the activation of a G-protein (G), and subsequent activation of phospholipase C (PLC). This cleaves phosphotidylinositol bisphosphate into diacylglycerol and inositol trisphosphate (IP_3), which is free to diffuse through the cell cytoplasm. When IP_3 binds to an IP_3 receptor (IPR) on the endoplasmic reticulum (ER) membrane it causes the release of Ca^{2+} from the ER, and this Ca^{2+} in turn modulates the open probability of the IPR and ryanodine receptors (RyR). Calcium fluxes are denoted by solid arrows. Calcium can be released from the ER through IPR (J_{IPR}) or RyR (J_{RyR}), can be pumped from the cytoplasm into the ER (J_{serca}) or to the outside (J_{pm}), can be taken up into (J_{uni}), or released from (J_{mito}), the mitochondria, and can be bound to (J_{on}), or released from (J_{off}), Ca^{2+} buffers. Entry from the outside (J_{in}) is controlled by a variety of possible channels, including store-operated channels (SOC), Ca^{2+}-release-activated channels (I_{crac}), and arachidonic-acid-operated channels (AAOC).

membrane, leading to the opening of these receptors and subsequent release of Ca^{2+} from the ER. Similar to ryanodine receptors, IP_3 receptors are modulated by the cytoplasmic Ca^{2+} concentration, with Ca^{2+} both activating and inactivating Ca^{2+} release, but at different rates. Thus, Ca^{2+}-induced Ca^{2+} release occurs through IP_3 receptors also.

As an additional control for the cytoplasmic Ca^{2+} concentration, Ca^{2+} is heavily buffered (i.e., bound) by large proteins, with estimates that at least 99% of the total cytoplasmic Ca^{2+} is bound to buffers. The Ca^{2+} in the ER and mitochondria is also heavily buffered.

7.1   Calcium Oscillations and Waves

One of the principal reasons that modelers have become interested in Ca^{2+} dynamics is that the concentration of Ca^{2+} shows highly complex spatiotemporal behavior (Fig. 7.2). Many cell types respond to agonist stimulation with oscillations in the concentration of Ca^{2+}. These oscillations can be grouped into two major types: those that are dependent on periodic fluctuations of the cell membrane potential and the associated periodic entry of Ca^{2+} through voltage-gated Ca^{2+} channels (for example in cardiac cells), and those that occur in the presence of a voltage clamp. The focus in this chapter is on the latter type, within which group further distinctions can be made by whether the oscillatory Ca^{2+} flux is through ryanodine or IP_3 receptors. We consider models of both types here, beginning with models of IP_3-dependent Ca^{2+} oscillations. In many cell types one important feature of IP_3-dependent Ca^{2+} oscillations is that they persist for some time in the absence of extracellular Ca^{2+} (although often with a changed shape and period), and must thus necessarily involve Ca^{2+} transport to and from the internal stores.

Calcium oscillations have been implicated in a vast array of cellular control processes. The review by Berridge et al. (2003) mentions, among other things, oocyte activation at fertilization, axonal growth, cell migration, gene expression, formation of nodules in plant root hairs, development of muscle, and release of cytokines from epithelial cells. In many cases, the oscillations are a frequency-encoded signal that allows a cell to use Ca^{2+} as a second messenger while avoiding the toxic effects of prolonged high Ca^{2+} concentration. For example, exocytosis in gonadotropes is known to be dependent on the frequency of Ca^{2+} oscillations (Tse et al., 1993), and gene expression can also be modulated by Ca^{2+} spike frequency (Dolmetsch et al., 1998; Li et al., 1998). However, there are still many examples for which the signal carried by a Ca^{2+} oscillation has not been unambiguously decoded.

The period of IP_3-dependent oscillations ranges from a few seconds to a few minutes. In some cell types these oscillations appear to occur at a constant concentration of IP_3, while in other cell types they appear to be driven by oscillations in $[IP_3]$. Modulation of the IP_3 receptor by other factors such as phosphatases and kinases also plays an important role in setting the oscillation period, while Ca^{2+} influx from outside the cell is another important regulatory mechanism. Thus, there is a tremendous variety of mechanisms that control Ca^{2+} oscillations, and it is not realistic to expect that a single model can capture the important behaviors in all cell types. Nevertheless, most of the models have much in common, and a great deal can be learned about the overall approach by the study of a small number of models. The concept of a cellular

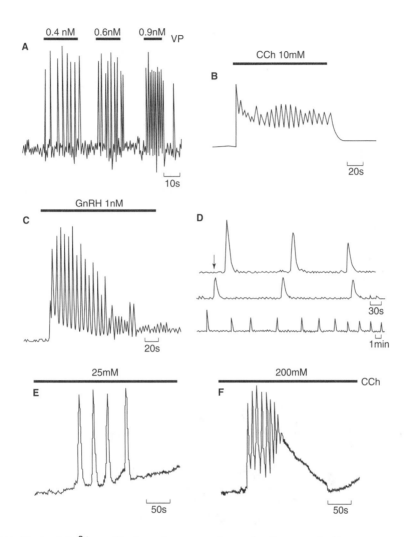

Figure 7.2 Typical Ca^{2+} oscillations from a variety of cell types. A: Hepatocytes stimulated with vasopressin (VP). B: Rat parotid gland stimulated with carbachol (CCh). C: Gonadotropes stimulated with gonadotropin-releasing hormone (GnRH). D: Hamster eggs after fertilization. The time of fertilization is denoted by the arrow. E and F: Insulinoma cells stimulated with two different concentrations of carbachol. (Berridge and Galione, 1988, Fig. 2.)

Ca^{2+}-signaling "toolkit" has been introduced by Berridge et al. (2003). There is a large number of components in the toolkit (receptors, G proteins, channels, buffers, pumps, exchangers, and so on); by expressing those components that are needed, each cell can fine-tune the spatiotemporal properties of its intracellular Ca^{2+}. Models of Ca^{2+} dynamics follow a similar modular approach. Each member of the cellular toolkit has a corresponding module in the modeling toolkit (or if one doesn't exist it can be constructed). To construct whole-cell models one then combines appropriate components from the modeling toolkit.

Often, Ca^{2+} oscillations do not occur uniformly throughout the cell, but are organized into repetitive intracellular waves (Rooney and Thomas, 1993; Thomas et al., 1996; Røttingen and Iversen, 2000; Falcke, 2004). One of the most visually impressive examples of this occurs in *Xenopus* oocytes. In 1991, Lechleiter and Clapham and their coworkers discovered that intracellular Ca^{2+} waves in immature *Xenopus* oocytes showed remarkable spatiotemporal organization. By loading the oocytes with a Ca^{2+}-sensitive dye, releasing IP_3, and observing Ca^{2+} release patterns with a confocal microscope, Lechleiter and Clapham (1992; Lechleiter et al., 1991b) observed that the intracellular waves develop a high degree of spatial organization, forming concentric circles, plane waves, and multiple spirals. Typical experimental results are shown in Fig. 7.3A. The feature of *Xenopus* oocytes that makes these observations possible is their large size. *Xenopus* oocytes can have a diameter larger than 600 μm, an order of magnitude greater than most other cells. In a small cell, a typical Ca^{2+} wave (often with a width of close to 100 μm) cannot be observed in its entirety, and there is not enough room for a spiral to form. However, in a large cell it may be possible to observe both the wave front and the wave back, as well as spiral waves, and this has made the *Xenopus* oocyte an important system for the study of Ca^{2+} waves. Of course, what is true for *Xenopus* oocytes is not necessarily true for other cells, and so one must be cautious about extrapolating to other cell types. The physiological significance of these spatiotemporal patterns is also not apparent.

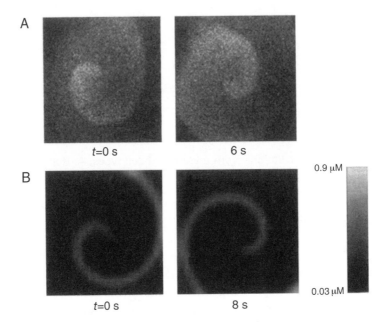

Figure 7.3   A: Spiral Ca^{2+} wave in the *Xenopus* oocyte. The image size is 420 $\times$ 420 μm. The spiral has a wavelength of about 150 μm and a period of about 8 seconds. B: A model spiral wave simulated on a domain of size 250 $\times$ 250 μm, with $[IP_3] = 95$ nM. See Section 7.3.1. (Atri et al., 1993, Fig. 11.)

Another well-known example of an intracellular Ca^{2+} wave occurs across the cortex of an egg directly after fertilization (Ridgway et al., 1977; Nuccitelli et al., 1993). In fact, this wave motivated the first models of Ca^{2+} wave propagation, which appeared as early as 1978 (Gilkey et al., 1978; Cheer et al., 1987; Lane et al., 1987). However, because they do not account for the underlying physiology, these early models have since been superseded (Wagner et al., 1998; Bugrim et al., 2003).

In addition to traveling across single cells, Ca^{2+} waves can be transmitted between cells, forming intercellular waves that can travel over distances of many cell lengths. Such intercellular waves have been observed in intact livers, slices of hippocampal brain tissue, epithelial and glial cell cultures (see Fig. 7.4), and many other preparations

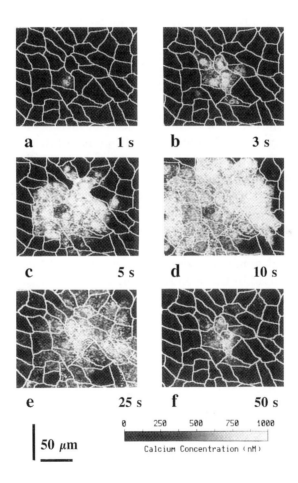

Figure 7.4 Mechanically stimulated intercellular wave in airway epithelial cells. The solid white lines denote the cell borders, and the white dot in the approximate center of frame **a** denotes the place of mechanical stimulation. After mechanical stimulation a wave of increased $[Ca^{2+}]$ can be seen spreading from the stimulated cell through other cells in the culture. The time after mechanical stimulation is given in seconds in the lower right corner of each panel (Sneyd et al., 1995b, Fig. 4A). A model of this wave is discussed in Section 7.7.

(Sanderson et al., 1990, 1994; Charles et al., 1991, 1992; Cornell-Bell et al., 1990; Kim et al., 1994; Robb-Gaspers and Thomas, 1995). Not all intercellular coordination is of such long range; synchronized oscillations are often observed in small groups of cells such as pancreatic or parotid acinar cells (Yule et al., 1996) or multiplets of hepatocytes (Tordjmann et al., 1997, 1998).

Although there is controversy about the exact mechanisms by which Ca^{2+} waves propagate (and it is true that the mechanisms differ from cell type to cell type), it is widely believed that in many cell types, intracellular Ca^{2+} waves are driven by the diffusion of Ca^{2+} between Ca^{2+} release sites. According to this hypothesis, the Ca^{2+} released from one group of release sites (usually either IPR or RyR) diffuses to neighboring release sites and initiates further Ca^{2+} release from them. Repetition of this process can generate an advancing wave front of high Ca^{2+} concentration, i.e., a Ca^{2+} wave. Since they rely on the active release of Ca^{2+} via a positive feedback mechanism, such waves are actively propagated. The theory of such waves was presented in Chapter 6. However, when the underlying Ca^{2+} kinetics are oscillatory (i.e., there is a stable limit cycle), waves can propagate by a kinematic, or phase wave, mechanism, in which the wave results from the spatially ordered firing of local oscillators. These waves do not depend on Ca^{2+} diffusion for their existence, but merely on the fact that one end of the cell is oscillating with a different phase from that of the other end. In this case, Ca^{2+} diffusion serves to synchronize the local oscillators, but the phase wave persists in the absence of Ca^{2+} diffusive coupling.

The most common approach to the study of Ca^{2+} oscillations and waves is to assume that the underlying mechanisms are deterministic. However, even the most cursory examination of experimental data shows that Ca^{2+} dynamics are inherently stochastic in nature. The most prominent of the stochastic events underlying Ca^{2+} oscillations and waves are small localized release events called *puffs*, or *sparks*, and these elementary events, caused by the opening of single, or a small number of, Ca^{2+} release channels, are the building blocks from which global events are built. Calcium puffs, caused by localized release through IPR, have been extensively studied in *Xenopus* oocytes and HeLa cells (Marchant et al., 1999; Sun et al., 1998; Callamaras et al., 1998; Marchant and Parker, 2001; Thomas et al., 2000; Bootman et al., 1997a,b), while Ca^{2+} sparks, caused by localized release through RyR and occurring principally in muscle cells, were discovered by Cheng et al. (1993) and studied by a multitude of authors since (Smith et al., 1998; Izu et al., 2001; Sobie et al., 2002; Soeller and Cannell, 1997, 2002). To study the stochastic properties of puffs and sparks, stochastic models are necessary. In general, such models are based on stochastic simulations of Markov models for the IPR and RyR. We discuss such models briefly in Section 7.6.

Many different models have been constructed to study Ca^{2+} oscillations and waves in different cell types, and there is not space enough here to discuss them all. Thus, after discussing particular models of each of the most important Ca^{2+} fluxes, we discuss in detail only a few simple models of oscillations and waves. The most comprehensive review of the field is Falcke (2004); this review is almost 200 pages long and is the best yet written of models of Ca^{2+} dynamics.

7.2 Well-Mixed Cell Models: Calcium Oscillations

If we assume the cell is well-mixed, then the concentration of each species is homogeneous throughout. We write c for the concentration of free Ca^{2+} ions in the cytoplasm and note that $c = c(t)$, i.e., c has no spatial dependence. Similarly, we let c_e denote the homogeneous concentration of Ca^{2+} in the ER.

The differential equations for c and c_e follow from conservation of Ca^{2+}. In words,

Rate of change of total calcium = net flux of calcium into the compartment,

or in mathematical notation,

$$\frac{d(vc)}{dt} = \tilde{J}_{net}, \tag{7.1}$$

where v is the volume of the compartment, and $\tilde{J}_{net}$ is the net Ca^{2+} flux into the compartment, in units of number of moles per second. Usually it is assumed that the volumes of the cell and its internal compartments are constant, in which case the conservation equation is written as

$$\frac{dc}{dt} = J_{net} = \tilde{J}_{net}/v. \tag{7.2}$$

Here, J_{net} is the net flux into the cytoplasm per cytoplasmic volume, and has units of concentration per second.

In applications in which the volume is changing (as, for example, in models of cell volume control, or models of the control of fluid secretion by Ca^{2+} oscillations), it is necessary to use $\tilde{J}_{net}$ instead of J_{net}.

More specifically, in view of the fluxes shown in Fig. 7.1, we have

$$\frac{dc}{dt} = J_{IPR} + J_{RyR} + J_{in} - J_{pm} - J_{serca} - J_{on} + J_{off} + J_{uni} - J_{mito}. \tag{7.3}$$

For the equation for ER Ca^{2+}, the fact that the cytoplasmic volume is different from the ER volume must be taken into account. Again, in view of the fluxes shown in Fig. 7.1, we have

$$\frac{dc_e}{dt} = \gamma(J_{serca} - J_{IPR} - J_{RyR}) + J_{off,e} - J_{on,e}, \tag{7.4}$$

where $\gamma = \frac{v_{cyt}}{v_{ER}}$ is the ratio of the cytoplasmic volume to the ER volume.

Each of the fluxes in these equations corresponds to a component of the Ca^{2+}-signaling toolkit, of which there are many possibilities; here, we discuss only those that appear most often in models. In general, these equations are coupled to additional differential equations that describe the gating of the IPR or RyR, or the dynamics of the pumps and exchangers. An example of such a model is discussed in Section 7.2.6. However, before we see what happens when it is all put together, we first discuss how each toolkit component is modeled.

7.2.1 Influx

In general, the influx of Ca^{2+} into a cell from outside is voltage-dependent. However, when Ca^{2+} oscillations occur at a constant voltage (as is typical in nonexcitable cells), the voltage dependence is unimportant. This influx is certainly dependent on a host of other factors, including Ca^{2+}, IP_3, arachidonic acid, and the Ca^{2+} concentration in the ER. For example, there is evidence that, in some cell types, depletion of the ER causes an increase in Ca^{2+} influx through store-operated channels, or SOCs (Clapham, 1995). There is also evidence that SOCs play a role only at high agonist concentration (when the ER is highly depleted), but that at lower agonist concentrations Ca^{2+} influx is controlled by arachidonic acid (Shuttleworth, 1999). However, the exact mechanism by which this occurs is unknown. What is known is that J_{in} increases as agonist concentration increases; if this were not so, the steady-state level of Ca^{2+} would be independent of IP_3 (see Exercise 1), which it is not. One common approach is to assume that J_{leak} is a linear increasing function of p, the IP_3 concentration, with

$$J_{in} = \alpha_1 + \alpha_2 p, \tag{7.5}$$

for some constants α_1 and α_2. Although this extremely simple model does not take into account any dependence of influx on the loading of the ER, it is probably acceptable for lower agonist concentrations. Our current state of knowledge of what controls Ca^{2+} influx is not sufficient to allow the construction of much more detailed models.

7.2.2 Mitochondria

Mitochondrial Ca^{2+} handling is a highly complex process, and a number of detailed models have been constructed. For the sake of brevity we do not discuss these models at all; the interested reader is referred to Colegrove et al. (2000), Friel (2000), Falcke et al. (2000), Grubelnik et al. (2001), Marhl et al. (2000), Schuster et al. (2002) and Selivanov et al. (1998). It seems that one function of the mitochondria is to take up and release large amounts of Ca^{2+}, but relatively slowly. Thus, the mitochondria tend to modulate the trailing edges of the waves, reduce wave amplitude, and change the long-term oscillatory behavior. However, this is certainly an oversimplification.

7.2.3 Calcium Buffers

At least 99% of Ca^{2+} in the cytoplasm is bound to large proteins, called Ca^{2+} buffers. Typical buffers are calsequestrin, calbindin, fluorescent dyes, and the plasma membrane itself. A detailed discussion of Ca^{2+} buffering, and its effect on oscillations and waves, is given in Section 7.4.

7.2.4   Calcium Pumps and Exchangers

Calcium ATPases

Early models of the Ca^{2+} ATPase pump were of the Hill equation type (Section 1.4.4). For example, data from Lytton et al. (1992) showed that the flux through the ATPase was approximately a sigmoidal function of c, with Hill coefficient of about 2. Thus, a common model is

$$J_{serca} = \frac{V_p c^2}{K_p^2 + c^2}. \tag{7.6}$$

Such simple models are known to have a number of serious flaws. For example, this flux has no dependence on ER Ca^{2+} concentration and is always of one sign. However, it is known that with high enough ER Ca^{2+} concentration, it is possible for the pump to reverse, generating ATP in the process.

MacLennan et al. (1997) constructed a more detailed model, shown schematically in Fig. 7.5. The pump can be in one of two basic conformations: E_1 and E_2. In the E_1 conformation the pump binds two Ca^{2+} ions from the cytoplasm, whereupon it exposes a phosphorylation site. Once phosphorylated, the pump switches to the E_2 conformation in which the Ca^{2+} binding sites are exposed to the ER lumen and have a much lower affinity for Ca^{2+}. Thus, Ca^{2+} is released into the ER, the pump is dephosphorylated,

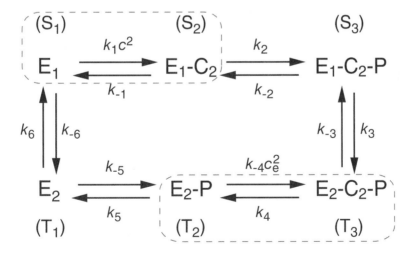

Figure 7.5 Schematic diagram of the SERCA model of MacLennan et al. (1997). E_1 is the conformation in which the Ca^{2+} binding sites are exposed to the cytoplasm, and E_2 is the conformation in which they are exposed to the lumen and have a much lower affinity for Ca^{2+}. The P denotes that the pump has been phosphorylated; for simplicity ATP and ADP have been omitted from the diagram. The cotransport of H^+ has also been omitted. By assuming fast equilibrium between the pairs of states in the dashed boxes, a simplified model, shown in Fig. 7.6, can be derived.

and completes the cycle by switching back to the E_1 conformation. For each Ca^{2+} ion transported from the cytoplasm to the ER, one proton is cotransported from the ER to the cytoplasm. This makes the similarity with the Na^+–K^+ ATPase more apparent. The rate-limiting step of the transport cycle is the transition from E_1-P-C_2 to E_2-P-C_2.

To calculate the flux in this model is relatively straightforward, and is similar to examples that were amply discussed in Chapter 2, so we leave this calculation as an exercise for the interested reader. It turns out that

$$J_{\text{serca}} = \frac{c^2 - K_1 K_2 K_3 K_4 K_5 K_6 c_e^2}{\alpha_1 c^2 + \alpha_2 c_e^2 + \alpha_3 c^2 c_e^2 + \alpha_4}, \tag{7.7}$$

where the α's are functions of the rate constants, too long and of too little interest to include in full detail. As always, $K_i = k_{-i}/k_i$. If the affinity of the Ca^{2+} binding sites is high when the pump is in conformation E_1, and low when the pump is in conformation E_2, then $\frac{k_{-4}}{k_4} \ll \frac{k_1}{k_{-1}}$, so that $K_1 K_4$ is much less than one. However, it is also reasonable to assume that $\frac{k_6}{k_{-6}} = \frac{k_{-3}}{k_3}$ and that $\frac{k_2}{k_{-2}} = \frac{k_5}{k_{-5}}$, in which case $K_2 K_3 K_5 K_6 = 1$. It follows that $K_1 K_2 K_3 K_4 K_5 K_6 \ll 1$, so that the pump can support a positive flux even when c_e is much greater than c.

Notice that if this were a model of a closed system, the law of detailed balance would require that $K_1 K_2 K_3 K_4 K_5 K_6 = 1$. This is not the case here since the cycle is driven by phosphorylation of the pump, so the reaction rates depend on the concentrations of ATP, ADP, and P, and energy is continually consumed (if $J_{\text{serca}} > 0$) or generated (if $J_{\text{serca}} < 0$). For a more detailed discussion, see Section 2.5.1.

In the original description of the model, MacLennan et al. assumed that the binding and release of Ca^{2+} occurs quickly. This results in a simpler model with a similar expression for the steady-state flux. The process of reducing a model in this way using a fast-equilibrium assumption is important and useful, and is explored in considerable detail in Chapter 2, and in Exercise 7. However, for convenience, we briefly sketch the derivation here.

In Fig. 7.5, states S_1 and S_2 have been grouped together by a box with a dotted outline, as have states T_2 and T_3. The assumption of fast equilibrium gives

$$s_1 = \frac{K_1}{c^2} s_2, \tag{7.8}$$

with a similar expression for t_2 and t_3. Here, we denote the fraction of pumps in state S_1 by s_1, and similarly for the other states. We now define two new variables; $\bar{s}_1 = s_1 + s_2$ and $\bar{t}_2 = t_2 + t_3$. From (7.8) it follows that

$$\bar{s}_1 = s_1 \left(1 + \frac{c^2}{K_1} \right) = s_2 \left(1 + \frac{K_1}{c^2} \right). \tag{7.9}$$

Hence, the rate at which $\bar{S}_1$ is converted to S_3 is $k_2 s_2 = \frac{c^2 k_2 \bar{s}_1}{c^2 + K_1}$. Similarly, the rate at which $\bar{S}_1$ is converted to T_1 is $k_{-6} s_1 = \frac{K_1 k_{-6} \bar{s}_1}{K_1 + c^2}$.

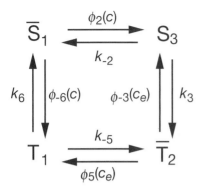

Figure 7.6 Schematic diagram of the simplified version of the SERCA model of MacLennan et al. (1997). By assuming fast equilibrium between the pairs of states shown grouped in Fig. 7.5 by the dashed boxes, this diagram, in which there are fewer states, but with Ca^{2+}-dependent transitions, can be derived. The functions in the transition rates are $\phi_2 = c^2 k_2/(c^2 + K_1)$, $\phi_{-3} = k_{-3} K_4 c_e^2/(1 + K_4 c_e^2)$, $\phi_5 = k_5/(1 + K_4 c_e^2)$ and $\phi_{-6} = K_1 k_{-6}/(K_1 + c^2)$.

Repetition of this process for each of the transitions results in the simplified model shown in Fig. 7.6. The form of the functions ϕ can be understood intuitively. For example, as c increases, the equilibrium between S_1 and S_2 is shifted further toward S_2. This increases the rate at which S_3 is formed, but decreases the rate at which S_1 is converted to T_1. Thus, the transition from $\bar{S}_1$ to S_3 is an increasing function of c, while the transition from $\bar{S}_1$ to T_1 is a decreasing function of c.

From this diagram, the steady-state flux can be calculated in the same way as before. Not surprisingly, the final result looks much the same, giving

$$J_{\text{serca}} = \frac{c^2 - K_1 K_2 K_3 K_4 K_5 K_6 c_e^2}{\beta_1 c^2 + \beta_2 c_e^2 + \beta_3 c^2 c_e^2 + \beta_4}, \tag{7.10}$$

for constants β_i that are limiting values of the constants α_i that appeared in (7.7).

The Ca^{2+} ATPases on the plasma membrane are similar to the SERCA ATPases, and are modeled in a similar way.

Calcium Exchangers

Another important way in which Ca^{2+} is removed from the cytoplasm is via the action of Na^+–Ca^{2+} exchangers, which remove one Ca^{2+} ion from the cytoplasm at the expense of the entry of three Na^+ ions. They are particularly important for the control of Ca^{2+} in cardiac cells and are discussed further in Chapter 12 in the context of models of excitation–contraction coupling.

7.2.5 IP$_3$ Receptors

A basic property of IP$_3$ receptors is that they respond in a time-dependent manner to step changes of Ca^{2+} or IP$_3$. Thus, in response to a step increase of IP$_3$ or Ca^{2+} the receptor open probability first opens to a peak and then declines to a lower plateau (see Fig. 7.7). This decline is called *adaptation* of the receptor, since it adapts to a maintained Ca^{2+} or IP$_3$ concentration. If a further step is applied on top of the first, the receptor responds with another peak, followed by a decline to a plateau. In this way the IPR responds to *changes* in $\left[Ca^{2+}\right]$ or [IP$_3$], rather than to absolute concentrations.

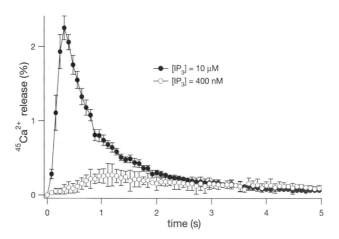

Figure 7.7 Experimental data from Marchant and Taylor (1998; Fig. 1a). In response to a maintained elevation of IP$_3$, the flux through a hepatocyte IPR (measured here as release of radioactive $^{45}Ca^{2+}$) first increases to a peak and then declines to a lower level.

Adaptation is a well-known feature of many physiological systems (see, for example, Chapters 16 and 19), and is often characterized by a fast activation followed by a slower inactivation, as seen in the Na^+ channel in the Hodgkin–Huxley equations (Chapter 5).

Adaptation of the IPR is now believed to result, at least in part, from the fact that Ca^{2+} not only stimulates its own release, but also inhibits it on a slower time scale. It is hypothesized that this sequential activation and inactivation of the IP$_3$ receptor by Ca^{2+} is one mechanism underlying IP$_3$-dependent Ca^{2+} oscillations and waves, and a number of models incorporating this hypothesis have appeared (reviewed by Sneyd et al., 1995b; Tang et al., 1996; Schuster et al., 2002; Falcke, 2004). However, as is described below, there are almost certainly other important mechanisms in operation; some of the most recent models have suggested that the IPR kinetics are of less importance than previously thought, while depletion of the ER might also play in important role. A detailed review of IPR models is given by Sneyd and Falcke (2005), while Sneyd et al. (2004a) compare a number of models to experimental data.

An Eight-State IP$_3$ Receptor Model

One of the earliest models of the IPR to incorporate sequential activation and inactivation by Ca^{2+} was that of De Young and Keizer (1992).

For this model, it is assumed that the IP$_3$ receptor consists of three equivalent and independent subunits, all of which must be in a conducting state for there to be Ca^{2+} flux. Each subunit has an IP$_3$ binding site, an activating Ca^{2+} binding site, and an inactivating Ca^{2+} binding site, each of which can be either occupied or unoccupied, and thus each subunit can be in one of eight states. Each state of the subunit is labeled

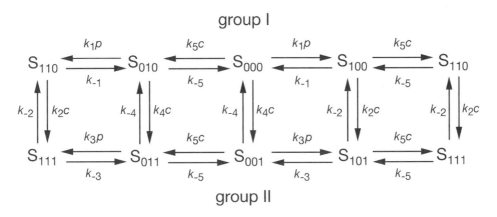

Figure 7.8 The binding diagram for the IP$_3$ receptor model. Here, c denotes $[Ca^{2+}]$, and p denotes $[IP_3]$.

S_{ijk}, where i, j, and k are equal to 0 or 1, with 0 indicating that the binding site is unoccupied and 1 indicating that it is occupied. The first index refers to the IP$_3$ binding site, the second to the Ca^{2+} activation site, and the third to the Ca^{2+} inactivation site. This is illustrated in Fig. 7.8. While the model has 24 rate constants, because of the requirement for detailed balance, these are not all independent. In addition, two simplifying assumptions are made to reduce the number of independent constants. First, the rate constants are assumed to be independent of whether activating Ca^{2+} is bound or not. Second, the kinetics of Ca^{2+} activation are assumed to be independent of IP$_3$ binding and Ca^{2+} inactivation. This leaves only 10 rate constants, $k_1, \ldots, k_5$ and $k_{-1}, \ldots, k_{-5}$. Notice that with these simplifying assumptions, detailed balance is satisfied.

The fraction of subunits in the state S_{ijk} is denoted by x_{ijk}. The differential equations for these are based on mass-action kinetics, and thus, for example,

$$\frac{dx_{000}}{dt} = -(V_1 + V_2 + V_3), \tag{7.11}$$

where

$$V_1 = k_1 p x_{000} - k_{-1} x_{100}, \tag{7.12}$$

$$V_2 = k_4 c x_{000} - k_{-4} x_{001}, \tag{7.13}$$

$$V_3 = k_5 c x_{000} - k_{-5} x_{010}, \tag{7.14}$$

where p denotes $[IP_3]$ and c denotes $[Ca^{2+}]$. V_1 describes the rate at which IP$_3$ binds to and leaves the IP$_3$ binding site, V_2 describes the rate at which Ca^{2+} binds to and leaves the inactivating site, and similarly for V_3. Since experimental data indicate that the receptor subunits act in a cooperative fashion, the model assumes that the IP$_3$ receptor passes Ca^{2+} current only when three subunits are in the state S_{110} (i.e., with

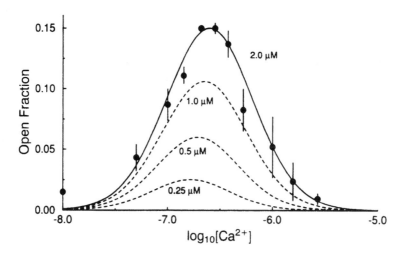

Figure 7.9 The steady-state open probability of the IP$_3$ receptor, as a function of [Ca^{2+}]. The symbols are the experimental data of Bezprozvanny et al. (1991), and the smooth curves are from the receptor model (calculated at four different IP$_3$ concentrations). (De Young and Keizer, 1992, Fig. 2A.)

one IP$_3$ and one activating Ca^{2+} bound), and thus the open probability of the receptor is x_{110}^3.

In Fig. 7.9 we show the open probability of the IP$_3$ receptor as a function of [Ca^{2+}], which is some of the experimental data on which the model is based. Bezprozvanny et al. (1991) showed that this open probability is a bell-shaped function of [Ca^{2+}]. Thus, at low [Ca^{2+}], an increase in [Ca^{2+}] increases the open probability of the receptor, while at high [Ca^{2+}] an increase in [Ca^{2+}] decreases the open probability. Parameters in the model were chosen to obtain agreement with these steady-state data. The kinetic properties of the IP$_3$ receptor are equally important: the receptor is activated quickly by Ca^{2+}, but inactivated by Ca^{2+} on a slower time scale. In the model, this is incorporated in the magnitude of the rate constants ($k_5 > k_2$ and $k_5 > k_4$).

Reduction of the Eight-State IP$_3$ Receptor Model

The complexity of the eight-state receptor model (seven differential equations and numerous parameters) provides ample motivation to seek a simpler model that retains its essential properties. Since IP$_3$ binds quickly to its binding site and Ca^{2+} binds quickly to the activating site, we can dispense with the transient details of these binding processes and assume instead that the receptor is in quasi-steady state with respect to IP$_3$ binding and Ca^{2+} activation (De Young and Keizer, 1992; Keizer and De Young, 1994; Li and Rinzel, 1994; Tang et al., 1996). Notice that this is implied by the parameter values for the detailed receptor model shown in Table 7.1, where k_1, k_3, and k_5 are substantially larger than k_2 and k_4, and k_{-1}, k_{-3}, and k_{-5} are also larger than k_{-2} and k_{-4}.

Table 7.1 Parameters of the eight-state IPR model (De Young and Keizer, 1992).

$k_1 = 400 \ \mu\text{M}^{-1}\text{s}^{-1}$	$k_{-1} = 52 \ \text{s}^{-1}$
$k_2 = 0.2 \ \mu\text{M}^{-1}\text{s}^{-1}$	$k_{-2} = 0.21 \ \text{s}^{-1}$
$k_3 = 400 \ \mu\text{M}^{-1}\text{s}^{-1}$	$k_{-3} = 377.2 \ \text{s}^{-1}$
$k_4 = 0.2 \ \mu\text{M}^{-1}\text{s}^{-1}$	$k_{-4} = 0.029 \ \text{s}^{-1}$
$k_5 = 20 \ \mu\text{M}^{-1}\text{s}^{-1}$	$k_{-5} = 1.64 \ \text{s}^{-1}$

As shown in Fig. 7.8, we arrange the receptor states into two groups: those without Ca^{2+} bound to the inactivating site ($S_{000}, S_{010}, S_{100}$, and S_{110}, shown in the upper line of Fig. 7.8; called group I states), and those with Ca^{2+} bound to the inactivating site ($S_{001}, S_{011}, S_{101}$, and S_{111}, shown in the lower line of Fig. 7.8; called group II states). Because the binding of IP$_3$ and the binding of Ca^{2+} to the activating site are assumed to be fast processes, within each group the receptor states are assumed to be at quasi-steady state. However, the transitions between group I and group II (between top and bottom in Fig. 7.8), due to binding or unbinding of the inactivating site, are slow, and so the group I states are not in equilibrium with the group II states.

Following what by now is a standard procedure (see, for instance, Section 2.4.1, or Section 7.2.4), we set

$$y = x_{001} + x_{011} + x_{101} + x_{111} \tag{7.15}$$

and find that

$$\frac{dy}{dt} = \left[\frac{(k_{-4}K_1K_2 + k_{-2}pK_4)c}{K_4K_2(p + K_1)} \right](1 - y) - \left(\frac{k_{-2}p + k_{-4}K_3}{p + K_3} \right)y. \tag{7.16}$$

The details are left as an exercise (Exercise 4). Equation (7.16) can be written in the form

$$\tau_y(c,p)\frac{dy}{dt} = y_\infty(c,p) - y, \tag{7.17}$$

which is useful for comparison with other models such as the Hodgkin–Huxley equations. The open probability is obtained from the equation

$$x_{110} = \frac{pc(1 - y)}{(p + K_1)(c + K_5)}. \tag{7.18}$$

Note that $1 - y$, which is the proportion of receptors that are not inactivated by Ca^{2+}, plays the role of an inactivation variable, similar in spirit to the variable h in the Hodgkin–Huxley equations (Chapter 5). To emphasize this similarity, the reduced model can be written in the form

$$x_{110} = \frac{pc}{(p+K_1)(c+K_5)}h, \tag{7.19}$$

$$\tau_h(c,p)\frac{dh}{dt} = h_\infty(c,p) - h, \tag{7.20}$$

where $h = 1 - y$, and τ_h and h_∞ are readily calculated from the corresponding differential equation for y.

A Model with Saturating Binding Rates

Recently it has become clear that the eight-state model has some serious flaws. Most importantly, it has been shown experimentally that the rate of opening of the IPR varies only over a single order of magnitude, while the concentrations of Ca^{2+} or IP_3 vary over several orders of magnitude. It follows that opening of the IPR cannot be governed by simple mass action kinetics of Ca^{2+} or IP_3 binding, but must follow some kinetic scheme that allows for saturation of the binding rate.

Given what we have learned in Section 3.5.4, there is one obvious way to do this. If, instead of assuming a single-step reaction of Ca^{2+} or IP_3 binding, we separate the binding step from the opening step, thus using the concepts of affinity and efficacy we saw in models of agonist-controlled channels, we obtain a reaction that has a saturating rate as the concentration of the agonist gets high. This approach is also essentially identical to models of enzyme kinetics, as discussed in Section 1.4. Because of the saturation of the reaction rate, simple mass action kinetics are not appropriate to model enzyme reactions, leading to the development of the Michaelis–Menten-type models.

To illustrate the idea, consider the reaction scheme

$$\tilde{A} \underset{k_{-1}}{\overset{k_1 c}{\rightleftarrows}} \bar{A} \underset{k_{-2}}{\overset{k_2}{\rightleftarrows}} I. \tag{7.21}$$

If the transitions between $\tilde{A}$ and $\bar{A}$ are faster than other reactions, so that these are in instantaneous equilibrium, then

$$c\tilde{A} = K_1\bar{A}, \tag{7.22}$$

where, as usual, $K_1 = k_{-1}/k_1$. Following what by now should be the routine method to find slow dynamics, we find that

$$\frac{dA}{dt} = k_{-2}I - \phi(c)A, \tag{7.23}$$

where $A = \bar{A} + \tilde{A}$, and

$$\phi(c) = \frac{k_2 c}{c + K_1}, \tag{7.24}$$

a rate that is saturating in c. In other words, the assumption of fast equilibrium lets us simplify (7.21) to

$$A \underset{k_{-2}}{\overset{\phi(c)}{\rightleftharpoons}} I. \tag{7.25}$$

Thus, in this simple way, saturating binding kinetics can be incorporated into a model.

Using saturating binding schemes of this type, Sneyd and Dufour (2002) constructed a model of the IPR that was based on the qualitative models of Taylor and his colleagues, and is consistent with the scheme of Hajnóczky and Thomas (1997). In addition to the saturating binding rates, the main features of the model are

1. The IPR can be opened by IP_3 in the absence of Ca^{2+}, but with a lower conductance.
2. The IPR can be inactivated by Ca^{2+} in the absence of IP_3.
3. Once IP_3 is bound, the IPR can spontaneously inactivate (to the shut state, S), independently of Ca^{2+}.
4. Once IP_3 is bound, the IPR can also bind Ca^{2+} to activate the receptor. Thus, there is an intrinsic competition between Ca^{2+}-mediated receptor activation and spontaneous inactivation.
5. Once the IPR is activated by Ca^{2+} binding, it can be inactivated by binding of additional Ca^{2+}. (This feature of the model is the principal point of disagreement with the qualitative models of Taylor, 1998.)
6. Binding of IP_3 and Ca^{2+} is sequential.

Following these assumptions, the model takes the form shown in Fig. 7.10, with corresponding equations

$$\frac{dR}{dt} = \phi_{-2}O - \phi_2 pR + k_{-1}I_1 - \phi_1 R, \tag{7.26}$$

$$\frac{dO}{dt} = \phi_2 pR - (\phi_{-2} + \phi_4 + \phi_3)O + \phi_{-4}A + k_{-3}S, \tag{7.27}$$

$$\frac{dA}{dt} = \phi_4 O - \phi_{-4}A - \phi_5 A + k_{-1}I_2, \tag{7.28}$$

$$\frac{dI_1}{dt} = \phi_1 R - k_{-1}I_1, \tag{7.29}$$

$$\frac{dI_2}{dt} = \phi_5 A - k_{-1}I_2, \tag{7.30}$$

where $R + O + A + S + I_1 + I_2 = 1$, and where

$$\phi_1(c) = \frac{\alpha_1 c}{\beta_1 + c}, \qquad \phi_3(c) = \frac{\alpha_3}{\beta_3 + c}, \qquad \phi_5(c) = \frac{\alpha_5 c}{\beta_5 + c},$$

$$\phi_2(c) = \frac{\alpha_2 + \beta_2 c}{\beta_1 + c}, \qquad \phi_{-2}(c) = \frac{\alpha_{-2} + \beta_{-2} c}{\beta_3 + c},$$

$$\phi_4(c) = \frac{\alpha_4 c}{\beta_3 + c}, \qquad \phi_{-4}(c) = \frac{\alpha_{-4}}{\beta_5 + c}.$$

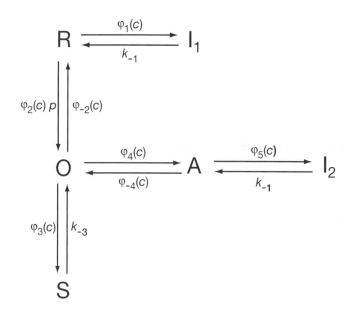

Figure 7.10   Sequential model of the IP$_3$ receptor with saturating binding rates. R — receptor; O — open; S — shut; A — activated; I$_1$ and I$_2$ — inactivated.

Table 7.2   Parameters of the IPR model with saturating binding rates.

α_1	$= 0.3\ s^{-1}$	β_1	$= 0.02\ \mu M$
α_2	$= 0.77\ s^{-1}$	β_2	$= 1.38\ \mu M^{-1} s^{-1}$
α_{-2}	$= 76.6\ \mu M\, s^{-1}$	β_{-2}	$= 137\ s^{-1}$
α_3	$= 6\ \mu M\, s^{-1}$	β_3	$= 54.7\ \mu M$
α_4	$= 4926\ s^{-1}$	α_{-4}	$= 1.43\ s^{-1}$
α_5	$= 1.78\ s^{-1}$	β_5	$= 0.12\ \mu M$
k_{-1}	$= 0.84\ s^{-1}$	k_{-3}	$= 29.8\ s^{-1}$

The parameters were determined by fitting to experimental data and are shown in Table 7.2. We assume that the IPR consists of four identical and independent subunits, and that it allows Ca^{2+} current when all four subunits are in either the O or the A state. We also assume that the more subunits in the A state, the greater the conductance. One simple way to express this is to write the open probability, P_o, as

$$P_o = (a_1 O + a_2 A)^4, \tag{7.31}$$

for some constants a_1 and a_2. In the original model $a_1 = 0.1$ and $a_2 = 0.9$, but these values can be changed without appreciably changing the fit or the model's behavior.

As a side issue, but an important one, note that the simple phrase "the parameters were determined by fitting to experimental data" covers a multitude of thorny problems. First, it is usually the case that not all the parameters in a model can be unambiguously determined from the available experimental data. This is certainly the case in the IPR model described above, for which, in fact, only a few of the parameters can be pinned down with any confidence. There is no way of getting around this unpleasant fact except either to simplify the model or to collect more experimental data of the appropriate type. Of course, determining which additional experimental data are needed is not a trivial problem.

Second, the process of determining the parameters from the data is itself complicated. Simple methods such as least-squares fits suffer from serious disadvantages, while more sophisticated methods such as Bayesian inference and Markov chain Monte Carlo can be more difficult to implement. A discussion of these issues is far beyond the scope of this text, and the interested reader is referred to Ball et al. (1999).

7.2.6 Simple Models of Calcium Dynamics

We now have constructed models of most of the fluxes that are important for the control of Ca^{2+}. To put them together into a model of Ca^{2+} dynamics is straightforward; simply choose your favorite model of the IPR (or RyR if appropriate), your favorite models of the ATPases and the influx, and put them all into (7.3) and (7.4).

To illustrate with a simple model, suppose there are only two fluxes, the IPR and SERCA fluxes, so that

$$\frac{dc}{dt} = (k_f P_O + J_{er})(c_e - c) - J_{serca}. \tag{7.32}$$

Since the only fluxes are between the ER and the cytoplasm, it follows that

$$\frac{dc_e}{dt} = -\gamma \frac{dc}{dt}, \tag{7.33}$$

so that $c + \frac{c_e}{\gamma} = c_t$ is unchanging. Such a model is called a *closed-cell* model (Section 7.2.7). Next, we use a Hill function (Chapter 1) to model the SERCA pump, giving

$$J_{serca} = \frac{V_p c^2}{K_p^2 + c^2}, \tag{7.34}$$

and to model the IPR we use the simplified version of the DeYoung–Keizer model (7.18),

$$P_O = \left(\frac{pc(1-y)}{(p+K_1)(c+K_5)} \right)^3, \tag{7.35}$$

where y satisfies (7.16).

Note that the flux through the IPR is assumed to be proportional to the concentration difference between the ER and the cytoplasm. This is appropriate only if there is

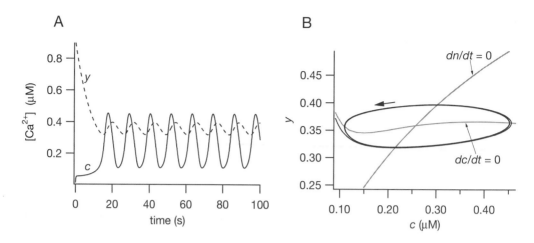

Figure 7.11 Calcium oscillations in the simple model of Ca^{2+} dynamics described by (7.32)–(7.35) and (7.16). The parameters are given in Tables 7.3 and 7.1. A: a typical oscillation (for $p = 0.5$). B: the phase portrait.

Table 7.3 Parameter values of the simple model of Ca^{2+} dynamics described by (7.32)–(7.35) and (7.16). The parameter values for the IPR model are given in Table 7.1. The value of γ is based on experimental estimates of the relative volumes of the cytoplasm and the ER, while k_f is a scaling factor that controls the total amount of Ca^{2+} efflux through the IPR; k_f can be loosely interpreted as the product of the IPR density and the single channel current. Although J_{er} is based on estimations of the rate of Ca^{2+} leak from the ER at steady state, there are no reliable data for this parameter.

$V_p = 0.9 \, \mu M \, s^{-1}$	$K_p = 0.1 \, s$
$k_f = 1.11 \, s^{-1}$	$\gamma = 5.5$
$J_{er} = 0.02 \, s^{-1}$	$c_t = 2 \, \mu M$

no potential difference across the ER membrane. Even though Ca^{2+} carries a charge, the flow of Ca^{2+} does not induce a potential difference because there are counterions that flow freely, so that the electrical balance is not upset.

Because c_t is fixed, this simple model can be reduced to a two-variable model for which a phase portrait can be constructed. The numerical solution of this model with $p = 0.5$ is shown in Fig. 7.11. The phase portrait is identical in structure to those of several two-variable excitable media models, such as the FitzHugh–Nagumo equations or the reduced Hodgkin–Huxley equations (Chapter 5). The nullcline for the inactivation variable y is a monotone increasing function of c, and the nullcline for c is an N-shaped function of c (although on this scale the N appears very shallow). There is a single intersection of the two nullclines. The stability of this fixed point is determined (roughly, but not precisely) by the branch of the c-nullcline on which it lies. On the leftmost branch, the fixed point is stable; on the middle branch it is unstable.

If the fixed point is unstable, there is a stable periodic limit cycle, corresponding to spontaneous Ca^{2+} oscillations.

A More Complex Example

A slightly more realistic model is as follows: If we

1. ignore RyR and mitochondrial fluxes,
2. use the saturating binding model of the IPR,
3. use a four-state Markov model for the SERCA pump, with the transport of a single Ca^{2+} ion for each pump cycle,
4. use a Hill function for the plasma membrane pump, and
5. assume that Ca^{2+} buffering is fast and linear (see Section 7.4),

then we get

$$\frac{dc}{dt} = (k_f P_O + J_{er})(c_e - c) - J_{serca} + J_{in} - J_{pm}, \tag{7.36}$$

$$\frac{dc_e}{dt} = \gamma [J_{serca} - (k_f P_O + J_{er})(c_e - c)], \tag{7.37}$$

where

$$J_{serca} = \frac{c - \alpha_1 c_e}{\alpha_2 + \alpha_3 c + \alpha_4 c_e + \alpha_5 c c_e}, \tag{7.38}$$

$$P_O = (0.1O + 0.9A)^4, \tag{7.39}$$

$$J_{pm} = \frac{V_p c^2}{K_p^2 + c^2}, \tag{7.40}$$

$$J_{er} = \text{constant}, \tag{7.41}$$

$$J_{in} = a_1 + a_2 p. \tag{7.42}$$

The constant J_{er} represents a constant leak from the ER that is necessary to balance the ATPase flux at steady state. Equations (7.36) and (7.37) are coupled to (7.26)–(7.30) which describe the IPR. We have used two different pump models here to emphasize the fact that it is possible to mix and match the individual models of the various fluxes. Since there are five parameters describing the SERCA ATPase, and only two parameters describing the plasma membrane ATPase, it is likely that the SERCA pump is overparameterized. If we were fitting the pump models to data this would be a concern, since the data may not contain sufficient information to determine all the parameters unambiguously. Here, however, we are content to use some function that can be justified by a mechanistic model and for which the parameters can be adjusted to give reasonable agreement with experimental data. The parameter values are given in Table 7.4.

Since p corresponds to the concentration of IP_3, and thus, indirectly, the agonist concentration, we describe the behavior of this model as p varies. The bifurcation diagram for this model is shown in Fig. 7.12A. As p increases, the steady-state Ca^{2+} concentration increases also (because J_{in} increases with p), and oscillations occur for

Table 7.4   Parameter values of the model of Ca^{2+} dynamics (7.36)–(7.42).

$\alpha_1 = 10^{-4}$	$\alpha_2 = 0.007$ s
$\alpha_3 = 0.06\ \mu M^{-1}s$	$\alpha_4 = 0.0014\ \mu M^{-1}s$
$\alpha_5 = 0.007\ \mu M^{-2}s$	$J_{er} = 0.002\ s^{-1}$
$V_p = 2.8\ \mu M\,s^{-1}$	$K_p = 0.425\ \mu M$
$a_1 = 0.003\ \mu M\,s^{-1}$	$a_2 = 0.02\ s^{-1}$
$k_f = 0.96\ s^{-1}$	$\gamma = 5.4$

a range of intermediate values of p. Typical oscillations for two different values of p are shown in Fig. 7.12B and C. As p increases, the oscillation frequency increases.

Given the variety of models of the IPR, the RyR, the ATPases, and the other fluxes involved in Ca^{2+} dynamics, it is important to think about what a model of Ca^{2+} dynamics can tell us. For example, by appropriate choice of IPR and ATPase models one can construct a whole-cell model that exhibits Ca^{2+} oscillations of an enormous variety of shapes and sizes. The oscillations in Fig. 7.12 give a reasonably accurate description of Ca^{2+} oscillations in pancreatic acinar cells (the cell type for which this model was originally designed); however, the parameters and the flux models could be adjusted, practically indefinitely, to obtain oscillations that mimic the behavior seen in other cell types. Hence, the simple fact that the model exhibits Ca^{2+} oscillations with reasonable properties tells us very little about the underlying mechanisms. A great deal more work is needed before we can conclude anything about specific mechanisms in actual cells. This extra work is beyond the scope of this book; the interested reader is referred to the extensive literature on modeling Ca^{2+} dynamics, which can most easily be accessed through Falcke's 2004 review.

7.2.7   Open- and Closed-Cell Models

The period of a Ca^{2+} oscillation is determined by a number of factors, including the dynamics of the IPR and RyR, the rate of formation and degradation of IP$_3$ (see Section 7.2.8), the rate at which the ER refills after a Ca^{2+} spike, and the rate of Ca^{2+} transport across the cell membrane. One good way to study these latter two effects is to rewrite the model in a slightly different form, one in which membrane Ca^{2+} fluxes can readily be separated from ER fluxes. To do this, we introduce a new variable

$$c_t = c + \frac{c_e}{\gamma}. \tag{7.43}$$

This new variable c_t is the total number of moles of Ca^{2+} in the cell (including both the cytoplasm and the ER) divided by the cytoplasmic volume, and is thus a measure

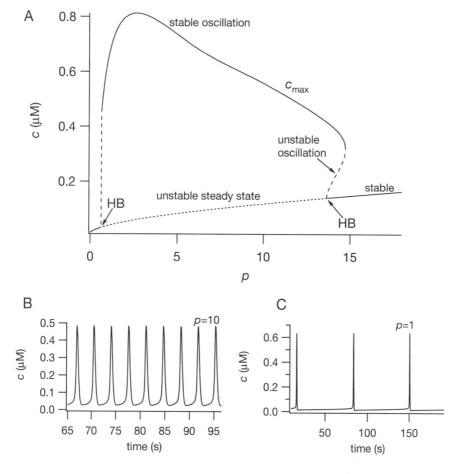

Figure 7.12 A: Simplified bifurcation diagram of the model of Ca^{2+} dynamics (7.36)–(7.42). A solid line denotes stability, and a dashed line, instability. HB denotes a Hopf bifurcation. Although the branch of oscillations shown here appears to begin at the lower Hopf bifurcation point, such is not the case. The bifurcation diagram is much more complicated in that region and is not shown in full detail here. B and C: typical oscillations in the model for two different values of p.

of the total Ca^{2+} content of the cell. Rewriting (7.36) and (7.37) in terms of c_t gives

$$\frac{dc_t}{dt} = \delta(J_{in} - J_{pm}),\tag{7.44}$$

$$\frac{dc_e}{dt} = \gamma[J_{serca} - (k_f P_O + J_{er})(c_e - (c_t - \gamma c_e))],\tag{7.45}$$

where, for convenience, we have introduced the scale factor δ into the c_t equation. This scale factor makes it easier to modify the ratio of membrane transport to ER transport without affecting anything else in the model. We could, of course, equally well have written the model in terms of c_t and c, rather than c_t and c_e.

If membrane fluxes are much smaller than ER fluxes, δ is small. This is the case for a number of cell types. When $\delta = 0$, Ca^{2+} neither enters nor leaves the cell, and c_t remains constant. Such a model is called a *closed-cell* model, a simple example of which was discussed in Section 7.2.6. When $\delta \neq 0$ we have an *open-cell* model, in which the total amount of Ca^{2+} in the cell can change over time.

In some cell types, c_t varies slowly enough that we can gain a reasonable understanding of the Ca^{2+} dynamics by first understanding the behavior of the closed-cell model. If δ is small enough, the behavior of the open-cell model is similar to that of the closed-cell model (since c_t remains almost constant over short time scales) but is slowly modulated as c_t slowly drifts toward its steady state. Such an analysis relies on the existence of two time scales; fast changes in c and c_e that are modulated by slow changes in c_t. This approach was first introduced by Rinzel (1985) to study a model of bursting oscillations in pancreatic β-cells, and is discussed in more detail in Chapter 9.

A particularly interesting use of this formulation arises from the ability to control both δ and c_t experimentally (Sneyd et al., 2004b). Application of high concentrations of La^{3+} to the outside of a cell blocks both Ca^{2+} influx and the plasma membrane ATPase, thus reducing δ effectively to zero. Furthermore, preloading of cells with photoreleasable Ca^{2+} allows the manipulation of c_t by flashing light. Thus, once La^{3+} has been applied, c_t can be viewed as a control parameter and increased at will (although, unfortunately, not so easily decreased) and the resultant cellular behavior observed. This allows for the detailed testing of model predictions.

7.2.8 IP$_3$ Dynamics

In the above models, Ca^{2+} oscillations occur at a constant concentration of IP$_3$. Thus, the role of IP$_3$ is to activate the IPR; once the receptor is activated, Ca^{2+} feedback takes over, and the period of oscillations is controlled by the kinetics of Ca^{2+} feedback on the IPR as well as the interaction of the membrane and ER fluxes.

This is certainly an oversimplification. It is known that the rate of production of IP$_3$ is dependent on Ca^{2+}, and that, in some cell types, oscillations in $[Ca^{2+}]$ are accompanied by oscillations in [IP$_3$] (Hirose et al., 1999; Nash et al., 2001; Young et al., 2003). It is not yet clear whether oscillations in [IP$_3$] are *necessary* for Ca^{2+} oscillations, since the former could merely be a passive follower of the latter. Experimental evidence is neither consistent nor conclusive.

There are a number of models of Ca^{2+} dynamics that incorporate the Ca^{2+} dependence of IP$_3$ production and degradation. Early models were those of Meyer and Stryer (1988, 1991), Swillens and Mercan (1990), De Young and Keizer (1992), and Cuthbertson and Chay (1991); more recent models have been constructed by Shen and Larter (1995), Borghans et al. (1997), Dupont and Erneux (1997), Houart et al. (1999), and Politi et al. (2006).

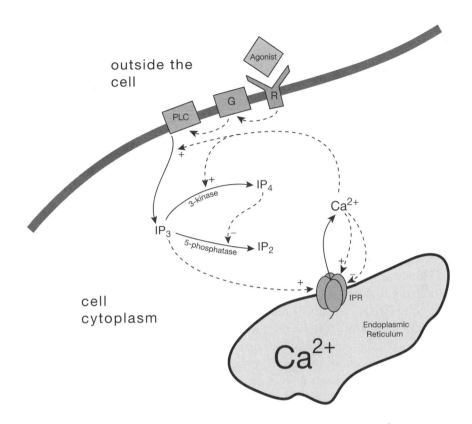

Figure 7.13 Schematic diagram of some of the interactions between Ca^{2+} and IP_3. Calcium can activate PLC, leading to an increase in the rate of production of IP_3, and it can also increase the rate at which IP_3 is phosphorylated by the 3-kinase. The end product of phosphorylation by the 3-kinase, IP_4, acts as a competitive inhibitor of dephosphorylation of IP_3 by the 5-phosphatase. Not all of these feedbacks are significant in every cell type.

There are two principal ways in which Ca^{2+} can influence the concentration of IP_3, and these are sketched in Fig. 7.13. First, activation of PLC by Ca^{2+} is known to occur in many cell types; this forms the basis of the models of Meyer and Stryer (1988) and De Young and Keizer (1992). Second, IP_3 is degraded in two principal ways, each of which is subject to feedback regulation; it is dephosphorylated by IP_3 5-phosphatase to inositol 1,4-bisphosphate (IP_2), or phosphorylated by IP_3 3-kinase to inositol 1,3,4,5-tetrakisphosphate (IP_4). However, since IP_4 is also a substrate for the 5-phosphatase, IP_4 acts as a competitive inhibitor of IP_3 dephosphorylation. In addition, Ca^{2+} (in the form of a Ca^{2+}-calmodulin complex) can enhance the activity of the 3-kinase.

Intuitively, it seems clear that an intricate array of feedbacks such as this could easily give rise to interdependent oscillations in $[IP_3]$ and Ca^{2+}, and the models mentioned above have confirmed that this intuitive expectation is correct. However, few of those models have been used to make predictions that were tested experimentally. Two notable exceptions are the models of Dupont et al. (2003) and of Politi et al. (2006).

In hepatocytes, oscillations of Ca^{2+} and IP_3 occur together. Each Ca^{2+} spike causes an increase in the rate of phosphorylation of IP_3, and thus a decline in $[IP_3]$, with a consequent decline in $[Ca^{2+}]$. However, although both species oscillate, it is not clear that oscillations in $[IP_3]$ are *necessary*. Dupont et al. first used a model (Dupont and Erneux, 1997) in which Ca^{2+} influences the rate of the 3-kinase. This model predicted that, if the rate of the 5-phosphatase is increased by a factor of 25, and the rate of IP_3 production also increased, oscillations with the same qualitative properties persist, with only a slight decrease in frequency. Under these conditions, little IP_3 is degraded by the 3-kinase, and thus Ca^{2+} feedback on the 3-kinase cannot play a major role. In other words, these oscillations arise because of the feedback of Ca^{2+} on the IPR.

Dupont et al. then devised an experimental test of this model prediction. They increased both the rate of the 5-phosphatase (by direct microinjection of the 5-phosphatase), and the rate of production of IP_3 (by increasing the concentration of agonist), and observed that the oscillations remained essentially unchanged, with the predicted decrease in frequency. Thus, under conditions whereby the degradation of IP_3 is principally through the 5-phosphatase pathway (in which case it is not directly affected by $[Ca^{2+}]$), there is little change in the properties of the oscillations. The conclusion is that, in hepatocytes, the oscillations are not a result of the interplay between $[Ca^{2+}]$ and the IP_3 dynamics, but rather a result of Ca^{2+} feedback on the IPR. This work illustrates an excellent interplay between modeling and experimentation.

Similar studies have been also been performed in CHO cells (Chinese hamster ovary cells, which are a cell line). Politi et al. constructed a model incorporating both positive and negative feedback of Ca^{2+} on IP_3, and then tested the consequences of the addition of exogenous IP_3 buffer. Their analysis is involved and subtle, too much so to be discussed here in detail. However, their conclusion was that Ca^{2+} oscillations in CHO cells are a result of positive feedback of Ca^{2+} on IP_3 production.

It is possible that the response to an exogenous pulse of IP_3 can be used to determine whether IP_3 oscillations are a necessary accompaniment to Ca^{2+} oscillations in any given cell type (Sneyd et al., 2006). If both IP_3 and Ca^{2+} are dynamic variables, necessarily oscillating together, then a pulse in IP_3 should perturb the solution off the stable limit cycle, leading to a delay before the oscillations resume. In other words, the phase response curve should exhibit a phase lag. (A phase advance, although theoretically possible, is rarely observed.) If the Ca^{2+} oscillations occur at a constant $[IP_3]$, then the pulse of IP_3 temporarily increases the frequency of the oscillations. These predictions have been tested in two cells types; in pancreatic acinar cells, the IP_3 pulse causes a delay before the next oscillation peak, while in airway smooth muscle cells, the IP_3 pulse causes a temporary increase in oscillation frequency. The model thus predicts that in pancreatic acinar cells the Ca^{2+} oscillations are necessarily accompanied by IP_3 oscillations, while in airway smooth muscle cells they occur for constant $[IP_3]$. However, since this approach yields inconclusive results in hepatocytes (Harootunian et al., 1988) and has not been applied widely to other cell types, it is not yet clear how useful it will prove to be.

7.2.9 Ryanodine Receptors

The second principal way in which Ca^{2+} can be released from intracellular stores is through ryanodine receptors, which are found in a variety of cells, including cardiac cells, smooth muscle, skeletal muscle, chromaffin cells, pituitary cells, neurons, and sea urchin eggs. Ryanodine receptors share many structural and functional similarities with IP_3 receptors, particularly in their sensitivity to Ca^{2+}. Just as Ca^{2+} can activate IP_3 receptors and increase the Ca^{2+} flux, so too can Ca^{2+} trigger Ca^{2+}-induced Ca^{2+} release (CICR) from the sarcoplasmic or endoplasmic reticulum through ryanodine receptors (Endo et al., 1970; Fabiato, 1983). Calcium can also inactivate ryanodine receptors *in vitro*, although it is unknown whether such inactivation plays any important physiological role, or even occurs to any significant extent *in vivo*.

Ryanodine receptors are so named because of their sensitivity to ryanodine, which decreases the open probability of the channel. On the other hand, caffeine increases the open probability of ryanodine receptors.

There are many models of the ryanodine receptor but since they are mostly adapted for use in models of excitation–contraction coupling, we discuss them in Section 12.2.4. Here we discuss one of the few models that has been developed for use in other cell types.

Calcium Oscillations in Bullfrog Sympathetic Neurons

Sympathetic neurons respond to caffeine, or mild depolarization, with robust and reproducible Ca^{2+} oscillations. Although these oscillations are dependent on external Ca^{2+}, they occur at a fixed membrane potential and involve the release of Ca^{2+} from the ER via ryanodine receptors, as is indicated by the fact that they are abolished by ryanodine. Typical oscillations are shown in Fig. 7.14.

A simple model of CICR (Friel, 1995) provides an excellent quantitative description of the behavior of these oscillations in the bullfrog sympathetic neuron. Despite the

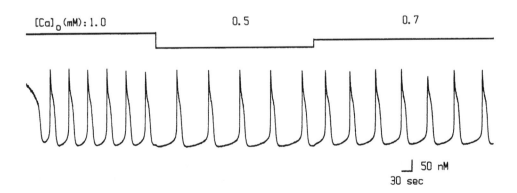

Figure 7.14 Caffeine-induced Ca^{2+} oscillations in sympathetic neurons, and their dependence on the extracellular Ca^{2+} concentration. $[Ca]_0$ stands for c_o. (Friel, 1995, Fig. 5a.)

A

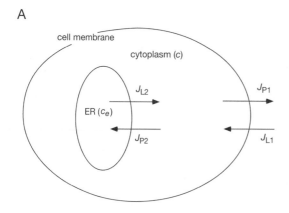

B

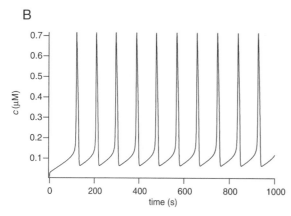

Figure 7.15 A: Schematic diagram of the CICR model of Ca^{2+} oscillations in bullfrog sympathetic neurons. B: Typical oscillation in the model, for $c_o = 1000$ μM (i.e. 1 mM), and calculated using the parameters in Table 7.5.

model's simplicity (or perhaps because of it), it is a superb example of how theory can supplement experiment, providing an interpretation of experimental results as well as quantitative predictions that can subsequently be tested.

A schematic diagram of the model is given in Fig. 7.15A. A single intracellular Ca^{2+} store exchanges Ca^{2+} with the cytoplasm (with fluxes J_{L2} and J_{P2}), which in turn exchanges Ca^{2+} with the external medium (J_{L1} and J_{P1}). Thus,

$$\frac{dc}{dt} = J_{L1} - J_{P1} + J_{L2} - J_{P2}, \tag{7.46}$$

$$\frac{dc_e}{dt} = \gamma(-J_{L2} + J_{P2}), \tag{7.47}$$

where c denotes $[\text{Ca}^{2+}]$ in the cytoplasm and c_e denotes $[\text{Ca}^{2+}]$ in the intracellular store as before. As before, γ is the ratio of the cytoplasmic volume to the ER volume. The fluxes are chosen in a simple way, as linear functions of the concentrations:

$$J_{L1} = k_1(c_o - c), \qquad \text{Ca}^{2+}\text{entry}, \tag{7.48}$$

$$J_{P1} = k_2 c, \qquad\qquad \text{Ca}^{2+}\text{extrusion}, \tag{7.49}$$

Table 7.5 Typical parameters of the model of Ca^{2+} oscillations in sympathetic neurons.

$k_1 = 2 \times 10^{-5}$ s^{-1}	$\kappa_2 = 0.58$ s^{-1}
$k_2 = 0.13$ s^{-1}	$K_d = 0.5$ μM
$k_4 = 0.9$ s^{-1}	$n = 3$
$\kappa_1 = 0.013$ s^{-1}	$\gamma = 4.17$

$$J_{L2} = k_3(c_e - c), \quad Ca^{2+}\text{release}, \tag{7.50}$$

$$J_{P2} = k_4 c, \quad Ca^{2+}\text{uptake}, \tag{7.51}$$

where c_o denotes the external $[Ca^{2+}]$, which is assumed to be fixed. (In the experiment shown in Fig. 7.14, c_o was fixed at 1, 0.5, and 0.7 mM.) Depolarization induced by the application of high external K^+ is modeled as an increase in k_1, the rate of Ca^{2+} entry from the outside, while the application of caffeine (which increases the rate of Ca^{2+} release from the internal store) is modeled by an increase in k_3.

We model CICR in a simple way by making k_3 an increasing sigmoidal function of c, i.e.,

$$k_3 = \kappa_1 + \frac{\kappa_2 c^n}{K_d^n + c^n}, \tag{7.52}$$

and then determine the parameters of the nonlinear model by fitting to the time course of an oscillation (Table 7.5). A typical result is shown in Fig. 7.15.

Not only does this model provide an excellent quantitative description of the Ca^{2+} oscillation (with a different parameter set from that shown in Table 7.5), it also predicts the fluxes that should be observed over the oscillatory cycle. Subsequent measurement of these fluxes confirmed the model predictions. It therefore appears that CICR (at least in bullfrog sympathetic neurons) can be well described by a relatively simple model. It is necessary only for the ryanodine receptors to be activated by Ca^{2+} to generate physiological oscillations; inactivation by Ca^{2+} is not necessary.

7.3 Calcium Waves

In some cell types, Ca^{2+} oscillations occur practically uniformly across the cell. In such a situation, measurement of the Ca^{2+} concentration at any point of the cell gives the same time course, and a well-mixed model is appropriate. More often, however, each oscillation takes the form of a wave moving across the cell; these intracellular "oscillations" are actually periodic intracellular waves. To model and understand such spatially distributed behavior, inclusion of Ca^{2+} diffusion is necessary. Furthermore, to study objects such as spiral waves of Ca^{2+} (Fig. 7.3), partial differential equation models are again necessary.

When attempting to model spatially distributed Ca^{2+} dynamics, it is tempting to generalize the whole-cell model by simply adding a diffusion term to (7.3). However, with a little reflection, one realizes that this cannot be correct for several reasons. First, most of the flux terms represent movement of Ca^{2+} across a boundary, whether the plasma membrane, the ER, or mitochondria; the only exceptions are the fluxes on and off the buffers. Second, since the cytoplasmic space is highly inhomogeneous, it is not clear that movement of Ca^{2+} in this space is governed by a standard diffusive process.

A more proper formulation of the model would be to assume that the cell is a three-dimensional structure comprised of two interconnected domains, the cytoplasm Ω_c, and the ER Ω_e (ignoring all other subdomains, such as mitochondria, Golgi apparati, etc, to make the problem simpler). In the cytoplasm and the ER, Ca^{2+} is assumed to move by normal diffusion and to react with buffers. Thus,

$$\frac{\partial c}{\partial t} = \nabla \cdot (D_c \nabla c) - J_{on} + J_{off}, \qquad \text{in } \Omega_c, \tag{7.53}$$

and

$$\frac{\partial c_e}{\partial t} = \nabla \cdot (D_e \nabla c) - J_{on,e} + J_{off,e}, \qquad \text{in } \Omega_e. \tag{7.54}$$

Fluxes across the plasma membrane into the cytoplasm lead to the boundary condition

$$D_c \nabla c \cdot \mathbf{n} = J_{in} - J_{pm}, \qquad \text{on } \partial\Omega_{c,m}, \tag{7.55}$$

while fluxes across the ER into the cytoplasm yield

$$D_c \nabla c \cdot \mathbf{n} = -D_e \nabla c_e \cdot \mathbf{n} = J_{IPR} + J_{RyR} - J_{serca}, \qquad \text{on } \partial\Omega_e, \tag{7.56}$$

where $\partial\Omega_{c,m}$ is the cell membrane, $\partial\Omega_e$ is the boundary of the ER, and $\mathbf{n}$ is the unit outward normal vector to the domain in question. Notice that here we have assumed that there is no direct communication between the ER and extracellular space (although this assumption, as with almost everything in biology, is not without controversy).

Because of the intricate geometry of the ER, it is immediately obvious that these models are far too complicated to make any progress at all, so some simplifying assumptions are needed. This is where homogenization comes to the rescue. The fundamental idea is that since diffusion is rapid over short distances, local variations are smoothed out quickly and it is only necessary to know the average, or mean field, behavior. Thus, one derives mean field equations that describe the behavior over larger space scales. In these descriptions, concentrations c and c_e are assumed to coexist at every point in space, and one finds (see Appendix 7.8)

$$\frac{\partial c}{\partial t} = \nabla \cdot (D_c^{eff} \nabla c) + \chi_c (J_{IPR} + J_{RyR} - J_{serca}) - J_{on} + J_{off}, \tag{7.57}$$

$$\frac{\partial c_e}{\partial t} = \nabla \cdot (D_e^{eff} \nabla c_e) - \chi_e (J_{IPR} + J_{RyR} - J_{serca}) - J_{on,e} + J_{off,e}, \tag{7.58}$$

where D_c^{eff} and D_e^{eff} are effective diffusion coefficients for the cytoplasmic space and the ER, respectively, and χ_c and χ_e are the surface-to-volume ratios of these two comingled spaces. It is usually assumed that the cellular cytoplasm is isotropic and homogeneous. Exceptions to the assumption of homogeneity are not uncommon, but exceptions to the isotropic assumption are rare. It is not known, however, how Ca^{2+} diffuses in the ER, or the extent to which the tortuosity of the ER plays a role in determining the effective diffusion coefficient of ER Ca^{2+}. It is typical to assume either that Ca^{2+} does not diffuse in the ER, or that it does so with a restricted diffusion coefficient, $D_e^{\text{eff}} \ll D_c^{\text{eff}}$. Henceforth we delete the superscript eff.

The surface-to-volume ratios are necessary to get the units correct. In fact, one should notice that a flux term, for example J_{IPR}, in a whole-cell model is not the same as in a spatially distributed model; in a whole-cell model the flux must be in units of concentration per unit time, while in a spatially distributed model the flux must be in units of moles per unit time per surface area. Multiplication by a surface-to-volume ratio changes the units to number of moles per volume per unit time. However, it is typical to scale all fluxes to be in units of moles per unit time per cytoplasmic volume, in which case (7.57) and (7.57) become (after rescaling)

$$\frac{\partial c}{\partial t} = \nabla \cdot (D_c \nabla c) + J_{\text{IPR}} + J_{\text{RyR}} - J_{\text{serca}} - J_{\text{on}} + J_{\text{off}} \qquad (7.59)$$

and

$$\frac{\partial c_e}{\partial t} = \nabla \cdot (D_e \nabla c_e) - \gamma (J_{\text{IPR}} + J_{\text{RyR}} - J_{\text{serca}}) - J_{\text{on,e}} + J_{\text{off,e}}, \qquad (7.60)$$

where γ is the ratio of cytoplasmic to ER volumes.

Additional approximations can be made when one dimension of the cell is very much smaller than the diffusion length scale of Ca^{2+}. For instance, many cultured cells are relatively thin, in which case spatial variation in the direction of the cell thickness can be ignored. However, to derive the correct model equations requires a little care, as is discussed in detail in Appendix 7.8.

Consider, for example, a long, thin cylindrical cell, with boundary fluxes (on the cylindrical wall) J_{in} and J_{pm}, in units of moles per surface area per time. If the cell radius is small compared to the diffusion length scale of Ca^{2+}, then Ca^{2+} is homogeneously distributed in each cross-section, with the fluxes across the (cylindrical) wall included as source terms in the governing partial differential equation. In this case, we can model the cell as a one-dimensional object. For example, for a "one-dimensional" cell of length L with no Ca^{2+} flux at the ends, the boundary conditions are $\frac{\partial c}{\partial x} = 0$ at $x = L$ and $x = 0$, while at each internal point we have

$$\frac{\partial c}{\partial t} = D_c \frac{\partial^2 c}{\partial x^2} + J_{\text{IPR}} + \frac{\rho}{A}(J_{\text{in}} - J_{\text{pm}}) + J_{\text{RyR}} - J_{\text{serca}} - J_{\text{on}} + J_{\text{off}}, \qquad (7.61)$$

where ρ is the cell circumference and A is the cell cross-sectional area. Note that the original boundary fluxes, with units of moles per surface area per time, have been converted to fluxes with units of moles per length per time, which are appropriate for a one-dimensional model.

There are a number of ways to study reaction–diffusion models of this type, but the two most common (at least in the study of Ca^{2+} waves) are numerical simulation or bifurcation analysis of the traveling wave equations.

7.3.1   Simulation of Spiral Waves in *Xenopus*

One common experimental procedure for initiating waves in *Xenopus* oocytes is to photorelease a bolus of IP_3 inside the cell and observe the subsequent Ca^{2+} activity (Lechleiter and Clapham, 1992). After sufficient time, Ca^{2+} wave activity disappears as IP_3 is degraded, but in the short term, the observed Ca^{2+} activity is the result of Ca^{2+} diffusion and IP_3 diffusion. Another technique is to release IP_3S_3, a nonhydrolyzable analogue of IP_3, which has a similar effect on IP_3 receptors but is not degraded by the cell. In this case, after sufficient time has passed, the IP_3S_3 is at a constant concentration in all parts of the cell.

When a bolus of IP_3S_3 is released in the middle of the domain, it causes the release of a large amount of Ca^{2+} at the site of the bolus. The IP_3S_3 then diffuses across the cell, releasing Ca^{2+} in the process. Activation of IP_3 receptors by the released Ca^{2+} can lead to periodic Ca^{2+} release from the stores, and the diffusion of Ca^{2+} between IP_3 receptors serves to stabilize the waves, giving regular periodic traveling waves. These periodic waves are the spatial analogues of the oscillations seen in the temporal model, and arise from the underlying oscillatory kinetics. If the steady $[IP_3S_3]$ is in the appropriate range (see, for example, Fig. 7.12, which shows that limit cycles exist for $[IP_3]$ in some intermediate range) over the entire cell, every part of the cell cytoplasm is in an oscillatory state. It follows from the standard theory of reaction–diffusion systems with oscillatory kinetics (see, for example, Kopell and Howard, 1973; Duffy et al., 1980; Neu, 1979; Murray, 2002) that periodic and spiral waves can exist for these values of $[IP_3S_3]$. When IP_3, rather than IP_3S_3, is released, the wave activity lasts for only a short time, which is consistent with the theoretical results. When the wave front is broken, a spiral wave of Ca^{2+} often forms (Fig. 7.3A). These results have all been reproduced by numerical simulation; although Atri et al. (1993) were the first to do so, most models of Ca^{2+} waves behave in the same qualitative fashion. Depending on the initial conditions, these spiral waves can be stable or unstable. In the unstable case, the branches of the spiral can intersect themselves and cause breakup of the spiral, in which case a region of complex patterning emerges in which there is no clear spatial structure (McKenzie and Sneyd, 1998).

A more detailed understanding of the stability of spiral Ca^{2+} waves has been developed by Falcke et al. (1999, 2000), who showed that an increased rate of Ca^{2+} release from the mitochondria can dramatically change the kinds of waves observed, and that, in extreme cases, the cytoplasm can be made bistable, with an additional steady state with a high resting $[Ca^{2+}]$. When this happens, the spiral waves become unstable, resulting in the emergence of more complex spatiotemporal patterns.

7.3.2 Traveling Wave Equations and Bifurcation Analysis

The second principal way in which Ca^{2+} waves are studied is via the traveling wave equations. This is most useful for the study of waves in one spatial dimension. If we introduce the traveling wave variable $\xi = x + st$, where s is the wave speed (see Chapter 6, Section 6.2), we can write (7.61) as the pair of equations

$$c' = d, \tag{7.62}$$

$$D_c d' = sd - \sum J, \tag{7.63}$$

where $\sum J$ denotes all the fluxes on the right-hand side of (7.61), and where a prime denotes differentiation with respect to ξ. Thus, a single reaction–diffusion equation is converted to two ordinary differential equations. In general, these two equations are coupled to the other equations for c_e, p, and the states of the various receptors.

Traveling pulses, traveling fronts, and periodic waves correspond to, respectively, homoclinic orbits, heteroclinic orbits, and limit cycles of the traveling wave equations. Thus, by studying the bifurcations of the traveling wave equations we can gain considerable insight into what kinds of waves exist in the model, and for which parameter values. However, such an approach does not give information about the *stability* of the wave solutions of the original reaction–diffusion equations. Stability is much more difficult to determine. Numerical simulation of the reaction–diffusion equations can begin to address the question of stability.

Here we briefly illustrate the method using the model of Section 7.2.6. Adding diffusion of c (but not of c_e) to that model gives

$$\frac{\partial c}{\partial t} = D_c \frac{\partial^2 c}{\partial x^2} + \left(k_f P_O + J_{er}\right)(c_e - c) - J_{serca} + J_{in} - J_{pm}, \tag{7.64}$$

$$\frac{\partial c_e}{\partial t} = \gamma [J_{serca} - \left(k_f P_O + J_{er}\right)(c_e - c)], \tag{7.65}$$

and these two equations are coupled, as before, to the five equations of the six-state IPR model (Section 7.2.5). (Here, J_{in} and J_{pm} are scaled to be in units of flux per unit cellular length.) Rewriting these equations in the traveling wave variable gives

$$c' = d, \tag{7.66}$$

$$D_c d' = sd - \left(k_f P_O + J_{er}\right)(c_e - c) + J_{serca} - J_{in} + J_{pm}, \tag{7.67}$$

$$sc_e' = \gamma [J_{serca} - \left(k_f P_O + J_{er}\right)(c_e - c)], \tag{7.68}$$

coupled to the five receptor equations. Note that, given the units of the parameters in Table 7.4, the natural units for the speed s are $\mu m\,s^{-1}$.

The result of a two-parameter numerical bifurcation analysis of these equations (using AUTO), with s and p as the bifurcation parameters, is shown in Fig. 7.16. The behavior as $s \to \infty$ is that of the model in the absence of diffusion, as expected from the general theory (Maginu, 1985). Thus, for large values of s, the behavior mirrors that seen in Fig. 7.12; there are two Hopf bifurcations, with a branch of periodic solutions for

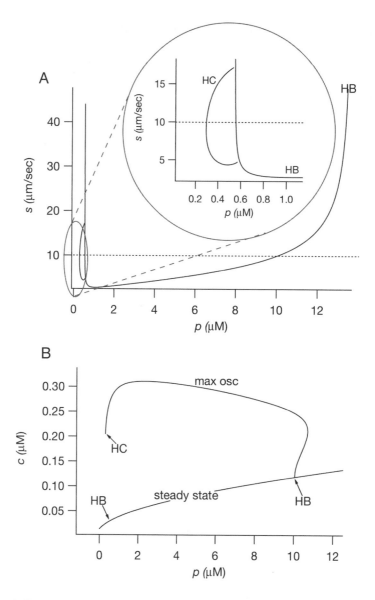

Figure 7.16 A: Two-parameter bifurcation diagram of the traveling wave equations of the model of Ca^{2+} wave propagation. HC — homoclinic bifurcation; HB — Hopf bifurcation. The inset shows a blowup of the branch of homoclinic bifurcations. B: bifurcation diagram obtained by taking a cross-section at $s = 10 \ \mu ms^{-1}$, as shown by the dotted line in panel A. Both panels were computed using the parameter values in Table 7.4, and using $D_c = 25 \ \mu m^2 s^{-1}$.

intermediate values of p. If we track these Hopf bifurcations in the s, p plane they form a U-shaped curve (Fig. 7.16A). To find homoclinic bifurcations corresponding to traveling waves, we take a cross-section of this diagram for a constant value of $s = 10 \ \mu m s^{-1}$, i.e., for that value of s we plot the bifurcation diagram of c against p (as in Fig. 7.12).

The result is shown in Fig. 7.16B. The two Hopf bifurcation points correspond to where the U-shaped curve in panel A intersects the dotted line at $s = 10 \ \mu m \, s^{-1}$. Tracking the branch of periodic orbits that emerges from the rightmost Hopf bifurcation gives a branch that ends in a homoclinic orbit. Tracking this homoclinic orbit in the s, p plane gives the C-shaped curve of homoclinics shown in the inset to panel A. More precisely, only branches of large period (10,000 seconds in this case) can be tracked, so the curve labeled HC in panel A is actually a branch of large-period orbits. However, tracking the branch of orbits of period 1,000 seconds or of period 30,000 seconds gives a nearly identical numerical result, so we can be relatively confident that we are tracking a homoclinic orbit.

This branch of homoclinic orbits is the one that gives stable traveling wave solutions of the original reaction–diffusion equation (i.e., of the partial differential equation). This is easily checked numerically by direct simulation; with $p = 0.5$ fixed, and following a large Ca^{2+} stimulus at one end of the (one-dimensional) domain, the wave that results travels at approximately speed $15 \ \mu m \, s^{-1}$ (consistent with Fig. 7.16A), and has the same shape as the homoclinic orbit on that C-shaped branch.

The branch of homoclinic orbits predicts traveling waves with speeds of around $10–17 \ \mu m \, s^{-1}$, within the physiological range. Furthermore, as p increases, so does the wave speed, consistent with what has been observed experimentally. The ends of the C-shaped homoclinic branch are physiologically important, since the upper branch corresponds to the waves that would be observed as p is slowly increased. At some point there is a transition from single traveling pulses to periodic waves, but exactly how that transition occurs is not completely understood (Champneys et al., 2007).

The basic structure of the bifurcation diagram shown in Fig. 7.16, with a C-shaped branch of homoclinic orbits and a U-shaped branch of Hopf bifurcations, seems to be generic to models of excitable systems. The FitzHugh–Nagumo equations have the same basic structure, as do the Hodgkin–Huxley equations. All models of Ca^{2+} wave propagation (at least all the ones for which this question has been studied) also have the same basic C-U structure in the s, p plane.

7.4 Calcium Buffering

Calcium is heavily buffered in all cells, with at least 99% (and often more) of the available Ca^{2+} bound to large Ca^{2+}-binding proteins, of which there are about 200 encoded by the human genome (Carafoli et al., 2001). For example, calsequestrin and calreticulin are major Ca^{2+} buffers in the endoplasmic and sarcoplasmic reticula, while in the cytoplasm Ca^{2+} is bound to calbindin, calretinin, and parvalbumin, among many others. Calcium pumps and exchangers and the plasma membrane itself are also major Ca^{2+} buffers. In essence, a free Ca^{2+} ion in solution in the cytoplasm cannot do much, or go far, before it is bound to something.

The basic chemical reaction for Ca^{2+} buffering can be represented by the reaction

$$P + Ca^{2+} \underset{k_-}{\overset{k_+}{\rightleftharpoons}} B, \tag{7.69}$$

where P is the buffering protein and B is buffered Ca^{2+}. Letting b denote the concentration of buffer with Ca^{2+} bound, and c the concentration of free Ca^{2+}, a simple model of Ca^{2+} buffering is

$$\frac{\partial c}{\partial t} = D_c \nabla^2 c + f(c) + k_- b - k_+ c(b_t - b), \tag{7.70}$$

$$\frac{\partial b}{\partial t} = D_b \nabla^2 b - k_- b + k_+ c(b_t - b), \tag{7.71}$$

where k_- is the rate of Ca^{2+} release from the buffer, k_+ is the rate of Ca^{2+} uptake by the buffer, b_t is the total buffer concentration, and $f(c)$ denotes all the other reactions involving free Ca^{2+} (release from the IP_3 receptors, reuptake by pumps, etc.).

7.4.1   Fast Buffers or Excess Buffers

If the buffer has fast kinetics, its effect on the intracellular Ca^{2+} dynamics can be analyzed simply (Section 2.2.5). If k_- and k_+ are large compared to the time constant of Ca^{2+} reaction, we take b to be in the quasi-steady state

$$k_- b - k_+ c(b_t - b) = 0, \tag{7.72}$$

and so

$$b = \frac{b_t c}{K + c}, \tag{7.73}$$

where $K = k_-/k_+$. Adding (7.70) and (7.71), we find the "slow" equation

$$\frac{\partial}{\partial t}(c + b) = D_c \nabla^2 c + D_b \nabla^2 b + f(c), \tag{7.74}$$

which, after using (7.73) to eliminate b, becomes

$$\frac{\partial c}{\partial t} = \frac{1}{1 + \theta(c)} \left(\nabla^2 \left(D_c c + D_b b_t \frac{c}{K + c} \right) + f(c) \right) \tag{7.75}$$

$$= \frac{D_c + D_b \theta(c)}{1 + \theta(c)} \nabla^2 c - \frac{2 D_b \theta(c)}{(K + c)(1 + \theta(c))} |\nabla c|^2 + \frac{f(c)}{1 + \theta(c)}, \tag{7.76}$$

where

$$\theta(c) = \frac{b_t K}{(K + c)^2}. \tag{7.77}$$

Note that we assume that b_t is a constant, and does not vary in either space or time.

Nonlinear buffering changes the model significantly. In particular, Ca^{2+} obeys a nonlinear diffusion–advection equation, where the advection is the result of Ca^{2+}

transport by a mobile buffer (Wagner and Keizer, 1994). The effective diffusion coefficient

$$D_{\text{eff}} = \frac{D_c + D_b \theta(c)}{1 + \theta(c)} \tag{7.78}$$

is a convex linear combination of the two diffusion coefficients D_c and D_b, so lies somewhere between the two. Since buffers are large molecules, $D_{\text{eff}} < D_c$. If the buffer is not mobile, i.e., $D_b = 0$, then (7.76) reverts to a reaction–diffusion equation. Also, when Ca^{2+} gradients are small, the nonlinear advective term can be ignored (Irving et al., 1990). Finally, the buffering also affects the qualitative nature of the nonlinear reaction term, $f(c)$, which is divided by $1 + \theta(c)$. This may change many properties of the model, including oscillatory behavior and the nature of wave propagation.

If the buffer is not only fast, but also of low affinity, so that $K \gg c$, it follows that

$$b = \frac{b_t c}{K}, \tag{7.79}$$

in which case

$$\theta = \frac{b_t}{K}, \tag{7.80}$$

a constant. Thus, D_{eff} is constant also.

It is commonly assumed that the buffer has fast kinetics, is immobile, and has a low affinity. With these assumptions we get the simplest possible model of Ca^{2+} buffers (short of not including them at all), in which

$$\frac{\partial c}{\partial t} = \frac{K}{K + b_t} (D_c \nabla^2 c + f(c)), \tag{7.81}$$

wherein both the diffusion coefficient and the fluxes are scaled by the constant factor $K/(K + b_t)$; each flux in the model can then be interpreted as an *effective* flux, i.e., that fraction of the flux that contributes to a change in free Ca^{2+} concentration.

In 1986, Neher observed that if the buffer is present in large excess then $b_t - b \approx b_t$, in which case the buffering reaction becomes linear, the so-called *excess buffering approximation*:

$$\frac{\partial c}{\partial t} = D_c \nabla^2 c + f(c) + k_- b - k_+ c b_t, \tag{7.82}$$

$$\frac{\partial b}{\partial t} = D_b \nabla^2 b - k_- b + k_+ c b_t. \tag{7.83}$$

If we now assume the buffers are fast we recover (7.79) and thus (7.81). In other words, the simple approach to buffering given in (7.81) can be obtained in two ways; either by assuming a low affinity buffer or by assuming that the buffer is present in excess. It is intuitively clear why these two approximations lead to the same result—in either case the binding of Ca^{2+} does little to change the fraction of unbound buffer.

Typical parameter values for three different buffers are given in Table 7.6. BAPTA is a fast high-affinity buffer, and EGTA is a slow high-affinity buffer, both of which

Table 7.6 Typical parameter values for three different buffers, taken from Smith et al., (2001). BAPTA and EGTA are commonly used as exogenous buffers in experimental work, while Endog refers to a typical endogenous buffer. Typically $b_t = 100\ \mu M$ for an endogenous buffer.

Buffer	D_b $\mu m^2\,s^{-1}$	k_+ $\mu M^{-1}s^{-1}$	k_- s^{-1}	K μM
BAPTA	95	600	100	0.17
EGTA	113	1.5	0.3	0.2
Endog	15	50	500	10

are used as exogenous Ca^{2+} buffers in experimental work. Parameters for a typical endogenous buffer are also included.

Many studies of Ca^{2+} buffering have addressed the problem of the steady-state Ca^{2+} distribution that arises from the flux of Ca^{2+} through a single open channel (Naraghi and Neher, 1997; Stern, 1992; Smith, 1996; Smith et al., 1996, 2001; Falcke, 2003b). Such studies are motivated by the fact that, when a channel opens, the steady-state Ca^{2+} distribution close to the opening of the channel is reached within microseconds. In that context, Smith et al. (2001) derived both the rapid buffering approximation and the excess buffering approximation as asymptotic solutions of the original equations.

Despite the complexity of (7.76), it retains the advantage of being a single equation. However, if the buffer kinetics are not fast relative to the Ca^{2+} kinetics, the only way to proceed is with numerical simulations of the complete system, a procedure followed by a number of groups (Backx et al., 1989; Sala and Hernández-Cruz, 1990; Nowycky and Pinter, 1993; Falcke, 2003a).

Of particular current interest are recent observations that buffers with different kinetics have remarkably different effects on the observed Ca^{2+} responses. For example, Dargan and Parker (2003) have shown that in oocytes, the addition of EGTA, a slow buffer with high affinity, "balkanizes" the Ca^{2+} response so that local regions tend to respond independently of their neighbors. However, the addition of BAPTA, a fast buffer, has quite different effects, giving slower Ca^{2+} responses that occur globally rather than locally. These results are possibly the result of the different effects that these two buffers have on the diffusion of Ca^{2+} between individual release sites (see Section 7.6). Similar studies in pancreatic acinar cells (Kidd et al., 1999) have shown that the addition of EGTA increases the frequency, but decreases the amplitude, of Ca^{2+} spikes, while addition of BAPTA has the opposite effect, decreasing the frequency but maintaining a larger spike amplitude. Furthermore, EGTA broke the response into a number of independent spatially separate release events, as in oocytes. Kidd et al. concluded that EGTA disrupts long-range Ca^{2+} diffusion between release sites, thus decreasing the global coordination of Ca^{2+} release, while BAPTA is fast enough to disrupt Ca^{2+} inactivation of the IPR, thus leading to large Ca^{2+} spikes with a lower frequency. However, no detailed modeling studies have yet confirmed these qualitative explanations.

7.4.2 The Existence of Buffered Waves

Since the presence of fast Ca^{2+} buffers changes the nature of the Ca^{2+} transport equation, it is of considerable interest to determine how Ca^{2+} buffering affects the properties of waves. For example, can the addition of a buffer eliminate wave activity? How much do buffers affect the speed of traveling waves? Does the addition of exogenous buffer, such as a fluorescent Ca^{2+} dye, affect the existence or the speed of the Ca^{2+} waves?

Tsai and his colleagues (Tsai and Sneyd, 2005, 2007a,b; Guo and Tsai, 2006) have done a great deal of work on these questions, and their results can be summarized simply: immobile buffers have no effect on the existence, stability, or uniqueness of traveling waves in the buffered bistable equation, while mobile buffers can eliminate traveling waves when present in sufficient quantity. However, when waves exist in the presence of mobile buffers, they remain unique and stable.

The proofs of these results are too technical to present here. Instead, we discuss some of the simpler results from earlier work (Sneyd et al., 1998). First, we address the question of whether buffers can eliminate wave activity.

The form of (7.75) suggests the change of variables

$$w = D_c c + D_b b_t \frac{c}{K+c}, \tag{7.84}$$

so that w is a monotone increasing function of c, since

$$\frac{dw}{dc} = D_c + D_b \theta(c) \tag{7.85}$$

is positive. The unique inverse of this function is denoted by

$$c = \phi(w). \tag{7.86}$$

In terms of w, (7.75) becomes

$$\frac{\partial w}{\partial t} = \frac{D_c + D_b \Theta}{1 + \Theta} \left(\nabla^2 w + f(\phi(w)) \right), \tag{7.87}$$

where $\Theta = \frac{b_t K}{(K + \phi(w))^2}$.

Now we assume that $f(c)$ is of bistable form, with three zeros, $C_1 < C_2 < C_3$, of which C_1 and C_3 are stable. It immediately follows that $f(\phi(w))$ has three zeros $W_1 < W_2 < W_3$, with W_1 and W_3 stable. The proof of existence of a one-dimensional traveling wave solution for (7.87) uses exactly the same arguments as those for the bistable equation presented in Chapter 6 (Sneyd et al., 1998). It follows that a traveling wave solution that provides a transition from W_1 to W_3 exists if and only if

$$\int_{W_1}^{W_3} f(\phi(w)) \, dw > 0. \tag{7.88}$$

(If this inequality is reversed, a traveling wave solution still exists, but it moves "backward", providing a transition from W_3 to W_1.) Using (7.85), we write this condition in terms of c as

$$\int_{C_1}^{C_3} f(c)(D_c + D_b\theta(c))\, dc > 0. \tag{7.89}$$

In general, this integral cannot be evaluated explicitly. However, for the simple case of cubic bistable kinetics $f(c) = c(1 - c)(c - a)$, $0 < a < 1/2$, explicit evaluation of the integral (7.89) shows that traveling waves exist if and only if

$$a < a_c = \frac{1}{2}\frac{D_c - 12D_b b_t K[(3K^2 + 2K)\ln(\frac{K+1}{K}) - (3K + \frac{1}{2})]}{D_c + 12D_b b_t K[(K + \frac{1}{2})\ln(\frac{K+1}{K}) - 1]}. \tag{7.90}$$

One conclusion that can be drawn immediately from (7.89) is that a stationary buffer (i.e., one with $D_b = 0$) has no effect on the existence of traveling waves in the bistable equation. For when $D_b = 0$, the condition (7.89) for the existence of the traveling wave reduces to

$$\int_0^1 f(c)\, dc > 0, \tag{7.91}$$

which is exactly the condition for the existence of a wave in the absence of a buffer.

Note that a_c is a monotonically decreasing function of $D_b b_t / D_c$, and

$$a_c \to a_{c,\min}(K) = -\frac{1}{2}\frac{[(3K^2 + 2K)\ln(\frac{K+1}{K}) - (3K + \frac{1}{2})]}{[(K + \frac{1}{2})\ln(\frac{K+1}{K}) - 1]} \qquad \text{as } D_b \to \infty. \tag{7.92}$$

In Fig. 7.17 we give a plot of $a_{c,\min}(K)$ against K. When K is large, the minimum value of a_c is close to 0.5, and thus wave existence is insensitive to D_b. However, when K is small, a_c also becomes small as D_b increases, and so in this case, a mobile buffer can easily stop a wave.

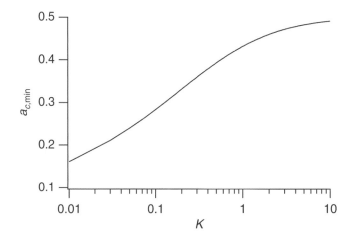

Figure 7.17 Plot of $a_{c,\min}$ against K.

7.5 Discrete Calcium Sources

In all of the models presented so far, Ca^{2+} release from the ER was assumed to be homogeneous in space. In fact, this is not the case. In *Xenopus* oocytes, for example, IPR are arranged in clusters with a density of about 1 per 30 μm^2, with each cluster containing about 25 IPR. Furthermore, the propagation of Ca^{2+} waves is saltatory, with release jumping from one cluster of release sites to another.

To explore the properties of such saltatory waves, and to see the effect of this discrete structure, we assume that Ca^{2+} is released from discrete release sites but is removed continuously. Thus

$$\frac{\partial c}{\partial t} = D_c \frac{\partial^2 c}{\partial x^2} - k_s c + L \sum_n \delta(x - nL) f(c), \tag{7.93}$$

where the function f is your favorite description of Ca^{2+} release, and L is the spatial separation between release sites. Two examples of the release function $f(c)$ that are useful are

$$f(c) = Ac^2(c_e - c) \tag{7.94}$$

and

$$f(c) = AH(c - c^*)(c_e - c), \tag{7.95}$$

where H is the usual Heaviside function. Both of these release functions have Ca^{2+}-dependent release, representing CICR.

The corresponding continuous space model can be found by homogenization (Exercise 16). It is not difficult to show (and it makes intuitive sense) that in the limit that $\frac{L^2 k_s}{D_c} \ll 1$, the spatially inhomogeneous problem

$$\frac{\partial c}{\partial t} = D_c \frac{\partial^2 c}{\partial x^2} - k_s c + g(x)f(c), \tag{7.96}$$

where $g(x)$ is periodic with period L, can be replaced by its average

$$\frac{\partial c}{\partial t} = D_c \frac{\partial^2 c}{\partial x^2} - k_s c + Gf(c), \tag{7.97}$$

where $G = \frac{1}{L} \int_0^L g(x)\,dx$.

Since the effective release function $F(c) = Gf(c) - k_s c$ is of bistable type, with three zeros $0 < c_1 < c_2$, one would expect there to be traveling waves of Ca^{2+} release, provided the medium is sufficiently excitable, i.e., provided

$$\int_0^{c_2} F(c)\,dc > 0. \tag{7.98}$$

However, this condition fails to be correct if $\frac{L^2 k_s}{D_c} \neq 0$.

Waves fail to propagate if there is a standing wave solution. Standing waves are stationary solutions of (7.93), i.e., solutions of

$$0 = D_c \frac{\partial^2 c}{\partial x^2} - k_s c + L \sum_n \delta(x - nL) f(c). \tag{7.99}$$

On the intervals $nL < x < (n+1)L$, this becomes

$$0 = D_c \frac{\partial^2 c}{\partial x^2} - k_s c. \tag{7.100}$$

We find jump conditions at $x = nL$ by integrating from nL^- to nL^+ to obtain

$$D_c c_x |_{nL^-}^{nL^+} + L f(c_n) = 0, \tag{7.101}$$

where $c_n = c(nL)$.

Now we solve (7.100) to obtain

$$c(x) = (c_{n+1} - c_n \cosh \beta) \frac{\sinh(\frac{\beta}{L}(x - nL))}{\sinh \beta} + c_n \cosh\left(\frac{\beta}{L}(x - nL)\right), \tag{7.102}$$

for $nL < x < (n+1)L$, where $c_n = c(nL)$, $\beta^2 = \frac{k_s L^2}{D_c}$, so that

$$c_x(nL^+) = (c_{n+1} - c_n \cosh \beta) \frac{\beta}{L \sinh \beta}. \tag{7.103}$$

Similarly,

$$c_x(nL^-) = -(c_{n-1} - c_n \cosh \beta) \frac{\beta}{L \sinh \beta}. \tag{7.104}$$

It follows that (7.101) is the difference equation

$$\frac{k_s}{\beta \sinh \beta}(c_{n+1} - 2c_n \cosh \beta + c_{n-1}) + f(c_n) = 0, \tag{7.105}$$

which is a difference equation for c_n.

Finding solutions of nonlinear difference equations is nontrivial in general. However, if $f(c)$ is piecewise linear, as it is with release function (7.95), analytical solutions can be found. We seek solutions of (7.105) of the form

$$c_n = \begin{cases} a\mu_0^{-n}, & n \le 0, \\ C - b\mu_f^n, & n > 0, \end{cases} \tag{7.106}$$

where the steady solution C satisfies

$$\frac{k_s}{\beta \sinh \beta}(C - 2C \cosh \beta + C) + A(c_e - C) = 0, \tag{7.107}$$

so that

$$C = c_e \frac{A_f}{2\cosh \beta - 2 + A_f}, \qquad A_f = A \frac{\beta \sinh \beta}{k_s}. \tag{7.108}$$

The numbers μ_0 and μ_f are both less than one and satisfy the quadratic equations

$$\mu_j - 2\lambda_j + \frac{1}{\mu_j} = 0, \qquad j = 0, f, \tag{7.109}$$

where

$$\lambda_0 = \cosh \beta, \qquad \lambda_f = \cosh \beta + \frac{A_f}{2}, \tag{7.110}$$

so that

$$\mu_j = \lambda_j - \sqrt{\lambda_j^2 - 1}. \tag{7.111}$$

It is easy to determine that $\mu_0 = \exp(-\beta)$; the expression for μ_f is more complicated.

We determine the scalars a and b by examining the difference equation for $n = 0$,

$$\frac{k_s}{\beta \sinh \beta}(c_1 - 2c_0 \cosh \beta + c_{-1}) + f(c_0) = 0, \tag{7.112}$$

and for $n = 1$,

$$\frac{k_s}{\beta \sinh \beta}(c_2 - 2c_1 \cosh \beta + c_0) + f(c_1) = 0. \tag{7.113}$$

Since $c_0 < c^*$ and $c_1 > c^*$ by assumption, $f(c_0) = 0$ and $f(c_1) = A(c_e - C + b\mu_f)$. After some algebraic manipulation we find that

$$a = C\left(\frac{\mu_f - 1}{\mu_f - \frac{1}{\mu_0}}\right), \qquad b = C\left(\frac{1 - \frac{1}{\mu_0}}{\mu_f - \frac{1}{\mu_0}}\right). \tag{7.114}$$

The condition for the existence of these standing waves is that

$$a \le c^*, \qquad C - b\mu_f \ge c^*, \tag{7.115}$$

so a plot of a and $C - b\mu_f$ is revealing. In Fig. 7.18 are shown the two curves $\frac{a}{c_e}$ (lower curve) and $\frac{C}{c_e} - \frac{b}{c_e}\mu_f$ (upper curve) plotted as functions of $\beta = \sqrt{\frac{k_s L^2}{D_c}}$ for $\frac{A}{k} = 5$. The interpretation of this plot is that for a fixed value of β, if $\frac{c^*}{c_e}$ lies in the region below the curve $\frac{a}{c_e}$, then there is a traveling wave, whereas if $\frac{c^*}{c_e}$ and β lies between the two curves, there is a standing wave, which precludes the possibility of propagation. Thus, in general, the larger is β, the more excitable the release sites must be ($\frac{c^*}{c_e}$ must be smaller) for propagation to occur. Said another way, discrete release from clumped receptors makes propagation less likely than if the same amount of Ca^{2+} were released in a spatially continuous and homogeneous fashion. Notice also that for fixed $\frac{c^*}{c_e}$, increasing β sufficiently always leads to propagation failure. Increasing β corresponds to increasing the distance between release sites L or the rate of uptake k_s, or decreasing the Ca^{2+} diffusion coefficient D_c.

A similar approach to this problem was explored by Sneyd and Sherratt (1997).

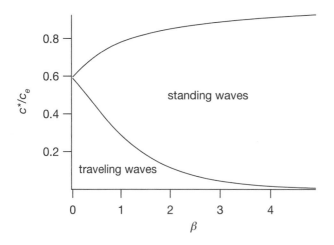

Figure 7.18 Plot of $\frac{a}{c_e}$ (lower curve) and $\frac{C}{c_e} - \frac{b}{c_e}\mu_f$ (upper curve) plotted as functions of $\beta = \sqrt{\frac{k_s L^2}{D}}$ for $\frac{A}{k_s} = 5$.

7.5.1 The Fire–Diffuse–Fire Model

It is difficult to determine when propagation failure occurs, and it is even more difficult to find the speed of propagation in a discrete release model. One approach, which uses techniques that are beyond the scope of this text, was used in Keener (2000b). Another approach is to use a different model. One model that has attracted a lot of attention is the *fire–diffuse–fire* model (Pearson and Ponce-Dawson, 1998; Keizer et al., 1998; Ponce-Dawson et al., 1999; Coombes, 2001; Coombes and Bressloff, 2003; Coombes and Timofeeva, 2003; Coombes et al., 2004).

In this model, once $[\text{Ca}^{2+}]$ reaches a threshold value, c^*, at a release site, that site fires, instantaneously releasing a fixed amount, σ, of Ca^{2+}. Thus, a Ca^{2+} wave is propagated by the sequential firing of release sites, each responding to the Ca^{2+} diffusing from neighboring release sites. Hence the name fire–diffuse–fire.

We assume that Ca^{2+} obeys the reaction–diffusion equation

$$\frac{\partial c}{\partial t} = D_c \frac{\partial^2 c}{\partial x^2} + \sigma \sum_n \delta(x - nL)\delta(t - t_n), \tag{7.116}$$

where, as before, L is the spacing between release sites. Although this equation looks linear, appearances are deceptive. Here, t_n is the time at which c first reaches the threshold value c^* at the nth release site. When this happens, the nth release site releases the amount σ. Thus, t_n depends in a complicated way on c.

The Ca^{2+} profile resulting from the firing of a single site, site i, say, is

$$c_i(x, t) = \sigma \frac{H(t - t_i)}{\sqrt{4\pi D_c(t - t_i)}} \exp\left(-\frac{(x - iL)^2}{4D_c(t - t_i)}\right), \tag{7.117}$$

where H is the Heaviside function. This is the fundamental solution of the diffusion equation with a delta function input at $x = i$, $t = t_i$, and can be found in any standard book on analytical solutions to partial differential equations (see, for example,

Keener, 1998, or Kevorkian, 2000). If we superimpose the solutions from each site, we get

$$c(x,t) = \sum_i c_i(x,t) = \sigma \sum_i \frac{H(t-t_i)}{\sqrt{4\pi D_c(t-t_i)}} \exp\left(-\frac{(x-iL)^2}{4D_c(t-t_i)}\right). \qquad (7.118)$$

Notice that because of the instantaneous release, $c(x,t)$ is not a continuous function of time at any release site.

Now suppose that sites $i = N, N-1, \ldots$ have fired at known times $t_N > t_{N-1} > \ldots$. The next firing time t_{N+1} is determined by when c at x_{N+1} first reaches the threshold, c^*, that is,

$$c((N+1)L, t_{N+1}^-) = c^*, \qquad \frac{\partial}{\partial t} c((N+1)L, t_{N+1}^-) > 0. \qquad (7.119)$$

Thus, t_{N+1} must satisfy

$$c^* = \sigma \sum_{i \leq N} \frac{1}{\sqrt{4\pi D_c(t_{N+1}-t_i)}} \exp\left(-\frac{L^2(N+1-i)^2}{4D_c(t_{N+1}-t_i)}\right). \qquad (7.120)$$

A steadily propagating wave corresponds to having $t_i - t_{i-1} = \text{constant} = \tau$ for all i, i.e., each site fires a fixed time after its leftward neighbor fires. Note that the resulting wave does not propagate with a constant profile, but has a well-defined wave speed L/τ. If such a τ exists, then $t_{N+1} - t_i = \tau(N+1-i)$ and τ is a solution of the equation

$$\frac{c^* L}{\sigma} = \sum_{n=1}^{\infty} \frac{1}{\sqrt{4\pi n\eta}} \exp\left(-\frac{n}{4\eta}\right) \equiv g(\eta), \qquad (7.121)$$

where $\eta = \frac{D_c \tau}{L^2}$ is the dimensionless delay.

To find η we need to invert this equation. A plot of $g(\eta)$ is shown in Fig. 7.19. It can be shown that $0 \leq g(\eta) \leq 1$ and that g is monotonic with $g \to 0$ as $\eta \to 0$, and $g \to 1$ as $\eta \to \infty$. It follows that a solution to (7.121) exists only if $\frac{c^* L}{\sigma} < 1$. Thus, when the intercluster distance or the threshold is too large, or the amount of release is too

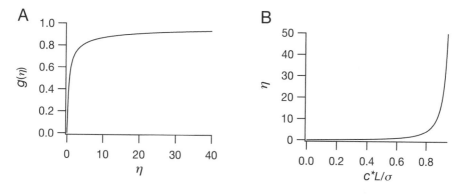

Figure 7.19 A: Plot of $g(\eta)$, where $g(\eta)$ is defined by (7.121). B: Plot of the delay η using (7.121).

small, there is no propagation. However, when $\frac{c^*L}{\sigma} < 1$, a unique solution of (7.121) is guaranteed, and thus there is a propagating wave.

It is an easy matter to plot the delay as a function of $\frac{c^*L}{\sigma}$ by appropriately reversing the axes of Fig. 7.19A. Thus, in Fig. 7.19B is plotted the dimensionless delay η as a function of $\frac{c^*L}{\sigma}$. It is equally easy to plot the dimensionless velocity $\frac{1}{\eta}$ as a function of $\frac{c^*L}{\sigma}$ (not shown). The result is that the velocity is infinitely large as $\frac{c^*L}{\sigma} \to 0$ and is zero for $\frac{c^*L}{\sigma} \geq 1$.

If a solution of (7.121) exists, then the speed of the wave is proportional to D_c. This is because the velocity is $\frac{L}{\tau} = \frac{D}{\eta L}$. This result is disconcerting for two reasons. First, the speed of propagation of waves in spatially homogeneous reaction–diffusion systems usually scales with the square root of D. It follows from simple scaling arguments that since the units of the diffusion coefficient, D_c, are (length)^2 per time, the distance variable can be scaled by $\sqrt{D_c k}$, where k is a typical time constant, to remove all distance units, so that the wave speed scales with $\sqrt{D_c k}$. Second, this disagrees with the result of the previous section showing that if D_c is sufficiently small, there is propagation failure. For the fire–diffuse–fire model, propagation success or failure is independent of D_c.

This mismatch is explained by the fact that the fire–diffuse–fire model allows only for release of Ca^{2+}, but no uptake, and thus the Ca^{2+} transient is unrealistically monotone increasing.

This deficiency of the fire–diffuse–fire model is easily remedied by adding a linear uptake term that is homogeneous in space (Coombes, 2001), so that the model becomes

$$\frac{\partial c}{\partial t} = D_c \frac{\partial^2 c}{\partial x^2} - k_s c + \sigma \sum_n \delta(x - nL)\delta(t - t_n). \tag{7.122}$$

The analysis of this modified model is almost identical to the case with $k_s = 0$. The fundamental solution is modified slightly by k_s, with the Ca^{2+} profile resulting from the firing at site i given by

$$c_i(x, t) = \sigma \frac{H(t - t_i)}{\sqrt{4\pi D_c(t - t_i)}} \exp\left(-\frac{(x - iL)^2}{4D_c(t - t_i)} - k_s(t - t_i)\right). \tag{7.123}$$

Following the previous arguments, we learn that a propagating solution exists if there is a solution of the equation

$$\frac{c^*L}{\sigma} = \sum_{n=1}^{\infty} \frac{1}{\sqrt{4\pi n\eta}} \exp\left(-\frac{n}{4\eta} - \beta^2 n\right) \equiv g_\beta(\eta), \tag{7.124}$$

where $\eta = \frac{D_c \tau}{L^2}$ is the dimensionless delay, and $\beta^2 = \frac{k_s L^2}{D_c}$.

Estimates of the size of β^2 can vary substantially. For example, using $k_s = 143/s$ ($= 1/\alpha_2$ from Table 7.4), $L^2 = 30 \ \mu m^2$, and $D_c = 20 \ \mu m^2/s$, we find $\beta^2 = 172$, whereas with $k_s = 3.78/s$ ($= k_4$ from the Friel model, see Table 7.5), $L^2 = 4 \ \mu m^2$ (appropriate

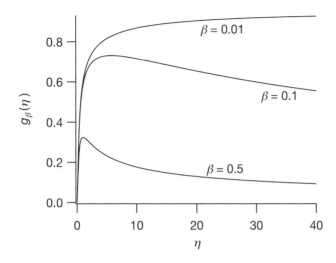

Figure 7.20 Plots of $g_\beta(\eta)$ for three different values of β. Note that g_β is a decreasing function of β for all positive values of η.

for Ca^{2+} release in cardiac cells), and $D_c = 25\ \mu\text{m}^2/\text{s}$, we find $\beta^2 = 0.6$. Regardless, the effect of β is significant. Plots of $g_\beta(\eta)$ are shown for several values of β in Fig. 7.20. In particular, if $\beta \neq 0$, the function $g_\beta(\eta)$ is not monotone increasing, but has a maximal value, say $g_{\max}(\beta)$, which is a decreasing function of β. Furthermore, $g_\beta(\eta) \to 0$ as $\eta \to 0$ and as $\eta \to \infty$. If $\frac{c^*L}{\sigma} > g_{\max}(\beta)$, then no solution of (7.124) exists; there is propagation failure. On the other hand, if $\frac{c^*L}{\sigma} < g_{\max}(\beta)$, then there are two solutions of (7.124); the physically meaningful solution is the smaller of the two, corresponding to the first time that $c(x,t)$ reaches c^*.

For larger values of β, the curve $g_{\max}(\beta)$ agrees very well with the exponential function $\exp(-\beta)$. This provides us with an approximate criterion for propagation failure, namely, if $\frac{c^*L}{\sigma} > \exp(-\beta)$, propagation fails. Recall that $\beta = \sqrt{\frac{k_s L^2}{D_c}}$ to see the $\sqrt{D_c}$ dependence in this criterion.

By reversing the axes in Fig. 7.20, one obtains a plot of the dimensionless delay as a function of $\frac{c^*L}{\sigma}$. Notice the significant qualitative difference when $\beta \neq 0$ compared to $\beta = 0$. With $\beta \neq 0$, propagation ceases at a finite, not infinite, delay. This implies that propagation fails at a positive, not zero, velocity.

7.6   Calcium Puffs and Stochastic Modeling

In all the models discussed above, the release of Ca^{2+} was modeled as deterministic. However, it is now well known that this is not always appropriate. In fact, each Ca^{2+} oscillation or wave is built up from a number of stochastic elementary release events, called *puffs*, each of which corresponds to Ca^{2+} release from a single, or a small group of, IPR. At low concentrations of IP$_3$, punctate release from single clusters occurs, while at high [IP$_3$] these local release events are coordinated into global intracellular waves (Yao et al., 1995; Parker et al., 1996a,b; Parker and Yao, 1996). Detailed studies

A

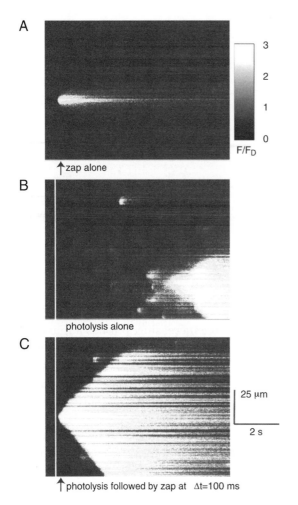

3

2

1

0

F/F_D

↑zap alone

B

photolysis alone

C

25 μm

2 s

↑photolysis followed by zap at Δt=100 ms

Figure 7.21 Calcium waves and puffs caused by release of Ca^{2+} and IP_3. Lighter colors correspond to higher Ca^{2+} concentrations. A: release of Ca^{2+} by a UV laser zap results in a single localized Ca^{2+} response. B: photolysis of IP_3 across the entire region causes several Ca^{2+} puffs (see, for example, the puff at top center), followed by an abortive wave about 5 seconds after the release of IP_3. C: photolysis of IP_3 followed by release of Ca^{2+} by a UV laser zap. A global wave is initiated immediately by the additional Ca^{2+} release. (Adapted from Marchant et al., 1999, Fig. 3. We thank Ian Parker for providing the original figure of this experimental data.)

of puffs have been done principally in *Xenopus* oocytes (Marchant et al., 1999; Sun et al., 1998; Callamaras et al., 1998; Marchant and Parker, 2001) and HeLa cells (Thomas et al., 2000; Bootman et al., 1997a,b).

Some typical experimental results are shown in Fig. 7.21. In the top panel is shown the release of Ca^{2+} in response to a single laser pulse. (The cell was first loaded with a photoreleasable form of Ca^{2+}.) The Ca^{2+} pulse is localized, and does not spread to form a global wave. This is because the background IP_3 concentration is too low to support a wave. However, when the background IP_3 concentration is raised by the photorelease of IP_3 (panel B), the cytoplasm becomes more excitable, and spontaneous Ca^{2+} puffs occur (for instance, the white dot at the top and center of the panel). If, by chance, the spontaneous release is large enough, a traveling wave will form, spreading from release site to release site, as shown in the bottom right of the panel. Panel C shows the response when the Ca^{2+} pulse is applied on top of a higher background IP_3 concentration. In this case the medium is excitable enough for the Ca^{2+} pulse to initiate a global traveling wave.

These experimental results raise several important modeling questions. First, when is it appropriate to use deterministic models and when must stochastic behavior be incorporated? Second, how should one best model stochastic Ca^{2+} release through a small number of IPR, and, finally, how can one model the coordination of such local release into global events such as intracellular waves? Although such questions are most obvious from the experimental work in oocytes and HeLa cells, they are also important for the study of Ca^{2+} oscillations in other cell types. As Falcke (2004) has pointed out, stochastic effects appear to be so fundamental and widespread that they raise questions about the applicability of deterministic approaches in general.

7.6.1 Stochastic IPR Models

The basic assumption behind the modeling of Ca^{2+} puffs is that each IPR release event can be modeled as a stochastic Markov process, while the diffusion of Ca^{2+}, and the action of the other pumps and Ca^{2+} fluxes can be modeled deterministically. Given a Markov state model of the IPR (for example, as shown in Fig. 7.8 or Fig. 7.10), we can simulate the model by choosing a random number at each time step and using that random number to determine the change of state for that time step. This is done most efficiently using the Gillespie method, as described in Section 2.9.3.

In general, each transition rate is a function of c and p, and thus the transition probabilities are continually changing as the concentrations change. It follows that c and p must be updated at each time step. Such updates are done deterministically, by solving the reaction–diffusion equations for c and p. When the simulated IPR model is in the open state, the reaction–diffusion equation for Ca^{2+} has an additional flux through the IPR. When the simulated IPR model is in some other state, this flux is absent. Thus, we obtain a reaction–diffusion equation for c that is driven by a stochastically varying input.

The first stochastic model of an IPR was due to Swillens et al. (1998). Their model is an 18-state model in which the IPR can have 0 or 1 IP_3 bound, 0, 1, or 2 activating Ca^{2+} bound, and 0, 1, or 2 inactivating Ca^{2+} bound. The receptor is open only when it has one IP_3 bound, two Ca^{2+} bound to activating sites, and no Ca^{2+} bound to inactivating sites. The steady-state open probability of the model is constrained to fit the experimental data of Bezprozvanny et al. (1991), and it is assumed that Ca^{2+} diffuses radially away from the mouth of the channel. Calcium can build up to high concentrations at the mouth of the channel, and these local concentrations are used in the stochastic simulation of the IPR model.

Simulations of this model show two things in particular. First, channel openings occur in bursts, as Ca^{2+} diffuses away from the mouth of the channel slowly enough to allow rebinding to an activating site. Second, by comparing to experimentally observed distributions of puff amplitudes, Swillens et al. (1999) showed that a typical cluster contains approximately 25 receptors, and that within the cluster the IPR are probably separated by no more than 12 nm.

Because of the intensive computations involved in direct stochastic simulation of gating schemes for the IPR, Shuai and Jung approximated a stochastic version of the eight-state model (Section 7.2.5) by a Langevin equation (Shuai and Jung, 2002a,b, 2003), an approach that had already been used to study the Hodgkin–Huxley equations by Fox and Lu (1994). Although 25 receptors in each cluster makes a Langevin equation approach less accurate than direct stochastic simulation, the qualitative behavior agrees well with direct simulation.

Local Concentrations

In any stochastic model of the IPR it is impossible to ignore the local high concentrations that occur at the mouth of the channel. Neglect of this factor results in either an IPR model that bears no relation to reality, or a model that does not exhibit realistic Ca^{2+} oscillations. Thus, for a stochastic model of the IPR to be incorporated into a whole-cell model of Ca^{2+} oscillations, it is necessary to somehow relate the Ca^{2+} concentration in the microdomain at the channel mouth to the bulk Ca^{2+} concentration in the cytoplasm of the cell. Two approaches to this problem are those of Huertas and Smith (2007) and Bentele and Falcke (2007).

This problem is similar to the problem of how best to model Ca^{2+} release in cardiac cells (Section 12.2.4). There, the ryanodine receptors release Ca^{2+} into a very restricted domain, the diadic cleft, and experience Ca^{2+} concentrations much higher than those in the remainder of the cytoplasm. Thus, many recent models of Ca^{2+} dynamics in cardiac cells do not use a single cytoplasmic Ca^{2+} domain, but incorporate various Ca^{2+} microdomains, with greater or lesser complexity.

7.6.2   Stochastic Models of Calcium Waves

Of all the results from stochastic modeling, the most intriguing are those of Falcke (2003a,b), who studied the transition from puffs to waves in *Xenopus* oocytes as [IP$_3$] is increased. At low [IP$_3$] only puffs are observed; there is not enough Ca^{2+} released from each cluster to stimulate Ca^{2+} release from neighboring clusters, and thus the responses are purely local. However, as [IP$_3$] increases, both the sensitivity of, and the amount of Ca^{2+} released from, each IPR increases. This allows for the development of global waves that emerge from a *nucleation* site. However, until [IP$_3$] gets considerably larger, these global events are rare, and in many cases form only abortive waves that progress a short distance before dying out. Both the interwave time interval, T_{av}, and its standard deviation ΔT_{av}, decrease as [IP$_3$] increases. Finally, at high [IP$_3$] global waves occur regularly with a well-defined period.

Falcke (2003a,b) showed that all these behaviors can be reproduced by a stochastic version of the eight-state model of Section 7.2.5. The long time interval between

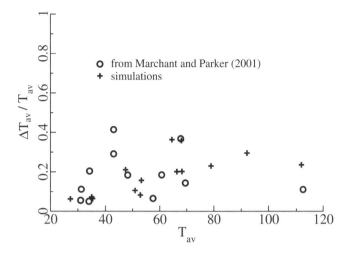

Figure 7.22 Plot of the relative standard deviation $\Delta T_{av}/T_{av}$. The experimental data (open circles) are from Marchant and Parker (2001), while the simulations are from Falcke (2003a).

successive waves when [IP$_3$] is low is a result almost entirely of the stochastic dynamics. In each time interval there is a chance that one IPR fires, stimulating the firing of the entire cluster, and thus initiating a global wave, but because of the intercluster separation and the low sensitivity of each IPR, such events are rare. Hence in this regime both T_{av} and ΔT_{av} are large. Conversely, when [IP$_3$] is large, each IPR is much more sensitive to Ca^{2+}, and the flux through each receptor is larger. Thus, firing of a single IPR is nearly always sufficient to stimulate a global wave. In this case, the interwave period is set, not by stochastic effects, but by the intrinsic dynamics of the IPR, i.e., the time required for the receptor to reactivate and be ready to propagate another wave. These results are most easily seen in Fig. 7.22, which shows experimental data from Marchant and Parker (2001) and the corresponding simulations of Falcke (2003a). The ratio of $\Delta T_{av}/T_{av}$ remains approximately constant over a wide range of T_{av}.

What is particularly interesting about these results is that in the simulations the periodic global waves occur for concentrations of IP$_3$ for which the deterministic version of the model is nonoscillatory. In other words, oscillatory waves are not necessarily the result of oscillatory kinetics. They can result from a stochastic process that, every so often, causes a cluster to fire sufficiently strongly that it initiates a global wave. If the standard deviation of the average time between firings is sufficiently small the resulting almost-periodic responses can appear to be the result of an underlying limit cycle, even when no such limit cycle exists. Such results call into serious question the relevance of deterministic approaches to Ca^{2+} waves and oscillations that are caused by small numbers of stochastic IPR. However, the implications have yet to be fully digested (Keener, 2006).

7.7 Intercellular Calcium Waves

Not only do Ca^{2+} waves travel across individual cells, they also travel from cell to cell to form intercellular waves that can travel across many cells. One of the earliest examples of such an intercellular wave was discovered by Sanderson et al. (1990, 1994), who discovered that in epithelial cell cultures, a mechanical stimulus (for example, poking a single cell with a micropipette) can initiate a wave of increased intracellular Ca^{2+} that spreads from cell to cell to form an intercellular wave. Typical experimental results from airway epithelial cells are shown in Fig. 7.4. The epithelial cell culture forms a thin layer of cells, connected by gap junctions. When a cell in the middle of the culture is mechanically stimulated, the Ca^{2+} in the stimulated cell increases quickly. After a time delay of a second or so, the neighbors of the stimulated cell also show an increase in Ca^{2+}, and this increase spreads sequentially through the culture. An intracellular wave moves across each cell, is delayed at the cell boundary, and then initiates an intracellular wave in the neighboring cell. The intercellular wave moves via the sequential propagation of intracellular waves. Of particular interest here is the fact that in the absence of extracellular Ca^{2+}, the stimulated cell shows no response, but an intercellular wave still spreads to other cells in the culture. It thus appears that a rise in Ca^{2+} in the stimulated cell is not necessary for wave propagation. Neither is a rise in Ca^{2+} sufficient to initiate an intercellular wave. For example, epithelial cells in culture sometimes exhibit spontaneous intracellular Ca^{2+} oscillations, and these oscillations do not spread from cell to cell. Nevertheless, a mechanically stimulated intercellular wave does spread through cells that are spontaneously oscillating.

Intercellular Ca^{2+} waves in glial cultures were also studied by Charles et al. (1991) as well as by Cornell-Bell et al. (1990). Over the past few years there has accumulated an increasing body of evidence to show that such intercellular communication between glia, or between glia and neurons, plays an important role in information processing in the brain (Nedergaard, 1994; Charles, 1998; Vesce et al., 1999; Fields and Stevens-Graham, 2002; Lin and Bergles, 2004).

Just as there is a wide variety of intercellular Ca^{2+} waves in different cell types, so is there a corresponding variety in their mechanism of propagation. Nevertheless, two basic mechanisms are predominant: propagation by the diffusion of an extracellular messenger, and propagation by the diffusion of an intracellular messenger through gap junctions. Sometimes both mechanisms operate in combination to drive an intercellular wave (see, for example, Young and Hession, 1997). Most commonly the intracellular messenger is IP_3 or Ca^{2+} (or both), but a much larger array of extracellular messengers, including ATP, ADP, and nitric oxide, has been implicated.

There have been few models of intercellular Ca^{2+} waves. The earliest were due to Sneyd et al. (1994, 1995a, 1998) who studied the mechanisms underlying mechanically induced waves, while another early model was that of Young (1997). More recent versions of this basic model have been used to study intercellular coupling in hepatocytes (Höfer, 1999; Höfer et al., 2001, 2002; Dupont et al., 2000) and pancreatic acinar cells

(Tsaneva-Atanasova et al., 2005), while a different approach was taken by Jung et al. (1998) and Ullah et al. (2006). There have been even fewer studies of the interactions between intracellular and extracellular messengers (two recent examples are Bennett et al., 2005, and Iacobas et al., 2006), and much remains to be discovered about how such models behave.

7.7.1 Mechanically Stimulated Intercellular Ca²⁺ Waves

Sanderson and his colleagues (Boitano et al., 1992; Sanderson et al., 1994; Sneyd et al., 1994, 1995a,b) proposed a model of mechanically stimulated intercellular Ca^{2+} waves in epithelial cells (Fig. 7.23). They proposed that mechanical stimulation causes the production of large amounts of IP_3 in the stimulated cell, and this IP_3 moves through the culture by passive diffusion, moving from cell to cell through gap junctions. Since IP_3 releases Ca^{2+} from the ER, the diffusion of IP_3 from cell to cell results in a corresponding intercellular Ca^{2+} wave. Experimental results indicate that the movement of Ca^{2+} between cells does not play a major role in wave propagation, and thus the model assumes that intercellular movement of Ca^{2+} is negligible. Relaxation of this assumption makes little difference to the model behavior, as it is the movement of IP_3 through gap junctions that determines the intercellular wave properties.

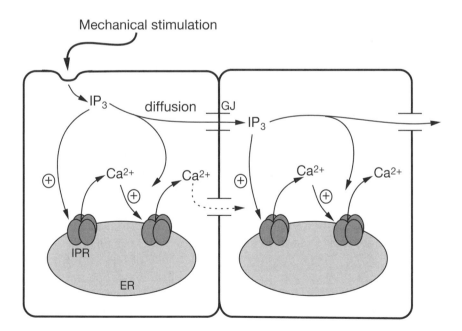

Figure 7.23 Schematic diagram of the model of intercellular Ca^{2+} waves. GJ: gap junction; ER: endoplasmic reticulum; IPR: IP_3 receptor.

In the model, the epithelial cell culture is modeled as a grid of square cells. It is assumed that IP$_3$ moves by passive diffusion and is degraded with saturable kinetics. Thus, if p denotes [IP$_3$], then

$$\frac{\partial p}{\partial t} = D_p \nabla^2 p - \frac{V_p p k_p}{k_p + p}. \tag{7.125}$$

When $p \ll k_p$, p decays with time constant $1/V_p$. Ca^{2+} is also assumed to move by passive diffusion, but it is released from the ER by IP$_3$ and pumped back into the ER by Ca^{2+} ATPases. The equations are

$$\frac{\partial c}{\partial t} = D_c \nabla^2 c + J_{\text{IPR}} - J_{\text{serca}} + J_{\text{in}}, \tag{7.126}$$

$$\tau_h \frac{dh}{dt} = \frac{k_2^2}{k_2^2 + c^2} - h, \tag{7.127}$$

$$J_{\text{IPR}} = k_f \mu(p) h \left[b + \frac{(1-b)c}{k_1 + c} \right], \tag{7.128}$$

$$J_{\text{serca}} = \frac{\gamma c^2}{k_\gamma^2 + c^2}, \tag{7.129}$$

$$J_{\text{in}} = \beta, \tag{7.130}$$

$$\mu(p) = \frac{p^3}{k_\mu^3 + p^3}. \tag{7.131}$$

This was an early model of Ca^{2+} dynamics, which is reflected in its relative simplicity. The model of the IPR was based on the model of Atri et al. (1993); J_{IPR} is a function of p, c, and a slow variable h, which denotes the proportion of IP$_3$ receptors that have not been inactivated by Ca^{2+}. Opening by IP$_3$ is assumed to be instantaneous, as is activation by Ca^{2+}; hence the term $\mu(p) \left[b + \frac{(1-b)c}{k_1+c} \right]$, that is an increasing function of both p and c. The inactivation variable, h, has a steady state which is a decreasing function of Ca^{2+}, and it reaches this steady state with time constant τ_h. Thus, the property of fast activation by Ca^{2+} followed by slower inactivation is built into the model equations.

As usual, J_{serca} denotes the removal of Ca^{2+} from the cytoplasm by Ca^{2+} ATPases in the ER membrane, and is modeled as a Hill equation with coefficient 2 (based on the data of Lytton et al., 1992); J_{in} is an unspecified leak of Ca^{2+} into the cytoplasm, either from outside the cell or from the ER. Values of the model parameters are given in Table 7.7.

There is one important difference between this model and others discussed earlier in this chapter; here there is no variable describing the Ca^{2+} concentration in the ER, i.e., no c_e variable. This is equivalent to assuming that depletion of the ER is negligible, so that c_e is constant.

Table 7.7   Parameters of the model of intercellular Ca^{2+} waves. One important point to note is the difference between D_p and D_c. The value of D_c used here is the effective diffusion coefficient of Ca^{2+}. Although IP_3 is a much larger molecule than Ca^{2+}, it diffuses faster through the cell because it is not buffered (Allbritton et al., 1992).

$k_f = 3\ \mu M\,s^{-1}$	$b = 0.11$
$k_1 = 0.7\ \mu M$	$k_2 = 0.7\ \mu M$
$\tau_h = 0.2\ s$	$k_\gamma = 0.27\ \mu M$
$\gamma = 1\ \mu M\,s^{-1}$	$\beta = 0.15\ \mu M\,s^{-1}$
$k_\mu = 0.01\ \mu M$	$V_p = 0.08\ s^{-1}$
$k_p = 1\ \mu M$	$F = 2\ \mu m\,s^{-1}$
$D_c = 20\ \mu m^2 s^{-1}$	$D_p = 300\ \mu m^2 s^{-1}$

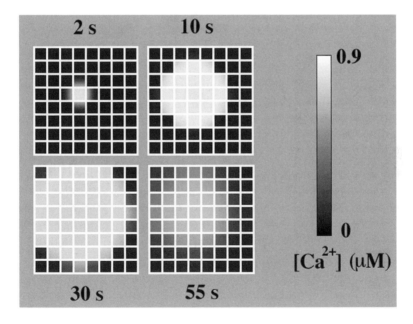

Figure 7.24   Density plot of a two-dimensional intercellular Ca^{2+} wave, computed numerically from the intercellular Ca^{2+} wave model. (Sneyd et al., 1995b, Fig. 4.)

Finally, the internal boundary conditions are given in terms of the flux of IP_3 from cell to cell. If cell n has $[IP_3] = p_n$, it is assumed that the flux of IP_3 from cell n to cell $n + 1$ is given by $F(p_n - p_{n+1})$ for some constant F, called the permeability.

Initially, a single cell was injected with IP_3, which was then allowed to diffuse from cell to cell, thereby generating an intercellular Ca^{2+} wave. Figure 7.24 shows a density plot of a numerical solution of the model equations in two dimensions. An intercellular wave can be seen expanding across the grid of cells and then retreating as the IP$_3$ degrades. As expected for a process based on passive diffusion, the intracellular wave speed (i.e., the speed at which the intercellular wave moves across an individual cell) decreases with distance from the stimulated cell, and the arrival time and the intercellular delay increase exponentially with distance from the stimulated cell. For the values chosen for F, ranging from 1 to 8 μm s^{-1}, the model agrees well with experimental data from epithelial, endothelial, and glial cells (Demer et al., 1993; Charles et al., 1992). However, the most important model prediction is the value of F needed to obtain such agreement. If F is lower than about 1 μm s^{-1}, the intercellular wave moves too slowly to agree with experimental data. Since the value of F is unknown, this prediction provides a way to test the underlying hypothesis of passive diffusion of IP$_3$.

7.7.2 Partial Regeneration

One of the major questions raised about the previous model was whether an IP$_3$ molecule is able to diffuse through multiple cells without being degraded. It is unlikely that an IP$_3$ molecule could survive long enough to cause intercellular waves propagating distances of up to several hundred micrometers. However, if the production of IP$_3$ is regenerative, for example, by the activation of PLC by increased $[Ca^{2+}]$, then the waves could potentially propagate indefinitely, as do action potentials in the axon (Chapter 6). Since in most cell types the waves eventually stop, this would seem, at first sight, to preclude a regenerative mechanism for IP$_3$ production.

This question was resolved by Höfer et al. (2002), who showed that a partially regenerative mechanism could propagate intercellular Ca^{2+} waves much farther than would be possible by simple diffusion, but not indefinitely; the waves still eventually terminate. Their model was used to study long-range intercellular Ca^{2+} waves in astrocytes cultured from rat striatum, a preparation in which it is known that the intercellular waves propagate principally through gap-junctional diffusion of an intracellular messenger.

The model equations are similar to those in the previous section, with two major exceptions. First, they include the dynamics of c_e, and second, they assume that the rate of production of IP$_3$ by one of the isoforms of PLC, PLCδ, is an increasing function of $[Ca^{2+}]$. Thus,

$$\frac{\partial p}{\partial t} = D_p \nabla^2 p + \frac{v_7 c^2}{K_{Ca}^2 + c^2} - k_9 p, \qquad (7.132)$$

where $K_{Ca} = 0.3$ μM, $k_9 = 0.08$ s^{-1}, and v_7 varies between 0 and 0.08.

Results are shown in Fig. 7.25. When the positive feedback from Ca^{2+} to IP$_3$ production is small (i.e., when the maximal rate of PLCδ activity, v_7, is small), the intercellular Ca^{2+} wave behaves as if it resulted from passive diffusion of IP$_3$ from the stimulated cell, as in the model of mechanically stimulated intercellular Ca^{2+} waves. Conversely,

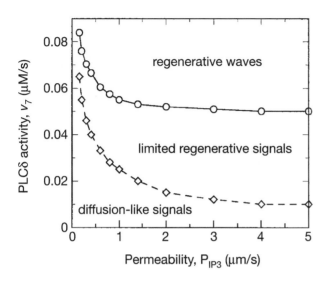

Figure 7.25 Types of intercellular waves in the partially regenerative model as a function of the gap junction permeability of IP_3 and the maximal activity of PLCδ (Höfer et al., 2002, Fig. 4). The intercellular permeability to IP_3, here called P_{IP_3}, was called F in Table 7.7.

when v_7 is large the waves become fully regenerative, propagating indefinitely. However, numerical computations show an intermediate range of v_7 values that give limited regenerative waves; waves that propagate further than they would have, had they been governed solely by passive diffusion of IP_3, but that eventually terminate and thus are not fully regenerative.

7.7.3 Coordinated Oscillations in Hepatocytes

Periodic intercellular Ca^{2+} waves propagate across entire lobules of the liver (Nathanson et al., 1995; Robb-Gaspers and Thomas, 1995), and this behavior is mirrored in smaller clusters (doublets and triplets) of coupled hepatocytes. The mechanism for the propagation of these waves appears to be quite different from that proposed for mechanically stimulated waves. First, the waves in hepatocytes propagate only if each cell is stimulated, i.e., only if each cell has increased IP_3. Second, the waves are time periodic. Despite these differences, it is known that the coordination of the waves depends on gap-junctional diffusion of an intracellular messenger; in the absence of gap-junctional coupling no coordinated wave activity appears, with each cell oscillating independently of its neighbors (Tordjmann et al., 1997). It is also known that there is an increasing gradient of hormone receptor density as one moves from the periportal to the perivenous zone of the liver lobule. This gradient is mirrored in triplets of hepatocytes; when stimulated with noradrenaline, the cells in the triplet respond in order, with the cell with the highest receptor density responding first. Since each cell in the triplet responds in synchrony to intracellular photorelease of IP_3, it follows that the different latencies are the result of different rates of production of IP_3.

Based on these observations, it has been proposed (Dupont et al., 2000) that the coordinated waves in hepatocyte clusters are phase waves, in which each cell oscillates with a phase slightly different from that of its neighbors. In that model, intercellular synchronization is provided by the diffusion of IP$_3$ through gap junctions. Since each cell produces IP$_3$ at a slightly different rate, the intercellular diffusion of IP$_3$ serves to smooth out the intercellular IP$_3$ gradients, allowing for greater synchronization of the oscillations and the appearance of phase waves. Such synchronization does not persist for a long time, since the cells are essentially uncoupled—in the model there are no oscillations in IP$_3$ driving the Ca^{2+} oscillations, and thus, in the absence of intercellular Ca^{2+} diffusion, the cells act as uncoupled oscillators over a longer time scale. Nevertheless, synchronization persists for long enough to explain experimental observations. A slightly different model, in which the cells are coupled by the intercellular diffusion of Ca^{2+}, was proposed by Höfer (1999) and also tested by Dupont and her colleagues. According to the latter group, a model based on intercellular diffusion of Ca^{2+} is insufficient to explain all the experimental data.

Similar studies have been done of intercellular waves in pancreatic acinar cells (Tsaneva-Atanasova et al., 2005), in which it was concluded that the intercellular diffusion of Ca^{2+} alone was able to synchronize intercellular phase waves over a long time period, and that the function of intercellular IP$_3$ diffusion was to minimize the intercellular IP$_3$ gradients, thus making it easier for Ca^{2+} diffusion to synchronize the oscillators.

However, these results are all still attended by considerable controversy. It is not yet clear how important IP$_3$ oscillations are for driving Ca^{2+} oscillations in either hepatocytes or pancreatic acinar cells. Much work remains to be done to elucidate the detailed mechanisms underlying these intercellular waves and coupled oscillations.

7.8 Appendix: Mean Field Equations

In Section 7.3, we added diffusion to our model equations in order to study Ca^{2+} dynamics in a spatially distributed system. However, in doing so, we made use of some very important approximation techniques, which we describe in this appendix.

7.8.1 Microdomains

Suppose that a substance u is diffusing and reacting in a three-dimensional region in which one or two of the dimensions are small compared to the diffusion length scale. (If all dimensions are small compared to a typical diffusion length scale, then it is reasonable to assume that the region is well mixed, and a whole-cell model is appropriate.) Such a region could be, for example, the region between the junctional SR and the plasma membrane in cardiac cells (the diadic cleft), a narrow region of the cytoplasm between two planar sheets of ER, or a narrow region of the cytoplasm between the ER and the mitochondria. Such microdomains are known to be very important in cardiac cells and skeletal muscle, and are believed to be important (although the details remain uncertain) in cells such as the interstitial cells of Cajal, smooth muscle, and neurons.

In this domain,

$$\frac{\partial u}{\partial t} = D\nabla^2 u + f,\qquad(7.133)$$

with boundary conditions $\mathbf{n} \cdot D\nabla u = J$ on the boundary of the domain, where $\mathbf{n}$ is the unit outward normal. For this discussion, it is most convenient to assume that (7.133) is in units of dimensionless time, so that D has units of length squared.

Now suppose, for example, that the domain is a long cylindrical tube with cross-sectional area A that is small compared to D. Because A is small compared to D, we expect that the solution should be nearly uniform in each cross-section, varying only slightly from its average value. Thus, we seek an equation describing the evolution of the average value, also called a mean field equation.

To exploit the difference in length scales, we split the Laplacian operator into two, writing

$$\nabla^2 u = u_{xx} + \nabla_y^2 u,\qquad(7.134)$$

where x is the coordinate along the length of the tube, and y represents the coordinates of the cross-section of the tube. The boundary conditions along the sides of the tube are also expressed as $\mathbf{n} \cdot D\nabla_y u = J$, using the coordinate system of the cross-section. Here we ignore the boundary conditions at the ends of the tube.

The quick and dirty derivation of the mean field equation is to define $\bar{u}$ as the average cross-sectional value of u,

$$\bar{u} = \frac{1}{A}\int_\Omega u \, dA,\qquad(7.135)$$

where Ω is the cross-sectional domain, and then to integrate the governing equation (7.133) over Ω, to find

$$\frac{\partial \bar{u}}{\partial t} = D\bar{u}_{xx} + \frac{D}{A}\int_\Omega \nabla_y^2 u \, dA + \frac{1}{A}\int_\Omega f \, dA.\qquad(7.136)$$

Applying the divergence theorem,

$$\int_\Omega \nabla_y^2 u \, dA = \int_{\partial\Omega} \mathbf{n} \cdot \nabla_y u \, dS,\qquad(7.137)$$

yields

$$\frac{\partial \bar{u}}{\partial t} = D\bar{u}_{xx} + \frac{S}{A}\bar{J} + \bar{f},\qquad(7.138)$$

where

$$\bar{J} = \frac{1}{S}\int_{\partial\Omega} J \, dS,\qquad(7.139)$$

with S the circumference of the cross-section. Equation (7.138) is the equation governing the behavior of $\bar{u}$.

While (7.138) gives the correct answer, some of us may wish for a more systematic derivation. Furthermore, if either J or f depends on u, then it is not clear how to determine $\bar{J}$ or $\bar{f}$.

To this end we introduce a small parameter ϵ where $\epsilon^2 = \frac{A}{D}$, and introduce the scaled coordinate $y = \sqrt{A}\xi$, in terms of which (7.133) becomes

$$\frac{\partial u}{\partial t} = Du_{xx} + \frac{1}{\epsilon^2}\nabla^2_\xi u + f, \tag{7.140}$$

subject to boundary conditions $\mathbf{n} \cdot \nabla_\xi u = \epsilon \frac{J}{\sqrt{D}}$. Now we seek a solution of (7.140) of the form

$$u = \bar{u} + \epsilon w, \tag{7.141}$$

where w is required to have zero average in each cross-section (i.e., $\int_{\Omega_\xi} w\, dA_\xi = 0$), and $\bar{u}$ is independent of ξ. It is important to note that $\bar{u}$ and ω are not independent of ϵ, and this is not a power series representation of the solution. Instead, this is a splitting of the solution by a projection operator, say

$$Pu = \int_{\Omega_\xi} u\, dA_\xi, \tag{7.142}$$

and in this notation $\bar{u} = Pu$ and $\epsilon w = u - Pu$.

Now we find equations for $\bar{u}$ and w by substituting (7.141) into (7.140) and then applying the projection operators P and $I - P$. First, applying P (i.e., integrating over Ω_ξ), we find

$$\frac{\partial \bar{u}}{\partial t} = D\bar{u}_{xx} + \frac{1}{\epsilon}\frac{S_\xi}{\sqrt{D}}\bar{J} + \bar{f}, \tag{7.143}$$

where $\bar{J} = \frac{1}{S_\xi}\int_{\partial\Omega_\xi} J\, dS_\xi$, and S_ξ is the circumference of the tube measured in units of ξ. Next, applying the operator $I - P$ we find

$$\frac{1}{\epsilon}\nabla^2_\xi w = \frac{1}{\epsilon}\frac{S_\xi}{\sqrt{D}}\bar{J} + \bar{f} - f + \epsilon\frac{\partial w}{\partial t} - \epsilon Dw_{xx}. \tag{7.144}$$

The function w must also satisfy the boundary condition $\mathbf{n} \cdot \nabla_\xi w = \frac{J}{\sqrt{D}}$ along the sides of the tube.

If both J and f are known functions, with no dependence on u, then we are done, because (7.143) is the exact mean field equation. No further approximation is needed.

However, if either J or f depends on u, as is typical, then it is necessary to know more about w. In particular, if w is bounded and of order one, then

$$
\begin{aligned}
\bar{J} &= \frac{1}{S_\xi} \int_{\partial\Omega_\xi} J(u)\, dS_\xi \\
&= \frac{1}{S_\xi} \int_{\partial\Omega_\xi} J(\bar{u} + \epsilon w)\, dS_\xi \\
&= \frac{1}{S_\xi} \int_{\partial\Omega_\xi} \left(J(\bar{u}) + \epsilon J_u(\bar{u})w + O(\epsilon^2) \right) dS_\xi \\
&= J(\bar{u}) + O(\epsilon^2),
\end{aligned}
\tag{7.145}
$$

and similarly, $\bar{f} = f(\bar{u}) + O(\epsilon^2)$. Thus, the mean field equation is

$$
\frac{\partial \bar{u}}{\partial t} = D\bar{u}_{xx} + \frac{1}{\epsilon}\frac{S_\xi}{\sqrt{D}}J(\bar{u}) + f(\bar{u}) + O(\epsilon),
\tag{7.146}
$$

or, in terms of the original parameters,

$$
\frac{\partial \bar{u}}{\partial t} = D\bar{u}_{xx} + \frac{S}{A}J(\bar{u}) + f(\bar{u}) + O\left(\sqrt{\frac{A}{D}}\right).
\tag{7.147}
$$

It remains to establish that w is bounded and of order one. To this end, we seek a solution w of (7.144) as a power series in ϵ, with $\bar{u}$ fixed (that is, ignoring the implicit ϵ dependence of $\bar{u}$),

$$
w = w_1 + \epsilon w_2 + O(\epsilon^2).
\tag{7.148}
$$

It is immediate that w_1 must satisfy

$$
\nabla_\xi^2 w_1 = \frac{S_\xi}{\sqrt{D}}\bar{J},
\tag{7.149}
$$

subject to the boundary condition $n \cdot \nabla_\xi w_1 = \frac{J(\bar{u})}{\sqrt{D}}$. Now we let $W(\xi)$ be the fundamental solution of the boundary value problem

$$
\nabla_\xi^2 W = S_\xi \qquad \text{on } \Omega,
\tag{7.150}
$$

subject to the boundary condition $n \cdot \nabla_\xi W = 1$ on $\partial\Omega$, and the condition $\int_\Omega W(\xi)\, dA_\xi = 0$ (which, from the standard theory of Poisson's equation, is known to exist). It follows that

$$
w_1 = \frac{J(\bar{u})}{\sqrt{D}}W(\xi).
\tag{7.151}
$$

This establishes, at least to our satisfaction, that w is a well-behaved, bounded function of ϵ, and we are done.

Suppose that instead of a long thin cylinder, the region of interest lies between two flat two-dimensional membranes separated by the distance L, where $L^2 \ll D$. Suppose also that $Du_z = -J_0$ and $Du_z = J_1$ on the lower and upper membranes, respectively, where z represents the vertical spatial coordinate. In this case, one can show using the same methodology (Exercise 19) that the mean field equation is

$$\frac{\partial \bar{u}}{\partial t} = D\nabla^2 \bar{u} + \frac{1}{L}(J_1(\bar{u}) + J_0(\bar{u})) + f(\bar{u}) + O\left(\frac{L}{\sqrt{D}}\right), \tag{7.152}$$

where $\bar{u} = \frac{1}{L}\int_0^L u\, dz$, and ∇^2 represents the two-dimensional Laplacian operator.

7.8.2 Homogenization; Effective Diffusion Coefficients

As noted in Section 7.3, to take the details of the microstructure of the cytoplasmic and ER boundaries into account is both impractical and of little use. Instead, there is a need to find a mean field description of the Ca^{2+} concentrations that uses an effective diffusion coefficient. This need to avoid the details of the microstructure is evident in many other contexts. For example, as is described in Chapter 12, there is a need for equations of action potential propagation in cardiac tissue that do not rely on the details of cellular structure and its interconnectedness.

Homogenization is the very powerful technique by which this is accomplished. In this section we show how homogenization is used to find averaged, or mean field, equations with an effective diffusion coefficient. This same technique is invoked in Chapter 12 to find effective conductances for cardiac tissue. Finally, this technique allows for the derivation of the bidomain equations for Ca^{2+}, and the bidomain equations for the cardiac action potential.

As a warmup problem, suppose a substance is reacting and diffusing along a one-dimensional region, and that the diffusion coefficient is rapidly varying in space. To be specific, suppose u is governed by the reaction–diffusion equation

$$\frac{\partial u}{\partial t} = \frac{\partial}{\partial x}\left(D\left(\frac{x}{\epsilon}\right)\frac{\partial u}{\partial x}\right) + f(u). \tag{7.153}$$

Here x is dimensionless, $D(x)$ is a periodic function of period one and of order one, and ϵ is small. We expect that u should have some average or mean field behavior with a characteristic length scale of order one, with small variations from this mean field that are of order ϵ.

To explore this possibility, we introduce two variables,

$$z = x, \qquad \xi = \frac{x}{\epsilon}, \tag{7.154}$$

which we treat as independent variables. From the chain rule,

$$\frac{\partial}{\partial x} = \frac{\partial}{\partial z} + \frac{1}{\epsilon}\frac{\partial}{\partial \xi}, \tag{7.155}$$

and the original partial differential equation (7.153) becomes

$$\frac{\partial u}{\partial t} = \frac{\partial}{\partial z}\left(D(\xi)\left(\frac{\partial u}{\partial z} + \frac{1}{\epsilon}\frac{\partial u}{\partial \xi}\right)\right) + \frac{1}{\epsilon}\frac{\partial}{\partial \xi}\left(D(\xi)\left(\frac{\partial u}{\partial z} + \frac{\partial u}{\partial \xi}\right)\right) + f(u). \tag{7.156}$$

While this equation is clearly more intricate than (7.134), its structure is essentially the same. Thus the calculation that follows is the same as that of the previous section. That is, we seek a solution of (7.156) of the form

$$u = \bar{u} + \epsilon w, \tag{7.157}$$

where $\bar{u}$ is independent of ξ, and w is a periodic function of ξ with zero mean value, $\int_0^1 w \, d\xi = 0$. However, rather than applying the projection operators and then finding power series solutions of these, it is slightly more convenient (and in this case equivalent) to seek power series solutions of (7.156) directly. That is, we set $\bar{u} = u_0 + \epsilon u_1 + O(\epsilon^2)$, and $w = w_1 + \epsilon w_2 + O(\epsilon^2)$, substitute these into (7.156), collect terms of like powers of ϵ, resulting in a hierarchy of equations to be solved,

$$\frac{\partial}{\partial \xi} \left(D(\xi) \left(\frac{\partial w_1}{\partial \xi} + \frac{\partial u_0}{\partial z} \right) \right) = 0, \tag{7.158}$$

and

$$\frac{\partial}{\partial \xi} \left(D(\xi) \frac{\partial w_2}{\partial \xi} \right) = \frac{\partial u_0}{\partial t} - \frac{\partial}{\partial z} \left(D(\xi) \left(\frac{\partial u_0}{\partial z} + \frac{\partial w_1}{\partial \xi} \right) \right)$$
$$- \frac{\partial}{\partial \xi} \left(D(\xi) \left(\frac{\partial w_1}{\partial z} + \frac{\partial u_1}{\partial z} \right) \right)$$
$$- f(u_0). \tag{7.159}$$

The first of these is readily solved by direct integration. We define $W(\xi)$ to be the periodic function with zero mean that satisfies the differential equation

$$\frac{dW}{d\xi} = \frac{R}{\bar{R}} - 1, \tag{7.160}$$

where $R = \frac{1}{D}$ and $\bar{R} = \int_0^1 R \, d\xi$. Then

$$w_1 = -W(\xi) \frac{\partial u_0}{\partial z}. \tag{7.161}$$

The function $W(\xi)$ determines the small-scale structure of the solution.

Next, we examine (7.159) and observe that w_2 can be periodic only if the right-hand side of this equation has zero average with respect to ξ. Thus, it must be that

$$\frac{\partial u_0}{\partial t} = \int_0^1 \frac{\partial}{\partial z} \left(D(\xi) \left(\frac{\partial u_0}{\partial z} + \frac{\partial w_1}{\partial \xi} \right) \right) d\xi + f(u_0)$$
$$= D_{\text{eff}} \frac{\partial^2 u_0}{\partial z^2} + f(u_0), \tag{7.162}$$

where

$$D_{\text{eff}} = \frac{1}{\bar{R}}. \tag{7.163}$$

Equation (7.162) is the mean field equation we seek, and D_{eff} is the effective diffusion coefficient.

A

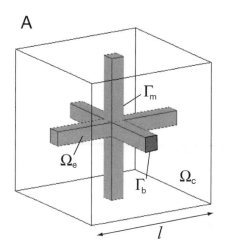

B

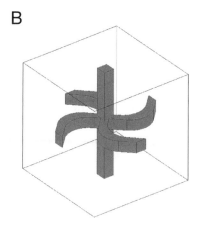

Figure 7.26 Two possible ER microscale structures. The more complicated structure in B results in a nonisotropic effective diffusion tensor.

Homogenization for the full three-dimensional problem follows the same menu. We suppose that a substance is reacting and diffusing in a subregion of three-dimensional space that is subdivided into small periodic subunits Ω, called the microstructure, which are each contained in a small rectangular box. The region Ω is further subdivided into a cytoplasmic region, Ω_c, and an ER region, Ω_e, as shown in Fig. 7.26A. Γ_m is the membrane boundary between the ER and the cytoplasm, and Γ_b is the intersection of the ER with the walls of the box. Because we assume that the microdomain is repeated periodically, the ER of one box connects, through Γ_b, to the ER of the neighboring boxes.

We assume that the rectangular box is of length l (although the box need not be a cube), where l is much less than the natural length scale of the problem, in this case $\Lambda = \sqrt{D}$, the diffusion length. Thus, there are two natural length scales, l and Λ, and so $\epsilon = l/\Lambda \ll 1$ is a natural small parameter. Since l is small compared to the diffusion length scale, it is reasonable to assume that the solution is nearly homogeneous in each microstructural unit, and that the microstructure causes only small perturbations to the background solution.

Now we suppose that the substance u is diffusing and reacting according to (7.133) in Ω_c, and that the flux across Γ_m is $\mathbf{n} \cdot D\nabla u = J$. Another variable, v say, will be reacting and diffusing inside the region Ω_e, but it is unnecessary to carry the analysis through for both variables. Again, it is convenient to assume that these equations are in units of dimensionless time, so that D has units of length^2. We further assume that J is proportional to l. Notice that the total flux across Γ_m is proportional to the surface area, so that the total flux per unit volume is proportional to J/l. In order that this remain bounded in the limit that $\epsilon \to 0$, we assume that $J = lj$.

The standard procedure is to introduce two variables

$$z = \frac{x}{\Lambda}, \qquad \xi = \frac{z}{\epsilon}, \tag{7.164}$$

so that ξ denotes the space variable on the microscale, and z denotes the space variable on the original, or long, scale. Treating z and ξ as independent variables, we write

$$\nabla_x = \frac{1}{\Lambda} \left(\frac{1}{\epsilon} \nabla_\xi + \nabla_z \right), \tag{7.165}$$

where the subscript denotes the space variable with which the gradient derivatives are taken. In these variables, (7.133) becomes

$$\frac{\partial u}{\partial t} = \nabla_z^2 u + \frac{2}{\epsilon} \nabla_\xi \nabla_z u + \frac{1}{\epsilon^2} \nabla_\xi^2 u + f, \tag{7.166}$$

subject to the boundary condition

$$\mathbf{n} \cdot \left(\nabla_z u + \frac{1}{\epsilon} \nabla_\xi u \right) = \epsilon j \qquad \text{on } \Gamma_m. \tag{7.167}$$

Notice that $\mathbf{n}$ is a dimensionless quantity, containing only directional information about the normal to the boundary and so does not change with changes in space scale.

Once again we have a partial differential equation with the same structure as (7.134). Thus, as before, we look for a solution of (7.166) in the form

$$u = \bar{u}(z, t, \epsilon) + \epsilon w(\xi, z, t, \epsilon), \tag{7.168}$$

where $\bar{u}$ is the background, or mean field, solution, varying only on the large space scale and independent of ξ, while ϵw is a correction term periodic in ξ and having zero average value in the variable ξ. Next, we seek power series representations for $\bar{u}$ and w of the form

$$\bar{u} = u_0 + \epsilon u_1 + O(\epsilon^2), \qquad w = w_1 + \epsilon w_2 + O(\epsilon^2). \tag{7.169}$$

Substituting these into (7.166) and (7.167), and equating coefficients of powers of ϵ, gives a hierarchy of equations to be solved sequentially. The leading-order equation is

$$\nabla_\xi^2 w_1 = 0 \qquad \text{in } \Omega_c, \tag{7.170}$$

subject to boundary conditions

$$\mathbf{n} \cdot (\nabla_z u_0 + \nabla_\xi w_1) = 0 \qquad \text{on } \Gamma_m, \tag{7.171}$$

and the next equation in the hierarchy is

$$\frac{\partial u_0}{\partial t} = \nabla_z^2 u_0 + 2 \nabla_\xi \nabla_z w_1 + \nabla_\xi^2 w_2 + f, \qquad \text{in } \Omega_c, \tag{7.172}$$

with boundary conditions

$$\mathbf{n} \cdot (\nabla_z w_1 + \nabla_\xi w_2) = j \qquad \text{on } \Gamma_m. \tag{7.173}$$

The first of these equations can be solved in the way derived by Bensoussan et al. (1978) by letting

$$w_1 = W(\xi) \cdot \nabla_z u_0, \tag{7.174}$$

where $W(\xi)$ is a fundamental solution satisfying the vector differential equation

$$\nabla_\xi^2 W = 0 \qquad \text{in } \Omega_c, \tag{7.175}$$

$$\mathbf{n} \cdot (\nabla_\xi W + I) = 0 \qquad \text{on } \Gamma_m. \tag{7.176}$$

In addition, to specify $W(\xi)$ uniquely, we require that

$$\int_{\Omega_c} W(\xi) \, dV_\xi = 0. \tag{7.177}$$

The mean-field equation for u_0 can be found by integrating (7.172) over Ω_c, which gives

$$
\begin{aligned}
V_c \frac{\partial u_0}{\partial t} &= \int_{\Omega_c} \nabla_z \cdot (\nabla_z u_0 + \nabla_\xi w_1) \, dV_\xi + \int_{\Omega_c} \nabla_\xi \cdot (\nabla_z w_1 + \nabla_\xi w_2) \, dV_\xi + \int_{\Omega_c} f \, dV_\xi \\
&= \int_{\Omega_c} \nabla_z \cdot (\nabla_z u_0 + \nabla_\xi W \nabla_z u_0) \, dV_\xi + \int_{\partial\Omega_c} \mathbf{n} \cdot (\nabla_\xi w_2 + \nabla_z w_1) \, dS_\xi + \int_{\Omega_c} f \, dV_\xi \\
&= \nabla_z \cdot \left[\int_{\Omega_c} (I + \nabla_\xi W) \right] \nabla_z u_0 \, dV_\xi + \int_{\Gamma_m} j(u_0) \, dS_\xi + V_{\Omega_c} f(u_0).
\end{aligned} \tag{7.178}
$$

Note that the integral over $\partial\Omega_c$ reduces to the integral over Γ_m because we require w_1 and w_2 to be periodic in ξ.

Equation (7.178) is a consistency condition that is required in order that w_2 exist. In other words, (7.172) is a differential equation for w_2 that, if w_2 is to be periodic, can be solved only if a solvability or consistency condition holds (this is an application of the Fredholm alternative). The required consistency condition is found by integrating (7.172) over Ω_c to give (7.178).

It thus follows that

$$\frac{\partial u_0}{\partial t} = \nabla_z \cdot D_{\text{eff}} \nabla_z u_0 + \frac{S_c}{V_c} j(u_0) + f(u_0), \tag{7.179}$$

where V_c and S_c denote the (dimensionless) volume and surface area of Ω_c, respectively, and D_{eff}, the (dimensionless) effective diffusion tensor, is given by

$$D_{\text{eff}} = \frac{1}{V_c} \int_{\Omega_c} (I + \nabla_\xi W) \, dV_\xi. \tag{7.180}$$

Hence, for any given periodic geometry, the effective diffusion tensor (7.180) can be calculated by solving (7.175) and (7.176). Goel et al. (2006) calculated effective diffusion coefficients for a variety of possible ER microstructures. For the regular structure of Fig. 7.26A, the effective diffusion coefficient of Ca^{2+} in the cytoplasm decreases by 60% as the volume fraction of the ER increases from 0 to 0.9. The more complicated structure of Fig. 7.26B gives nonisotropic diffusion, with the diffusion coefficient in the vertical direction about 50% larger than the diffusion coefficient in the other two directions.

7.8.3 Bidomain Equations

The homogenization technique described here can be applied to any spatial domain with a periodic microstructure. Thus, for example, it can be applied to intertwined intracellular and extracellular spaces, or to intertwined cytoplasmic and ER spaces. In each case, one finds descriptions of the mean fields that are defined everywhere in space, so that the two spaces are effectively comingled, hence viewed as a bidomain. The only (slight) complication is that fluxes across common boundaries must be equal in amplitude and opposite in direction. Thus, a flux into the intracellular space is a flux out of the extracellular space, or a flux into the cytoplasm is also a flux out of the ER. Furthermore, even though the domains share a common interface, their volumes are complementary so that the surface-to-volume ratios for the two spaces differ.

7.9 EXERCISES

1. Show that in a general model of intracellular Ca^{2+} dynamics in a well-mixed cell, the resting cytoplasmic Ca^{2+} concentration is independent of Ca^{2+} exchange rates with the internal pools.

2. Murray (2002) discusses a simple model of CICR that has been used by a number of modelers (Cheer et al., 1987; Lane et al., 1987). In the model, Ca^{2+} release from the ER is an increasing sigmoidal function of Ca^{2+}, and Ca^{2+} is removed from the cytoplasm with linear kinetics. Thus,

$$\frac{dc}{dt} = L + \frac{k_1 c^2}{K + c^2} - k_2 c,$$

   where L is a constant leak of Ca^{2+} from the ER into the cytoplasm.

   (a) Nondimensionalize this equation. How many nondimensional parameters are there?

   (b) Show that when $L = 0$ and $k_1 > 2k_2$, there are two positive steady states and determine their stability. For the remainder of this problem assume that $k_1 > 2k_2$.

   (c) How does the nullcline $dc/dt = 0$ vary as the leak from the internal store increases? Show that there is a critical value of L, say L_c, such that when $L > L_c$, only one positive solution exists.

   (d) Fix $L < L_c$ and suppose the solution is initially at the lowest steady state. How does c behave when small perturbations are applied to c? How does c behave when large perturbations are applied? How does c behave when L is raised above L_c and then decreased back to zero? Plot the bifurcation diagram in the L, c phase plane, indicating the stability or instability of the branches. Is there hysteresis in this model?

3. One of the earliest models of Ca^{2+} oscillations was the two-pool model of Goldbeter, Dupont, and Berridge (1990). They assumed that IP_3 causes an influx, r, of Ca^{2+} into the cell and that this influx causes additional release of Ca^{2+} from the ER via an IP_3-independent

Table 7.8 Typical parameter values for the two-pool model of Ca^{2+} oscillations (Goldbeter et al., 1990).

$k = 10\ s^{-1}$	$K_1 = 1\ \mu M$
$K_2 = 2\ \mu M$	$K_3 = 0.9\ \mu M$
$V_1 = 65\ \mu M\,s^{-1}$	$V_2 = 500\ \mu M\,s^{-1}$
$k_f = 1\ s^{-1}$	$m = 2$
$n = 2$	$p = 4$

mechanism. Thus,

$$\frac{dc}{dt} = r - kc - f(c, c_e), \tag{7.181}$$

$$\frac{dc_e}{dt} = f(c, c_e), \tag{7.182}$$

$$f(c, c_e) = J_{\text{uptake}} - J_{\text{release}} - k_f c_e, \tag{7.183}$$

where

$$J_{\text{uptake}} = \frac{V_1 c^n}{K_1^n + c^n}, \tag{7.184}$$

$$J_{\text{release}} = \left(\frac{V_2 c_e^m}{K_2^m + c_e^m} \right) \left(\frac{c^p}{K_3^p + c^p} \right). \tag{7.185}$$

Here, $k_f c_e$ is a leak from the ER into the cytoplasm. Typical parameter values are given in Table 7.8. (All the concentrations in this model are with respect to the total cell volume, and thus there is no need for the correction factor γ (cf. Equation 7.4) to take into account the difference in the ER and cytoplasmic volumes.)

(a) Nondimensionalize these equations. How many nondimensional parameters are there?

(b) Show that in a closed cell (i.e., one without any interaction with the extracellular environment) the two-pool model cannot exhibit oscillations.

(c) How does the steady-state solution depend on influx?

(d) Use a bifurcation tracking program such as AUTO to plot the bifurcation diagram of this model, using r as the bifurcation parameter. Find the Hopf bifurcation points and locate the branch of stable limit cycle solutions. Plot some typical limit cycle solutions for different values of r.

4. Complete the details of the reduction of the IP_3 receptor model (Section 7.2.5).

5. Write down the equations for the reduced IP_3 receptor model (Section 7.2.5) when $k_4 = k_2$ and $k_{-4} = k_{-2}$. Let $h = 1 - y$. What is the differential equation for h? Write it in the form

$$\tau_h \frac{dh}{dt} = h_\infty - h. \tag{7.186}$$

Derive this simplified model directly from the state diagram in Fig. 7.8.

6. Write down a reaction scheme like that of Fig. 7.8, but assuming that two Ca^{2+} ions inactivate the receptor in a cooperative fashion. Assume a simple model of cooperativity,

$$S_{ij0} \underset{k_{-2}}{\overset{k_2 c^2}{\rightleftarrows}} S_{ij1}, \qquad (7.187)$$

   for $i, j = 0$ or 1, and assume that the group I and group II states are each in quasi-steady state. Derive the model equations.

7. Starting with a basic scheme like that of Fig. 7.5 one can construct a large number of variants, each of which has a similar steady-state flux. The purpose of this exercise is to investigate two simple variants of the basic model, and show that the flux is similar for each. To distinguish between these models on the basis of experimental data is, in general, quite difficult.

   (a) First, calculate the expressions for α_1 to α_4 in (7.7) and for β_1 to β_4 in (7.7). (Use of a symbolic manipulation program such as Maple is, as usual, recommended.)

   (b) Extend the ATPase model of Section 7.2.4 by assuming that the two Ca^{2+} ions bind sequentially, not simultaneously, and calculate the flux.

   (c) Modify the model still further by assuming that the binding of the second Ca^{2+} ion is much faster than binding of the first, and calculate the flux.

   (d) Compare these three different expressions for the steady-state flux. Are experimental data of the steady-state flux sufficient to distinguish between these models?

8. Does the binding diagram in Fig. 7.8 satisfy the principle of detailed balance (Section 1.3)?

9. In 2006 Domijan et al. constructed a model of Ca^{2+} oscillations that uses a simple model of the IPR (based on an earlier model of Atri et al., 1993):

$$\frac{dc}{dt} = J_{\text{release}} - J_{\text{serca}} + \delta(J_{\text{in}} - J_{\text{pm}}), \qquad (7.188)$$

$$\frac{dc_e}{dt} = \gamma(J_{\text{serca}} - J_{\text{release}}) \qquad (7.189)$$

$$\tau_n \frac{dn}{dt} = \frac{k_2^2}{k_2^2 + c^2} - n, \qquad (7.190)$$

$$\frac{dp}{dt} = \nu\left(\frac{c + (1 - \alpha)k_4}{c + k_4}\right) - \beta p, \qquad (7.191)$$

where

$$J_{\text{release}} = \left[k_{\text{flux}} \left(\mu_0 + \frac{\mu_1 p}{k_\mu + p} \right) n \left(b + \frac{V_1 c}{k_1 + c} \right) \right] (c_e - c), \qquad (7.192)$$

$$J_{\text{serca}} = \frac{V_e c}{k_e + c}, \qquad (7.193)$$

$$J_{\text{pm}} = \frac{V_p c^2}{k_p^2 + c^2}, \qquad (7.194)$$

$$J_{\text{in}} = \alpha_1 + \alpha_2 \frac{\nu}{\beta}. \qquad (7.195)$$

Table 7.9 Parameters of the model of Ca^{2+} oscillations in Exercise 9.

δ	$= 0.01$	k_1	$= 1.1\ \mu M$
γ	$= 5.405$	k_2	$= 0.7\ \mu M\,s^{-1}$
k_{flux}	$= 6.0\ \mu M\,s^{-1}$	k_μ	$= 4\ \mu M$
V_p	$= 24\ \mu M\,s^{-1}$	μ_0	$= 0.567$
k_p	$= 0.4\ \mu M$	μ_1	$= 0.433$
V_e	$= 20\ \mu M\,s^{-1}$	b	$= 0.111$
k_e	$= 0.06\ \mu M\,s^{-1}$	V_1	$= 0.889$
α_1	$= 1\ \mu M\,s^{-1}$	α_2	$= 0.2\ s^{-1}$
β	$= 0.08\ s^{-1}$	k_4	$= 1.1\ \mu M$

Typical parameter values are given in Table 7.9.

(a) Explain the physiological basis of each term in the model.

(b) What does the parameter ν represent?

(c) The model takes quite different forms when $\alpha = 0$ or $\alpha = 1$. What are the physiological assumptions behind each of these forms?

(d) Using AUTO, plot the bifurcation diagram of the model using ν as the bifurcation parameter. How are the bifurcation diagrams different for $\alpha = 0$ and $\alpha = 1$? (Do not plot every detail of the bifurcation diagrams, but try to find the main branches.)

(e) How would you model a pulse of IP$_3$ applied by the experimentalist?

(f) How does the model respond to an external pulse of IP$_3$? What is the difference between the responses when $\alpha = 0$ and $\alpha = 1$?

10. Suppose that a model of Ca^{2+} oscillations has two variables, c and n. Suppose further that the bifurcation diagram of this model has the structure shown in Fig. 7.27, as a function of some parameter, μ. (This basic structure is common to many models of Ca^{2+} dynamics.) Sketch the phase planes for μ below, at, and above each bifurcation point.

11. (a) Plot the nullclines of the model of CICR in bullfrog sympathetic neurons (Section 7.2.9) and show that they are both N-shaped curves.

(b) Rewrite the model in terms of c and $c_t = c + \frac{c_e}{\gamma}$ and draw the phase portrait and nullclines of the model in these new variables.

(c) Plot the bifurcation diagram of the model of CICR in bullfrog sympathetic neurons (Section 7.2.9) using the external Ca^{2+} concentration c_o as the bifurcation parameter. Verify the behavior shown in Fig. 7.14, that the period but not the amplitude of the oscillations is sensitive to c_o.

12. Generalize (7.76) to the case of multiple buffers, both mobile and immobile.

13. Prove that (7.87) has a traveling wave solution that is a transition between W_1 and W_3 if and only if

$$\int_{W_1}^{W_3} f(\phi(w))\,dw > 0. \tag{7.196}$$

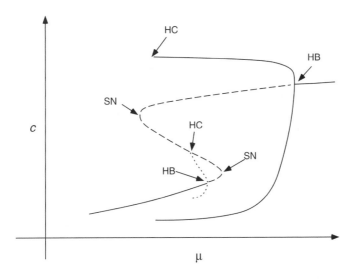

Figure 7.27 Typical bifurcation diagram of the model of Ca^{2+} oscillations in Exercise 10 (not drawn to scale). HB denotes a Hopf bifurcation, SN denotes a saddle-node bifurcation, HC denotes a homoclinic bifurcation.

14. By looking for solutions to (7.70) and (7.71) of the form $c = A \exp(\xi/\lambda)$, $b = B \exp(\xi/\lambda)$, where λ is the space constant of the wave front and $\xi = x + st$, show that the speed of the wave s is related to the space constant of the wave front by

$$s = \frac{D_c}{\lambda} + \lambda \left[f'(0) - \frac{k_+ b_t(\lambda s - D_b)}{\lambda^2 k_- + \lambda s - D_b} \right]. \tag{7.197}$$

What is the equation in the limit $k_+, k_- \to \infty$, with $k_-/k_+ = K$? Hence show that for the generalized bistable equation,

$$\lambda s < D_{\text{eff}}. \tag{7.198}$$

15. Repeat the analysis of Section 7.4.1, without assuming that b_t is constant. Under what conditions may we assume that b_t is constant?

16. Develop a formal asymptotic expansion (following the method used in the appendix) to homogenize the equation

$$\frac{\partial c}{\partial t} = D_c \frac{\partial^2 c}{\partial x^2} - kc + g\left(\frac{x}{L}\right) f(c), \tag{7.199}$$

where $g(y)$ is periodic with period 1, in the limit that $\frac{L^2 k}{D_c} \ll 1$. Show that the averaged equation is

$$\frac{\partial c}{\partial t} = D_c \frac{\partial^2 c}{\partial x^2} - kc + Gf(c), \tag{7.200}$$

where $G = \frac{1}{L} \int_0^L g(x)\, dx$.

17. (a) Find conditions on A guaranteeing that there are traveling waves of Ca^{2+} release in the spatially homogenized equation (7.97) with the cubic release function (7.94).

   (b) Find conditions on A guaranteeing that there are traveling waves of Ca^{2+} release in the spatially homogenized equation (7.97) with the piecewise-linear release function (7.95).

18. Find the solution of (7.150) for a circular domain of radius 1. In other words, solve the boundary value problem

$$\nabla^2 W = 2, \tag{7.201}$$

on a circle of radius 1, subject to the boundary condition $n \cdot \nabla W = 1$, and requiring W to have zero average value. (Why is 2 the correct right-hand side for this partial differential equation?)

19. Suppose that a substance u reacts and diffuses in a region between two flat two-dimensional membranes separated by the distance L, where $L^2 \ll D$ (where D is in units of length^2). Suppose also that $Du_z = -J_0$ and $Du_z = J_1$ on the lower and upper membranes, respectively, where z represents the vertical spatial coordinate. Derive the mean field equation for the evolution of u (7.152), and find the leading-order correction to the mean field.

CHAPTER 8

Intercellular Communication

For multicellular organisms to form and operate, cellular behavior must be vastly more complex than what is seen at the single-cell level. Not only must cells regulate their own growth and behavior, they must also communicate and interact with their neighbors to ensure the correct behavior of the entire organism. Intercellular communication occurs in a variety of ways, ranging from hormonal communication on the level of the entire body to localized interactions between individual cells. The discussion in this chapter is limited to cellular communication processes that occur between cells or over a region of a small number of cells. Other forms of communication and control, such as hormone feedback systems, are described in other chapters.

There are two primary ways that cells communicate with neighbors. Many cells (muscle and cardiac cells for example) are connected to their immediate neighbors by gap junctions in the cell membrane that form a relatively nonselective, low-resistance pore through which electrical current or chemical species can flow. Hence, a gap junction is also called an *electrical synapse*. The second means of communication is through a *chemical synapse*, in which the message is mediated by the release of a chemical from one cell and detected by receptors on its neighbor. Electrically active cells such as neurons typically communicate via chemical synapses, which are thus a crucial feature of the nervous system. Chemical synapses are considerably more complex than electrical synapses, and as a result, most of this chapter is devoted to them.

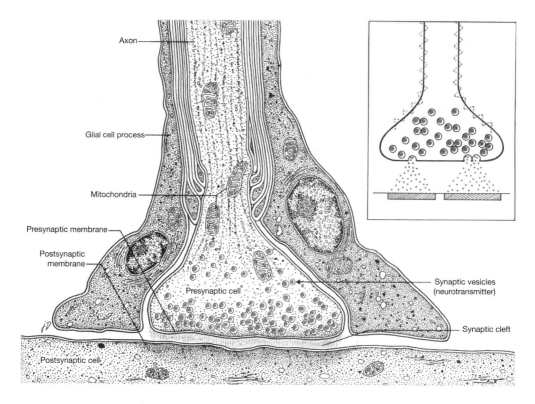

Figure 8.1 Schematic diagram of a chemical synapse. The inset shows vesicles of neurotransmitter being released into the synaptic cleft. (Davis et al., 1985, Fig. 8-11, p. 135.)

8.1 Chemical Synapses

At a chemical synapse (Fig. 8.1) the nerve axon and the postsynaptic cell are in close apposition, being separated by the *synaptic cleft*, which is about 500 angstroms wide. When an action potential reaches the presynaptic nerve terminal, it opens voltage-gated Ca^{2+} channels, leading to an influx of Ca^{2+} into the nerve terminal. Increased $[Ca^{2+}]$ causes the release of a neurotransmitter, which diffuses across the synaptic cleft, binds to receptors on the postsynaptic cell, and initiates changes in its membrane potential. The neurotransmitter is then removed from the synaptic cleft by diffusion and hydrolysis.

There are over 40 different types of synaptic transmitters, with differing effects on the postsynaptic membrane. For example, acetylcholine (ACh) binds to ACh receptors, which in skeletal muscle act as cation channels. Thus, when they open, the flow of the cation causes a change in the membrane potential, either depolarizing or hyperpolarizing the membrane. If the channel is a Na^+ channel, then the flow is inward and depolarizing, whereas if the channel is a K^+ channel, then the flow is outward and hyperpolarizing. Other receptors, such as those for γ-aminobutyric acid (GABA), open anion channels (mainly chloride), thus hyperpolarizing the postsynaptic membrane, rendering it less excitable. Synapses are classified as excitatory or inhibitory according to whether they depolarize or hyperpolarize the postsynaptic membrane. Other important neurotransmitters include adrenaline (often called epinephrine), noradrenaline (often called norepinephrine), dopamine, glycine, glutamate, and serotonin. Of these, dopamine, glycine, and GABA are usually inhibitory, while glutamate and ACh are usually, but not always, excitatory.

Many of these neurotransmitters work by gating an ion channel directly, as described for ACh above. Such channels are called *agonist-controlled* or *ligand-gated* ion channels, and a simple model of an agonist-controlled channel is described in Section 3.5.4. However, not all neurotransmitters work only in this way. For example, receptors for glutamate, the principal excitatory neurotransmitter in the brain, come in two types—*ionotropic* receptors, which are agonist-controlled, and *metabotropic* receptors, which are linked to G-proteins, and act via the production of the second messenger inositol trisphosphate and the subsequent release of Ca^{2+} from the endoplasmic reticulum (as described in Chapter 7).

The loss of specific neurotransmitter function corresponds to certain diseases. For example, Huntington's disease, a hereditary disease characterized by flicking movements at individual joints progressing to severe distortional movements of the entire body, is associated with the loss of certain GABA-secreting neurons in the brain. The resulting loss of inhibition is believed to allow spontaneous outbursts of neural activity leading to distorted movements.

Similarly, Parkinson's disease results from widespread destruction of dopamine-secreting neurons in the basal ganglia. The disease is associated with rigidity of much of the musculature of the body, involuntary tremor of involved areas, and a serious difficulty in initiating movement. Although the causes of these abnormal motor effects are uncertain, the loss of dopamine inhibition could lead to overexcitation of many muscles, hence rigidity, or to lack of inhibitory control of feedback circuits with high feedback gain, leading to oscillations, i.e., muscular tremor.

8.1.1 Quantal Nature of Synaptic Transmission

Chemical synapses are typically small and inaccessible, crowded together in very large numbers in the brain. However, neurons also make synapses with skeletal muscle cells, and these are usually much easier to isolate and study. For this reason, a great deal of the early experimental and theoretical work on synaptic transmission was performed on the neuromuscular junction, where the axon of a motorneuron forms a chemical synapse with a skeletal muscle fiber. The response of the muscle cell to a neuronal stimulus is called an *end-plate potential*, or epp.

In 1952 Fatt and Katz discovered that when the concentration of Ca^{2+} in the synaptic cleft was very low, an action potential stimulated only a small end-plate potential (Fig. 8.2). Further, these miniature end-plate potentials appeared to consist of multiples of an underlying minimum epp of the same amplitude as an epp arising spontaneously,

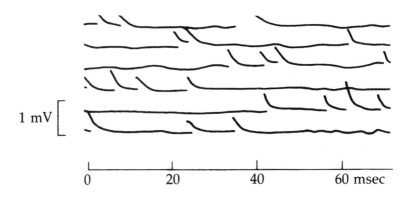

1 mV

0   20   40   60 msec

Figure 8.2 Miniature end-plate potentials (epp's) in the frog neuromuscular junction. Each epp has an amplitude of around 1 mV and results from the independent release of a single quantum of ACh. (Kuffler et al., 1984, p. 251, reproducing a figure of Fatt and Katz, 1952.)

i.e., due to random activity other than an action potential. Their findings suggested that an epp is made up of a large number of identical building blocks each of which is of small amplitude.

It is now known that quantal synaptic transmission results from the packaging of ACh into discrete vesicles. Each nerve terminal contains a large number of synaptic vesicles that contain ACh. Upon stimulation, these vesicles fuse with the cell membrane, releasing their contents into the synaptic cleft. Even in the absence of stimulation, background random activity can cause vesicles to fuse with the cell membrane and release their contents. The epp seen in spontaneous activity results from the release of the contents of a single vesicle, while the miniature epp's result from the fusion of a small integer number of vesicles, and thus appear in multiples of the spontaneous epp.

Based on their observations in frog muscle, del Castillo and Katz (1954) proposed a probabilistic model of ACh release. Their model was later applied to mammalian neuromuscular junctions by Boyd and Martin (1956). The model is based on the assumption that the synaptic terminal of the neuron consists of a large number, say n, of releasing units, each of which releases a fixed amount of ACh with probability p. If each releasing site operates independently, then the number of quanta of ACh that is released by an action potential is binomially distributed. The probability that k releasing sites fire (i.e., release a quantum of ACh) is the probability that k sites fire and the remaining sites do not, and so is given by $p^k(1-p)^{n-k}$. Since k sites can be chosen from n total sites in $n!/[(k!(n-k)!)]$ ways, it follows that

$$\text{Probability } k \text{ sites fire} = P(k) = \frac{n!}{k!(n-k)!}p^k(1-p)^{n-k}. \tag{8.1}$$

Under normal conditions, p is large (and furthermore, the assumption of independent release sites is probably inaccurate). However, under conditions of low external Ca^{2+} and high Mg^{2+}, p is small. This is because Ca^{2+} entry into the synapse is required for the release of a quantum of ACh. If only a small amount of Ca^{2+} is able to enter (because the external Ca^{2+} concentration is low), the probability of transmitter release is small. If n is correspondingly large, while $np = m$ remains fixed, the binomial distribution can be approximated by a Poisson distribution. That is,

$$
\begin{aligned}
\lim_{n \to \infty} P(k) &= \lim_{n \to \infty} \left[\frac{n!}{k!(n-k)!} \left(\frac{m}{n} \right)^k \left(1 - \frac{m}{n} \right)^{n-k} \right] \\
&= \frac{m^k}{k!} \lim_{n \to \infty} \left[\frac{n!}{n^k(n-k)!} \left(1 - \frac{m}{n} \right)^{n-k} \right] \\
&= \frac{m^k}{k!} \lim_{n \to \infty} \left(1 - \frac{m}{n} \right)^n \\
&= \frac{e^{-m} m^k}{k!},
\end{aligned}
\tag{8.2}
$$

which is the Poisson distribution with mean, or expected value, m.

There are two ways to estimate m. First, notice that $P(0) = e^{-m}$, so that

$$
e^{-m} = \frac{\text{number of action potentials with no epp's}}{\text{total number of action potentials}}.
\tag{8.3}
$$

Second, according to the assumptions of the model, a spontaneous epp results from the release of a single quantum of ACh, and a miniature epp is a linear sum of spontaneous epp's. Thus, m can be calculated by dividing the mean amplitude of a miniature epp response by the mean amplitude of a spontaneous epp, giving the mean number of quanta in a miniature epp, which should be m. As del Castillo and Katz showed, these two estimates of m agree well, which confirms the model hypotheses.

The spontaneous epp's are not of constant amplitude, because the amounts of ACh released from each vesicle are not identical. In the inset to Fig. 8.4 is shown the amplitude distribution of spontaneous epp's. To a good approximation, the amplitudes of single-unit release, denoted by $A_1(x)$, are normally distributed (i.e., a Gaussian distribution), with mean μ and variance σ^2. From this it is possible to calculate the amplitude distribution of the miniature epp's, as follows. If k vesicles are released, the amplitude distribution, denoted by $A_k(x)$, will be normally distributed with mean $k\mu$ and variance $k\sigma^2$, being the sum of k independent, normal distributions each of mean μ and variance σ^2 (Fig. 8.3). Summing the distributions for $k = 1, 2, 3, \ldots$, and noting that the probability of $A_k(x)$ is $P(k)$, gives the amplitude distribution

$$
\begin{aligned}
A(x) &= \sum_{k=1}^{\infty} P(k) A_k(x) \\
&= \frac{1}{\sqrt{2\pi\sigma^2}} \sum_{k=1}^{\infty} \frac{e^{-m} m^k}{k!\sqrt{k}} \exp \left[\frac{-(x - k\mu)^2}{2k\sigma^2} \right],
\end{aligned}
\tag{8.4}
$$

which is graphed in Fig. 8.4. There are clear peaks corresponding to 1, 2, or 3 released quanta, but these peaks are smeared out and flattened by the normal distribution of amplitudes. There is excellent agreement between the theoretical prediction and the

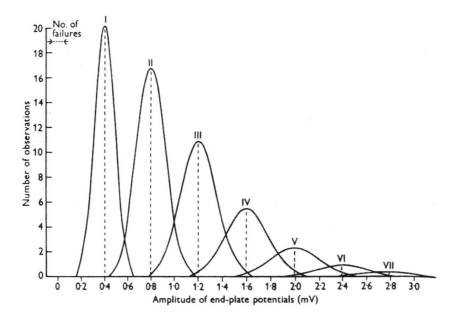

Figure 8.3 Theoretical distributions for epp's consisting of integer multiples of the sponta-neous epp, which is the basic building block, or quantum, of the epp. Summation of these curves for all integral numbers of quanta gives the theoretical prediction for the overall amplitude distribution, (8.4), which is plotted in the next figure. (Boyd and Martin, 1956, Fig. 9.)

experimental observations, lending further support to the quantal model of synaptic transmission (del Castillo and Katz, 1954; Boyd and Martin, 1956).

8.1.2 Presynaptic Voltage-Gated Calcium Channels

Chemical synaptic transmission begins when an action potential reaches the nerve terminal and opens voltage-gated Ca^{2+} channels, leading to an influx of Ca^{2+} and consequent neurotransmitter release. Based on voltage-clamp data from the squid giant synapse, Llinás et al. (1976) constructed a model of the Ca^{2+} current and its relation to synaptic transmission.

When the presynaptic voltage is stepped up and clamped at a constant level, the presynaptic Ca^{2+} current I_{Ca} increases in a sigmoidal fashion. To model these data, we assume that the voltage-gated Ca^{2+} channel consists of n identical, independent subunits, each of which can be in one of two states, S and O. Only when all subunits are in the state O does the channel admit Ca^{2+} current. Hence

$$S \underset{k_2}{\overset{k_1}{\rightleftarrows}} O, \tag{8.5}$$

and the probability that a channel is open is proportional to o^n where o is the probability that a subunit is in the open state. To incorporate the voltage dependence of the channel,

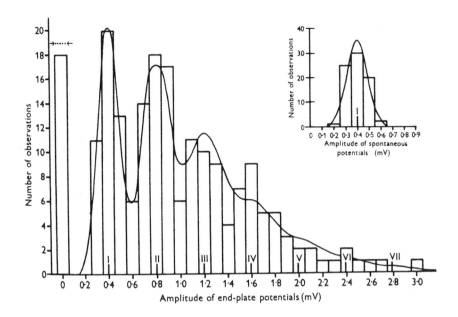

Figure 8.4 Amplitude distribution of miniature epp's. The histogram gives the frequency of the miniature epp as a function of its amplitude, as measured experimentally. The smooth curve is a fit of the theoretical prediction (8.4). The inset shows the amplitude distribution of spontaneous epp's: the smooth curve is a fit of the normal distribution to the data. (Boyd and Martin, 1956, Fig. 8.)

the opening and closing rate constants k_1 and k_2 are assumed to be functions of voltage of the form

$$k_1 = k_1^0 \exp\left(\frac{qz_1 V}{kT}\right), \qquad k_2 = k_2^0 \exp\left(\frac{qz_2 V}{kT}\right), \tag{8.6}$$

where k is Boltzmann's constant, T is the absolute temperature, V is the membrane potential, q is the positive elementary electric charge, z_1 and z_2 are the number of charges that move across the width of the membrane as S $\to$ O and O $\to$ S respectively, and k_1^0 and k_2^0 are constants. This is the same type of expression as that for the rate constant seen in Chapter 3 (for example, (3.44) and (3.45)). In Chapter 3, z referred to the number of charges on each ion crossing the membrane by passing through a channel. In this model, z_1 and z_2 denote charges that cross the membrane as a result of a change in the conformation of the channel as it opens or closes. In either case, the result of z charges crossing the membrane is the same, and we have a simple and plausible way to incorporate voltage dependence into the rate constants.

From (8.5) it follows that

$$\frac{do}{dt} = k_1(V)(1 - o) - k_2(V)o. \tag{8.7}$$

We denote the steady state of o by $\hat{o}$,

$$\hat{o}(V) = \frac{k_1(V)}{k_1(V) + k_2(V)}, \tag{8.8}$$

and assume that the membrane potential jumps instantaneously from V_0 to V_1 at time $t = 0$. We also set $o(t = 0) = \hat{o}(V_0)$. Then the solution of (8.7) is

$$o(t) = \frac{k_1}{k_1 + k_2}(1 - \exp[-(k_1 + k_2)t]) + \hat{o}(V_1)\exp[-(k_1 + k_2)t], \tag{8.9}$$

where k_1 and k_2 are evaluated at V_0. Now we assume that the single-channel current for an open Ca^{2+} channel, i, is given by the Goldman–Hodgkin–Katz current equation (2.123). Then

$$i = P_{Ca} \cdot \frac{4F^2}{RT} \cdot V \cdot \frac{c_i - c_e \exp(\frac{-2FV}{RT})}{1 - \exp(\frac{-2FV}{RT})}, \tag{8.10}$$

where c_i and c_e are the internal and external Ca^{2+} concentrations respectively, and P_{Ca} is the permeability of the Ca^{2+} channel. Note that an inward flux of Ca^{2+} gives a negative current. Finally, I_{Ca} is the product of the number of open channels with the single-channel current, and so

$$I_{Ca} = s_0 i o^n, \tag{8.11}$$

where s_0 is the total number of channels.

By fitting such curves to experimental data, Llinás et al. determined that the best-fit values for the unknowns are $n = 5$, $k_1^0 = 2$ ms^{-1}, $k_2^0 = 1$ ms^{-1}, $z_1 = 1$, and $z_2 = 0$. Hence, the best-fit parameters imply that the Ca^{2+} channel consists of 5 independent subunits, that the conversion of O to S is independent of voltage ($z_2 = 0$), but that the conversion of S to O involves the movement of a single charge across the membrane ($z_1 = 1$) and is thus dependent on the membrane potential.

Typical responses are shown in Fig. 8.5. To plot the curves shown here we used the fixed parameters $c_i = 0.1~\mu$M, $c_e = 40$ mM, $s_0 = 20$, $P_{Ca} = 10~\mu$m s^{-1}, $V_0 = -70$ mV. (The internal Ca^{2+} concentration is much smaller than the external concentration, so that the exact number used makes no essential difference to the result. See Exercise 1.) Note that the responses speed up as the voltage step increases, but the plateau level reaches a maximum (in absolute value) and then declines as the size of the voltage step is increased further. This is because of two competing effects that are discussed below.

Because the conversion of a closed subunit to an open subunit involves the movement of a charge across the cell membrane, there must be a current associated with channel opening, i.e., a gating current. This is generally the case when the rate constants for conformational changes of a channel protein are voltage-dependent, and these gating currents have been measured experimentally. We do not discuss gating currents any further; the interested reader is referred to Hille (2001) and Armstrong and Bezanilla (1973, 1974, 1977).

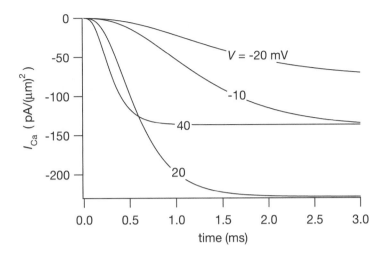

Figure 8.5 Presynaptic Ca^{2+} currents in response to presynaptic voltage steps in a model of squid stellate ganglion.

Synaptic Suppression

At steady state, the percentage of open channels is

$$(o(t = \infty))^5 = \left(\frac{k_1}{k_1 + k_2}\right)^5, \tag{8.12}$$

which is an increasing function of V. However, the single-channel current (8.10) is a decreasing function of V. Thus, the steady-state I_{Ca}, being a product of these two functions, is a bell-shaped function of V, as illustrated in Fig. 8.6. There are two time scales in the model; the single channel current depends instantaneously on the voltage, while the number of open channels is controlled by the voltage on a slower time scale. When the voltage is stepped up, the single-channel current decreases instantaneously. However, since there are so few channels open, the instantaneous decrease in the single-channel current has little effect on the total current. On a longer time scale, the channels gradually open in response to the increase in voltage, and this results in the slow monotonic responses to a positive step seen in Fig. 8.5. Of course, if the single-channel current is reduced to zero, no increase in the current is seen as the channels begin to open.

In response to a step *decrease* in voltage, the single-channel Ca^{2+} current increases instantaneously, but in contrast to the previous case where there were few channels open before the stimulus, there are now many open channels. Hence, the instantaneous increase in the single-channel current results in a large and fast increase in the total current. Over a longer time scale, the decrease in the voltage leads to a slow decrease in the number of open channels, and thus a slow decrease in the total current.

These responses are illustrated in Fig. 8.7. When a small positive step is turned on and then off, the Ca^{2+} current I_{Ca} responds with a monotonic increase followed by

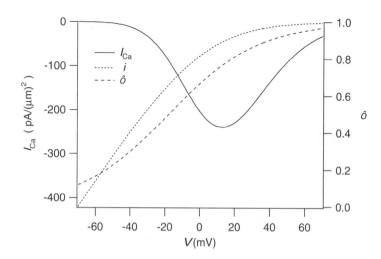

Figure 8.6 Steady-state I_{Ca}, i and $\hat{o}$ as functions of V. Note that the scale for I_{Ca}, and i are shown on the left while the scale for $\hat{o}$ is shown on the right. $\hat{o}$ increases with V while the magnitude of i decreases, their product (the steady-state Ca^{2+} current) is a bell-shaped curve.

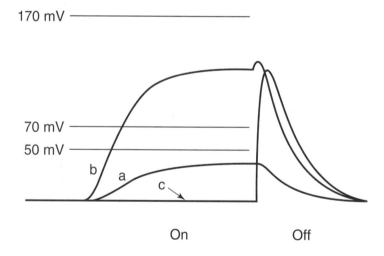

Figure 8.7 Synaptic suppression of the Ca^{2+} current. These curves are not calculated from the model in the text (they are reproduced from Llinás et al., 1976, Fig. 2D) and are used merely to demonstrate the qualitative behavior. Note that here the Ca^{2+} current is plotted as positive, rather than negative as in Fig. 8.5. Numerical solution of the model in the text (8.7) is left for Exercise 2.

a monotonic decrease (curve a). When the step is increased to 70 mV, the increase is still monotonic, but the decrease is preceded by a small bump as the current initially increases slightly (curve b). For a large step of 150 mV, the initial response is suppressed completely as the single-channel current is essentially zero, but when this suppression

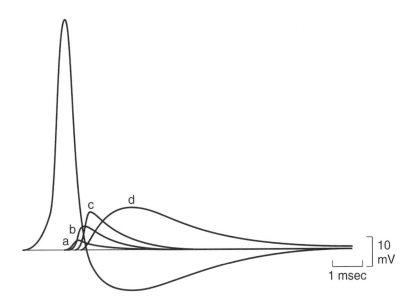

Figure 8.8 Theoretical responses to an action potential. Using the experimentally measured action potential (the leftmost curve in the figure) as input, the model can be used to predict the time courses of (a) the proportion of open channels, (b) the Ca^{2+} current, (c) the postsynaptic current, and (d) the postsynaptic potential. Details of how curve d is calculated are given in the text. Curve c is obtained by assuming that it has the same form as curve b, delayed by 200 ms and amplified. (Llinás et al., 1976, Fig. 2C.)

is released, a large voltage response is seen, which finally decreases to the resting state (curve c). This phenomenon is called *synaptic suppression*.

Response to an Action Potential

So far we have analytically calculated the response of the model to a voltage step. This was possible because under voltage clamp conditions, the voltage is piecewise constant. However, a more realistic stimulus would be a time-varying voltage corresponding to an action potential at the nerve terminal. It is easiest to find the solution of (8.7) numerically, and this is shown in Fig. 8.8. Given an input $V(t)$ that looks like an action potential, the number of open channels (curve a) and the presynaptic Ca^{2+} current (curve b) can be calculated. Figure 8.8 also includes theoretical predictions of the postsynaptic current (curve c) and the postsynaptic membrane potential (curve d). The postsynaptic current is obtained by assuming that it has the same form as the presynaptic current, delayed by 200 ms and amplified appropriately, assumptions that are justified by experimental evidence not discussed here. The postsynaptic membrane potential is obtained by using the postsynaptic current as an input into a model electrical circuit that is described later.

Although the Llinás model provides a detailed description of the initial stages of synaptic transmission and the voltage-gated Ca^{2+} channels, its picture of the postsynaptic response is oversimplified. There are a number of steps between the presynaptic Ca^{2+} current and the postsynaptic current that in this model are assumed to be linearly related. Thus, a decrease in the postsynaptic current is the direct result of a decrease in the presynaptic Ca^{2+} current, leading to a decrease in the concentration of neurotransmitter in the synaptic cleft. However, more detailed models of neurotransmitter kinetics show that, at least at the neuromuscular junction, this is not an accurate description.

8.1.3 Presynaptic Calcium Dynamics and Facilitation

One of the fundamental assumptions of the above model (and of the others in this chapter) is that neurotransmitter release is caused by the entry of Ca^{2+} through voltage-sensitive channels into the presynaptic neuron. However, although there is much evidence in favor of this hypothesis, there is also evidence that cannot be easily reconciled with this model. This has led to considerable controversy; some favor the *calcium hypothesis*, in which transmitter release is the direct result of the influx of Ca^{2+} and the only role of voltage is to cause Ca^{2+} influx (Fogelson and Zucker, 1985; Zucker and Fogelson, 1986; Zucker and Landò, 1986; Yamada and Zucker, 1992). Others favor the *calcium–voltage hypothesis*, in which transmitter release can be triggered directly by the presynaptic membrane potential, with Ca^{2+} playing a regulatory role (Parnas and Segel, 1980; Parnas et al., 1989; Aharon et al., 1994).

The calcium/voltage controversy is particularly interesting because of the role mathematical models have played. In 1985, Fogelson and Zucker proposed a model in which the diffusion of Ca^{2+} from an array of single channels was used to explain the duration of transmitter release and the decay of facilitation. This model was later used as the basis for a large number of other modeling studies, some showing that Ca^{2+} diffusion could not by itself explain all the experimental data, others showing how various refinements of the basic model could result in better agreement with experiment. Experimental and theoretical groups alike used the model as a basis for discussion, and thus, irrespective of the final verdict concerning its accuracy, the Fogelson–Zucker model is an excellent example of the value and use of modeling.

The controversy has shown no signs of abating (Parnas et al., 2002; Felmy et al., 2003), and mathematical models are playing an increasingly important role in the study of the presynaptic terminal. Two Nobel Laureates, Erwin Neher and Bert Sakmann, have both been actively involved in the construction and study of mathematical models of Ca^{2+} entry, binding and diffusion in the presynaptic cell (see, for example, Meinrenken et al., 2003, and Neher, 1998a,b), and current models are now both highly detailed and closely based on experimental data.

Facilitation

One of the major challenges presented by presynaptic Ca^{2+} dynamics is the phenomenon of *facilitation*, which occurs when the amount of neurotransmitter release caused by an action potential is increased by an earlier action potential, provided that the time interval between the action potentials is not too great. There is a lot of evidence that facilitation is primarily a presynaptic mechanism, and thus quantitative explanations have focused on the presynaptic terminal.

One of the earliest hypotheses (Katz and Miledi, 1968) was that facilitation is caused by the buildup of Ca^{2+} in the presynaptic terminal, since the Ca^{2+} introduced by one action potential is not completely removed before the next one comes along, and this hypothesis, in its general form, is still widely accepted.

However, although there is general agreement that buildup of Ca^{2+} is, one way or another, responsible for facilitation, the devil is in the details; there is still no agreement as to where, or in what form, the Ca^{2+} is building up. The *residual free calcium* hypothesis claims that it is an increase in free presynaptic Ca^{2+} that underlies facilitation, and there is considerable evidence in support of this claim (Zucker and Regehr, 2002). For example, Kamiya and Zucker (1994) tested how facilitation responds to a rapid increase of the concentration of Ca^{2+} buffer in the cytoplasm. This rapid increase was accomplished by the photorelease of a Ca^{2+} buffer, diazo-2. They found that the fast release of diazo-2 into the presynaptic terminal causes a decrease of the synaptic response within milliseconds. Since, on this time scale, the main effect of the added buffers should be to decrease the free Ca^{2+} concentration, it implies that it is the resting free Ca^{2+} concentration that is responsible for facilitation.

Mathematical models of the residual free Ca^{2+} hypothesis (Tang et al., 2000; Matveev et al., 2002; Bennett et al., 2004, 2007) have shown that this hypothesis is capable of explaining many experimental results. These models require that the Ca^{2+} facilitation binding site (to the vesicle release machinery) be situated at least 150 nm away from the release binding site, for if this were not so, then every action potential would saturate the facilitation binding site, leading to no increase in neurotransmitter release for subsequent action potentials. In addition, these models require low diffusion coefficients of Ca^{2+} close to the Ca^{2+} channel, and low diffusion coefficients of the exogenous buffer.

A similar hypothesis is that facilitation is caused by the local saturation of Ca^{2+} buffers close to the mouth of the Ca^{2+} channel, leading to the accumulation of cytoplasmic Ca^{2+} during a train of action potentials, and thus facilitation (Klingauf and Neher, 1997). Although a mathematical model of this *buffer saturation* hypothesis (Matveev et al., 2004) can explain many experimental results (at least from the crayfish neuromuscular junction), some questions remain. Thus, although there is strong evidence that buffer saturation plays a major role in some experimental preparations (Blatow et al., 2003), in other situations the evidence is inconclusive.

The third hypothesis discussed here was, historically, the first to be proposed. In 1968, Katz and Miledi proposed that facilitation was the result of residual Ca^{2+} bound to the vesicle release site; i.e., if the action potentials followed upon one another too closely there would be not enough time for Ca^{2+} to unbind, making it easier for the subsequent action potential to release neurotransmitter. Although this *residual bound calcium* hypothesis has fallen somewhat out of favor in recent years, due principally to the experiments described briefly above, in which the application of an exogenous buffer was shown to decrease the synaptic response quickly, more recent detailed models have shown that it is, after all, consistent with this experiment (Matveev et al., 2006). Indeed, the residual bound Ca^{2+} hypothesis is superior in some ways, as it does not suffer from the necessity of somewhat artificial requirements, such as the 150 nm separation between the release and the facilitation binding sites, or the small diffusion coefficients around the mouth of the Ca^{2+} channel.

Nevertheless, the question of what causes facilitation is still far from completely resolved. As is often the case, the most likely explanation is a combination of all the above three hypotheses, each acting with different strengths depending on the exact situation.

A Model of the Residual Bound Calcium Hypothesis

A simple model of the residual bound Ca^{2+} hypothesis does not include any spatial information, but is based entirely on the kinetics of Ca^{2+} binding to the vesicle binding sites (Bertram et al., 1996). This simplification is justified by experimental results showing that the minimum latency between Ca^{2+} influx and the onset of transmitter release can be as short as 200 μs. Since the Ca^{2+} binding site must thus be close to the Ca^{2+} channel, in a simple model we can neglect Ca^{2+} diffusion and assume that Ca^{2+} entering through the channel is immediately available for binding to the vesicle binding site.

It is also assumed that transmitter release is the result of Ca^{2+} entering through a single channel, the so-called Ca^{2+}-domain hypothesis. If the Ca^{2+} channels are far enough apart, or if only few open during each action potential, the Ca^{2+} domains of individual channels are independent.

Our principal goal here is to provide a plausible explanation for the intriguing experimental observation that facilitation increases in a steplike fashion as a function of the frequency of the conditioning action potential train.

We assume that Ca^{2+} entering through the Ca^{2+} channel is immediately available to bind to the transmitter release site, which itself consists of four independent, but not identical, gates, denoted by S_1 through S_4. Gate S_j can be either closed (with probability C_j) or open (with probability O_j), and thus

$$Ca^{2+} + C_j \underset{k_{-j}}{\overset{k_j}{\rightleftarrows}} O_j. \tag{8.13}$$

Hence,

$$\frac{dO_j}{dt} = k_j c - \frac{O_j}{\tau_j(c)}, \tag{8.14}$$

where $\tau_j(c) = 1/(k_j c + k_{-j})$, and c is the Ca^{2+} concentration. Finally, the probability R that the release site is activated is

$$R = O_1 O_2 O_3 O_4. \tag{8.15}$$

The rate constants were chosen to give good agreement with experimental data and are shown in Table 8.1. Note that the rates of closure of S_3 and S_4 are much greater than

Table 8.1 Parameter values for the binding model of synaptic facilitation (Bertram et al., 1996).

$$
\begin{array}{ll}
k_1 = 3.75 \times 10^{-3} \text{ ms}^{-1}\mu\text{M}^{-1} & k_{-1} = 4 \times 10^{-4} \text{ ms}^{-1} \\
k_2 = 2.5 \times 10^{-3} \text{ ms}^{-1}\mu\text{M}^{-1} & k_{-2} = 1 \times 10^{-3} \text{ ms}^{-1} \\
k_3 = 5 \times 10^{-4} \text{ ms}^{-1}\mu\text{M}^{-1} & k_{-3} = 0.1 \text{ ms}^{-1} \\
k_4 = 7.5 \times 10^{-3} \text{ ms}^{-1}\mu\text{M}^{-1} & k_{-4} = 10 \text{ ms}^{-1}
\end{array}
$$

for S_1 and S_2, and thus Ca^{2+} remains bound to S_1 and S_2 for a relatively long time, providing the possibility of facilitation.

To demonstrate how facilitation works in this model, we suppose that a train of square pulses of Ca^{2+} (each of width t_p and amplitude c_p) arrives at the synapse. We want to calculate the level of activation at the end of each pulse and show that this is an increasing function of time. The reason for this increase is clear from the governing differential equation, (8.14). If a population of gates is initially closed, then a Ca^{2+} pulse begins to open them, but when Ca^{2+} is absent, the gates close. If the interval between pulses is sufficiently short and the decay time constant sufficiently large, then when the next pulse arrives, some gates are already open, so the new pulse activates a larger fraction of transmitter release sites than the first, and so on.

To quantify this observation we define t_n to be the time at the end of the nth pulse,

$$t_n = t_p + (n - 1)T, \tag{8.16}$$

where $T = t_p + t_I$ is the period and t_I is the interpulse interval. For any gate (temporarily omitting the subscript j) with $O(0) = 0$, the open probability at the end of the first pulse is

$$O(t_1) = O_\infty(1 - e^{-t_p/\tau_p}), \tag{8.17}$$

where $O_\infty = kc_p\tau_p$ is the steady-state probability corresponding to a steady concentration of Ca^{2+}, c_p, and $\tau_p = \tau(c_p) = 1/(kc_p + k_-)$.

Suppose that $O(t_{n-1})$ is the open probability at the end of the $(n - 1)$st Ca^{2+} pulse. During the interpulse period, O decays with rate constant $\tau(0)$. Thus, at the start of the nth pulse,

$$O(t_{n-1} + t_I) = O(t_{n-1})e^{-t_I/\tau(0)}, \tag{8.18}$$

and so at the end of the nth pulse,

$$\begin{aligned}
O(t_n) &= O(t_{n-1})e^{-t_I/\tau(0)}e^{-t_p/\tau_p} + O_\infty(1 - e^{-t_p/\tau_p}) \\
&= \alpha O(t_{n-1}) + O(t_1),
\end{aligned} \tag{8.19}$$

where $\alpha = \exp(-(t_I/\tau(0) + t_p/\tau_p)) = \exp(-k_-(T + t_p\frac{c_p}{K}))$ and $K = k_-/k_+$. This is a simple difference equation for $O(t_n)$, which can be solved by setting $O(t_n) = A\alpha^n + B$

and substituting into (8.19), from which we find that

$$\frac{O(t_n)}{O(t_1)} = \frac{1 - \alpha^n}{1 - \alpha}. \tag{8.20}$$

Notice that as the interpulse interval gets large ($t_I \to \infty$), we have $\alpha \to 0$, so that $O(t_n)$ is independent of n. On the other hand, α increases if the Ca^{2+} pulses are shortened (t_p is decreased).

Now we define facilitation as the ratio

$$F_n = \frac{R(t_n)}{R(t_1)}, \tag{8.21}$$

and find that

$$F_n = \left(\frac{1 - \alpha_1^n}{1 - \alpha_1}\right)\left(\frac{1 - \alpha_2^n}{1 - \alpha_2}\right)\left(\frac{1 - \alpha_3^n}{1 - \alpha_3}\right)\left(\frac{1 - \alpha_4^n}{1 - \alpha_4}\right), \tag{8.22}$$

where α_j is the α corresponding to gate j. For the numbers shown in Table 8.1, α_4 is nearly zero in the physiologically relevant range of frequencies, so it can be safely ignored. A plot of F_n against the pulse train frequency shows a steplike function, as is observed experimentally. In Fig. 8.9 is shown the maximal facilitation,

$$F_{max} = \lim_{n \to \infty} F_n = \left(\frac{1}{1 - \alpha_1}\right)\left(\frac{1}{1 - \alpha_2}\right)\left(\frac{1}{1 - \alpha_3}\right), \tag{8.23}$$

which also has a steplike appearance.

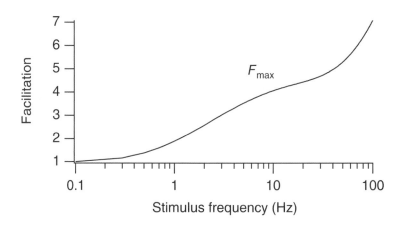

Figure 8.9 Facilitation as a function of stimulus frequency in the binding model of synaptic facilitation, calculated using $t_p c_p = 200\ \mu M\,ms$. Here $F_{max} = \lim_{n \to \infty} F_n$ is the maximal facilitation produced by a pulse train.

A More Complex Version

A more complex model of the residual bound Ca^{2+} hypothesis takes into account diffusion of Ca^{2+} from the channel to the binding site and the presence of Ca^{2+} buffers (Matveev et al., 2006).

We assume that each of n Ca^{2+} channels is located at a discrete position r_j, $j = 1, \ldots, n$, and that there are two Ca^{2+} buffers, one with slower kinetics. Thus, following the discussion of buffers and discrete Ca^{2+} release in Chapter 7, we get

$$\frac{\partial c}{\partial t} = D_c \nabla^2 c + \sum_{i=1}^{2} k_{-,i} b_i - k_{+,i} c(b_{t,i} - b_i) + \frac{1}{2F} I_{Ca} \sum_{j=1}^{n} \delta(r - r_j), \quad (8.24)$$

$$\frac{\partial b_i}{\partial t} = D_b \nabla^2 b_i + \sum_{i=1}^{2} k_{-,i} b_i - k_{+,i} c(b_{t,i} - b_i). \quad (8.25)$$

These equations are solved numerically on the domain illustrated in Fig. 8.10. Calcium channels are clustered into active zones, with 16 channels per zone, and the symmetry of the domain around each active zone is used to reduce the problem to one on the quarter box bordered by the dashed lines. In each quarter box there is a single Ca^{2+} binding domain, denoted by the filled circle, which contains multiple Ca^{2+} binding sites, all colocalized. Calcium is pumped out only on the top and bottom surfaces (shaded), while the other surfaces have no-flux boundary conditions, which assumes a regular array of active sites. Calcium pumping is assumed to be by a simple saturating mechanism, with Hill coefficient 1, and is balanced by a constant leak into the cell.

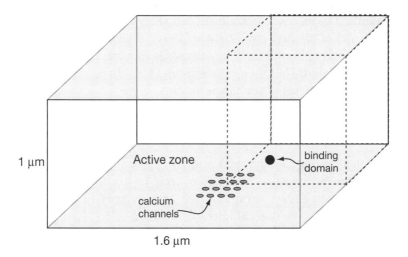

Figure 8.10 Schematic diagram of a presynaptic active zone, the Ca^{2+} channels, and the Ca^{2+} binding site (adapted from Matveev et al., 2006, Fig. 1). The binding domain is situated 130 nm away from the center of the active zone, and about 55 nm from the nearest Ca^{2+} channel.

Thus, on the shaded boundaries,

$$\nabla c \cdot \mathbf{n} = J_{\text{leak}} - \frac{V_p c}{K_p + c},\qquad(8.26)$$

where $\mathbf{n}$ is the normal to the boundary.

The binding domain is assumed to contain two different types of binding sites, one faster (type X), one slower (type Y). The X binding site can bind two Ca^{2+} ions, and thus

$$X \underset{k_{-x}}{\overset{2ck_x}{\rightleftarrows}} CaX \underset{2k_{-x}}{\overset{ck_x}{\rightleftarrows}} Ca_2X,\qquad(8.27)$$

while binding sites Y_1 and Y_2 can bind one Ca^{2+} ion each, and thus

$$Y_1 \underset{k_{-1}}{\overset{ck_1}{\rightleftarrows}} CaY_1,\qquad(8.28)$$

$$Y_2 \underset{k_{-2}}{\overset{ck_2}{\rightleftarrows}} CaY_2.\qquad(8.29)$$

If R denotes the rate of release neurotransmitter, then the rate of increase of R is assumed to be proportional to $[Ca_2X][CaY_1][CaY_2]$:

$$\frac{dR}{dt} = k_R[Ca_2X][CaY_1][CaY_2] - k_I R.\qquad(8.30)$$

Typical model simulations are shown in Fig. 8.11. Significantly, these results show that, in the model, the release of an exogenous buffer by a UV flash does indeed decrease the synaptic response within milliseconds (Fig. 8.11, panels C and D). Thus, the experimental results of Kamiya and Zucker (1994), which have often been taken as evidence that the residual bound Ca^{2+} hypothesis is incorrect, show no such thing. Although we cannot conclude that the residual bound Ca^{2+} hypothesis is correct, neither can it be rejected on the basis of such experiments.

8.1.4 Neurotransmitter Kinetics

When the end-plate voltage is clamped and the nerve stimulated (so that the end-plate receives a stimulus of ACh, of undetermined form), the end-plate current first rises to a peak and then decays exponentially, with a decay time constant that is an exponential function of the voltage. Magleby and Stevens (1972) constructed a detailed model of end-plate currents in the frog neuromuscular junction that gives a mechanistic explanation of this observation and shows how a simple model of the receptor kinetics can quantitatively reproduce the observed end-plate currents.

First, Magleby and Stevens showed that the instantaneous end-plate current–voltage relationship is linear, and thus, for a fixed voltage, the end-plate current is proportional to the end-plate conductance. Because of this, it is sufficient to study the

A

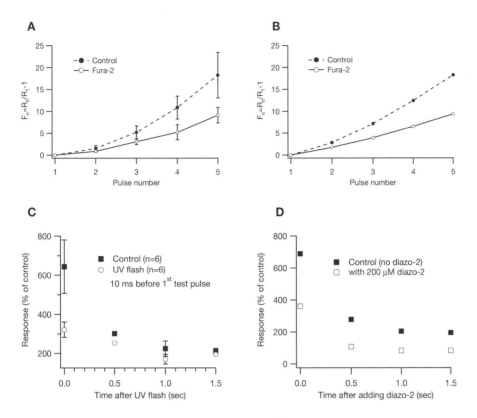

Figure 8.11   Comparison of the residual bound Ca^{2+} model to experimental data. A and B: Facilitation as a function of pulse number, in the presence and absence of exogenous buffer. Panel A shows the experimental results of Tang et al. (2000), while panel B shows model simulations. In both the experimental data and the model, facilitation increases with pulse number, and has an increasing slope. Addition of exogenous buffer decreases facilitation. C and D; time dependence of facilitation after the fast release of exogenous buffer by a UV flash. Panel C is the experimental data of Kamiya and Zucker (1994), while panel D are model simulations. Adapted from Matveev et al., (2006), Figs. 3 and 4.

end-plate conductance rather than the end-plate current. Since the end-plate conductance is a function of the concentration of ACh, we restrict our attention to the kinetics of ACh in the synaptic cleft.

We assume that ACh reacts with its receptor, R, in enzymatic fashion,

$$\text{ACh} + \text{R} \underset{k_2}{\overset{k_1}{\rightleftarrows}} \text{ACh} \cdot \text{R} \underset{\alpha}{\overset{\beta}{\rightleftarrows}} \text{ACh} \cdot \text{R}^*, \qquad (8.31)$$

and that the ACh-receptor complex passes current only when it is in the open state $\text{ACh} \cdot \text{R}^*$. We let $c = [\text{ACh}]$, $y = [\text{ACh} \cdot \text{R}]$, and $x = [\text{ACh} \cdot \text{R}^*]$, and then it follows from the law of mass action that

$$\frac{dx}{dt} = -\alpha x + \beta y, \tag{8.32}$$

$$\frac{dy}{dt} = \alpha x + k_1 c(N - x - y) - (\beta + k_2)y, \tag{8.33}$$

$$\frac{dc}{dt} = f(t) - k_e c - k_1 c(N - x - y) + k_2 y, \tag{8.34}$$

where N is the total concentration of ACh receptor, which is assumed to be conserved, and ACh decays by a simple first-order process at rate k_e. The postsynaptic conductance is assumed to be proportional to x, and the rate of formation of ACh is some given function $f(t)$. One option for $f(t)$ is to use the output of the single-domain/bound-calcium model described in the previous section (Exercise 8).

The model equations, as given, are too complicated to solve analytically, and so we proceed by making some simplifying assumptions. First, we assume that the kinetics of ACh binding to its receptor are much faster than the other reactions in the scheme, so that y is in instantaneous equilibrium with c. To formalize this assumption, we introduce dimensionless variables $X = x/N, Y = y/N, C = k_1 c/k_2$, and $\tau = \alpha t$, in terms of which (8.33) becomes

$$\epsilon \frac{dY}{d\tau} = \epsilon X + C(1 - X - Y) - \left(\epsilon \frac{\beta}{\alpha} + 1\right)Y, \tag{8.35}$$

where $\epsilon = \alpha/k_2 \ll 1$. Upon setting ϵ to zero, we find the quasi-steady approximation

$$Y = \frac{C(1 - X)}{1 + C}, \tag{8.36}$$

or in dimensioned variables,

$$y = \frac{c(N - x)}{K + c}, \tag{8.37}$$

where $K = k_2/k_1$. Now we can eliminate y from (8.32) to obtain

$$\frac{dx}{dt} = -\alpha x + \beta \frac{c(N - x)}{K + c}. \tag{8.38}$$

Next we observe that

$$\frac{dx}{dt} + \frac{dy}{dt} + \frac{dc}{dt} = f(t) - k_e c. \tag{8.39}$$

In dimensionless variables this becomes

$$\frac{N}{K}\left(\frac{dX}{d\tau} + \frac{dY}{d\tau}\right) + \frac{dC}{d\tau} = F(\tau) - K_e C, \tag{8.40}$$

where $F(\tau) = \frac{f(t)}{\alpha K}$ and $K_e = k_e/\alpha$ are assumed to be of order one. If we suppose that $N \ll K$, then setting N/K to zero in (8.40), we find (in dimensioned variables)

$$\frac{dc}{dt} = f(t) - k_e c. \tag{8.41}$$

One further simplification is possible if we assume that $\beta \ll \alpha$. Notice from (8.38) that

$$\frac{dx}{dt} < -\alpha x + \beta(N - x),\tag{8.42}$$

so that

$$x(t) \leq x(0)e^{-(\alpha+\beta)t} + \frac{\beta N}{\alpha + \beta}(1 - e^{-(\alpha+\beta)t}).\tag{8.43}$$

Once this process has been running for some time, so that effects of initial data can be ignored, x is of order $\frac{\beta N}{\alpha+\beta}$. If $\beta \ll \alpha$, then $x \ll N$, and (8.38) simplifies to

$$\frac{dx}{dt} = -\alpha x + \beta \frac{cN}{K + c}.\tag{8.44}$$

For any given input $f(t)$, (8.41) can be solved for $c(t)$, and then (8.44) can be solved to give $x(t)$, the postsynaptic conductance.

As mentioned above, the decay of the postsynaptic current has a time constant that depends on the voltage. This could happen in two principal ways. First, if the conformational changes of the receptor were much faster than the decay of ACh in the synaptic cleft, x would be in quasi-equilibrium, and we would have

$$x = \frac{\beta c(t)N}{\alpha[K + c(t)]}.\tag{8.45}$$

Thus, if c is small, x would be approximately proportional to c. In this case an exponential decrease of c caused by the decay term $-k_e c$ would cause an exponential decrease in the postsynaptic conductance. An alternative possibility is that ACh degrades quickly in the synaptic cleft, so that c quickly approaches zero, but that the decay of the end-plate current is due to conformational changes of the ACh receptor. According to this hypothesis, the release of ACh into the cleft would cause an increase in x, which then would decay according to

$$\frac{dx}{dt} = -\alpha x,\tag{8.46}$$

(since c is nearly zero). In this case, the exponential decrease of end-plate current would be governed by the term $-\alpha x$.

Magleby and Stevens argued that the latter hypothesis is preferable. Assuming therefore that the rate-limiting step in the decay of the end-plate current is the decay of x, α can be estimated directly from experimental measurements of end-plate current decay to be

$$\alpha(V) = Be^{AV},\tag{8.47}$$

where $A = 0.008$ mV^{-1} and $B = 1.43$ ms^{-1}.

To calculate the complete time course of the end-plate current from (8.44), it remains to determine $c(t)$. In general this is not known, as it is not possible to measure synaptic cleft concentrations of ACh accurately.

A method to determine $c(t)$ from the experimental data was proposed by Magleby and Stevens. First, suppose that β is also a function of V, as is expected, since α is a function of V. Then (8.44) can be written as

$$\frac{dx}{dt} = -\alpha(V)x + \beta(V)W(t), \qquad (8.48)$$

where $W(t) = Nc(t)/[K + c(t)]$. Since for any fixed voltage the time course of x can be measured experimentally (recall that the experiments were done under voltage clamp conditions), it follows that $\beta(V)W(t)$ can be determined from

$$\beta(V)W(t) = \frac{dx}{dt} + \alpha(V)x. \qquad (8.49)$$

Although this requires numerical differentiation (which is notoriously unstable), the experimental records are smooth enough to permit a reasonable determination of W from the time course of x. Since W is assumed to be independent of V, it can be determined (up to an arbitrary scale factor) from a time course of x obtained for any fixed voltage. Further, if the model is valid, then we expect the same result no matter what voltage is used to obtain W. A typical result for W, shown in Fig. 8.12, rises and falls in a way reminiscent of the responses calculated from the Llinás model described above (Fig. 8.8).

The final unknown is $\beta(V)$, the scale factor in the determination of W. Relative values of β can be obtained by comparing time courses taken at different voltages. If $\beta(V_1)W(t)$ and $\beta(V_2)W(t)$ are time courses obtained from (8.49) at two different voltages, the ratio $\beta(V_1)/\beta(V_2)$ is obtained from the ratio of the time courses. However, because of experimental variability or invalid model assumptions, this ratio may not be constant as a function of time, in which case the ratio cannot be determined unambiguously. Magleby and Stevens used the ratio of the maximum amplitudes of the time courses, in which case $\beta(V_1)/\beta(V_2)$ can be obtained uniquely. They determined that β,

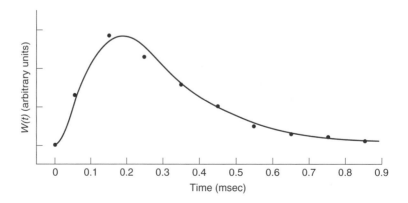

Figure 8.12 $W(t)$ calculated from the time course of x using (8.49). (Magleby and Stevens, 1972, Fig. 4.)

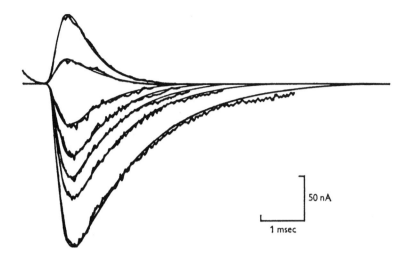

Figure 8.13 End-plate currents from the Magleby–Stevens model. Equation (8.48) was solved numerically, using as input the function plotted in Fig. 8.12. The functions $\alpha(V)$ and $\beta(V)$ are given in the text. The corresponding values for V are, from top to bottom, 32, 20, −30, −56, −82, −106 and −161 mV. The wavy lines are the experimental data; the smooth curves are the model fit. (Magleby and Stevens, 1972, Fig. 6.)

like α, is an exponential function of V,

$$\beta(V) = be^{aV}, \tag{8.50}$$

where $a = 0.00315 \text{ mV}^{-1}$ and b is an arbitrary scaling factor.

Equation (8.48) can now be solved numerically to determine the time course of the end-plate current for various voltages. Typical results are shown in Fig. 8.13. Although the model construction guarantees the correct peak response (because that is how β was determined) and also guarantees the correct time course at one particular voltage (because that is how W was determined), the model responses agree well with the experimental records over all times and voltages. This confirms the underlying assumption that W is independent of voltage.

Although the approach of Magleby and Stevens of determining $W(t)$ directly from the data leads to excellent agreement with the experimental data, it suffers from the disadvantage that no mechanistic rationale is given for the function W. It would be preferable to have a derivation of the behavior of W from fundamental assumptions about the kinetics of ACh release and degradation in the synaptic cleft, but such is not presently available.

Neurotransmitter Diffusion

A different style of model is that of Smart and McCammon (1998) or Tai et al. (2003). These groups constructed highly detailed finite element models of the neuromuscular junction, with the model of Tai et al. based directly on anatomical structural information (Stiles and Bartol, 2000). In this region they solved a simple diffusion equation for ACh, with degradation of ACh on selected parts of the boundary. These detailed structural models are still somewhat preliminary, as most of the efforts to date have been in the development of the necessary computational techniques; the models have provided, as yet, little additional physiological insight. However, they are ideally placed to provide insight into the comparative behavior of different neuromuscular junctions, and are likely to be a highly important style of model in the future.

8.1.5 The Postsynaptic Membrane Potential

Acetylcholine acts by opening ionic channels in the postsynaptic membrane that are permeable to Na^+ and K^+ ions. A schematic diagram of the electrical circuit model of the postsynaptic membrane is given in Fig. 8.14. This model is based on the usual assumptions (see, for example, Chapter 2) that the membrane channels can be modeled as ohmic resistors and that the membrane acts like a capacitor, with capacitance C_m.

The ACh-sensitive channels have a reversal potential V_s of about -15 mV and a conductance that depends on the concentration of ACh. The effects of all the other ionic channels in the membrane are summarized by a resting conductance, g_r and a resting potential V_r of about -70 mV. In the usual way, the equation for the membrane potential V is

$$C_m \frac{dV}{dt} + g_r(V - V_r) + g_s(V - V_s) = 0. \tag{8.51}$$

In general, g_s is a function of the number of ACh receptors with ACh bound, i.e., in the notation of the previous section, $g_s = g_s(x)$. Since x is a function of time, g_s is also a function of time. Hence,

$$C_m \frac{dV}{dt} + [g_r + g_s(t)]V = g_r V_r + g_s(t) V_s. \tag{8.52}$$

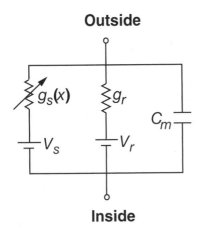

Outside

Inside

Figure 8.14 Electrical circuit model of the postsynaptic membrane.

This equation can be reduced to quadratures by using the integrating factor $I = \exp(\frac{1}{C_m} \int (g_r + g_s) \, dt)$, so in principle, the response of the postsynaptic membrane can be calculated from knowledge of the time course of x. However, since $g_s(t)$ is time-dependent, the quadratures can essentially never be evaluated explicitly, so exact formulas are effectively useless.

Suppose, however, that $g_s(t)$ is small compared to g_r. We set $V_1 = V - V_r$ so that

$$\frac{C_m}{g_r} \frac{dV_1}{dt} + V_1 = -\frac{g_s}{g_r}(V_1 + V_r - V_s). \tag{8.53}$$

Now we expect V_1 to be on the order of $\frac{g_s}{g_r}$, so that the term $\frac{g_s}{g_r} V_1$ can be ignored, and we are left with the equation

$$\frac{C_m}{g_r} \frac{dV_1}{dt} + V_1 = -\frac{g_s}{g_r}(V_r - V_s), \tag{8.54}$$

which can be solved exactly.

For example, a simple solution is obtained when g_s is taken to be proportional to x and the input $f(t)$ to the Magleby and Stevens model is assumed to be $\gamma \delta(t)$ for $\frac{\gamma}{K}$ small. In this case,

$$g_s(t) = x(t) = \frac{\gamma \beta N}{K(\alpha - k_e)} \left(e^{-k_e t} - e^{-\alpha t} \right), \tag{8.55}$$

as derived in Exercise 5. The solution of (8.54) is now readily found using standard solution techniques (Exercise 7).

8.1.6 Agonist-Controlled Ion Channels

The acetylcholine receptor/channel is an example of an agonist-controlled channel, a simple model for which was discussed in Section 3.5.4. In Section 8.1.5 above, we described an even simpler model of the ACh receptor/channel. There are many other more complex models of the postsynaptic agonist-controlled ion channels. A useful review is that of Destexhe et al. (1998), which briefly discusses a number of the major models of AMPA, NMDA, and GABA receptors.

A typical example is the Markov model of the AMPA receptor (Patneau and Mayer, 1991) shown in Fig. 8.15. This model has three closed states, two desensitized states, and one open state. The receptor can open only from the closed state with two transmitter molecules bound, C_2. The response of this model to a delta function input (i.e., if $T(t) = \delta(t)$) is shown in Fig. 8.16.

This multistate Markov model of the AMPA receptor can be simplified to a two-state model, without changing the qualitative properties of the impulse response. The two-state model

$$C + T \underset{\beta}{\overset{\alpha}{\rightleftharpoons}} O \tag{8.56}$$

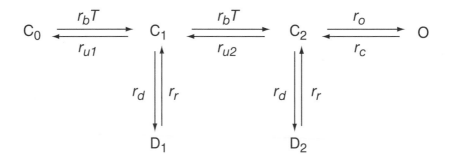

Figure 8.15 Markov model of the AMPA receptor (Patneau and Mayer, 1991). T is the concentration of transmitter.

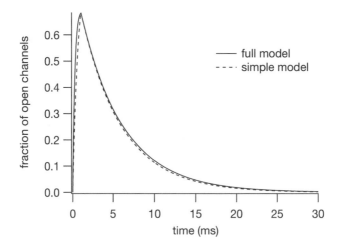

Figure 8.16 Response of the AMPA receptor model to a delta function input. The solid line is from the model shown in Fig. 8.15, with parameter values $r_b = 13 \times 10^6$ M^{-1}s^{-1}, $r_{u1} = 5.9$ s^{-1}, $r_{u2} = 8.6 \times 10^4$ s^{-1}, $r_d = 900$ s^{-1}, $r_r = 64$ s^{-1}, $r_o = 2.7 \times 10^3$ s^{-1}, $r_c = 200$ s^{-1}. The dotted line is from the simplified version of the model, (8.56), with parameters $\alpha = 1.35$ M^{-1}s^{-1}, $\beta = 200$ s^{-1}. The initial condition for both simulations is $c_0 = 1$.

responds to an impulse in much the same as the full model, as can be seen from the dotted line in Fig. 8.16.

Clearly, the Markov model shown in Fig. 8.15 is not well determined by its impulse response, a fact that, given the discussion in Section 3.6, should come as no surprise. It is, however, a fact that should inspire caution. In order to determine the rate constants unambiguously we would need to use, at the very least, single-channel data; even then, the task, for such a complex model, would not be trivial, and might not even be possible.

8.1.7 Drugs and Toxins

The foregoing models are sufficient to piece together a crude model of synaptic transmission. However, many features were ignored, and there are many situations that can change the behavior of this system. Primary among these are drugs and toxins that affect specific events in the neurotransmission process. For example, the influx of Ca^{2+} is reduced by divalent metal ions, such as Pb^{2+}, Cd^{2+}, Hg^{2+}, and Co^{2+}. By reducing the influx of Ca^{2+}, these cations depress or abolish the action-potential-evoked transmitter release. Certain toxins, including tetanus and clostridial botulinus, are potent inhibitors of transmitter exocytosis, an action that is essentially irreversible. Botulinus neurotoxin is selective for cholinergic synapses and is one of the most potent neuroparalytic agents known. Tetanus toxin is taken up by spinal motor nerve terminals and transported retrogradely to the spinal cord, where it blocks release of glycine at inhibitory synapses. Spread of the toxin throughout the brain and spinal cord can lead to severe convulsions and death. The venom from black widow spider contains a toxin (α-latrotoxin) that causes massive transmitter exocytosis and depletion of synaptic vesicles from presynaptic nerve terminals.

Agents that compete with the transmitter for receptor binding sites, thereby preventing receptor activation, are called receptor antagonists. An example of an antagonist of the ACh receptors of the skeletal neuromuscular junction is curare. By inhibiting ACh binding at receptor sites, curare causes progressive decrease in amplitude and shortening of epp's. In severe curare poisoning, transmission is blocked. Selective antagonists exist for most transmitter receptors. For example, bicuculline is an antagonist of GABA receptors, and is a well-known convulsant.

Agents that mimic the action of natural transmitters are known as receptor agonists. A well-known agonist of ACh receptors in neuromuscular junction is nicotine. Nicotine binds to the ACh receptor and activates it in the same manner as ACh. However, nicotine causes persistent receptor activation because it is not degraded, as is ACh, by ACh-esterase. On the other hand, diisopropylphosphofluoridate (commonly known as nerve gas) is an example of an anticholinesterase, because it inhibits the activity of ACh-esterase, so that ACh persists in the synaptic cleft. Similarly, one effect of cocaine is to prolong the activity of dopamine, by blocking the uptake of dopamine from the synaptic cleft.

Other agents interfere with receptor-gated permeabilities by interfering with the channel itself. Thus, picrotoxin, which blocks GABA-activated Cl^- channels, and strychnine, which blocks glycine-activated Cl^- channels, are potent blockers of inhibitory synapses and known convulsants.

8.2 Gap Junctions

Gap junctions are small nonselective channels (with diameters of about 1.2 nm) that form direct intercellular connections through which ions or other small molecules can flow. They are formed by the joining of two *connexons*, hexagonal arrays of *connexin*

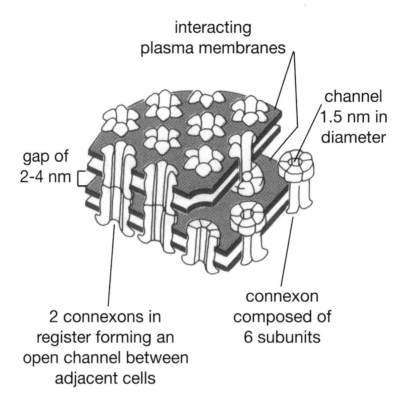

Figure 8.17 Diagram of a region of membrane containing gap junctions, based on electron microscope images and X-ray diffraction data. Each connexon is composed of six gap-junction proteins, called connexins, arranged hexagonally. Connexons in apposed membranes meet in the intercellular space to form the gap junction. (Alberts et al., 1994, Fig. 19-15.)

protein molecules (Fig. 8.17). Despite being called electrical synapses, in this chapter we concentrate on models in which membrane potential plays no role, focusing instead on the interaction between intracellular diffusion and intercellular permeability. Electrical aspects of gap junctions are important for the function of cardiac cells and are discussed in that context in Chapter 12. An example of how gap junctions are used for intercellular signaling via second messengers is described in Chapter 7, where we discuss a model of intercellular Ca^{2+} wave propagation.

8.2.1 Effective Diffusion Coefficients

We first consider a one-dimensional situation where a species u diffuses along a line of cells which are connected by gap junctions. Because of their relatively high resistance to flow (compared to cytoplasm), the gap junctions decrease the rate at which u diffuses along the line. Since this is a one-dimensional problem, we assume that each intercellular membrane acts like a single resistive pore with a given permeability F.

The effect of gap-junction distribution within the intercellular membrane is discussed later in this chapter.

We assume that Fick's law holds and thus the flux, J, of u is proportional to the gradient of u; i.e.,

$$J = -D\frac{\partial u}{\partial x}, \tag{8.57}$$

where D is the diffusion coefficient for the intracellular space. From the conservation of u it follows that in the interior of each cell,

$$\frac{\partial u}{\partial t} = D\frac{\partial^2 u}{\partial x^2}. \tag{8.58}$$

However, u need not be continuous across the intercellular boundary. In fact, if there is a cell boundary at $x = x_b$, the flux through the boundary is assumed to be proportional to the concentration difference across the boundary. Then, conservation of u across the boundary implies that

$$-D\frac{\partial u(x_b^-, t)}{\partial x} = -D\frac{\partial u(x_b^+, t)}{\partial x} = F[u(x_b^-, t) - u(x_b^+, t)], \tag{8.59}$$

for some constant F, called the *permeability coefficient*, with units of distance/time. The $+$ and $-$ superscripts indicate that the function values are calculated as limits from the right and left, respectively.

When the cells through which u diffuses are short compared to the total distance that u moves, the movement of u can be described by an effective diffusion coefficient. The effective diffusion coefficient is defined and is measurable experimentally by assuming that the analogue of Ohm's law holds. Thus, in a preparation of N cells, each of length L, with $u = U_0$ at $x = 0$ and $u = U_1$ at $x = NL$, the effective diffusion coefficient D_e is defined by

$$J = \frac{D_e}{NL}(U_0 - U_1), \tag{8.60}$$

where J is the steady-state flux of u.

To calculate D_e, we look for a function $u(x)$ that satisfies $u_{xx} = 0$ when $x \neq (j + \frac{1}{2})L$ and satisfies (8.59) at $x = (j + \frac{1}{2})L, j = 0, \ldots, N-1$. Further, we require that $u(0) = U_0$, and $u(NL) = U_1$. Note that we are assuming that the cell boundaries occur at $L/2, 3L/2, \ldots$, and thus the boundary conditions at $x = 0$ and $x = NL$ occur halfway through a cell.

A typical solution u that satisfies these conditions must be linear within each cell, and piecewise continuous with jumps at the cell boundaries. Suppose that the slope of u within each cell is $-\lambda$, and that the jump in u between cells is $u(x_b^+) - u(x_b^-) = -\Delta$. Then, since there are $N-1$ whole cells, two half cells (at the boundaries), and N interior cell boundaries, we have

$$(N - 1)\lambda L + 2\lambda \left(\frac{L}{2}\right) + N\Delta = U_0 - U_1. \tag{8.61}$$

Furthermore, it follows from (8.59) that

$$D\lambda = F\Delta. \tag{8.62}$$

We find from (8.60) and (8.61) that

$$D\lambda = -D\frac{\partial u}{\partial x} = J = \frac{D_e}{NL}(U_0 - U_1) \tag{8.63}$$

$$= \frac{D_e}{L}(L\lambda + \Delta) \tag{8.64}$$

$$= \lambda D_e\left(1 + \frac{D}{FL}\right), \tag{8.65}$$

from which it follows that

$$\frac{1}{D_e} = \frac{1}{D} + \frac{1}{LF}. \tag{8.66}$$

8.2.2   Homogenization

The above calculation of the effective diffusion coefficient can be formalized by the process of *homogenization*. Homogenization is an important technique that was described in the Appendix to Chapter 7 and which is seen again in Chapter 12. Homogenization is useful when there are two spatial scales, a microscopic and a macroscopic scale in determining the behavior of the solution on the macroscopic scale while accounting for influences from the microscopic scale, without calculating the full details of the solution on the microscopic scale.

The basic assumption here is that Fick's law holds, but that the diffusion coefficient D is a periodic, rapidly varying function, so that the flux is

$$J = -D\left(\frac{x}{\epsilon}\right)\frac{\partial u}{\partial x}. \tag{8.67}$$

The dimensionless parameter ϵ is small, indicating that the variations of D are rapid compared to other spatial scales of the problem, in particular, the diffusion length. Thus, cells are assumed to be short compared to the diffusion length scale.

In Section 7.8.2 we calculated that the effective diffusion coefficient is the inverse of the average resistance

$$D_e = \frac{1}{\bar{R}}. \tag{8.68}$$

where

$$\bar{R} = \int_0^1 R(s)\,ds = \int_0^1 \frac{1}{D(s)}\,ds \tag{8.69}$$

is the average resistance.

To apply this result to the specific problem of gap junctions, we take $R(x) = r_c + r_g \sum_k \delta(x - kL)$ to reflect the periodic occurrence of gap junctions with resistance r_g evenly spaced at the ends of cells of length L. Notice that $D = 1/r_c$ is the diffusion coefficient for the intracellular space, while $F = 1/r_g$ is the intercellular permeability. It follows easily that

$$\bar{R} = r_c + \frac{r_g}{L},$$ (8.70)

which is the same as (8.66).

8.2.3 Measurement of Permeabilities

Although an effective diffusion coefficient is useful when the species of interest diffuses through a large number of cells, in some experimental situations one is interested in how a dye molecule (or a second messenger such as IP_3) diffuses through a relatively small number of cells. In this case the effective diffusion coefficient approximation cannot always be used, and it is necessary to solve the equations with internal boundary conditions (Brink and Ramanan, 1985; Ramanan and Brink, 1990). By calculating exact solutions to the linear diffusion equation with internal boundary conditions (using transform methods, for example) and fitting them to experimental measurements on the movement of fluorescent probes, it is possible to obtain estimates of the intracellular diffusion coefficient as well as the permeability of the intercellular membrane.

The analytic solutions of Brink and Ramanan are useful only when the underlying equations are linear. In many cases, however, the species of interest are also reacting in a nonlinear way. This results in a system of nonlinear diffusion equations coupled by jump conditions at the gap junctions, a system that can only be solved numerically. Two groups have used numerical methods to study problems of this kind. Christ et al. (1994) studied the problem of diffusion through gap junctions, assuming that the diffusing species u decreases the permeability of the gap junction in a nonlinear fashion. A similar model was used by Sneyd et al. (1995a) to study the spread of a Ca^{2+} wave through a layer of cells coupled by gap junctions, and this model is described in Chapter 7.

8.2.4 The Role of Gap-Junction Distribution

It is not always appreciated that the intercellular permeability is strongly influenced by the distribution of gap junctions in the intercellular membrane, although it is a common observation in introductory biology texts that there is a similar relationship between the distribution of stomata on leaves and the rate of evaporation of water through the leaf surface. Individual gap junctions are usually found in aggregates forming larger junctional plaques, as individual gap-junction particles are not easily distinguished from other nonjunctional particles. However, numerical simulations show that the permeability of the intercellular membrane decreases as the gap junction particles aggregate in larger groupings. This raises the intriguing possibility that intercellular permeability may be lowest when the gap-junctional plaques are easiest to see. This in turn provides a possible explanation for the fact that it has been difficult to establish a direct link between the number of recognizable gap junctions and the intercellular permeability.

Chen and Meng (1995) constructed a cubic lattice model of a two-cell system with a common border. A number of gap-junction particles, with varying degrees of aggregation, were placed on the border lattice points. Marker particles were placed in one of the cubes and followed a random walk over the lattice points of the cube. When they encountered a gap-junction lattice point on the boundary, there was an assigned probability that the marker particle would move across to the other cell. By measuring the time required for a certain percentage of marker particles to cross from one cell to the other, Chen and Meng obtained a quantitative estimate of the efficiency of intercellular transport as a function of gap-junction aggregation. Their results are summarized in Fig. 8.18. When the gap junctions are clumped together into a single junctional plaque, 10,000 time steps were required for the transfer of about 10% of the marker particles. However, when the gap-junction particles were randomly scattered, only 1,000 time steps were required for the same transfer. The magnitude of this discrepancy emphasizes the fact that gap junction distribution can have a huge effect on the rate of intercellular transport.

This result makes intuitive sense. The transfer rate is related to the time it takes for a molecule undergoing a random walk to find an exit. It stands to reason that it is

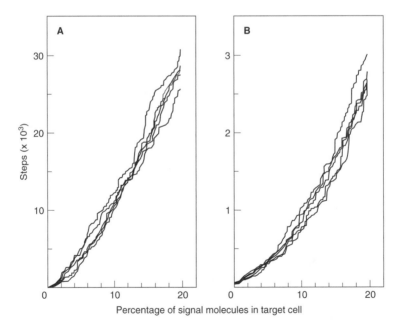

Figure 8.18 Simulation results of the cubic-lattice gap-junction model on a 50 × 50 × 50 lattice with 1000 signal molecules in the source cell at time 0. In A, 100 gap-junction particles are arranged in a compact junctional plaque, while in B they are scattered randomly on the intercellular interface. The random scattering of gap-junction particles results in a greatly increased intercellular transfer rate (note the different scales for the two panels). (Chen and Meng, 1995, Fig. 1.)

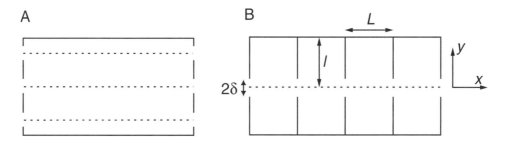

Figure 8.19 A: Sketch of a single rectangular cell with gap-junctional openings in the end faces. B: Sketch of cell array, reduced by symmetry to a single half-channel.

easier to find an exit if there are many of them scattered around as opposed to having all of them clumped together in one place.

To get a more analytical understanding of how the distribution of gap junctions affects the diffusion coefficient, we solve a model problem, similar to the one-dimensional problem solved in Section 8.2.1. We consider cells to be two-dimensional rectangles, with a portion of their ends open for diffusive transport (the gap junctions) and the remainder closed (Fig. 8.19A). The dashed lines in this figure are lines of symmetry across which there is no flux in a steady-state problem, so we can reduce the cell configuration to that shown in Fig. 8.19B.

To study diffusion in the x-coordinate direction, we assume that the vertical walls are separated by length L and have regularly spaced openings of width 2δ with centers separated by length $2l$. The fraction of the vertical separator that is open between cells is $\Delta = \delta/l$. To study how the distribution of gap junctions affects the diffusion coefficient, we hold Δ fixed while varying l. When l is small, the gap junctions are small and uniformly distributed, while when l is large, the gap junctions are clumped together into larger aggregates; in either case the same fraction (Δ) of the intercellular membrane is occupied by gap junctions.

Suppose that there are a large number of cells (say N) each of length L connected end to end. We impose a fixed concentration gradient across the array and use the definition (8.60) to define the effective diffusion coefficient for this array.

To find the flux, we solve Laplace's equation subject to no-flux boundary conditions on the horizontal lines $y = 0$ and $y = l$ and on the vertical lines $\delta < y < l, x = pL, p = 0, \ldots, N$. We further divide this region into two subregions, one for $y \geq \delta$ and one for $y \leq \delta$.

Consider first the solution on the upper region. The solution for a single cell $0 \leq x \leq L$ can be found by separation of variables to be

$$u(x,y) = \sum_{n=0}^{\infty} a_n \cos\left(\frac{n\pi x}{L}\right) \cosh\left(\frac{n\pi(y-l)}{L}\right). \qquad (8.71)$$

This solution satisfies no-flux boundary conditions at $y = l$ and at $x = 0, L$. Notice also that this solution is periodic, so it is a contender for the solution for any cell.

Now recall that in the one-dimensional case, the solution is piecewise linear, with jumps at the cell boundaries, and that the slope of the solution within each cell is the same (Section 8.2.1). This suggests that the derivative of the solution in the two-dimensional case should be the same in each cell, or equivalently, the solution in each cell should be the same up to an additive constant. Thus,

$$u(x, y) = \sum_{n=1}^{\infty} A_n \frac{\cosh(n\pi(y - l)/L)}{\cosh(n\pi(\delta - l)/L)} \cos\left(\frac{n\pi}{L}(x - pL)\right) + \alpha_p, \tag{8.72}$$

for $pL < x < (p + 1)L$, $\delta < y < l$, and $p = 0, \ldots, N - 1$. We have scaled the unknown constants, A_n, by $\cosh(n\pi(\delta - l)/L)$ for convenience.

On the lower region, a similar argument gives

$$u(x, t) = (U_1 - U_0)\frac{x}{NL} + U_0 + \sum_{n=1}^{\infty} \frac{\cosh(2n\pi y/L)}{\cosh(2n\pi\delta/L)}\left(C_n \sin\frac{2n\pi x}{L}\right), \tag{8.73}$$

for $0 < x < NL, 0 < y < \delta$. Notice that this solution satisfies a no-flux boundary condition at $y = 0$ and has the correct overall concentration gradient.

Now, to make these into a smooth solution of Laplace's equation we require that $u(x, y)$ and $u_y(x, y)$ be continuous at $y = \delta$. This gives two conditions,

$$\sum_{n=1}^{\infty} A_n \cos\left(\frac{n\pi}{L}(x - pL)\right) = \sum_{n=1}^{\infty}\left(C_n \sin\frac{2n\pi x}{L}\right) + (U_1 - U_0)\frac{x}{nL} + U_0 - \alpha_p, \tag{8.74}$$

and

$$\sum_{n=1}^{\infty} nA_n \tanh\left(\frac{n\pi}{L}(\delta - l)\right) \cos\left(\frac{n\pi}{L}(x - pL)\right) = \sum_{n=1}^{\infty} 2n \tanh\frac{2n\pi\delta}{L}C_n \sin\frac{2n\pi x}{L}, \tag{8.75}$$

on the interval $pL < x < (p + 1)L$.

We determine α_p by averaging (8.74) over cell p. Integrating (8.74) from $x = (p-1)L$ to $x = pL$ gives

$$\alpha_p = \frac{1}{L}\int_{(p-1)L}^{pL} (U_1 - U_0)\frac{x}{NL}\, dx + U_0, \tag{8.76}$$

since all the trigonometric terms integrate to zero. Hence,

$$\alpha_p = U_0 - (U_0 - U_1)\frac{p - 1/2}{N}. \tag{8.77}$$

Finally, for convenience, we choose $U_0 = N/2$ and $U_1 = -N/2$, which gives $\alpha_p = p + (1 + N)/2$. Since this is a linear problem, the values chosen for U_0 and U_1 have no effect on the effective diffusion coefficient.

To obtain equations for the coefficients, we project each of these onto $\cos \frac{k\pi x}{L}$ by multiplying by $\cos \frac{k\pi x}{L}$ and integrating from 0 to L. We find that

$$A_k \frac{L}{2} = F_k + \sum_{n=1}^{\infty} C_n I_{2n,k}, \tag{8.78}$$

and

$$k A_k \tanh \left(\frac{k\pi}{L} (\delta - l) \right) \frac{L}{2} = \sum_{n=1}^{\infty} 2n \tanh \frac{2n\pi\delta}{L} \left(C_n I_{2n,k} \right), \tag{8.79}$$

where

$$F_k = \int_0^L \left(\frac{x}{L} - \frac{1}{2} \right) \cos \frac{k\pi x}{L} \, dx = \frac{L}{n^2 \pi^2} ((-1)^k - 1), \tag{8.80}$$

$$I_{n,k} = \int_0^L \sin \frac{n\pi x}{L} \cos \frac{k\pi x}{L} \, dx = \frac{Ln}{\pi} \left(\frac{1 - (-1)^{n+k}}{n^2 - k^2} \right). \tag{8.81}$$

There is an immediate simplification possible. Notice that $I_{2n,k} = 0$ and $F_k = 0$ when k is even. Thus, $A_k = 0$ for all even k. Now we eliminate the coefficients A_k from (8.78) and (8.79) to obtain

$$\sum_{n=1}^{\infty} C_n \left(\frac{2n}{k} \frac{\tanh \frac{2n\pi l}{L} \Delta}{\tanh \frac{k\pi l}{L} (1 - \Delta)} + 1 \right) \frac{n}{4n^2 - k^2} = \frac{1}{2\pi k^2}, \tag{8.82}$$

for all odd k, with $\Delta = \delta/l$. Since k can take on any odd positive integer value, (8.82) is an infinite set of equations for the coefficients C_n. Since the solution of the differential equation converges, we can truncate this system of equations and solve the resulting finite linear system numerically.

In terms of this solution, the average flux is

$$J = \frac{D}{l} \int_0^\delta \frac{\partial u}{\partial x} \bigg|_{x=0} dy = D \left(\frac{\Delta}{L} + \frac{1}{l} \sum_{n=1}^{\infty} C_n \tanh \frac{2n\pi l \Delta}{L} \right), \tag{8.83}$$

and the effective diffusion coefficient is

$$D_e = D \left(\frac{L}{l} \sum_{n=1}^{\infty} C_n \tanh \frac{2n\pi l \Delta}{L} + \Delta \right). \tag{8.84}$$

Typical results are shown in Fig. 8.20A, where the ratio D_e/D is shown plotted as a function of l/L for different values of fixed $\Delta = \delta/l$, and in Fig. 8.20B, where D_e/D is shown plotted as a function of Δ for fixed l/L.

There are a number of important observations that can be made. First, notice that in the limit $\Delta \to 1$, or $l/L \to \infty$, $\frac{L}{l} C_n \tanh \frac{2n\pi l \Delta}{L} \to 0$. Thus,

$$\lim_{\Delta \to 1} D_e = D, \tag{8.85}$$

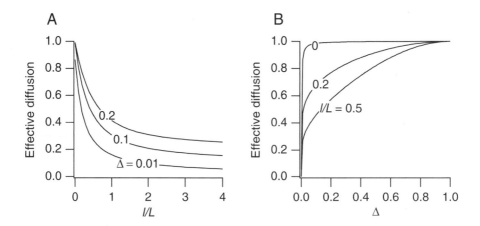

Figure 8.20 A: Effective diffusion ratio D_e/D as a function of the distribution ratio l/L for fixed gap-junction fraction Δ. B: Effective diffusion ratio D_e/D as a function of the gap-junction fraction Δ for fixed distribution ratio l/L.

and

$$\lim_{\frac{l}{L} \to \infty} D_e = D\Delta. \tag{8.86}$$

Finally, from the numerical solution, it can be seen that D_e is a decreasing function of l/L. Thus, clumping of gap junctions lowers the effective diffusion coefficient compared with spreading them out uniformly, in agreement with numerical experiments described above.

From Fig. 8.20B we see that when gap junctions are small but uniformly spread, there is little decrease in the effective diffusion coefficient, unless Δ is quite small. Thus, for example, with $\Delta = 0.01$ (so that gap junctions comprise 1% of the membrane surface area) and $l = 0$, the effective diffusion coefficient is about 86% of the cytoplasmic diffusion. On the other hand, with only one large gap junction with $\Delta = 0.01$ in the end membrane of a square cell ($l = 0.5$), the effective diffusion coefficient is reduced to about 27% of the original.

It is interesting to relate these results to the one-dimensional solution (8.66). This two-dimensional problem becomes effectively one-dimensional in the limit $\frac{l}{L} \to 0$, with a piecewise-linear profile in the interior of the cells and small boundary or corner layers at the end faces. In this limit, the effective diffusion coefficient satisfies

$$\frac{D}{D_e} = 1 + \mu \frac{1 - \Delta}{\Delta}, \tag{8.87}$$

with $\mu = 0.0016$. This formula was found by plotting the curve $\Delta(\frac{D}{D_e} - 1)$ against Δ, which, remarkably, is numerically indistinguishable from the straight line $\mu(1 - \Delta)$. Comparing this with the one-dimensional result, we find that the end-face permeability,

F, can be related to the fraction of gap junctions Δ through

$$F = \frac{D}{L}\frac{\Delta}{\mu(1-\Delta)}. \tag{8.88}$$

8.3 EXERCISES

1. Use that c_i is small compared to c_e to simplify the Llinás model of presynaptic Ca^{2+} channels (Peskin, 1991). Show that this simplification gives little difference to the behavior of the full model shown in Figs. 8.5 and 8.6.

2. (a) The model of the presynaptic Ca^{2+} current discussed in the text cannot give the curves shown in Fig. 8.7 in response to a voltage jump. Why not?

   (b) Calculate and plot the solution to (8.7) if $V = V_0 + (V_1 - V_0)(H(t) - H(t-4))$, where H is the Heaviside function.

3. Calculate the analytic solution to (8.7) when V is a given function of t.

4. (a) In Fig. 8.13, why are some of the curves positive and some negative?

   (b) Construct a simple function $F(t)$ with the same qualitative shape as the function $W(t)$ used in the Magleby and Stevens model of end-plate currents (Fig. 8.12).

   (c) Calculate the analytic solution to (8.48) for this function F. Compare to the curves shown in Fig. 8.13.

5. In the Magleby and Stevens model of end-plate currents, a simple choice for the release function $f(t)$ results in end-plate conductances with considerable qualitative similarity with those in Fig. 8.13. Suppose there is an instantaneous release of ACh into the synaptic cleft at time $t = 0$. Take $f(t) = \gamma\delta(t)$, where δ is the Dirac delta function and find the conductance x which is the solution of (8.44), to leading order in $\frac{\gamma}{K}$ in the limit that $\frac{\gamma}{K}$ is small. Show that $x(t)$ is always positive. Plot $x(t)$.

6. Peskin (1991) presented a more complex version of the Magleby and Stevens model. His model is based on the reaction scheme

$$\longrightarrow ACh, \qquad \text{rate } r_T \text{ per unit volume}, \tag{8.89}$$

$$ACh + R \underset{k_2}{\overset{k_1}{\rightleftarrows}} ACh \cdot R \underset{\alpha}{\overset{\beta}{\rightleftarrows}} ACh \cdot R^*, \tag{8.90}$$

$$ACh + E \underset{k_4}{\overset{k_3}{\rightleftarrows}} ACh \cdot E \overset{\gamma}{\longrightarrow} E, \tag{8.91}$$

where E is some enzyme that degrades ACh in the synaptic cleft. (The Peskin model differs from the Magleby and Stevens model in two ways, with the assumption of enzymatic degradation of ACh and the assumption that the amount of ACh bound to its receptor is not negligible.)

   (a) Write down the equations for the 6 dependent variables. Use conservation laws to eliminate two of the equations.

   (b) Assume that the reactions involving ACh with R, and ACh with E (with reaction rates $k_i, i = 1, 4$), are fast to obtain expressions for [R] and [E] in terms of the other variables.

Substitute these expressions into the differential equations for [ACh · R*] and [ACh] − [R] − [E] to find two differential equations for [ACh · R*] and [ACh].

(c) Solve these equations when the stimulus is a small sudden release of ACh (i.e., assume that $r_T = \epsilon \delta(t)$ where ϵ is small - small compared to what?). Compare this solution with that of Exercise 5.

Remark: After appropriate rescaling and approximation, the equations are of the form

$$\frac{dx}{dt} = a_1 y - \alpha x \tag{8.92}$$

$$\frac{dx}{dt} + a_2 \frac{dy}{dt} = \delta(t) - a_3 y, \tag{8.93}$$

for some constants a_1, a_2 and a_3.

7. Solve (8.54) with $g_s(t) = x(t)$ as found in Exercise 5 and plot the solution. Compare to $x(t)$ found in Exercise 5 (i.e., (8.55)).

8. The purpose of this exercise is to link together the models of this chapter to construct a unified model of the synaptic cleft that connects the presynaptic action potential to the postsynaptic voltage via the concentration of ACh in the synaptic cleft.

(a) First, use the FitzHugh–Nagumo equations (Section 5.2) to construct an action potential that serves as the input to the overall model. Remark: if we let V_1 denote the presynaptic membrane potential, and let $v = V_1 - V_r$, the equations

$$\frac{dv}{dt} = 0.01 v (70 - v)(v - 7) - 100 w, \tag{8.94}$$

$$\frac{dw}{dt} = 0.25(v - 5w), \tag{8.95}$$

do a reasonable job. What is V_r? What value should be chosen for V_r? Solve these equations and plot the presynaptic action potential.

(b) Link this presynaptic voltage to the presynaptic Ca^{2+} concentration by constructing a differential equation for the Ca^{2+} concentration. (Hint: Take the inward Ca^{2+} current given by the Llinás model of Section 8.1.2 $P_{Ca} = 0.03 \ \mu ms^{-1}$ and a linear Ca^{2+} removal with time constant 1 ms works well. Be careful with the units.) Plot the Ca^{2+} transient that results from the presynaptic action potential calculated in part (a) above.

(c) Link to the model of Ca^{2+}-mediated release of neurotransmitter (Section 8.1.3). Plot, as a function of time, the probability that the release site is activated.

(d) Next, link to the model of neurotransmitter kinetics in the synaptic cleft (Section 8.1.4). (Hint: The rate of release of neurotransmitter into the cleft should be proportional to the number of activated release sites. Choose this constant of proportionality to be 3×10^6.) Then link this to the model of the postsynaptic membrane potential, V_2, by letting $g_s(t) = x(t)$. Plot the concentration of ACh and the concentration of open ACh-sensitive channels as functions of time.

(e) Finally, plot V_1 and V_2 on the same graph to compare the input and output voltages of the model.

9. Incorporate the effects of nicotine into a model of ACh activation of receptors.

CHAPTER 9

Neuroendocrine Cells

There are many hormones that circulate through the body, controlling a diverse array of functions, from appetite to body temperature to blood pH. These hormones are secreted from specialized cells in various glands, such as the hypothalamus and pituitary, the pancreas, or the thyroid. Models of hormone physiology at the level of the entire body are discussed in Chapter 16. Here, we consider models of the cells that secrete the hormones, the *neuroendocrine* cells. They are called neuroendocrine (or sometimes neurosecretory) as they have many of the hallmarks of neurons, such as membrane excitability, but are specialized, not to secrete neurotransmitter into a synaptic cleft, but to secrete hormones into the blood. However, not only is there a fine line between hormones and neurotransmitters, there is also little qualitative difference between secretion into a synaptic cleft, and secretion into the bloodstream. Thus it does not pay to draw too rigid a distinction between neurons and neuroendocrine cells.

Although, unsurprisingly, there is a great variety of neuroendocrine cells, they have certain characteristics that serve to unify their study. First, they are excitable and therefore have action potentials. Second, the electrical activity usually is not a simple action potential, or periodic train of action potentials (Chapter 5). Instead, the action potentials are characterized by bursts of rapid oscillatory activity interspersed with quiescent periods during which the membrane potential changes only slowly. This behavior is called *bursting*. Third, bursting is often closely regulated by the intracellular Ca^{2+} concentration (Chapter 7). Thus, models of neurosecretory cells typically combine models of membrane electrical excitability and Ca^{2+} excitability, leading to a fascinating array of dynamic behaviors.

Other factors can also influence bursting; for example, as is described later in this chapter, recent models of bursting in the pancreatic β cell include models of the glycolytic pathway, thus leading to models that incorporate, in a single cell, many of the complexities described in Chapters 1, 5 and 7. Neuroendocrine cells are thus wonderful examples of how multiple oscillatory mechanisms can interact, and provide an enormous range of interesting behaviors to explore.

9.1   Pancreatic β Cells

In response to glucose, β cells of the pancreatic islet secrete insulin, which causes the increased use or uptake of glucose in target tissues such as muscle, liver, and adipose tissue. When blood levels of glucose decline, insulin secretion stops, and the tissues begin to use their energy stores instead. Interruption of this control system results in diabetes, a disease that, if left uncontrolled, can result in kidney failure, heart disease, and death. It is believed that bursting, a typical example of which is shown in Fig. 9.1, plays an important (but not exclusive) role in the release of insulin from β cells.

9.1.1   Bursting in the Pancreatic β Cell

Bursting in the pancreatic β cell occurs with a wide variety of periods, ranging from a few seconds to a few minutes. Typically, these are divided into three groups; fast bursting, with periods of around 2 to 5 s; medium bursting, with periods of around 10 to 60 s; and slow bursting, with periods of around 2 to 4 minutes.

Although bursting has been studied extensively for many years, most mathematical studies are based on the pioneering work of Rinzel (1985, 1987), which was in turn based on one of the first biophysical models of a pancreatic β cell (Chay and Keizer, 1983). Rinzel's interpretation of bursting in terms of nonlinear dynamics provides an excellent example of how mathematics can be used to understand complex biological dynamical systems.

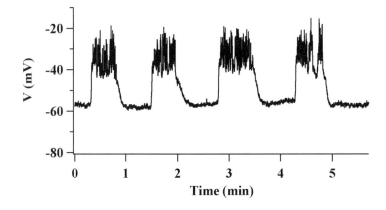

Figure 9.1   Bursting oscillations in the pancreatic β cell. Provided by Les Satin, Min Zhang and Richard Bertram.

Models of bursting in pancreatic β cells can be divided into two major groups (well summarized by de Vries, 1995). Earlier models were generally based on the assumption that bursting was caused by an underlying slow oscillation in the intracellular Ca^{2+} concentration (Chay, 1986, 1987; Chay and Cook, 1988; Chay and Kang, 1987; Himmel and Chay, 1987; Keizer and Magnus, 1989). In light of more recent experimental evidence showing that Ca^{2+} is not the slow variable underlying bursting, more recent models have modified this assumption, relying on alternative mechanisms to produce the underlying slow oscillation (Keizer and Smolen, 1991; Smolen and Keizer, 1992; Bertram and Sherman, 2004a,b; Nunemaker et al., 2006).

One of the first models of bursting was proposed by Atwater et al. (1980). It was based on extensive experimental data, incorporating the important cellular mechanisms that were thought to underlie bursting, and was later developed into a mathematical model by Chay and Keizer (1983). Although the mathematical model includes only those processes believed to be essential to the bursting process and thus omits many features of the cell, it is able to reproduce many of the basic properties of bursting. The ionic currents in the model are:

1. A Ca^{2+}-activated K^+ channel with conductance an increasing function of $c = [Ca^{2+}]$ of the form

$$g_{K,Ca} = \bar{g}_{K,Ca} \frac{c}{K_d + c}, \tag{9.1}$$

   for some constant $\bar{g}_{K,Ca}$.
2. A voltage-gated K^+ channel modeled in the same way as in the Hodgkin–Huxley equations, with

$$g_K = \bar{g}_K n^4, \tag{9.2}$$

   where n obeys the same differential equation as in the Hodgkin–Huxley equations (Chapter 5), except that the voltage is shifted by V^*, so that V in (5.28) and (5.29) is replaced by $V + V^*$. For example, $\beta_n(V) = 0.125 \exp[(-V - V^*)/80]$.
3. A voltage-gated Ca^{2+} channel, with conductance

$$g_{Ca} = \bar{g}_{Ca} m^3 h, \tag{9.3}$$

   where m and h satisfy Hodgkin–Huxley differential equations for Na^+ gating, shifted along the voltage axis by V'. That is, the inward Ca^{2+} current is modeled by the Na^+ current of the Hodgkin–Huxley equations.

Combining these ionic currents and adding a leak current gives

$$C_m \frac{dV}{dt} = -(g_{K,Ca} + g_K)(V - V_K) - 2g_{Ca}(V - V_{Ca}) - g_L(V - V_L), \tag{9.4}$$

where C_m is the membrane capacitance.

To complete the model, there is an equation for the regulation of intracellular Ca^{2+},

$$\frac{dc}{dt} = f(-k_1 I_{Ca} - k_c c), \tag{9.5}$$

where the Ca^{2+} current is $I_{Ca} = \bar{g}_{Ca} m^3 h (V - V_{Ca})$ and where k_1 and k_c are constants. The constant f is a scale factor relating total changes in $[Ca^{2+}]$ to the changes in free $[Ca^{2+}]$ (as discussed in the section on Ca^{2+} buffering in Chapter 7) and is usually a small number, while k_c is the rate at which Ca^{2+} is removed from the cytoplasm by the membrane ATPase pump.

For this model it is assumed that glucose regulates the rate of removal of Ca^{2+} from the cytoplasm. Thus, k_c is assumed to be an (unspecified) increasing function of glucose concentration. However, the concentration of glucose is not a dynamic variable of the model, so that k_c can be regarded as fixed, and the behavior of the model can be studied for a range of values of k_c.

A numerically computed solution of this model, shown in Fig. 9.2, exhibits bursts that bear a qualitative resemblance to those seen experimentally. It is also readily seen that there is a slow oscillation in c underlying the bursts, with bursting occurring during the peak of the Ca^{2+} oscillation. The fact that Ca^{2+} oscillations occur on a slower time scale is built into the Ca^{2+} equation (9.5) explicitly by means of the parameter f. As f becomes smaller, the Ca^{2+} equation evolves more slowly, and thus the relative speeds of

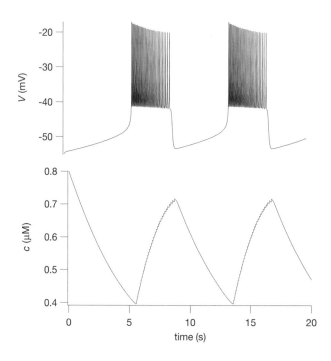

Figure 9.2 Bursting oscillations in the Chay–Keizer β cell model, calculated using the parameter values in Table 9.1.

Table 9.1 Parameters of the model of electrical bursting in pancreatic β cells.

$C_m = 1\ \mu F/cm^2$	$\bar{g}_{K,Ca} = 0.02\ mS/cm^2$
$\bar{g}_K = 3\ mS/cm^2$	$\bar{g}_{Ca} = 3.2\ mS/cm^2$
$\bar{g}_L = 0.012\ mS/cm^2$	$V_K = -75\ mV$
$V_{Ca} = 100\ mV$	$V_L = -40\ mV$
$V^* = 30\ mV$	$V' = 50\ mV$
$K_d = 1\ \mu M$	$f = 0.007$
$k_1 = 0.0275\ \mu M\ cm^2/nC$ $\quad$ $k_c = 0.02\ ms^{-1}$	

the voltage and Ca^{2+} equations are directly controlled. It therefore appears that there are two oscillatory processes interacting to give bursting, with a fast oscillation in V superimposed on a slower oscillation in c. This is the basis of the phase-plane analysis that we describe next.

Phase-Plane Analysis

The β cell model can be simplified by ignoring the dynamics of m and h, thus removing the time dependence (but not the voltage dependence) of the Ca^{2+} current (Rinzel and Lee, 1986). The simplified model equations are

$$C_m \frac{dV}{dt} = -I_{Ca}(V) - \left(\bar{g}_K n^4 + \frac{\bar{g}_{K,Ca} c}{K_d + c}\right)(V - V_K) - \bar{g}_L(V - V_L), \qquad (9.6)$$

$$\tau_n(V)\frac{dn}{dt} = n_\infty(V) - n, \qquad (9.7)$$

$$\frac{dc}{dt} = f(-k_1 I_{Ca}(V) - k_c c), \qquad (9.8)$$

where $I_{Ca} = \bar{g}_{Ca} m_\infty^3(V) h_\infty(V)(V - V_{Ca})$.

Since f is small, this β cell model separates into a fast subsystem (the V and n equations) and a slow equation for c. The fast subsystem can be studied using phase-plane methods, and then the behavior of the full system can be understood as slow variations of the fast phase plane system.

We first consider the structure of the fast subsystem as a function of c, treating c as a fixed parameter.

When c is low, the Ca^{2+}-activated K^+ channel is not activated, and the fast subsystem has a unique fixed point with V high. Conversely, when c is high, the Ca^{2+}-activated K^+ channel is fully activated, and the fast subsystem has a unique fixed point with V low, as the high conductance of the Ca^{2+}-activated K^+ channels pulls the membrane potential closer to the K^+ Nernst potential, about -75 mV. However, for intermediate values of c there are three fixed points, and the phase plane is much more intricate.

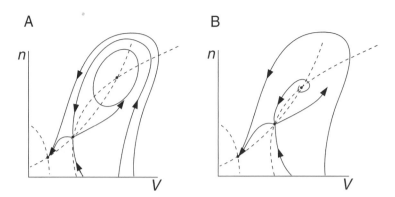

Figure 9.3 Phase planes of the fast subsystem of the Chay–Keizer β cell model, for two different values of c, both in the intermediate range. The phase planes are sketched, not drawn to scale. Nullclines are denoted by dashed lines, and the intersections of the nullclines show the positions of the fixed points. For both values of c there are three fixed points, of which the middle one is a saddle point. However, in A (with $c_{hb} < c < c_{hc}$; see Fig. 9.4) the unstable node is surrounded by a stable limit cycle, while in B (corresponding to $c > c_{hc}$) the limit cycle has disappeared via a homoclinic bifurcation.

Phase planes of the V, n subsystem for two different intermediate values of c are shown in Fig. 9.3.

In both cases, the lower fixed point is stable, the middle fixed point is a saddle point, and the upper fixed point is unstable. For some values of c the upper fixed point is surrounded by a stable limit cycle, which in turn is surrounded by the stable manifold of the saddle point (Fig. 9.3A). However, as c increases (still in the intermediate range), the limit cycle "hits" the saddle point and forms a homoclinic connection (a homoclinic bifurcation). Increasing c further breaks the homoclinic connection, and the stable manifold of the saddle point forms a heteroclinic connection with the upper, unstable, critical point (Fig. 9.3B). There is now no limit cycle.

This sequence of bifurcations can be summarized in a bifurcation diagram, with V plotted against the control parameter c (Fig. 9.4A). The Z-shaped curve is the curve of fixed points, and as usual, the stable oscillation around the upper steady state is depicted by the maximum and minimum of V through one cycle. As c increases, oscillations appear via a Hopf bifurcation (c_{hb}) and disappear again via a homoclinic bifurcation (c_{hc}). For a range of values of c the fast subsystem is bistable, with a lower stable fixed point and an upper stable periodic orbit. This bistability is crucial to the appearance of bursting.

We now couple the dynamics of the fast subsystem to the slower dynamics of c. Included in Fig. 9.4A is the curve defined by $dc/dt = 0$, i.e., the c nullcline. When V is above the c nullcline, $dc/dt > 0$, and so c increases, but when V is below the c nullcline, c decreases. Now suppose V starts on the lower fixed point for a value of c that is greater than c_{hc}. Since V is below the c nullcline, c starts to decrease, and

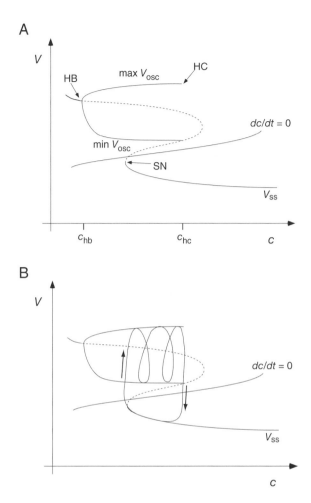

Figure 9.4 A: Sketch of the bifurcation diagram of the simplified Chay–Keizer β cell model, with c as the bifurcation parameter. V_{ss} denotes the curve of steady states of V as a function of c. A solid line indicates a stable steady state; a dashed line indicates an unstable steady state. The two branches of V_{osc} denote the maximum and minimum of V over one oscillatory cycle. HB denotes a Hopf bifurcation, HC denotes a homoclinic bifurcation, and SN denotes a saddle-node bifurcation. B: A burst cycle projected on the (V, c) plane. (Adapted from Rinzel and Lee, 1986, Fig. 3.)

V follows the lower branch of fixed points. However, when c becomes too small, this lower branch of fixed points disappears in a saddle-node bifurcation (SN), and so V must switch to the upper branch of the Z-shaped curve. Since this upper branch is unstable and surrounded by a stable limit cycle, V begins to oscillate. However, since V now lies entirely above the c nullcline, c begins to increase. Eventually, c increases enough to cross the homoclinic bifurcation at c_{hc}, the stable limit cycles disappear, and V switches back to the lower branch, completing the cycle. Repetition of this process causes bursting. The quiescent phase of the bursting cycle is when V is on the lower branch of the Z-shaped curve, and during this phase V increases slowly. A burst of oscillations occurs when V switches to the upper branch, and disappears again after passage through the homoclinic bifurcation. Clearly, in this scenario, bursting relies on the coexistence of both a stable fixed point and a stable limit cycle, and the bursting cycle is a hysteresis loop that switches between branches of the Z-shaped curve. Bursting also relies on the c nullcline intersecting the Z-shaped curve in the right location. For example, if the c nullcline intersects the Z-shaped curve on its lower branch, there is a unique stable fixed point for the whole system, and bursting does not occur. A projection of the bursting cycle on the (V, c) phase plane is shown in Fig. 9.4B. The periods of the oscillations in the burst increase through the burst, as the limit cycles get closer to the homoclinic trajectory, which has infinite period.

The relationship between bursting patterns and glucose concentration can also be deduced from Fig. 9.4. Notice that the $\frac{dc}{dt} = 0$ nullcline, given by $c = -\frac{k_1}{k_c} I_{Ca}(V)$, is inversely proportional to k_c. Increasing k_c moves the $\frac{dc}{dt} = 0$ nullcline to the left while decreasing k_c moves it to the right. Thus, when k_c is sufficiently small, the nullcline intersects the lower branch of the V nullcline. On the other hand, if k_c is extremely large, the c nullcline intersects the upper branch of the V nullcline, possibly to the left of c_{hb}. At intermediate values of c, the c nullclines intersects the middle branch of the V nullcline.

Under the assumption that k_c is monotonically related to the glucose concentration, when the glucose concentration is low, the system is at a stable rest point on the lower V nullcline; there is no bursting. If glucose is increased so that the c nullcline intersects the middle V nullcline with $c < c_{hc}$, there is bursting. However, the length of the bursting phase increases and the length of the resting phase decreases with increasing glucose, because Ca^{2+} increases at a slower rate and decreases at a faster rate when k_c is increased. For large enough k_c the bursting is sustained with no rest phase, as c becomes stalled below c_{hc}. Finally, at extremely high k_c values, bursting is replaced by a permanent high membrane potential, with $c < c_{hb}$. This dependence of the bursting phase on glucose is confirmed by experiments.

9.1.2   ER Calcium as a Slow Controlling Variable

There are two major problems with the above model. First, it does not reproduce the wide variety of periods and bursting patterns actually seen in bursting pancreatic β cells, being limited to a narrow range of fast bursting frequencies.

Second, more recent experimental evidence has shown that Ca^{2+} oscillates much too fast to be viewed as a slow control variable. This is illustrated in Fig. 9.5, which shows simultaneous Ca^{2+} and voltage measurements. The bursting oscillations in the voltage are mirrored by bursting oscillations in the cytoplasmic Ca^{2+} concentration, and the rise in Ca^{2+} concentration is almost as fast as the rise in voltage.

So the question arises of what controls the length of bursting. One possibility is that the ER Ca^{2+} concentration varies much more slowly than the cytoplasmic Ca^{2+} and could provide the necessary control mechanism. This would be the case if most of the Ca^{2+} during active bursting were coming from the extracellular space through transmembrane ion channels, and only a small amount of Ca^{2+} flowed between the cytoplasm and the ER. If this were the case, ER Ca^{2+} would act like a low pass filter for cytoplasmic Ca^{2+}, and therefore could be used to detect and regulate the length of bursting activity. The possible usefulness of a low-pass filter is seen in the Ca^{2+} traces shown in Fig. 9.5, where, during a burst, Ca^{2+} concentration oscillates rapidly around

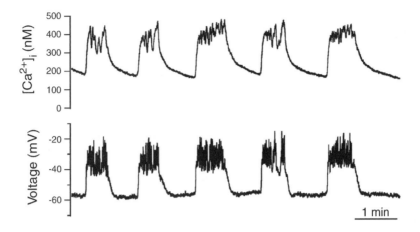

Figure 9.5 Simultaneous Ca^{2+} and voltage measurements from a bursting pancreatic β cell. (Adapted from Zhang et al., 2003, Fig. 4.)

a raised baseline. A low-pass filter would measure the length of time that the Ca^{2+} baseline is elevated, but filter out the rapid oscillations.

Thus, it was proposed by Chay (1996a,b, 1997) that slow variations in ER Ca^{2+} could be an important control mechanism, and could, in addition, generate a wider range of bursting periods. This proposal was analyzed in detail by Bertram and Sherman (2004a) who showed that the interaction of a (not very slow) cytoplasmic Ca^{2+} variable, with a (much slower) ER Ca^{2+} variable, could lead to bursting with a wide range of periods. They named this a *phantom bursting* model, as it can lead to bursting with a period intermediate between that of the slow variables.

The Membrane Voltage Submodel

The electrical part of the phantom bursting model is similar to the Chay–Keizer model, but with slight differences, so we present it in full here. As before, the current is the sum of a Ca^{2+} current, a K^+ current, and a Ca^{2+}-sensitive K^+ current, with an additional ATP-sensitive K^+ current, which is important below. The only current not assumed to be at pseudo-steady state is the K^+ current. Thus,

$$C_m \frac{dV}{dt} = -I_{Ca} - I_K - I_{K,Ca} - I_{K,ATP}, \tag{9.9}$$

$$\tau_n \frac{dn}{dt} = n_\infty(V) - n, \tag{9.10}$$

$$I_{Ca} = g_{Ca} m_\infty(V)(V - V_{Ca}), \tag{9.11}$$

$$I_K = g_K n(V - V_K), \tag{9.12}$$

$$I_{K,Ca} = g_{K,Ca} \omega(c)(V - V_K), \tag{9.13}$$

$$I_{K,ATP} = g_{K,ATP}(V - V_K), \tag{9.14}$$

where c denotes the free cytoplasmic Ca^{2+} concentration. The functions m_∞ and n_∞ are given by

$$m_\infty(V) = \frac{1}{1 + e^{(v_m - V)/s_m}},$$ (9.15)

$$n_\infty(V) = \frac{1}{1 + e^{(v_n - V)/s_n}}.$$ (9.16)

The variable ω is the fraction of open Ca^{2+}-sensitive K^+ channels, and is close to a step function,

$$\omega(c) = \frac{c^5}{c^5 + k_D^5}.$$ (9.17)

The Calcium Submodel

These equations for the electrical properties of the membrane must be coupled to equations that model the cytoplasmic and ER Ca^{2+} concentrations. The Ca^{2+} model is chosen to be relatively simple; more complex models of Ca^{2+} dynamics are discussed in Chapter 7.

As usual, let c and c_e denote the free concentrations of Ca^{2+} in the cytoplasm and ER respectively. The equations for c and c_e are simple balances of fluxes.

1. I_{Ca}, the transmembrane Ca^{2+} current as discussed above.
2. Pumping of Ca^{2+} across the plasma and ER membranes, denoted by J_{pm} and J_{serca} respectively. Both of these fluxes are assumed to be linear functions of Ca^{2+}, and thus

$$J_{serca} = k_{serca}c,$$ (9.18)

$$J_{pm} = k_{pm}c.$$ (9.19)

3. J_{leak}, a leak from the ER. This is assumed to be proportional to the difference between the ER and cytoplasmic concentrations, and thus

$$J_{leak} = k_{leak}(c_e - c).$$ (9.20)

Because of Ca^{2+} buffering, only a small fraction of each flux contributes to a change in the free Ca^{2+} concentration. If we assume that buffering is fast and unsaturated, then we need only multiply each flux by a scaling factor to get the change in free Ca^{2+} concentration (Section 7.4).

We put all these fluxes together to get

$$\frac{dc}{dt} = f_{cyt}(-\alpha I_{Ca} - J_{pm} + J_{leak} - J_{serca}),$$ (9.21)

$$\frac{dc_e}{dt} = -\gamma f_{er}(J_{leak} - J_{serca}).$$ (9.22)

Here, γ is the ratio of the cytoplasmic volume to the ER volume, while f_{cyt} and f_{er} are the buffering scaling factors for the cytoplasm and the ER.

Table 9.2 Parameters of the phantom bursting model. The values of $g_{K,Ca}$ and $g_{K,ATP}$ used for each simulation are given in the figure captions.

g_{Ca}	= 1200 pS	g_K	= 3000 pS
V_{Ca}	= 25 mV	V_K	= −75 mV
C_m	= 5300 fF	α	= $4.5 \times 10^{-6} \mu$M fA^{-1} ms^{-1}
τ_n	= 16 ms	f_{cyt}	= 0.01
k_{pm}	= 0.2 ms^{-1}	k_D	= 0.3 μM
v_n	= −16 mV	s_n	= 5 mV
v_m	= −20 mV	s_m	= 12 mV
k_{serca}	= 0.4 ms^{-1}	f_{er}	= 0.01
γ	= 5	p_{leak}	= 0.0005 ms^{-1}

The parameter values for this model are given in Table 9.2.

Fast Bursting

When the Ca^{2+}-sensitive K$^+$ conductance is high (900 pS), the model exhibits fast bursting, as shown in Fig. 9.6. The ER Ca^{2+} concentration varies little over the course of a burst, and indeed, if c_e is set to be a constant, the solution is nearly identical. As can be seen from the middle panel of Fig. 9.6, although c is slightly slower than V, it changes a lot faster than c_e, and is not obviously a slow variable.

A full analysis of this model requires examining a three-dimensional phase space, a difficult exercise. However, even though it is only an approximate analysis, it is useful to analyze this model in the same way as Rinzel's analysis of the Chay–Keizer model, as discussed previously. To do so, we pretend that c is a bifurcation parameter, and draw the bifurcation diagram of V against c (Fig. 9.7). This gives a diagram qualitatively similar to that shown in Fig. 9.4. A Z-shaped curve of steady-state solutions becomes unstable at a Hopf bifurcation, and the branch of stable limit cycles intersects the Z-shaped curve of steady states in a homoclinic bifurcation. This gives rise to bistability, where a stable steady state and a stable limit cycle exist simultaneously.

Bursting occurs in the same manner as the Chay–Keizer model. Above the $dc/dt = 0$ nullcline the solution trajectory moves to the right, and lives (approximately) on the branch of stable limit cycles, giving the active phase of the burst. When it moves far enough to the right, it falls off the branch of limit cycles and heads to the lower branch of stable steady states. Since this branch is below the c nullcline, the solution then moves to the left, eventually falling off the saddle-node to repeat the cycle. This way of interpreting the solution is only an approximate one, as it can be seen from Fig. 9.7 that the actual solution follows accurately neither the branch of periodic orbits nor

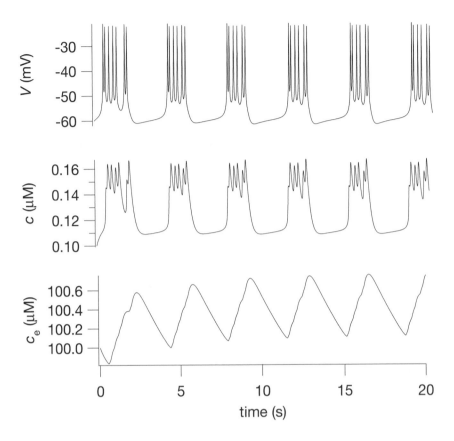

Figure 9.6 Fast bursting oscillations in the phantom bursting model. These solutions were computed with $g_{K,Ca} = 900$ pS and $g_{K,ATP} = 227.5$ pS. Note that the ER Ca^{2+} varies only slightly.

the lower branch of steady states. The upper turning point of the burst is close to osc$_{max}$, but the lower turning point is not close at all to osc$_{min}$. Similarly, when the solution falls off the branch of periodic orbits, it does so well before the homoclinic bifurcation, and then does not follow the lower branch of steady states closely. This happens because c is not really a slow parameter after all. In the limit as c becomes infinitely slow, the solution would track much more closely the bifurcation diagram of the fast subsystem. Despite these quantitative disagreements, the phase plane of the fast subsystem is nonetheless a useful way to interpret and understand the solutions of the full system.

However, there is one important difference between this model and the Chay–Keizer model. In this model there is a second slow variable, c_e, and thus the $dc/dt = 0$ nullcline moves as c_e varies (the Z-shaped curve, however, is independent of c_e). From (9.21) we see that the $dc/dt = 0$ nullcline is given by

$$c = \frac{p_{leak}c_e - \alpha I_{Ca}}{k_{pm} + p_{leak} + k_{serca}}, \qquad (9.23)$$

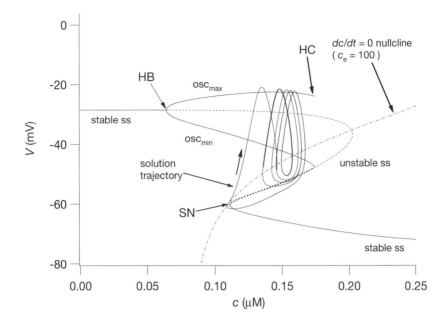

Figure 9.7 Fast bursting in the phantom bursting model (computed with $g_{K,Ca} = 900$ pS and $g_{K,ATP} = 227.5$ pS) superimposed on the bifurcation diagram of the fast subsystem, treating c as the bifurcation parameter. HB — Hopf bifurcation; HC — homoclinic bifurcation; SN — saddle-node bifurcation.

and thus, as c_e increases, the nullcline moves to the right. For the fast bursting shown in Fig. 9.6 the changes in c_e are so small that this movement of the nullcline has no effect on the bursting. However, for different parameter values, more interesting behaviors emerge.

Medium Bursting

One way to get a longer burst period would be to stretch the Z-shaped curve horizontally, so that the homoclinic bifurcation and the saddle-node are further apart. This can be accomplished by decreasing $g_{K,Ca}$. However, if the c nullcline remains unchanged, for small enough $g_{K,Ca}$ it intersects the Z-shaped curve on its upper branch, inside the branch of stable limit cycles. In this case, the limit cycle is a stable solution of the full system, and the solution remains stuck in the active phase.

   Now the slow dynamics of c_e come into play. During the active phase c_e increases (see Exercise 4), gradually moving the c nullcline to the right; the oscillations chase the c nullcline to the right, as shown in Fig. 9.8. When the c nullcline is moved far enough to the right, the solution falls off the limit cycles (i.e., leaves the active phase), moves toward the lower branch of steady-state solutions (thus starting the silent phase), and moves to the left toward the saddle-node. However, movement along this lower branch is very slow, as the c nullcline intersects the Z-shaped curve on its lower branch, giving

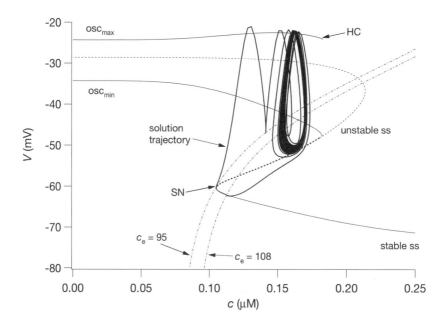

Figure 9.8 Medium bursting in the phantom bursting model (computed with $g_{K,Ca} = 700$ pS and $g_{K,ATP} = 300$ pS) superimposed on the bifurcation diagram of the fast subsystem, treating c as the bifurcation parameter. HC — homoclinic bifurcation; SN — saddle-node bifurcation. The Hopf bifurcation occurs at a negative value of c and so does not appear here. Two $dc/dt = 0$ nullclines are shown, for the maximum and minimum values of c_e over a burst.

a stable quasi-steady state. The solution is thus forced to move at the same speed as the slow variable c_e, tracking the quasi-steady state as it moves to the left. Eventually, the c nullcline moves sufficiently far to the left that the quasi-steady state disappears, the solution leaves the lower branch of the Z-shaped curve, and the burst cycle repeats.

Thus, in medium bursting the burst cycle relies on the slow movement of the c nullcline, which is caused by the slow increase of c_e during the active phase of the burst, and the slow decrease of c_e during the silent phase. Because the c nullcline moves slowly, this results in medium bursting with a longer period. Both the active and silent phases are longer than for fast bursting, giving a much longer burst period, and the ER Ca^{2+} varies much more over a cycle. Typical solutions are shown in Fig. 9.9.

The Effect of Agonists

Pancreatic β cells are also regulated by the nervous system. Secretion of acetylcholine from parasympathetic nerves increases the rate of insulin secretion, an effect due partly to changes in the electrical activity and Ca^{2+} dynamics of individual β cells. Since the action of acetylcholine is via the production of inositol trisphosphate (IP_3; see Chapter 7), and consequent release of Ca^{2+} from the ER, it can be modeled by including an additional term for a Ca^{2+} flux through IP_3 receptors (Exercise 3). When $[IP_3]$ is raised

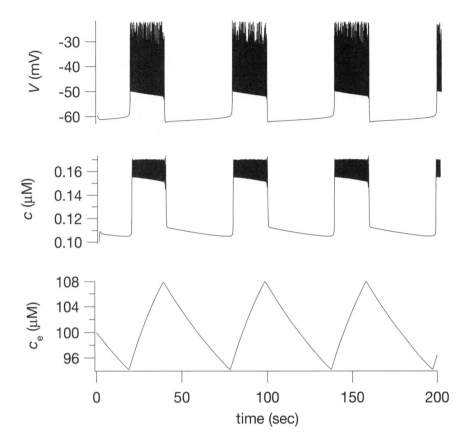

Figure 9.9 Medium bursting oscillations in the phantom bursting model. These solutions were computed with $g_{K,Ca} = 700$ pS and $g_{K,ATP} = 300$ pS. The active phase is not completely regular due to inadequate numerical resolution of the timestep.

from 0 to 0.3 μM the burst pattern changes from medium to fast, a change which is observed experimentally.

9.1.3 Slow Bursting and Glycolysis

Bursting in pancreatic β cells is highly sensitive to the level of glucose; as the concentration of glucose increases, so does the active fraction of the burst cycle. During the active fraction of the burst cycle the Ca^{2+} concentration is raised, and this in turn causes the secretion of insulin. Thus, the rate of insulin secretion is an increasing function of glucose concentration.

This effect of glucose is believed to be mediated by the ATP-sensitive K^+ channel, which is activated by ADP and inhibited by ATP. At low glucose concentrations the ATP/ADP ratio is low, the K^+ channel is open, hyperpolarizing the cell and thus preventing bursting. As glucose increases, so does the ATP/ADP ratio; this decreases the conductance of the ATP-sensitive K^+ channels, thus depolarizing the cell and allowing bursting to occur.

There are two important features of bursting and insulin secretion that we have not, as yet, addressed. First, bursting often occurs with a much longer period than that so far reproduced by the models above, and second, such slow bursting can take a more complex form than that shown in Figs. 9.6 and 9.9. In particular, slow bursting can occur as "compound" bursting, or bursts of bursts, in which each active phase itself consists of alternating active and silent subphases.

One hypothesis is that compound bursting and a long burst period arise from a slow oscillation in the glycolytic pathway (Chapter 1), which causes slow oscillations in the ATP/ADP ratio. One of the earliest quantitative models of this hypothesis was that of Wierschem and Bertram (2004), who showed that, by linking the Goldbeter–Lefever model of glycolytic oscillations (Chapter 1) to a simple bursting model, the burst pattern could be modulated on a long time scale, to obtain both compound and slow bursting. This initial model, more a proof of principle than a quantitative model, was quickly followed by much more detailed realizations, first using the Smolen model of glycolytic oscillations (Smolen, 1995; Bertram et al., 2004; Nunemaker et al., 2006), and then linking the Smolen model to the Magnus–Keizer model of mitochondrial metabolism (Magnus and Keizer, 1997, 1998a, 1998b; Bertram et al., 2006a, 2007a,b).

The later models being too complex to present in full here, we instead examine briefly the original model of Wierschem and Bertram (2004), as it contains most of the essentials of the more complex models.

The Glycolysis, Electrical and Calcium Submodels

To model glycolytic oscillations in a simple way we use the reduced Goldbeter–Lefever model discussed in Section 1.6. Thus,

$$\tau_c \frac{d[\text{ATP}]}{dt} = v_1 - F([\text{ATP}], [\text{ADP}]), \tag{9.24}$$

$$\tau_c \frac{d[\text{ADP}]}{dt} = F([\text{ATP}], [\text{ADP}]) - v_2[\text{ADP}], \tag{9.25}$$

where (as in (1.125))

$$F([\text{ATP}], [\text{ADP}]) = [\text{ATP}](1 + k_{\text{ADP}}[\text{ADP}])^2. \tag{9.26}$$

The only change introduced to this version of the glycolytic model is the time constant, τ_c, which is convenient for changing the period of the glycolytic oscillations.

The electrical submodel is almost the same as the models discussed previously in this chapter. The currents are described by (9.9)–(9.14) and (9.15), the only difference being that, instead of (9.17), we take

$$\omega(c) = \frac{c}{c + k_D}. \tag{9.27}$$

Table 9.3 Parameters of the compound bursting model.

g_{Ca}	$= 1200$ pS	g_K	$= 3000$ pS
$g_{K,Ca}$	$= 300$ pS	$g_{K,ATP}$	$= 350$ pS
V_{Ca}	$= 25$ mV	V_K	$= -75$ mV
C_m	$= 5300$ fF	α	$= 2.25 \times 10^{-6} \mu\text{M fA}^{-1} \text{ms}^{-1}$
τ_n	$= 16$ ms	f	$= 0.001$
k_c	$= 0.1$ ms^{-1}	k_D	$= 0.3 \mu$M
v_n	$= -16$ mV	s_n	$= 5.6$ mV
v_m	$= -20$ mV	s_m	$= 12$ mV
τ_c	$= 1.2 \times 10^{-6}$ ms	v_1	$= 10$ mM
v_2	$= 185$	k_{ADP}	$= 20$ mM^{-1}

Finally, the Ca^{2+} submodel is the same as that of the Chay–Keizer model. We thus ignore ER Ca^{2+} to get a single equation for c (as in (9.5)),

$$\frac{dc}{dt} = -f(\alpha I_{Ca} + k_c c). \tag{9.28}$$

All the parameters of the compound bursting model are given in Table 9.3.

Compound Bursting

A typical example of compound bursting in this model is shown in Fig. 9.10. The bursting occurs in clusters, with the duration of the active phase increasing and then decreasing through each cluster. As can be seen from the dotted curve, [ATP] is oscillating also; here these oscillations are independent of c and V, although this is not necessarily so in more complex models.

The reason for the compound bursting can be seen if we consider the fast–slow bifurcation structure of the electrical submodel, for various fixed values of [ATP]. As for the Chay–Keizer model, we treat c as a slow variable and construct the bifurcation diagrams using c as a bifurcation parameter.

At all values of [ATP], the curve of steady states (labeled ss in Fig. 9.10) is Z-shaped. The unstable branches of this Z-shaped curve are shown as a dashed line. For low values of c the steady state becomes unstable in a Hopf bifurcation, from which emerges a branch of stable periodic solutions (the dot-dash curve, labeled osc$_{max}$ and osc$_{min}$ in Fig. 9.10, panel **a**).

As [ATP] increases, the Z-shaped curve moves to the right, but the $dc/dt = 0$ nullcline remains unchanged. Thus, when [ATP] is low, the $dc/dt = 0$ nullcline intersects the Z-shaped curve on the lower, stable, branch, a situation that gives no bursting at all (panel **a**). Conversely, when [ATP] is high, the $dc/dt = 0$ nullcline intersects the branch

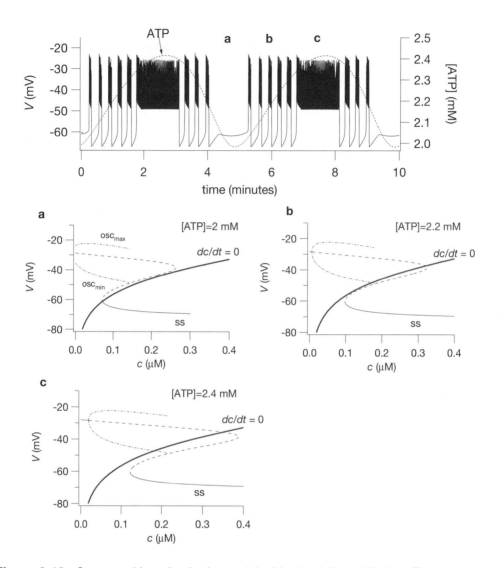

Figure 9.10 Compound bursting in the model with glycolytic oscillations. The upper panel shows a compound bursting pattern, in which the bursts are grouped in clusters that occur in a periodic manner. In panels **a**, **b**, and **c** are shown the bifurcation diagrams corresponding approximately to the regions labeled **a**, **b** and **c** in the top panel. The dashed lines are unstable branches of steady states, and the dot-dash lines are the maximum and minimum of stable periodic solutions.

of periodic orbits, which leads to a continuous active phase (panel **c**). When [ATP] takes intermediate values, the length of the active phase is either longer or shorter, depending on where the $dc/dt = 0$ nullcline intersects the Z-shaped curve (panel **b**).

Hence, as [ATP] varies over the course of an oscillation, so does the length of the active phase, thus giving compound bursting.

9.1.4 Bursting in Clusters

In the discussion so far, we have ignored the inconvenient fact that isolated pancreatic β cells usually do not burst in a regular fashion. It is not until several thousand cells are grouped into an islet and electrically coupled by gap junctions that regular spiking is seen in any cell. An isolated β cell behaves in a much more irregular fashion, with no discernible pattern of bursting. Indeed, blockage of gap junctions in an islet greatly reduces insulin secretion, and thus intercellular mechanisms that control bursting are of great physiological importance. Figure 9.11 shows how the behavior of an individual cell changes as a function of the number of other cells to which it is coupled.

Channel-Sharing

In 1983, Atwater et al. proposed a qualitative mechanism to account for the difference between the single-cell behavior and the behavior of the cell in a cluster. They

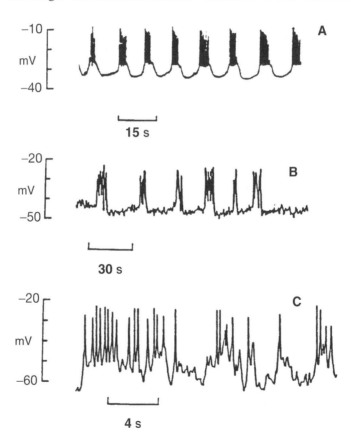

Figure 9.11 Bursting in clusters of β cells, compared to the electrical activity of an isolated β cell (Sherman et al., 1988, Fig. 1). A: Recording from a β cell in an intact cluster (Atwater and Rinzel, 1986), B: Recording from a cluster of β cells, with a radius of 70 μm (Rorsman and Trube, 1986, Fig. 1a). C: Recording from an isolated β cell (Rorsman and Trube, 1986, Fig. 2c).

proposed that an individual cell contains a small number of Ca^{2+}-sensitive K^+ (K^+–Ca^{2+}) channels each of which has a high conductance. At resting V and $[Ca^{2+}]$, a K^+–Ca^{2+} channel is open only infrequently, but the opening of a single channel passes enough current to cause a significant perturbation of the cell membrane potential. Thus, stochastic channel opening and closing causes the observed random fluctuations in V. However, when the cells are electrically coupled in a cluster, each K^+–Ca^{2+} channel has a much smaller effect on the potential of each individual cell, since the channel current is spread over the network of cells. Each cell integrates the effects of a large number of K^+–Ca^{2+} channels, each of which has only a small influence. The tighter the electrical coupling between cells, the better each cell is able to integrate the effects of all the K^+–Ca^{2+} channels in the cluster, and the more regular and deterministic is the overall behavior.

We can use this qualitative explanation as the basis for a quantitative model (Sherman et al., 1988; Chay and Kang, 1988). Initially, we assume infinitely tight coupling of the cells in the cluster, calling the cluster a "supercell," and show how the bursting becomes more regular as the size of the cluster increases.

The equations of the supercell model are similar to those of the β cell model. Recall that in the β cell model the conductance of the K^+–Ca^{2+} channel, $g_{K,Ca}$, is given by

$$g_{K,Ca} = \bar{g}_{K,Ca} \frac{c}{K_d + c}, \tag{9.29}$$

where c, as usual, denotes $[Ca^{2+}]$. This can be derived from a simple channel model in which the channel has one closed state and one open state, switching from closed to open upon the binding of a single Ca^{2+} ion. Thus

$$C + Ca^{2+} \underset{k_-}{\overset{k_+}{\rightleftharpoons}} O, \tag{9.30}$$

where C is a closed channel and O is an open one. If the rate constants k_+ and k_- are both large compared to the other kinetic parameters of the model, then

$$[O] = \frac{k_+[Ca^{2+}]}{k_-}[C] = \frac{[Ca^{2+}]}{K_d}(1 - [O]), \tag{9.31}$$

where $K_d = k_-/k_+$. Hence $[O] = [Ca^{2+}]/(K_d + [Ca^{2+}])$, from which (9.29) follows.

In the supercell model, (9.30) is interpreted as a Markov rather than a deterministic process, with k_+ denoting the probability per unit time that the closed channel switches to the open state, and similarly for k_-. Thus, the mean open and closed times are, respectively, $1/k_-$ and $1/k_+$. If we let $\langle N_o \rangle$ and $\langle N_c \rangle$ denote the mean number of open channels and closed channels respectively, then at equilibrium we have

$$\frac{\langle N_o \rangle}{\langle N_c \rangle} = \frac{k_+}{k_-}. \tag{9.32}$$

To incorporate the Ca^{2+} dependence of the channel we make

$$\frac{k_+}{k_-} = \frac{[Ca^{2+}]}{K_d},$$ (9.33)

which gives the steady-state mean proportion of open channels as

$$\langle p \rangle = \frac{\langle N_o \rangle}{\langle N_o \rangle + \langle N_c \rangle} = \frac{[Ca^{2+}]}{K_d + [Ca^{2+}]} = \frac{c}{K_d + c}.$$ (9.34)

The associated stochastic model is similar to (9.6)–(9.8), with the major difference being that the Ca^{2+}-sensitive K^+ current is governed by the above stochastic process. In other words, p is a random variable denoting the proportion of open channels in a single cell, and is calculated (as a function of time) by numerical simulation of the Markov process described by (9.30). Thus,

$$C_m \frac{dV}{dt} = -I_{Ca} - I_K - \bar{g}_{K,Ca} p (V - V_K),$$ (9.35)

$$\tau_n(V) \frac{dn}{dt} = \lambda(n_\infty(V) - n),$$ (9.36)

$$\frac{dc}{dt} = f(-\alpha I_{Ca} - k_c c),$$ (9.37)

where $I_{Ca} = \bar{g}_{Ca} m_\infty(V) h(V)(V - V_{Ca})$ and $I_K = \bar{g}_K n(V - V_K)$. The functions appearing in the supercell model are

$$m_\infty(V) = \frac{1}{1 + \exp\left[\frac{4-V}{14}\right]},$$ (9.38)

$$n_\infty(V) = \frac{1}{1 + \exp\left[\frac{-15-V}{5.6}\right]},$$ (9.39)

$$\tau_n(V) = \frac{\bar{\tau}_n}{\exp\left[\frac{V+75}{65}\right] + \exp\left[\frac{-V-75}{20}\right]},$$ (9.40)

$$h(V) = \frac{1}{1 + \exp\left[\frac{V+10}{10}\right]}.$$ (9.41)

The other parameters of the model are summarized in Table 9.4. Note that although the form of the model is similar to that of the β cell model, the details have been changed to agree with more recent experimental data. In particular, m_∞ appears only to the first power instead of the third, while n also appears to the first power. Further, the I–V curve of the open Ca^{2+} channel is assumed to be of the form $h(V)(V - V_{Ca})$, and $h(V)$ is chosen to fit the experimentally observed I–V curve. This has an effect similar to that of the function $h_\infty(V)$ that was used in the β cell model.

One of the most important features of the stochastic model is that the conductance of a single K^+–Ca^{2+} channel is an order of magnitude greater than the conductances of the other two channels (Table 9.4). However, each cell contains only a small number

Table 9.4 Parameters of the supercell model of electrical bursting in clusters of pancreatic β cells.

$C_m = 5.3$ pF	$\bar{g}_{K,Ca} = 30$ nS
$\bar{g}_K = 2.5$ nS	$\bar{g}_{Ca} = 1.4$ nS
$V_{Ca} = 110$ mV	$V_K = -75$ mV
$k_c = 0.03$ ms^{-1}	$f = 0.001$
$\alpha = 4.5\ \mu$M/C	$K_d = 100\ \mu$M
$\bar{\tau}_n = 60$ ms	

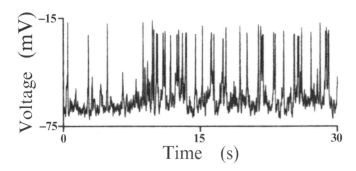

Figure 9.12 Behavior of an isolated cell in the supercell stochastic model. Because of the stochastic nature of each high-conductance K$^+$–Ca^{2+} channel, and because there are few channels in a single cell, no organized bursting appears (Sherman et al., 1988, Fig. 6).

of K$^+$–Ca^{2+} channels. Thus, the opening of a single K$^+$–Ca^{2+} channel in an isolated cell has a disproportionately large effect on the membrane potential of the cell; the stochastic nature of each K$^+$–Ca^{2+} channel then causes seemingly random fluctuations in membrane potential (Fig. 9.12).

However, when identical cells are coupled by gap junctions with zero resistance, different behavior emerges. First, since the gap junctions are assumed to have zero resistance, and thus the entire group of cells has the same membrane potential, it is not necessary to treat each cell explicitly. Second, the membrane capacitance and the ionic currents depend on the surface area of the cluster and are therefore proportional to the number of cells in the cluster, N_{cell}. Finally, the total number of K$^+$–Ca^{2+} channels is proportional to the total number of cells in the cluster, but the effect of each individual channel on the membrane potential of the cluster is proportional to $1/N_{cell}$. It follows that the cluster of cells behaves in the same way as a very large single cell with many K$^+$–Ca^{2+} channels, each with a smaller conductance. We expect the cluster of cells to behave in the same manner as a deterministic single-cell model in the limit as $N_{cell} \longrightarrow \infty$.

To see that this is what happens, we let $\hat{g}$ denote the conductance of a single K$^+$–Ca^{2+} channel and let N_o^i denote the number of open K$^+$–Ca^{2+} channels in the ith cell. Then

$$N_{\text{cell}}C_m\frac{dV}{dt} = -N_{\text{cell}}(I_{\text{Ca}} + I_{\text{K}}) - \hat{g}\sum_{i=1}^{N_{\text{cell}}} N_o^i(V - V_{\text{K}}), \qquad (9.42)$$

and so

$$C_m\frac{dV}{dt} = -(I_{\text{Ca}} + I_{\text{K}}) - \hat{g}\bar{N}\frac{1}{N_{\text{cell}}\bar{N}}\sum_{i=1}^{N_{\text{cell}}} N_o^i(V - V_{\text{K}})$$

$$= -(I_{\text{Ca}} + I_{\text{K}}) - \bar{g}_{\text{K,Ca}}p(V - V_{\text{K}}), \qquad (9.43)$$

where $\bar{N}$ is the number of K$^+$–Ca^{2+} channels per cell, or the channel density, and where, as before, $\bar{g}_{\text{K,Ca}} = \hat{g}\bar{N}$ is the total K$^+$–Ca^{2+} conductance per cell. Note that in the supercell model $p = \frac{1}{N_{\text{cell}}\bar{N}}\sum_{i=1}^{N_{\text{cell}}} N_o^i$ must be interpreted as the fraction of open channels in the cluster, rather than the fraction of open channels in a single cell. The mean of p is the same in both these cases, but as N_{cell} increases, the standard deviation of p decreases, leading to increasingly regular behavior. As before, p must be obtained by direct simulation of the Markov process. Simulations for different numbers of cells are shown in Fig. 9.13. Clearly, as the size of the cluster increases, bursting becomes more regular.

One obvious simplification in the supercell model is the assumption that the gap junctions have zero resistance and thus that every cell in the cluster has the same membrane potential. We can relax this assumption by modeling the cluster as individual cells coupled by gap junctions with finite conductance (Sherman and Rinzel, 1991). An individual cell, cell i say, satisfies a voltage equation of the form

$$C_m\frac{dV_i}{dt} = -I_{\text{Ca}}(V_i) - I_{\text{K}}(V_i, n_i) - \bar{g}_{\text{K,Ca}}p_i(V_i - V_{\text{K}}) - g_c\sum_j d_{ij}(V_i - V_j), \qquad (9.44)$$

where g_c is the gap junction conductance and where d_{ij} are coupling coefficients, with value one if cells i and j are coupled, and zero otherwise. As $g_c \to \infty$, the sum $\sum_j d_{ij}(V_i - V_j)$ must approach zero for every cell in the cluster. If all the cells are connected by some connecting path (so that there are no isolated cells or subclusters), then every cell must have the same voltage (see Exercise 9). Thus, in the limit of infinite conductance,

$$V_i \longrightarrow \bar{V} = \frac{1}{N_{\text{cell}}}\sum_{j=1}^{N_{\text{cell}}} V_j. \qquad (9.45)$$

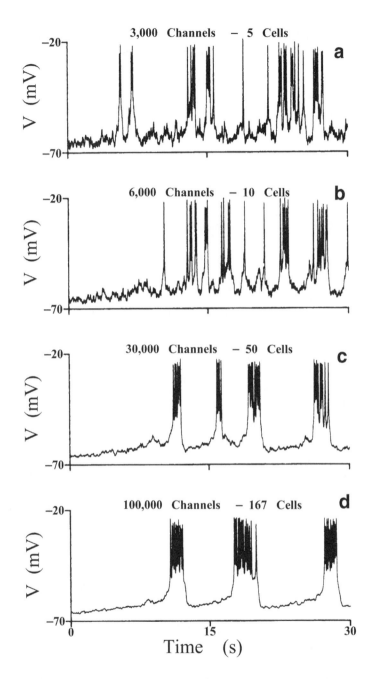

Figure 9.13 Numerical simulations of the supercell model of a cluster of cells ranging in size from 5 to 167 cells. As the size of the cluster increases, more organized bursting appears. (Sherman et al., 1988, Fig. 8.)

For large but finite coupling, $V_i = \bar{V} + O(\frac{1}{g_c})$. If we sum (9.44) over all the cells in the cluster and divide by N_{cell}, we find

$$C_m \frac{d\bar{V}}{dt} = -I_{\text{Ca}}(\bar{V}) - I_{\text{K}}(\bar{V}, n) - \bar{g}_{\text{K,Ca}} \frac{1}{N_{\text{cell}}} \sum_{j=1}^{N_{\text{cell}}} p_j(\bar{V} - V_{\text{K}}) + O\left(\frac{1}{N_{\text{cell}}}\right)$$

$$= -I_{\text{Ca}}(\bar{V}) - I_{\text{K}}(\bar{V}, n) - \bar{g}_{\text{K,Ca}}\bar{p}(\bar{V} - V_{\text{K}}) + O\left(\frac{1}{N_{\text{cell}}}\right), \qquad (9.46)$$

where $\bar{p} = \frac{\sum_i p_i}{N_{\text{cell}}} = \frac{\sum_i p_i \bar{N}}{N_{\text{cell}} \bar{N}}$ is the proportion of open K^+–Ca^{2+} channels in the cluster. Hence, the model with finite gap-junctional conductance (the *multicell* model) turns into the supercell model as the gap-junctional conductance and the number of cells in the cluster approach infinity.

As expected, synchronized bursting appears as the number of cells in the cluster increases and the coupling strength increases. Both strong coupling and a large cluster size are required to achieve regular bursting. This is illustrated in Fig. 9.14, where we

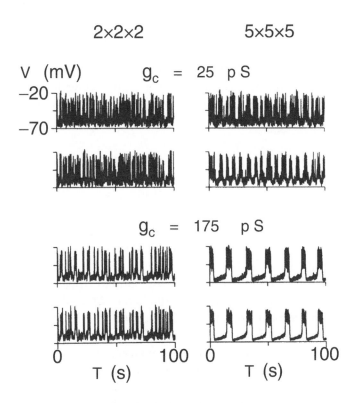

Figure 9.14 Numerical simulations of the multicell model, in which the cells in the cluster are coupled by gap junctions with finite conductance. Results are shown for two cells (upper and lower traces in each pair) from two different cluster sizes ($2 \times 2 \times 2$ cells and $5 \times 5 \times 5$ cells) and two different junctional conductances. (Sherman and Rinzel, 1991, Fig. 3.)

show numerical simulations for two different cluster sizes and two different coupling strengths. However, what is not expected (and is therefore particularly interesting) is that there is a coupling strength at which the length of the burst period is maximized. The reasons for this have been analyzed in depth in a simpler system consisting of two coupled cells (Sherman, 1994).

A different approach was taken by Tsaneva-Atanasova et al. (2006), who studied the effects on bursting of the intercellular diffusion of Ca^{2+} and metabolites. They concluded that intercellular diffusion of Ca^{2+} is not important for the synchronized bursting of a cluster, but that intercellular diffusion of metabolites can lead to either synchrony or the abolition of oscillations entirely.

9.1.5   A Qualitative Bursting Model

Just as the essential behaviors of the Hodgkin–Huxley equations can be distilled into the simpler FitzHugh–Nagumo equations, giving a great deal of insight into the mechanisms underlying excitability, so too can the Chay–Keizer model of bursting be simplified to a polynomial model, without losing the essential properties (Hindmarsh and Rose, 1982, 1984; Pernarowski, 1994).

We begin with the FitzHugh–Nagumo equations, and modify them slightly to make the period of the oscillations much longer. Suppose we let v denote the excitatory variable and w the recovery variable (as in Chapter 5). Then, the model equations are

$$\frac{dv}{dt} = \alpha(\beta w - f(v) + I), \tag{9.47}$$

$$\frac{dw}{dt} = \gamma(g(v) - \delta w), \tag{9.48}$$

where I is the applied current and α, β, γ, and δ are constants. As in the FitzHugh–Nagumo equations $f(v)$ is cubic, but unlike the FitzHugh–Nagumo equations $g(v)$ is not a linear function. In fact, as can be seen from Fig. 9.15, the w nullcline curves around to lie close to the v nullcline to the left of the oscillatory critical point. (Of course, this occurs only for a range of values of the applied current I for which oscillations occur.) As a result, between the peaks of the oscillation, the limit cycle trajectory lies close to both nullclines, and thus both $\dot{v}$ and $\dot{w}$ are small over that portion of the cycle. It follows that the intervals between the spikes are large.

With only a slight change, this modified FitzHugh–Nagumo model can be used as a model of bursting. Following the discussions in this chapter, it should come as no surprise that bursting can arise in this model when bistability is introduced. This can be done by deforming the $\dot{w} = 0$ nullcline so that it intersects the $\dot{v} = 0$ nullcline in three places rather than only one. Since the nullclines lie close to one another in the original model, only a slight deformation is required to create two new critical points.

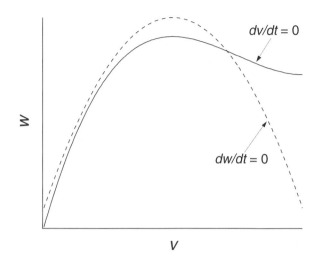

Figure 9.15 Sketch of typical null-clines in the modified FitzHugh–Nagumo equations.

With this change, the new model equations are

$$\frac{dv}{dt} = \alpha(\beta w - f(v) + I), \tag{9.49}$$

$$\frac{dw}{dt} = \gamma(g(v) + h(v) - \delta w), \tag{9.50}$$

where $h(v)$ is chosen such that the nullclines intersect in three places. For convenience, we scale and nondimensionalize the model by introducing the variables $T = \gamma \delta t$, $x = v$, and $y = \alpha \beta w/(\gamma \delta)$, in terms of which variables the model becomes

$$\frac{dx}{dT} = y - \tilde{f}(x), \tag{9.51}$$

$$\frac{dy}{dT} = \tilde{g}(x) - y, \tag{9.52}$$

where $\tilde{f}(x) = \alpha f(x)/(\gamma \delta)$ and $\tilde{g}(x) = \alpha \beta [g(x) + h(x)]/(\gamma \delta^2)$. Note that the form of the model is the same as that of the FitzHugh–Nagumo equations, although the functions appearing in the model are different. With appropriate choices for $\tilde{f}$ and $\tilde{g}$ the model exhibits bistability; we use the specific functions

$$\frac{dx}{dT} = y - x^3 + 3x^2 + I, \tag{9.53}$$

$$\frac{dy}{dT} = 1 - 5x^2 - y, \tag{9.54}$$

the phase plane of which is shown in Fig. 9.16 for $I = 0$.

There are three critical points: a stable node to the left at $x = -\frac{1}{2}(1 + \sqrt{5})$ (the resting state), a saddle point in the middle at $x = -1$, and an unstable node to the right at $x = \frac{1}{2}(-1 + \sqrt{5})$, which is surrounded by a stable limit cycle. As in the models of

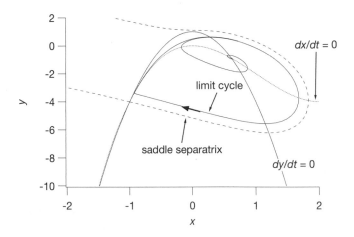

Figure 9.16 Phase plane of the polynomial bursting model.

bursting discussed above, the stable manifold of the saddle point acts as a threshold; if the perturbation from the resting state is large enough that the stable manifold is crossed, the trajectory approaches the stable limit cycle. Smaller perturbations die away to the resting state. This is called *triggered firing*. This bistable phase plane is essentially the same as the phase plane shown in Fig. 9.3A.

To generate bursting in addition to bistability, it is also necessary to have a slow variable so that the voltage can be moved in and out of the bistable regime. This is accomplished in this model by introducing a third variable, z, that modulates the applied current, I, on a slower time scale. Thus,

$$\frac{dx}{dT} = y - x^3 + 3x^2 + I - z, \tag{9.55}$$

$$\frac{dy}{dT} = 1 - 5x^2 - y, \tag{9.56}$$

$$\frac{dz}{dT} = r[s(x - x_1) - z], \tag{9.57}$$

where $x_1 = -\frac{1}{2}(1 + \sqrt{5})$ is the x-coordinate of the resting state in the two-variable model (9.53)–(9.54). When $r = 0.001$ and $s = 4$, (9.55)–(9.57) exhibit bursting (Fig. 9.17), arising via the same mechanism as in the Chay–Keizer model.

9.1.6   Bursting Oscillations in Other Cell Types

As can be seen in Fig. 9.18, bursting oscillations occur not only in pancreatic β cells, but also in a wide range of other types of neurons and neuroendocrine cells. The type of bursting that occurs is also widely variable. Because of the enormously complex dynamical behavior exhibited in these bursting patterns, they have been a popular topic

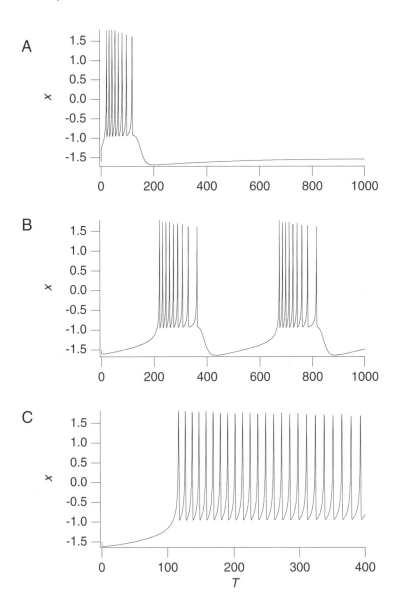

Figure 9.17 Bursting in the polynomial model, calculated numerically from (9.55)–(9.57) for three values of the applied current. A: $I = 0.4$; B: $I = 2$; C: $I = 4$.

of study among mathematicians, and the theoretical study of bursting oscillations has, by now, far outstripped our corresponding physiological knowledge.

Bursting in the Aplysia Neuron

Another well-studied example of bursting is found in the Aplysia R-15 neuron. Analysis of a detailed model by Plant (1981) shows that the mathematical structure of this

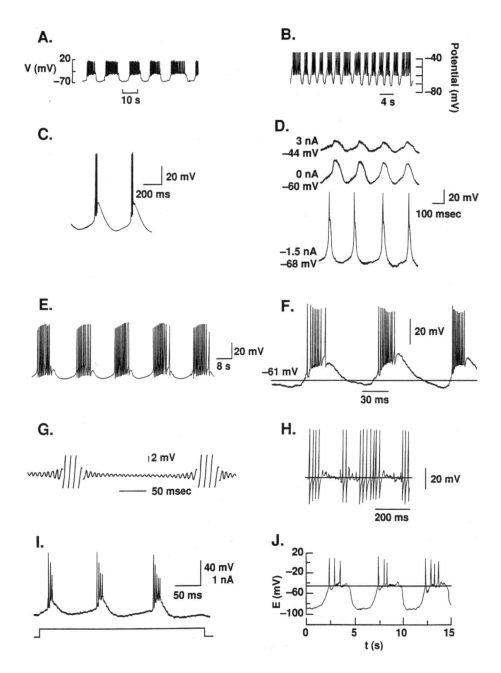

Figure 9.18 Electrical bursting in a range of different cell types. A: Pancreatic β cell. B: Dopamine-containing neurons in the rat midbrain. C: Cat thalamocortical relay neuron. D: Guinea pig inferior olivary neuron. E: *Aplysia* R15 neuron. F: Cat thalamic reticular neuron. G: *Sepia* giant axon. H: Rat thalamic reticular neuron. I: Mouse neocortical pyramidal neuron. J: Rat pituitary gonadotropin-releasing cell. (Wang and Rinzel, 1995, Fig. 2.)

bursting oscillator is different from that in the β cell model (Rinzel and Lee, 1987). The β cell model has two fast variables, one slow variable, bistability, and a hysteresis loop. At the end of a burst, a homoclinic bifurcation is crossed, leading to an increasing period through the burst. Plant's model, on the other hand, has no bistability, with bursting arising from the presence of two slow variables with their own oscillation. A homoclinic bifurcation is crossed at the beginning and the end of the burst, and so the instantaneous period of the burst oscillations starts high, decreases, and then increases again. The fact that the period is roughly a parabolic function of time has led to the name *parabolic* bursting.

Plant's parabolic bursting model is similar in some respects to the β cell model, incorporating Ca^{2+}-activated K^+ channels and voltage-dependent K^+ channels. However, it also includes a voltage-dependent Na^+ channel that activates and inactivates in typical Hodgkin–Huxley fashion and a slowly activating Ca^{2+} current. The Na^+, K^+, and leak currents form the fast subsystem

$$C_m \frac{dV}{dt} = -\bar{g}_{Na} m_\infty^3(V) h(V - V_{Na}) - \bar{g}_{Ca} x(V - V_{Ca})$$
$$- \left(\bar{g}_K n^4 + \frac{\bar{g}_{K,Ca} c}{0.5 + c} \right) (V - V_K) - \bar{g}_L(V - V_L), \tag{9.58}$$

$$\tau_h(V) \frac{dh}{dt} = h_\infty(V) - h, \tag{9.59}$$

$$\tau_n(V) \frac{dn}{dt} = n_\infty(V) - n, \tag{9.60}$$

while the Ca^{2+} current and its activation x form the slow subsystem

$$\tau_x \frac{dx}{dt} = x_\infty(V) - x, \tag{9.61}$$

$$\frac{dc}{dt} = f(k_1 x(V_{Ca} - V) - c). \tag{9.62}$$

The voltage dependence of the variables α_w and β_w with $w = m, n$, or h is of the form

$$\frac{C_1 \exp(\frac{V - V_0}{C_2}) + C_3(V - V_0)}{1 + C_4 \exp(\frac{V - V_0}{C_5})}, \tag{9.63}$$

in units of ms^{-1}, and the asymptotic values $w_\infty(V)$ and time constants $\tau_w(V)$ are of the form

$$w_\infty(V) = \frac{\alpha_w(\tilde{V})}{\alpha_w(\tilde{V}) + \beta_w(\tilde{V})}, \tag{9.64}$$

$$\tau_w(V) = \frac{1}{\lambda \left(\alpha_w(\tilde{V}) + \beta_w(\tilde{V}) \right)}, \tag{9.65}$$

Table 9.5 Parameters of the Plant model of parabolic bursting.

$C_m = 1\ \mu\text{F}/\text{cm}^2$	$\bar{g}_{\text{K,Ca}} = 0.03\ \text{mS}/\text{cm}^2$
$\bar{g}_{\text{Ca}} = 0.004\ \text{mS}/\text{cm}^2$	$V_{\text{Ca}} = 140\ \text{mV}$
$\bar{g}_{\text{Na}} = 4.0\ \text{mS}/\text{cm}^2$	$V_{\text{Na}} = 30\ \text{mV}$
$\bar{g}_{\text{K}} = 0.3\ \text{mS}/\text{cm}^2$	$V_{\text{K}} = -75\ \text{mV}$
$\bar{g}_{\text{L}} = 0.003\ \text{mS}/\text{cm}^2$	$V_{\text{L}} = -40\ \text{mV}$
$f = 0.0003\ \text{ms}^{-1}$	$k_1 = 0.0085\ \text{mV}^{-1}$
$\lambda = 1/12.5$	

Table 9.6 Defining values for the rate constants α and β in the Plant parabolic bursting model.

	C_1	C_2	C_3	C_4	C_5	V_0
α_m	0	—	0.1	−1	−10	50
β_m	4	−18	0	0	—	25
α_h	0.07	−20	0	0	—	25
β_h	1	10	0	1	10	55
α_n	0	—	0.01	−1	−10	55
β_n	0.125	−80	0	0	—	45

for $w = m$, n, or h (although $\tau_m(V)$ is not used), with $\tilde{V} = c_1 V + c_2$, $c_1 = 127/105$, $c_2 = 8265/105$. The constants $C_1, \ldots, C_5$ and V_0 are displayed in Table 9.6. Finally,

$$x_\infty(V) = \frac{1}{\exp\{-0.15(V + 50)\} + 1},\qquad (9.66)$$

and $\tau_x = 235$ ms.

The parameter values of the Plant model of parabolic bursting are given in Tables 9.5 and 9.6.

For a fixed $x = 0.7$, the bifurcation diagram of the fast subsystem as c varies is shown in Fig. 9.19. The curve of steady states is labeled V_{ss}, and is shaped like a Z, with two limit points. The rightmost saddle node (SN2) is not shown, being off to the right at $c = 8.634$. Oscillations arise via a subcritical Hopf bifurcation instead of a supercritical one as in the β-cell model. Note that in general, V_{ss} is a function of both c and x, and therefore the fast subsystem is properly described by a bifurcation surface. However, since surfaces are more difficult to draw and understand, we first examine a cross-section of the bifurcation surface for fixed x and then discuss how the important points behave as x varies. As usual, HB denotes a Hopf bifurcation, where a branch of periodic orbits appears, and HC a homoclinic bifurcation where the periodic solutions disappear.

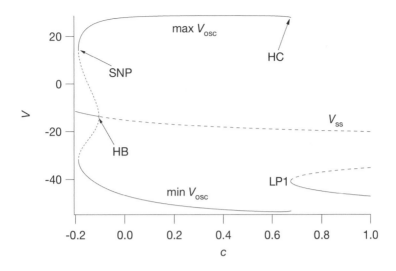

Figure 9.19 The bifurcation diagram of the fast subsystem of the parabolic bursting model, with $x = 0.7$. HB — Hopf bifurcation; HC — homoclinic bifurcation; SNP — saddle-node of periodic bifurcation; SN — saddle node. The maximum and minimum of V over an oscillation are labeled max V_{osc} and min V_{osc} respectively.

The bifurcation diagram is similar to that of the fast subsystem of the β-cell model (Fig. 9.4), except that the branch of periodic solutions around the upper branch of the Z-shaped curve does not extend past the lower "knee" of the Z-shaped curve. In fact, the homoclinic bifurcation coincides with the saddle-node bifurcation SN1. Hence, there is no bistability in the model, and a simple one-variable slow subsystem is insufficient to give bursting, because it is unable to move the fast subsystem in and out of the region where the oscillations occur, as in the β-cell model. However, because the parabolic bursting model has two slow variables, x and c, a slow oscillation in these variables moves c backward and forward across the homoclinic bifurcation, leading to bursts of fast oscillations during one portion of the slow oscillations.

In Fig. 9.20 we plot the positions of the various bifurcations as x varies. Note that the curves in Fig. 9.19 correspond to taking a cross-section of Fig. 9.20 for $x = 0.7$. In region J_{ss} there is a single stable fixed point of the fast subsystem, while in region J_{osc} (between the SNP and HC curves) the fast subsystem has stable oscillations. Now suppose that the slow subsystem has a periodic solution. In the parabolic bursting model, slow oscillations do not occur independently of the fast subsystem but rely on the interaction between the fast and slow variables, the details of which do not concern us here. Similar results are obtained when the slow variables oscillate independently of the fast variables, acting as a periodic driver of the fast subsystem (Kopell and Ermentrout, 1986). In any case, these oscillations correspond to closed curves in the (x, c) phase plane, two possible examples of which are shown in Fig. 9.20B. In case **a**, the dynamics of x and c are such that the slow periodic solution lies entirely within the

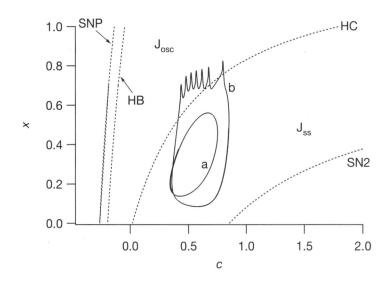

Figure 9.20 The Hopf and homoclinic bifurcations in the (x, c) plane, i.e., a projection of the bifurcation surface on the (x, c) plane. The SN1 bifurcation coincides with HC and is not shown separately. The rightmost curve, SN2, corresponds to the rightmost saddle node, not shown in Fig. 9.19. Two typical slow oscillations in x and c are shown as solid curves; curve a is with the parameters shown in Table 9.5, curve b is with the same parameters, but with f increased to 0.009.

region J_{ss}; i.e., the fast subsystem "lives" entirely on the lower branch of the Z-shaped curve and does not oscillate. In case **b**, the slow oscillation crosses the line of homoclinic bifurcations into the region J_{osc}, in which region the fast subsystem oscillates rapidly, leading to bursting.

Thus, by tuning the parameters of the underlying slow oscillation, different oscillatory patterns can be obtained from one model. We have already mentioned the parabolic nature of the period of the fast oscillations. This is easily understood in terms of the foregoing analysis. For a burst to occur, the slow oscillation must cross the line of homoclinic bifurcations both at the beginning and at the end of the burst. Since the period of the limit cycle tends to infinity at the homoclinic bifurcation, the interspike interval is large at both the beginning and end of the burst. However, as the slow periodic orbit penetrates further into region J_{osc}, away from the homoclinic bifurcation, the interspike interval decreases.

Classification of Bursting Oscillations

The earliest classification scheme for the different bursting mechanisms was proposed by Rinzel (1987) and later extended by Bertram et al. (1995). Originally, bursting oscillations were grouped into three classes: type I, with bursts arising from hysteresis and bistability as in the β cell model; type II, with bursts arising from an underlying slow oscillation, as in the Aplysia R-15 neuron; and type III, with bursts arising from a subcritical Hopf bifurcation. However, more recent studies (Izhikevich, 2000) have presented a huge array of different possible bursting behaviors, depending on the underlying bifurcations in the steady-state and periodic branches. Many of these possible bursters have no known biophysical implementation.

9.2 Hypothalamic and Pituitary Cells

In pancreatic β cells, the interaction of membrane-based electrical excitability with ER-based Ca^{2+} dynamics is crucial for the generation of fast bursting oscillations. Similar interactions are also known to be important in many cell types, including many types of hypothalamic neurons and pituitary cells. The close relationship between the hypothalamus and pituitary is discussed in Chapter 16. Briefly, hypothalamic neurons secrete hormones which typically travel a short distance along the inferior hypophyseal artery and the portal vein to the pituitary to act there on specific target cells (see Fig. 16.1). For example, gonadotropin-releasing hormone (GnRH) is secreted from GnRH cells in the hypothalamus; after traveling to the pituitary it stimulates the release of gonadotropin from pituitary gonadotrophs.

There are already a number of detailed models of electrical and Ca^{2+} excitability in hypothalamic or pituitary cells, and there are sure to be many more as our experimental understanding advances. For example, the release of corticotropin from pituitary corticotrophs in response to corticotropin-releasing hormone (CRH) has been studied in detail by LeBeau et al. (1997, 1998), Shorten et al. (2000) and Shorten and Wall (2000), while the interaction of Ca^{2+} and membrane potential in vasopressin hypothalamic neurons has been studied by Roper et al. (2003, 2004).

9.2.1 The Gonadotroph

One of the best models of this type is an earlier model of the gonadotroph (Li et al., 1994, 1995, 1997) and it is this one that we describe in more detail here. In gonadotrophs, Ca^{2+} spiking occurs in two different ways. First, spontaneous action potentials lead to the influx of Ca^{2+} through voltage-gated channels, and thus Ca^{2+} spikes. Second, in response to the agonist GnRH, gonadotrophs produce inositol trisphosphate (IP$_3$) (Chapter 7) which releases Ca^{2+} from the endoplasmic reticulum (ER) and leads to oscillations in the cytoplasmic Ca^{2+} concentration. Both the electrophysiological and agonist-induced Ca^{2+} signaling pathways are well-characterized experimentally (Stojilković et al., 1994).

Similarly to the model of the pancreatic β cell, the gonadotroph model consists of two pieces; a model of Hodgkin–Huxley type of the membrane electrical potential, and a model of the intracellular Ca^{2+} dynamics, including IP$_3$-induced Ca^{2+} release from the ER. Thus, the gonadotroph model is essentially a combination of the models of Chapters 5 and 7 (see Exercise 3). It is, however, more complicated in application, as Li et al. modeled the cell as a spherically symmetric region, with a gradient of Ca^{2+}

from the plasma membrane to the interior of the cell. Because the components of the model are no different from what has come before, and because a detailed reproduction of the results is complicated by the inclusion of Ca^{2+} diffusion, we do not present the model in complete detail here, but give instead only an overview of its construction and behavior.

The Membrane Model

The membrane potential is modeled in the usual way:

$$C\frac{dV}{dt} = -I_{Ca,T} - I_{Ca,L} - I_K - I_{K,Ca} - I_L, \tag{9.67}$$

where the Ca^{2+} current has been divided into T-type and L-type current, I_K is the current through K^+ channels, $I_{K,Ca}$ is the current through Ca^{2+}-sensitive K^+ channels, and I_L is a leak current.

Each of these currents is modeled in Hodgkin–Huxley fashion, in much the same way as in the model of the pancreatic β cell. Most importantly, the conductance of the Ca^{2+}-sensitive K^+ channel is assumed to be an increasing function of c_R, the concentration of Ca^{2+} at the plasma membrane. Thus,

$$I_{K,Ca} = g_{K,Ca}\frac{c_R^4}{c_R^4 + K_c^4}\phi_K(V), \tag{9.68}$$

where $\phi_K(V)$ denotes the Goldman–Hodgkin–Katz current–voltage relationship (Chapter 3) used here instead of the linear one used in the pancreatic β cell model.

The Calcium Model

Let c denote the intracellular free concentration of Ca^{2+}, and let c_e denote the free concentration of Ca^{2+} in the ER. Note that both c and c_e are functions of r, the radial distance from the center of the spherically symmetric cell. In the body of the cell, Ca^{2+} can enter the cytoplasm only from the ER (through IP_3 receptors), and can leave the cytoplasm only via the action of SERCA pumps in the ER membrane. However, on the boundary of the cell, Ca^{2+} can enter or leave via Ca^{2+} currents or plasma membrane Ca^{2+} ATPase pumps. Calcium buffering is assumed to be fast and linear (Section 7.4).

The equations for c and c_e are thus (Section 7.3)

$$\frac{\partial c}{\partial t} = D\nabla^2 c + J_{IPR} - J_{serca}, \tag{9.69}$$

$$\frac{\partial c_e}{\partial t} = D_e\nabla^2 c_e - \gamma(J_{IPR} - J_{serca}), \tag{9.70}$$

where, as usual, D and D_e are the effective diffusion coefficients of c and c_e respectively, and γ is the scale factor relating ER volume to cytoplasmic volume. On the boundary of the cell, i.e., at $r = R$, we have

$$D\frac{\partial c}{\partial r}\bigg|_{r=R} = -\alpha(I_{\text{Ca,T}} + I_{\text{Ca,L}}) - J_{\text{pm}}, \tag{9.71}$$

$$\frac{\partial c_e}{\partial r}\bigg|_{r=R} = 0, \tag{9.72}$$

where J_{pm} denotes the flux out of the cell due to the action of plasma membrane Ca^{2+} ATPase pumps. The scale factor $\alpha = 1/(2FA_{\text{cell}})$, where A_{cell} is the surface area of the cell, converts the current (in coulombs per second) to a mole flux density (moles per area per second).

The flux through the IP$_3$ receptor, J_{IPR}, is modeled by the Li–Rinzel simplification of the De Young–Keizer model (see Chapter 7 and Exercise 3), while the ATPase pump fluxes, J_{serca} and J_{pm}, are modeled by Hill functions. The exact form of the model equations and the parameter values can be found in the appendix of Li et al. (1997).

Results

When this model is solved for a range of IP$_3$ concentrations (corresponding to a range of GnRH concentrations), the agreement with experimental data is impressive (Fig. 9.21). Before the addition of agonist the cells exhibit continuous spiking; once agonist is applied the frequency of spiking drops dramatically, changing to a more complex burst pattern. At the highest agonist concentrations, bursting is initially suppressed by the agonist, although spiking eventually reappears.

In all cases, the spiking frequency is initially greatly decreased by the agonist, but recovers gradually. This recovery of the spike frequency is an interesting demonstration of the importance of the amount of Ca^{2+} in the ER.

In Fig. 9.22 we show V, c and c_e during the response to agonist stimulation (where c and c_e are averaged over the entire cell). During phase SS-1, before the addition of agonist, the cell exhibits tonic spiking. Upon addition of 0.42 μM IP$_3$, c starts to oscillate and the voltage spikes more slowly (phase T-1). The slowdown of the voltage spiking is caused by the much larger and slower Ca^{2+} spikes, which activate the Ca^{2+}-sensitive K$^+$ current to a much greater extent, thus making it more difficult for a voltage spike to occur. However, as time goes on, Ca^{2+} is lost from the cell by the action of the plasma membrane ATPase (recall that, because of the spatially distributed nature of the model, the concentration of Ca^{2+} is higher at the membrane than in the interior of the cell, which increases the rate at which Ca^{2+} is lost to the outside by the operation of the pump). This overall decline in $[Ca^{2+}]$ manifests itself in two principal ways; first, the concentration in the ER declines slowly, and secondly, the baseline of the Ca^{2+} spikes gradually decreases. It is this decrease that causes the slow increase in the frequency of the voltage spikes, as $I_{\text{K,Ca}}$ is gradually decreased.

Eventually, c_e decreases to such an extent that the ER can no longer support large Ca^{2+} spikes, and the oscillations in c become smaller and faster (SS-2 phase). This itself allows for a much greater frequency of the voltage spikes, which thus bring in more

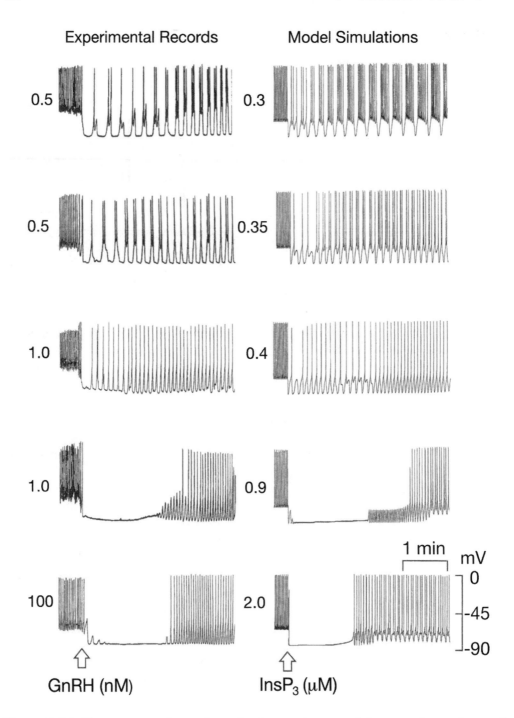

Figure 9.21 The response of gonadotrophs, in experiment and theory, to increasing applications of agonist. In the experiments (left panels), the indicated concentration of GnRH was added at the arrow, while this was simulated in the model (right panels) by addition of the indicated concentrations of IP$_3$. Li et al. (1997), Fig. 2.

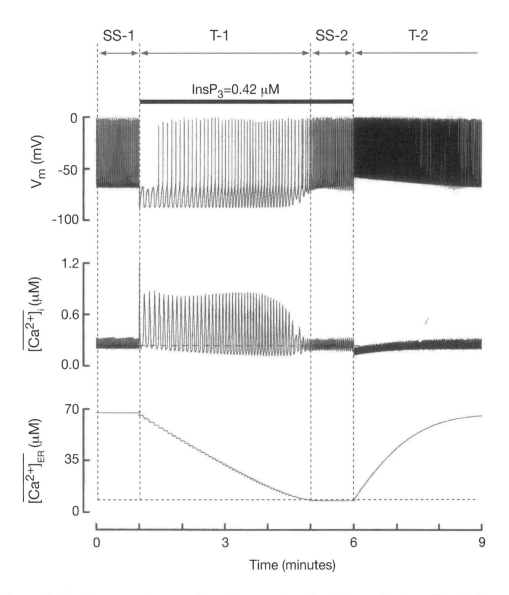

Figure 9.22 Response of a gonadotroph to agonist stimulation, with the different phases labeled. $\overline{[Ca^{2+}]}_i$ and $\overline{[Ca^{2+}]}_{ER}$ are, respectively, the average cytoplasmic and ER Ca^{2+} concentrations over the entire cell. Li et al. (1997), Fig. 3.

Ca^{2+} from the outside, allowing for the stabilization of the ER Ca^{2+} concentration. Upon removal of the agonist, c declines transiently, which causes an increase in the frequency of the voltage spikes. Because these voltage spikes bring in more Ca^{2+}, the ER gradually refills, and the system returns to basal levels (phase T-2).

This model demonstrates the complex interactions between ER Ca^{2+}, cytoplasmic Ca^{2+}, and membrane voltage spiking. It is clearly not the entire story, but it provides an excellent explanation for a wide variety of complex responses.

9.3 EXERCISES

1. (a) Numerically simulate the system of differential equations

$$\frac{dv}{dt} = f(v) - w - gs(v - v_\theta), \tag{9.73}$$

$$5\frac{dw}{dt} = w_\infty(v) - w, \tag{9.74}$$

$$\frac{ds}{dt} = f_s(s) + \alpha_s(x - 0.3), \tag{9.75}$$

$$\frac{dx}{dt} = \beta_x\left((1-x)H(v) - x\right), \tag{9.76}$$

where $f(v) = 1.35v(1 - \frac{1}{3}v^2), f_s(s) = -0.2(s - 0.05)(s - 0.135)(s - 0.21), w_\infty(v) = \tanh(5v)$, and $H(v) = \frac{3}{2}(1 + \tanh(5x - 2.5))$, and $v_\theta = -2, \alpha_s = 0.002, \beta_x = 0.00025, g = 0.73$.

   (b) Give a fast–slow analysis of this burster. Hint: The equations for v, w comprise the fast subsystem, while those for s, x comprise the slow subsystem.

   (c) Describe the bursting mechanism in this model. For what kind of burster might this be a reasonable model?

2. Compute some numerical solutions of the Rinzel–Lee simplification of the Chay–Keizer β cell model. How does the value of k_c affect the burst length?

3. To the Ca^{2+} submodel of phantom bursting in Section 9.1.2, add a Ca^{2+} flux, J_{IPR}, through IP_3 receptors (see Chapter 7). Assume that the IPR is at steady state, with an open probability given by the Li–Rinzel simplification (Section 7.2.5) of the De Young–Keizer model (Section 7.2.5). Thus

$$J_{IPR} = P_o(c_e - c), \tag{9.77}$$

   where

$$P_o = \left(\frac{c}{d_{act} + c}\right)^3 \left(\frac{p}{d_{IP_3} + p}\right)^3 \left(\frac{d_{inact}}{d_{inact} + c}\right)^3, \tag{9.78}$$

   where $d_{act} = 0.35\ \mu M$, $d_{inact} = 0.4\ \mu M$, and $d_{IP_3} = 0.5\ \mu M$. Let $g_{K,Ca} = 700$ and $g_{K,ATP} = 230$, so that the model exhibits medium bursting. Show that when p is increased from 0 to $0.3\ \mu M$ the bursting changes from medium to fast. Explain this in terms of changes to the $dc/dt = 0$ nullcline.

4. In the phantom bursting model of Section 9.1.2, c_e increases during the active phase and declines during the silent phase. Why? Explain this without using any mathematical equations. Then give a mathematical explanation.

5. Investigate the behavior of the model of compound bursting (Section 9.1.3) for the three fixed values of [ATP] shown in Fig. 9.10.

   (a) For each value of [ATP] solve the model equations numerically, and plot them in the V, c phase plane, as well as against time.

(b) Compute the bifurcation diagrams (as shown in Fig. 9.10), and superimpose the model solutions (with varying c) on the bifurcation diagrams (treating c as a bifurcation parameter).

(c) Show that, when [ATP] is small there is no bursting, while when [ATP] is large the active phase is continuous.

(d) Show that, for intermediate values of [ATP], the duration of the active phase depends on [ATP].

(e) Explain why these results should be so, using the bifurcation diagrams to illustrate your answers.

6. If the parameters of the model of compound bursting (Section 9.1.3) are changed slightly to $f = 0.0013$, $v_2 = 188$, $g_{K,ATP} = 357$, what kind of bursting occurs? (This kind of bursting has been called "accordion" bursting.) Explain this behavior by using a fast–slow analysis of the model, and plotting the bifurcation diagrams. Hint: What values are taken by [ATP] over the course of an oscillation, and how do these differ from the values taken in the top panel of Fig. 9.10?

7. Determine the value of I in the Hindmarsh–Rose fast subsystem (9.53), (9.54) for which the trajectory from the unstable node is also the saddle point separatrix.

8. (a) An equation of the form $\frac{d^2x}{dT^2} + F(x)\frac{dx}{dT} + x = 0$ is called an equation in the Liénard form (Minorsky, 1962; Stoker, 1950) and was important in the classical development of the theory of nonlinear oscillators. Show that the polynomial bursting model (9.55)–(9.57) can be written in the *generalized* Liénard form

$$\frac{d^2x}{dT^2} + F(x)\frac{dx}{dT} + G(x,z) = -\epsilon H(x,z), \tag{9.79}$$

$$\frac{dz}{dT} = \epsilon H(x,z), \tag{9.80}$$

where ϵ is a small parameter, and where F, G, and H are the polynomial functions

$$F(u) = a[(u - \hat{u})^2 - \eta^2], \tag{9.81}$$

$$G(u,z) = z + u^3 - \frac{3}{4}u - \frac{11}{27}, \tag{9.82}$$

$$H(u,z) = \beta(u - \bar{u}) - z. \tag{9.83}$$

(b) Construct the fast subsystem bifurcation diagram for this polynomial model for the following three different cases with $a = 0.25$, $\beta = 4$, $\hat{u} = -0.954$, and $\epsilon = 0.0025$.

i. $\eta = 0.7$, $\hat{u} = 1.6$. Show that in this case the model exhibits square-wave bursting.

ii. $\eta = 0.7$, $\hat{u} = 2.1$. Show that the model exhibits tapered bursting, resulting from passage through a supercritical Hopf bifurcation.

iii. $\eta = 1.2$, $\hat{u} = 1.0$. Describe the bursting pattern here.

Solve the model equations numerically to confirm the predictions from the bifurcation diagrams. A detailed analysis of this model has been performed by Pernarowski (1994) and de Vries (1995), with a perturbation analysis given by (Pernarowski et al., 1991, 1992).

9. Prove that if D is an irreducible matrix with nonnegative entries d_{ij}, then the only nontrivial solution of the system of equations $\sum_{j=1}^{n} d_{ij}(V_i - V_j) = 0, i = 1, \ldots, N$, is the constant vector. Remark: An irreducible nonnegative matrix is one for which some power of the matrix has no zero elements. To understand irreducibility, think of the elements of the matrix D as providing connections between nodes of a graph, and d_{ij} is positive if there is a path from node i to node j, but zero if not. Such a matrix is irreducible if between any two points there is a connecting path, with possibly multiple intermediate points. The smallest power of the matrix that is strictly positive is the smallest number of connections that is certain to connect any node with any other node.

    Hint: Represent the system of equations as $Av = 0$ and show that some shift $A + \lambda I$ of the matrix A is irreducible and has only nonnegative entries. Invoke the Perron–Frobenius theorem (a nonnegative irreducible matrix has a unique, positive, simple eigenvector with eigenvalue larger in magnitude than all other eigenvalues) to show that the null space of A is one-dimensional, spanned by the constant vector.

CHAPTER 10

Regulation of Cell Function

In all cells, the information necessary for the regulation of cell function is contained in strands of deoxyribose nucleic acid, or DNA. The nucleic acids are large polymers of smaller molecular subunits called *nucleotides*, which themselves are composed of three basic molecular groups: a *nitrogenous base*, which is an organic ring containing nitrogen; a 5-carbon (pentose) sugar, either *ribose* or *deoxyribose*; an inorganic phosphate group. Nucleotides may differ in the first two of these components, and consequently there are two specific types of nucleic acids: *deoxyribonucleic acid* (DNA) and *ribonucleic acid* (RNA).

There may be any one of five different nitrogenous bases present in the nucleotides: *adenine* (A), *cytosine* (C), *guanine* (G), *thymine* (T), and *uracil* (U). These are most often denoted by the letters A, C, G, T, and U, rather than by their full names.

The DNA molecule is a long double strand of nucleotide bases, which can be thought of as a twisted, or helical, ladder. The backbone (or sides of the ladder) is composed of alternating sugar and phosphate molecules, the sugar, deoxyribose, having one fewer oxygen atom than ribose. The "rungs" of the ladder are complementary pairs of nitrogenous bases, with G always paired with C, and A always paired with T. The bond between pairs is a weak hydrogen bond that is easily broken and restored during the replication process. In eukaryotic cells (cells that have a nucleus), the DNA is contained within the nucleus of the cell.

The ordering of the base pairs along the DNA molecule is called the genetic code, because it is this ordering of symbols from the four-letter alphabet of A, C, G, and T that controls all cellular biochemical functions. The nucleotide sequence is organized into code triplets, called *codons*, which code for amino acids as well as other signals, such as "start manufacture of a protein molecule" and "stop manufacture of a protein molecule." Segments of DNA that code for a particular product are called *genes*, of which there are about 30,000 in human DNA. Typically, a gene contains start and stop codons as well as the code for the gene product, and can include large segments of DNA whose function is unclear. One of the simplest known living organisms, *Mycoplasma genitalian*, has 470 genes and about 500,000 base pairs.

DNA itself, although not its structure, was first discovered in the late nineteenth century, and by 1943 it had been shown (although not widely accepted) that it is also the carrier of genetic information. How (approximately) it accomplishes this, and the structure of the molecule, was not established until the work of Maurice Wilkins and Rosalind Franklin at King's College in London, and James Watson and Francis Crick in Cambridge. Watson, Crick, and Wilkins received the 1962 Nobel Prize in Physiology or Medicine, Franklin having died tragically young some years previously, in 1958.

In recent years, the study of DNA and the genetic code has grown in a way that few would have predicted even 20 years ago. Nowadays, genetics and molecular biology have penetrated deeply into practically all aspects of life, from research and education to business and forensics. Mathematicians and statisticians have not been slow to join these advances. Departments and institutes of bioinformatics are springing up in all sorts of places, and there are few mathematics or statistics departments that do not have connections (some more extensive than others, of course) with molecular biologists. It is well beyond the scope of this book to provide even a cursory overview of this vast field. An excellent introduction to molecular biology is the book by Alberts et al. (1994); any reader who is seriously interested in learning about molecular biology will find this book indispensable. Waterman (1995), Mount (2001) and Krane and Raymer (2003) are good introductory bioinformatics texts, while for those who are more mathematically or statistically oriented there are Deonier et al. (2004), Ewens and Grant (2005), and Durrett (2002).

10.1   Regulation of Gene Expression

An RNA molecule is a single strand of nucleotides. It is different from DNA in that the sugar in the backbone is ribose, and the base U is substituted for T. Cells generally contain two to eight times as much RNA as DNA. There are three types of RNA, each of which plays a major role in cell physiology. For our purposes here, *messenger RNA* (mRNA) is the most important, since it carries the code for the manufacture of specific proteins. *Transfer RNA* (tRNA) acts as a carrier of one of the twenty amino acids that are to be incorporated into a protein molecule that is being produced. Finally, *ribosomal RNA* constitutes about 60% of the *ribosome*, a structure in the cellular cytoplasm on which proteins are manufactured.

The two primary functions that take place in the nucleus are the reproduction of DNA and the production of RNA. RNA is formed by a process called *transcription*, as follows. An enzyme called *RNA polymerase* (or, more precisely, a polymerase complex, since many other proteins are also needed) attaches to some starting site on the DNA, breaks the bonds between base pairs in that local region, and then makes a complementary copy of the nucleotide sequence for one of the DNA strands. As the RNA polymerase moves along the DNA strand, the RNA molecule is formed, and the DNA crossbridges reform. The process stops when the RNA polymerase reaches a transcriptional termination site and disengages from the DNA.

Proteins are manufactured employing all three RNA types. After a strand of mRNA that codes for some protein is formed in the nucleus, it is released to the cytoplasm. There it encounters ribosomes that "read" the mRNA much like a tape recording. As a particular codon is reached, it temporarily binds with the specific tRNA with the complementary codon carrying the corresponding amino acid. The amino acid is released from the tRNA and binds to the forming chain, leading to a protein with the sequence of amino acids coded for by the DNA.

Synthesis of a cellular biochemical product usually requires a series of reactions, each of which is catalyzed by a special enzyme. In prokaryotes, formation of the necessary enzymes is often controlled by a sequence of genes located in series on the DNA strand. This area of the DNA strand is called an *operon*, and the individual genes within the operon are called *structural genes*. At the beginning of the operon is a segment called a *promoter*, which is a series of nucleotides that has a specific affinity for RNA polymerase. The polymerase must bind with this promoter before it can begin to travel along the DNA strand to synthesize RNA. In addition, in the promoter region there is an area called a *repressor operator*, where a regulatory repressor protein can bind, preventing the attachment of RNA polymerase, thereby blocking the transcription of the genes of the operon. Repressor protein generally exists in two allosteric forms, one that can bind with the repressor operator and thereby repress transcription, and one that does not bind. A substance that changes the repressor so that it breaks its bond with the operator is called an *activator*, or *inducer*.

The original concept of the operon was due to Jacob et al. (1960), closely followed by mathematical studies (Goodwin, 1965; Griffith, 1968a,b; Tyson and Othmer, 1978). The interesting challenge is to understand how genes can be regulated by complex networks, and when, or how, gene expression can respond to the need of the organism or changes in the environment.

10.1.1 The *trp* Repressor

Tryptophan is an essential amino acid that cannot be synthesized by humans and therefore must be part of our diet. Tryptophan is a precursor for serotonin (a neurotransmitter), melatonin (a hormone), and niacin. Improper metabolism of tryptophan has been implicated as a possible cause of schizophrenia, since improper metabolism creates a waste product in the brain that is toxic, causing hallucinations and delusions. Tryptophan can, however, be synthesized by bacteria such as *E. coli*, and the regulation of tryptophan production serves as our first example of transcriptional regulation.

A number of models of the tryptophan (*trp*) repressor have been constructed, of greater or lesser complexity (Bliss et al., 1982; Sinha, 1988; Santillán and Mackey,

2001a,b; Mackey et al., 2004). Here, we present only a highly simplified version of these models, designed to illustrate some of the basic principles.

The *trp* operon comprises a regulatory region and a coding region consisting of five structural genes that code for three enzymes required to convert chorismic acid into tryptophan (Fig. 10.1A). Expression of the *trp* operon is regulated by the Trp repressor protein which is encoded by the *trpR* gene. In contrast to the *lac* operon, which is described in the next section, the *trpR* operon is independent of the *trp* operon, being located some distance on the DNA from the *trp* operon. TrpR protein is able to bind to the operator only when it is activated by the binding of two tryptophan molecules. Thus, we have a negative feedback loop; while tryptophan levels in the cell are low,

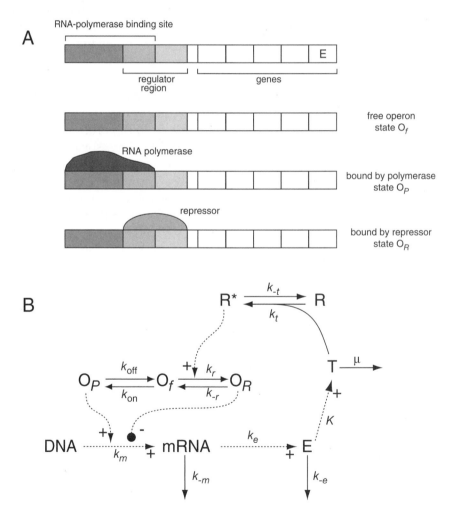

Figure 10.1 A: Control sites and control states of the *trp* operon. B: Feedback control of the *trp* operon. Dashed lines indicate reactions in which the reactants are not consumed.

production of tryptophan remains high. However, once the level of tryptophan builds up, the TrpR protein is activated, and represses further transcription of the operon. As a result, the synthesis of the three enzymes and consequently of tryptophan itself declines.

The ability to regulate production of a substance in response to its need is characteristic of negative feedback systems. Here, the negative feedback occurs because the product of gene activation represses the activity of the gene. Thus, the tryptophan operon is called a *repressor*.

A simple model of this network is sketched in Fig. 10.1B. We suppose that the operon has three states, either free O_f, bound by repressor O_R, or bound by polymerase O_P, and let $o_j, j = f, R, P$, be the probability that the operon is in state j. Messenger RNA (M) is produced only when polymerase is bound, so that

$$\frac{dM}{dt} = k_m o_P - k_{-m} M. \tag{10.1}$$

Note that the dashed lines in Fig. 10.1B correspond to reactions in which the reactants are not consumed. Thus, for example, the production of enzyme (E) does not consume mRNA, so that there is no consumption term $-k_e M$ in (10.1). The probabilities of being in an operon state are governed by the differential equations

$$\frac{do_P}{dt} = k_{on} o_f - k_{off} o_P, \qquad \frac{do_R}{dt} = k_r R^* o_f - k_{-r} o_R, \tag{10.2}$$

where $o_P + o_f + o_R = 1$, and R^* denotes activated repressor. Activation of repressor requires binding with two molecules of tryptophan (T) and so we take

$$\frac{dR^*}{dt} = k_t T^2 (1 - R^*) - k_{-t} R^* \tag{10.3}$$

(scaled so that $R + R^* = 1$).

According to Bliss et al. (1982), the most important of the enzymes is anthranilate synthase (denoted by E), which is the only enzyme concentration we track here. Enzyme (E) is produced from mRNA, and degraded,

$$\frac{dE}{dt} = k_e M - k_{-e} E, \tag{10.4}$$

and tryptophan production is proportional to the amount of enzyme, so that

$$\frac{dT}{dt} = KE - \mu T - 2\frac{dR^*}{dt}. \tag{10.5}$$

Here μ is the rate of tryptophan utilization and degradation. Note that the factor of 2 on the right-hand side comes from the fact that it takes two tryptophan molecules to activate the repressor.

For our purposes here, it is sufficient to examine the steady-state solutions of this system, which must satisfy the algebraic equation

$$F(T) \equiv \frac{k_e}{k_{-e}} \frac{k_m}{k_{-m}} \frac{k_{on}}{k_{off}} \frac{1}{1 + \frac{k_{on}}{k_{off}} + \frac{k_r R^*(T)}{k_{-r}}} = \frac{\mu}{K} T, \tag{10.6}$$

where

$$R^*(T) = \frac{T^2}{\frac{k_{-t}}{k_t} + T^2}.$$ (10.7)

The function $F(T)$ is a positive monotone decreasing function of T, and represents the steady-state rate of tryptophan production. The right-hand side of this equation is a straight line with slope μ/K. Thus, there is a unique positive intersection. Furthermore, as the utilization of tryptophan, quantified by μ/K, increases, the steady-state level of T decreases and the production rate $F(T)$ necessary to balance utilization increases, characteristic of negative feedback control. This is illustrated in Fig. 10.2, where we plot $F(T)$ and $\mu T/K$ for two different values of μ/K.

10.1.2 The *lac* Operon

When glucose is abundant, *E. coli* uses it exclusively as its food source, even when other sugars are present. However, when glucose is not available, *E. coli* is able to use other sugars such as lactose, a change that requires the expression of different genes by the bacterium. Jacob, Monod, and their colleagues (Jacob et al., 1960; Jacob and Monod, 1961) were the first to propose a mechanism by which this could happen, a mechanism that is now called a *genetic switch*. Forty years ago, the idea of a genetic switch was revolutionary, but the original description of this mechanism has withstood the test of time and is used, practically unchanged, in modern textbooks. Mathematicians were quick to see the dynamic possibilities of genetic switches, with the first model, by Goodwin, appearing in 1965, followed by that of Griffith (1968a,b). More recently, detailed models have been constructed by Wong et al. (1997), Yildirim and Mackey

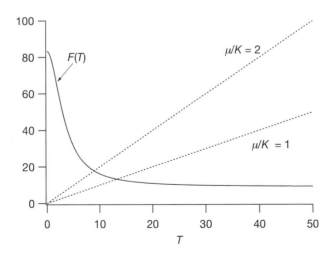

Figure 10.2 Plots of $F(T)$ and $\mu T/K$ from (10.6), for two different values of μ/K. Other parameter values were chosen arbitrarily: $\frac{k_e}{k_{-e}} \frac{k_m}{k_{-m}} \frac{k_{on}}{k_{off}} = 500$, $\frac{k_{on}}{k_{off}} = 5$, $\frac{k_r}{k_{-r}} = 50$, and $\frac{k_{-t}}{k_t} = 100$.

(2003), Yildirim et al. (2004), and Santillán and Mackey (2004). The structure and function of the *lac* repressor is reviewed by Lewis (2005), while an elegant blend of theoretical and experimental work was presented by Ozbudak et al. (2004). Mackey et al. (2004) review modeling work on both the *lac* operon and the tryptophan operon.

The *lac* operon consists of three structural genes and two principal control sites. The three genes are *lacZ*, *lacY*, and *lacA*, and they code for three proteins involved in lactose metabolism: *β-galactosidase*, *lac* permease, and *β-thiogalactoside acetyl transferase*, respectively. The permease allows entry of lactose into the bacterium. The *β*-galactosidase isomerizes lactose into *allolactose* (an allosteric isomer of lactose) and also breaks lactose down into the simple hexose sugars glucose and galactose, which can be metabolized for energy. The function of the transferase is not known.

Whether the operon is on or off depends on the two control sites. One of these control sites is a repressor, the other is an activator. If a repressor is bound to the repressor binding site, then RNA polymerase cannot bind to the operon to initiate transcription, and the three proteins cannot be produced. Preceding the promoter region of the *lac* operon, where the RNA polymerase must bind to begin transcription, there is another region, called a CAP site, which can be bound by a dimeric molecule CAP (*catabolic activator protein*). CAP by itself has no influence on transcription unless it is bound to cyclic AMP (cAMP), but when CAP is bound to cAMP the complex can bind to the CAP site, thereby promoting the binding of RNA polymerase to the promoter region, allowing transcription.

So, in summary, the three proteins necessary for lactose metabolism are produced only when CAP is bound and the repressor is not bound. This is illustrated in Fig. 10.3.

A bacterium is thus able to switch the *lac* operon on and off by regulating the concentrations of the repressor and of CAP, and this is how the requisite positive and negative feedbacks occur. Allolactose plays a central role here. In the absence of allolactose, the repressor is bound to the operon. However, allolactose can bind to the repressor protein, and prevent it binding to the repressor site. This, in turn, allows activation of the operon, the further production of allolactose (via the action of *β*-galactosidase), and increased entry of lactose (via the *lac* permease). Hence we have a positive feedback loop.

The second feedback loop operates through cAMP. The CAP protein is formed by a combination of cAMP with a cAMP receptor protein. When there is a large amount of cAMP in the cell, the concentration of CAP is high, CAP binds to the CAP binding site of the operon, thus allowing transcription. When cAMP concentration is low in the bacterium, the reverse happens, turning the operon off. A decrease in extracellular glucose leads to an increase in intracellular cAMP concentration (by an unknown mechanism), thus leading to activation of CAP and subsequent activation of the operon. Conversely, an increase in extracellular glucose switches the operon off.

To summarize, the operon is switched on only when lactose is present inside the cell, and glucose is not available outside (Fig. 10.3). Positive feedback is accomplished by allolactose preventing binding of the repressor. Negative feedback is accomplished by the control of CAP levels by extracellular glucose (Fig. 10.4).

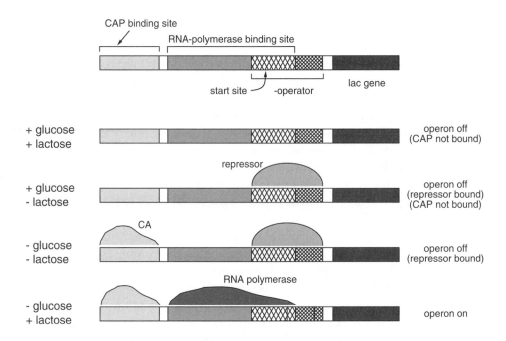

Figure 10.3   Control sites and control states for the *lac* operon.

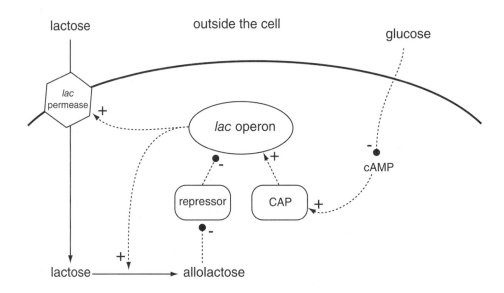

Figure 10.4   Feedback control of the *lac* operon. Indirect effects are denoted by dashed lines, with positive and negative effects denoted by different arrowheads and associated + or − signs.

Here we present a mathematical model of this process that is similar to the models of Griffith (1971) (see Exercise 1) and Yildirim and Mackey (2003). Our goal is to show how, when there is no lactose available, the operon is switched off, but that as the external lactose concentration increases, the operon is switched on (i.e., a genetic switch). Because of our limited goal we do not include the dynamics of CAP in our model.

Let A denote allolactose, with concentration A, and similarly for lactose (L), the permease (P), β-galactosidase (B), mRNA (M), and the repressor (R). We assume that the repressor, normally in its activated state R*, reacts with two molecules of allolactose to become inactivated (R), according to

$$R^* + 2A \underset{k_{-a}}{\overset{k_a}{\rightleftarrows}} R. \tag{10.8}$$

For simplicity, we assume that the operon can be in one of only two states, bound to (activated) repressor and therefore inactive (O_R), or bound by polymerase and therefore producing mRNA (O_P). Thus, the operon reacts with the repressor according to

$$O_P + R^* \underset{k_{-r}}{\overset{k_r}{\rightleftarrows}} O_R. \tag{10.9}$$

The probabilities for the operon to be in these states is governed by the equation

$$\frac{do_p}{dt} = k_{-r}(1 - o_p) - k_r R^* o_p, \tag{10.10}$$

since $o_P + o_R = 1$. Since effectively no repressor is consumed by binding with the operon, the repressor concentration is governed by the differential equation

$$\frac{dR^*}{dt} = k_{-a}R - k_a A^2 R^*, \tag{10.11}$$

where $R + R^* = R_t$.

Assuming each of these reactions is in steady state, we find that

$$R = K_a R^* A^2, \tag{10.12}$$

$$o_R = K_r R^* o_P, \tag{10.13}$$

where $K_i = k_i / k_{-i}$ for $i = a, r$. Thus,

$$R_t = R + R^* = R(1 + K_1 A^2) \tag{10.14}$$

and

$$o_P = \frac{1}{1 + K_r R^*} = \frac{1 + K_a A^2}{1 + K_r R_t + K_a A^2} = \frac{1 + K_a A^2}{K + K_a A^2}, \tag{10.15}$$

where $K = 1 + K_r R_t > 1$. Hence, the production of mRNA is described by the differential equation

$$\frac{dM}{dt} = \alpha_M o_P - \gamma_M M = \alpha_M \frac{1 + K_a A^2}{K + K_a A^2} - \gamma_M M, \tag{10.16}$$

where M is the concentration of mRNA that codes for the enzymes. The constant α_M is a proportionality constant that relates the probability of activated operon to the rate of mRNA production, while γ_M describes the degradation of mRNA. Note that, in the absence of allolactose, there is a residual production of mRNA. This is because the reaction in (10.9) has an equilibrium where O_P is nonzero, even at maximal concentrations of R.

We next assume that the enzymes are produced at a rate linearly proportional to available mRNA and are degraded, so that the concentrations of permease (denoted by P) and β-galactosidase (denoted by B) are determined by

$$\frac{dP}{dt} = \alpha_P M - \gamma_P P, \tag{10.17}$$

$$\frac{dB}{dt} = \alpha_B M - \gamma_B B. \tag{10.18}$$

Although it might appear that, since their codes are part of the same mRNA, the production rates of P and B are the same, this is not the case. First, mRNA reads the different genes within the operon (*lacZ* and *lacY*) in sequence, making β-galactosidase first and the permease second. Second, the permease must migrate to the cell membrane to be incorporated there. The different times of production, and the different time delays before these two enzymes can become effective, imply that they have different effective rates of production (see Table 10.1).

Lactose that is exterior to the cell, with concentration L_e, is brought into the cell to become the lactose substrate, with concentration L, at a Michaelis–Menten rate proportional to the permease P. Once inside the cell, lactose substrate is converted to allolactose, and then allolactose is converted to glucose and galactose via enzymatic reaction with β-galactosidase, so that

$$\frac{dL}{dt} = \alpha_L P \frac{L_e}{K_{Le} + L_e} - \alpha_A B \frac{L}{K_L + L} - \gamma_L L \tag{10.19}$$

and

$$\frac{dA}{dt} = \alpha_A B \frac{L}{K_L + L} - \beta_A B \frac{A}{K_A + A} - \gamma_A A. \tag{10.20}$$

Note that all the reactions here are modeled as unidirectional reactions. This is not strictly correct, as all the reactions are bidirectional, particularly a reaction such as the transport of lactose into the cell, which occurs by a passive mechanism. However, unidirectional reaction rates are adequate for our purpose, since they provide a reasonable description over a wide range of substrate concentrations. Ignoring the reverse reactions does not alter the conclusions in a model as simple as that presented here.

To summarize, the model is given by the five equations (10.16)–(10.20). A more complicated mechanism is studied by Wong et al. (1997), while Yildirim and Mackey (2003) include a number of time delays, rendering the model a system of delay–differential

Table 10.1 Parameters of the *lac* operon model.

$\alpha_A = 1.76 \times 10^4$ min^{-1}	$\gamma_A = 0.52$ min^{-1}
$\alpha_B = 1.66 \times 10^{-2}$ min^{-1}	$\gamma_B = 2.26 \times 10^{-2}$ min^{-1}
$\alpha_P = 10$ min^{-1}	$\gamma_P = 0.65$ min^{-1}
$\alpha_M = 9.97 \times 10^{-4}$ mM/min	$\gamma_M = 0.41$ min^{-1}
$\alpha_L = 2880$ min^{-1}	$\gamma_L = 2.26 \times 10^{-2}$ min^{-1}
$\beta_A = 2.15 \times 10^4$ min^{-1}	$\beta_L = 2.65 \times 10^3$ min^{-1}
$K = 6000$	$K_a = 2.52 \times 10^4$ (mM)$^{-2}$
$K_A = 1.95$ mM	$K_L = 9.7 \times 10^{-7}$ mM
$K_{L1} = 1.81$ mM	$K_{Le} = 0.26$ mM

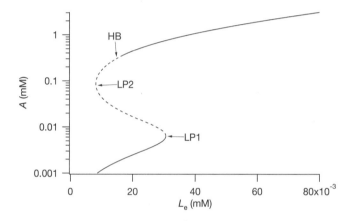

Figure 10.5 Steady states of the *lac* operon model as a function of the external lactose concentration, L_e. Unstable steady states are denoted by a dashed line; LP denotes a limit point and HB denotes a Hopf bifurcation point.

equations. However, this representation of the reaction mechanism is enough to show how a genetic switch can arise in a simple way.

Yildirim and Mackey (2003) devoted a great deal of effort to determine accurate values for the parameters, and slightly modified versions of these are given in Table 10.1, the modifications being necessary because our model is not exactly the same as theirs. The steady-state solution is shown in Fig. 10.5, plotted as a function of L_e, the external lactose concentration.

The curve of steady states shown in Fig. 10.5 was computed using AUTO (or, to be more precise, XPPAUT). However, if one does not have access to, or familiarity with, these kinds of sophisticated software packages, then the following do-it-yourself approach is recommended. The easiest method is to plot L_e as a function of A. That is, first use (10.16) to determine M as a function of A and then use (10.17) and (10.18) to obtain P and B as functions of A. Then use (10.20) to obtain L as a function of A, and finally use (10.19) to obtain L_e as a function of A.

When L_e is low there is only a single steady state, with a low concentration of allolactose, A; for an intermediate range of values of L_e there are two stable steady states, one (high A) corresponding to lactose usage, the other (low A) corresponding to negligible lactose usage; and for large values of L_e there is again only one stable steady-state solution, corresponding to lactose usage. Thus, if L_e increases past a critical value, here around 0.03 mM, the steady state switches from a low-usage state to the high-usage state as β-galactosidase is switched on. This discontinuous response to a gradually increasing L_e is characteristic of a genetic switch, and results from bistability in the model. This switching on of lactose usage is called *induction*, and this kind of operon is called an *inducer*.

The usage of lactose is switched off at a lower value of L_e than the value at which it is switched on. This feature of hysteretic switches is important because it prevents rapid cycling. That is, when allolactose usage is high, external lactose is consumed and therefore decreases. If the on switch and off switch were at the same value, then the lac operon would presumably be switched on and off rapidly, a strategy for resource allocation that seems unfavorable. In addition, separation of the on and off switches makes the system less susceptible to noise.

The Hopf bifurcation shown in Fig. 10.5 does not appear in more complicated models. It is thus of dubious relevance and we do not pursue it further.

10.2   Circadian Clocks

It has been known for a long time that many organisms have oscillators with a period of about 24 hours, hence "circadian" from the Latin *circa* = about and *dies* = a day. That humans have such a clock is apparent twice a year when the switch to or from daylight saving time occurs, or whenever one travels by air to a different time zone, and we experience "jet lag". Before the molecular basis of circadian clocks was known, the study of these rhythms focused on properties of generic autonomous oscillators, most typically the van der Pol oscillator, and its response to external stimuli either to entrain the oscillator or to reset the oscillator. There is a vast literature describing this endeavor, which we do not even attempt to summarize.

All of this changed in the 1980s when the first genes influencing the 24-hour cycle were discovered. These genes were the *per* (for *period*) gene in *Drosophila* and the *frq* (for *frequency*) gene in the fungus *Neurospora*.

Circadian clocks have been found in many organisms, including cyanobacteria, fungi, plants, invertebrate and vertebrate animals, but as yet, not in the archaebacteria. Wherever they are found, they employ biochemical loops that are self-contained within a single cell (requiring no cell-to-cell interaction). Further, the mechanism for their oscillation is always the same: there are positive elements that activate clock genes that yield clock proteins that act in some way to block the activity of the positive elements. Thus, the circadian clocks are all composed of a negative feedback loop (the gene product inactivates its own production) with a delay. An excellent review of circadian clocks is found in Dunlap (1999) (see also Dunlap, 1998).

Even before the details of the clocks were known, it was recognized that negative feedback loops with sufficient delay in the feedback could produce oscillatory behavior.

The first model of this type (Goodwin, 1965) was intended to model periodic enzyme synthesis in bacteria, and assumed that there were enzymes $X_1, X_1, \ldots, X_n$ such that

$$X_1 \rightarrow X_2 \rightarrow \cdots \rightarrow X_n,$$

i.e., X_n has an inhibitory effect on the production of X_1. The model equations for the Goodwin oscillator are

$$\frac{dX_1}{dt} = \frac{v_0}{1 + (\frac{X_n}{K_m})^p} - k_1 X_1, \tag{10.21}$$

$$\frac{dX_i}{dt} = v_{i-1} X_{i-1} - k_i X_i, \quad i = 2, \ldots, n. \tag{10.22}$$

Oscillations occur in this network only if $n \geq 3$, and if $p > 8$ when $n = 3$ (see Exercise 3).

So the question is not whether negative feedback loops with some source of time delay can produce oscillations. We know they can. Rather, the question is whether enough is known about the details of the biochemistry of circadian clocks to produce reasonably realistic models of their dynamics. There are several intriguing questions. What sets the intrinsic period of the oscillator; that is, what are the mechanisms that give a nearly 24 hour intrinsic cycle? How does phase resetting work?

Drosophila has a molecular circadian system whose investigation has been central to our understanding of how clocks work at the molecular level. Beginning in the morning, *per* and *tim* (for timeless) mRNA levels begin to rise, the result of activation of the clock gene promoters by a heterodimer of CLK (for CLOCK) and CYC (for CYCLE). The protein PER is unstable in the absence of TIM, but is stabilized by dimerization with TIM. Furthermore, the PER/TIM dimer is a target for nuclear translocation, and once it enters the nucleus (within three hours of dusk), it interacts with the CLK/CYC heterodimer to inhibit the activity of CLK/CYC, hence shutting down the production of *per* and *tim*. In the night, PER and TIM are increasingly phosphorylated, leading to their degradation. Once they are depleted, the CLK/CYC heterodimer is again activated and the cycle begins again.

Synchronization with the day/night light cycle and phase resetting occurs because the degradation of TIM is enhanced by light. Thus, exposure to light in the late day when TIM levels are rising delays the clock, while exposure to light in late night and early morning when TIM levels are falling advances the clock.

While the details are different in different organisms, this scenario is typical. However, as is also typical in biochemistry, the names of the main players are different in different organisms, even though their primary function is similar. In Table 10.2 we give a list of the main players in different organisms.

There are several mathematical models of circadian rhythms. An early model, due to Goldbeter (1995), is similar in structure to the enzyme oscillator of Goodwin. That is, the PER protein P_0 is (reversibly) phosphorylated into P_1 and then P_2

$$P_0 \rightleftarrows P_1 \rightleftarrows P_2. \tag{10.23}$$

Table 10.2   Circadian clock genes

System	Activator	Inhibitor
Synechoccus	*kaiA* (cycle in Japanese)	*kaiC*
Neurospora	*wc-1/wc-2* (WhiteCollar)	*frq* (Frequency)
Drosophila	*Clk/cyc* (Clock/Cycle)	*per tim* (Period/Timeless)
Mouse	*Clock/bmal1*	*per1,2,3* and *cry1,2* (Cryptochrome)

Table 10.3   Parameter values for the five-variable Goldbeter (1995) circadian clock model.

$v_s = 0.76 \ \mu\text{M h}^{-1}$	$k_s = 0.38 \ \text{h}^{-1}$
$v_m = 0.65 \ \mu\text{M h}^{-1}$	$k_1 = 1.9 \ \text{h}^{-1}$
$v_d = 0.95 \ \mu\text{M h}^{-1}$	$k_2 = 1.3 \ \text{h}^{-1}$
$V_1 = 3.2 \ \mu\text{M h}^{-1}$	$K_{1,2,3,4} = 2 \ \mu\text{M}$
$V_2 = 1.58 \ \mu\text{M h}^{-1}$	$K_I = 1 \ \mu\text{M}$
$V_3 = 5 \ \mu\text{M h}^{-1}$	$K_{M1} = 0.5 \ \mu\text{M}$
$V_4 = 2.5 \ \mu\text{M h}^{-1}$	$K_d = 0.2 \ \mu\text{M}$

The phosphorylated protein P_2 is then transported into the nucleus, where it (P_N) inhibits the production of *per* mRNA (M), closing the negative feedback loop, with a delay because of the phosphorylation steps and nuclear transport.

The equations for the Goldbeter model are as follows:

$$\frac{dM}{dt} = \frac{v_s}{1 + (\frac{P_n}{K_I})^4} - \frac{v_m M}{K_{m1} + M}, \tag{10.24}$$

$$\frac{dP_0}{dt} = k_s M - \frac{V_1 P_0}{K_1 + P_0} + \frac{V_2 P_1}{K_2 + P_1}, \tag{10.25}$$

$$\frac{dP_1}{dt} = \frac{V_1 P_0}{K_1 + P_0} - \frac{V_2 P_1}{K_2 + P_1} - \frac{V_3 P_1}{K_3 + P_1} + \frac{V_4 P_2}{K_4 + P_2}, \tag{10.26}$$

$$\frac{dP_2}{dt} = \frac{V_3 P_1}{K_3 + P_1} - \frac{V_4 P_2}{K_4 + P_2} - k_1 P_2 + k_2 P_N - \frac{v_d P_2}{K_d + P_2}, \tag{10.27}$$

$$\frac{dP_N}{dt} = k_1 P_2 - k_2 P_N. \tag{10.28}$$

Notice that the phosphorylation and dephosphorylation steps in this model are all at Michaelis–Menten rates, appropriate for enzymatic reactions, rather than linear, as in the original model of Goodwin. It is a straightforward matter to simulate these equations and verify that the solutions are periodic with a period of about 24 hours; we leave this verification to the interested reader.

More recently, Leloup and Goldbeter (1998, 2003, 2004) published a model that includes the two proteins PER and TIM, both of which undergo two phosphorylation steps before they form a dimer and then are transported into the nucleus, where they inhibit their own production. The model retains the basic structure of the Goodwin model, but it has the advantage that the effects of changes in the degradation rates of TIM can be studied as a model of how light interacts with the clock.

Another model of the circadian clock (Tyson et al., 1999) takes a different view of the post-translational regulation of PER and TIM, and is motivated by the more recent finding that another clock element *dbt* (doubletime) is important to the phosphorylation of PER. In this model, phosphorylation tags the protein for degradation, rather than activating it for nuclear translocation as in the Goldbeter model. In the Tyson et al. model, DBT protein phosphorylates monomers of PER at a faster rate than it does dimers, which means that PER monomers are much more likely to be degraded than its dimers.

As with the Goldbeter model, the Tyson et al. model does not include TIM, but assumes that dimers of PER inhibit the clock gene. The model equations are

$$\frac{dM}{dt} = \frac{v_m}{1 + (\frac{P_2}{P_{\text{crit}}})^2} - k_m M, \tag{10.29}$$

$$\frac{dP_1}{dt} = v_p M - \frac{k'_{p1} P_1}{J_p + P_1 + rP_2} - k_{p3} P_1 - 2k_a P_1^2 + 2k_d P_2, \tag{10.30}$$

$$\frac{dP_2}{dt} = k_a P_1^2 - k_d P_2 - \frac{k_{p2} P_2}{J_p + P_1 + rP_2} - k_{p3} P_2. \tag{10.31}$$

The most important assumptions entailed in these equations are, first, that both monomer (P_1) and dimers (P_2) bind to DBT, but P_1 is phosphorylated more rapidly (i.e., $k'_{p1} \gg k_{p2}$) and, second, that the DBT-catalyzed reaction is a saturating reaction and that the dimer is a competitive inhibitor of monomer phosphorylation. The extent of competitive inhibition is determined by r, the ratio of enzyme-substrate dissociation constants for the monomer and dimer (see Chapter 1). Here, the parameter r is chosen to be 2.

Next, we suppose that the dimerization reactions are fast (both k_a and k_d are large compared to other rate constants), so that P_1 and P_2 are in quasi-equilibrium. We let $P = P_1 + 2P_2$ be the total amount of PER protein, and observe that since $k_a^2 P_1^2 - k_d P_2 = 0$ (approximately),

$$P_1 = qP, \qquad P_2 = \frac{1}{2}(1 - q)P, \qquad q = \frac{2}{1 + \sqrt{1 + 8K_{\text{eq}}P}}. \tag{10.32}$$

Table 10.4 Parameter values for the two-variable Tyson et al. (1999) circadian clock model. Here C_p and C_m are typical concentrations of protein and mRNA, respectively.

$v_m = 1\ C_m\ \text{h}^{-1}$	$k_m = 0.1\ \text{h}^{-1}$
$v_p = 0.5\ C_p C_m^{-1}\ \text{h}^{-1}$	$k_{p1} = 10\ C_p \text{h}^{-1}$
$k_{p2} = 0.03\ C_p \text{h}^{-1}$	$k_{p3} = 0.1\ \text{h}^{-1}$
$K_{\text{eq}} = 200\ C_p^{-1}$	$P_{\text{crit}} = 0.1\ C_p$
$J_p = 0.5\ C_p$	$r = 2$

Now we add (10.30) and (10.31) to obtain a single equation for P,

$$\frac{dP}{dt} = v_p M - \frac{k_{p1} Pq + k_{p2} P}{J_p + P} - k_{p3} P. \tag{10.33}$$

This equation, coupled with (10.29) and the algebraic relationships (10.32), gives a two-variable model of the circadian clock.

A nice feature of two-variable systems is that they can be studied in the phase plane. The phase portrait for this system is shown in Fig. 10.6A, where it can be seen that this system is a canonical excitable system. That is, the $\frac{dM}{dt} = 0$ nullcline is a monotone decreasing curve having a single intersection with the N-shaped $\frac{dP}{dt} = 0$ nullcline. A typical solution is shown in Fig. 10.6B.

Research in circadian rhythms is quite active. Some recent work that relates to the topics of this text include work on the mammalian circadian clock (Leloup and Goldbeter, 2003; Forger and Peskin, 2003, 2004, 2005) and work describing the role of Ca^{2+} in plant circadian clocks (Dodd et al., 2005a,b).

10.3   The Cell Cycle

The *cell-division cycle* is the process by which a cell duplicates its contents and then divides in two. The adult human must manufacture many millions of new cells each second simply to maintain the status quo, and if all cell division is halted, the individual will die within a few days. On the other hand, abnormally rapid cell proliferation, i.e., *cancer*, can also be fatal, since rapidly proliferating cells interfere with the function of normal cells and organs. Control of the cell cycle involves, at a minimum, control of cell growth and replication of nuclear DNA in such a way that the size of the individual cells remains, on average, constant.

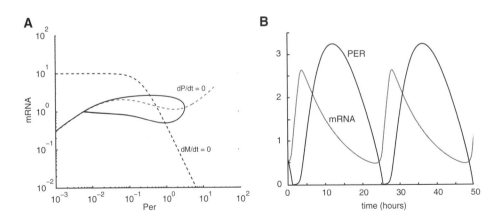

Figure 10.6   A: Phase portrait for the Tyson et al. circadian clock model. B: Solutions of the Tyson et al. circadian clock model.

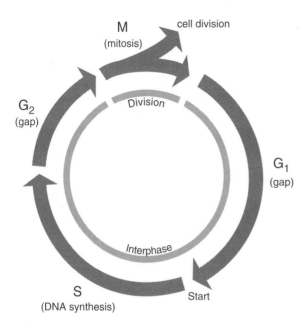

Figure 10.7 Schematic diagram of the cell cycle.

The cell cycle is traditionally divided into four distinct phases (shown schematically in Fig. 10.7), the most dramatic of which is mitosis, or *M phase*. Mitosis is characterized by separation of previously duplicated nuclear material, nuclear division, and finally the actual cell division, called *cytokinesis*. In most cells the whole of M phase takes only about an hour, a small fraction of the total cycle time. The much longer period of time between one M phase and the next is called *interphase*. In some cells, such as mammalian liver cells, the entire cell cycle can take longer than a year. The portion of interphase following cytokinesis is called G_1 *phase* (G for gap), during which cell growth occurs. When the cell is sufficiently large, DNA replication in the nucleus is initiated and continues during *S phase* (S for synthesis). Following S phase is G_2 *phase*, providing a safety gap during which the cell is presumably preparing for M phase, to ensure that DNA replication is complete before the cell plunges into mitosis.

There are actually two controlled growth processes. One is the chromosomal cycle, in which the genetic material is exactly duplicated and two nuclei are formed from one for every turn of the cycle. Accuracy is essential to this process, since each daughter nucleus must receive an exact replica of each chromosome. The second, less tightly controlled, process, the cytoplasmic cycle, duplicates the cytoplasmic material, including all of the structures (mitochondria, organelles, endoplasmic reticulum, etc.). This growth is continuous during the G_1, S, and G_2 phases, pausing briefly only during mitosis.

In mature organisms, these two processes operate in coordinated fashion, so that the ratio of cell mass to nuclear mass remains essentially constant. However, it is possible for these two to be uncoupled. For example, during *oogenesis*, a single cell (an *ovum*) grows in size without division. After fertilization, during *embryogenesis*, the egg undergoes twelve rapid synchronous mitotic divisions to form a ball consisting of 4096 cells, called the *blastula*.

The autonomous cell cycle oscillations seen in early embryos are unusual. Most cells proceed through the division cycle in fits and starts, pausing at "checkpoints" to ensure that all is ready for the next phase of the cycle. There are checkpoints at the end of the G_1, G_2, and M phases of the cell cycle, although not all cells use all of these checkpoints. During early embryogenesis, however, the checkpoints are inoperable, and cells divide as rapidly as possible, driven by the underlying limit cycle oscillation. The G_1 checkpoint is often called *Start*, because here the cell determines whether all systems are ready for S phase and the duplication of DNA. Before Start, newly born cells are able to leave the mitotic cycle and differentiate (into nondividing cells with specialized function). However, after Start, they have passed the point of no return and are committed to another round of DNA synthesis and division.

The cell cycle has been studied most extensively for frogs and yeast. Frog eggs are useful because they are large and easily manipulated. Yeast cells are much smaller, but are suitable for cloning and identification of the involved genes and gene products. Since both organisms use fundamentally similar mechanisms to regulate the cell cycle, insights gained from either may usefully be used to build up an overall picture of cell cycle control. The budding yeast *Saccharomyces cerevisiae*, used by brewers and bakers, divides by first forming a bud that is initiated and grows steadily during S and G_2 phases, and finally separates from its mother after mitosis. A similar organism, fission yeast *Schizosaccharomyces pombe*, is also used extensively in cell cycle studies. The cell cycle in mammalian cells is considerably more complex than in either frogs or yeast, and we do not consider it in any detail here.

Although there is a great deal of experimental work done on the cell cycle, there are few major modeling groups that specialize in the construction and analysis of cell cycle models. One of the most active is the group led by Bela Novak and John Tyson, who, over the last 15 years, have published a series of classic papers, beginning with relatively simple models of the cell cycle and progressing to their most recent, highly complex models. Despite the elegance of this work, there are substantial difficulties facing the novice to this field. Not the least of these difficulties is the proliferation of names. Although the basic mechanisms are similar, the genes (and proteins) that carry out analogous functions in frog, budding yeast, and fission yeast all have different names; when a simple model already contains seven or eight crucial proteins, and each of these proteins has a different name in different organisms, the potential for confusion is clear.

We begin by presenting a generic model of the eukaryotic cell cycle, and discuss the fundamental mechanism that is preserved in mammalian cells. We then specialize this generic model to the specific case of fission yeast. This requires a multitude of names; to keep track of them, the reader is urged to make frequent use of Table 10.6 where the analogous names for the generic model and for fission yeast are listed. We end with a brief discussion of cell division in frog eggs after fertilization.

10.3.1   A Simple Generic Model

The Fundamental Bistability

As with all cellular processes, the cell cycle is regulated by genes and the proteins that they encode. There are two classes of proteins that form the center of the cell cycle control system. The first is the family of *cyclin-dependent kinases* (Cdk), which induce a variety of downstream events by phosphorylating selected proteins. The second family are the *cyclins*, so named because the first members to be identified are cyclically synthesized and degraded in each division cycle of the cell. Cyclin binds to Cdk molecules and controls their ability to phosphorylate target proteins; without cyclin, Cdk is inactive. In budding yeast cells there is only one major Cdk and nine cyclins, leading to a possibility of nine active Cdk–cyclin complexes. In mammals, the story is substantially more complicated, as there are (at last count) six Cdks and more than a dozen cyclins.

Leland Hartwell and Paul Nurse received the 2001 Nobel Prize in Physiology or Medicine for their work in the 1970s that showed how the cyclin-dependent kinases Cdc2 (in fission yeast), Cdc28 (in budding yeast) and Cdk1 (in mammalian cells) control the cell cycle. Cyclins were discovered in 1982 by Tim Hunt who shared the Nobel Prize with Hartwell and Nurse.

Although the temptation for a modeler is to view the cycle in Fig. 10.7 as a limit cycle oscillator, it is more appropriate to view it as an alternation between two states, G_1 and S-G_2-M. This point of view was first proposed by Nasmyth (1995, 1996) and now forms the basis of practically all quantitative models. Transition between these two states is controlled by the concentration of the Cdk–cyclin complex. In the G_1 state, the concentration of Cdk–cyclin is low, due to the low concentration of cyclin. At Start (see Fig. 10.8), cyclin production is increased and cyclin degradation is inhibited. The concentration of Cdk–cyclin therefore rises (because there is always plenty of Cdk around), and the cell enters S state, beginning synthesis of DNA. At the end of S phase, each chromosome consists of a pair of chromatids. At the end of G_2 the nuclear envelope is broken down and the chromatid pairs are aligned along the metaphase spindle, shown as the lighter gray lines in Fig. 10.8. When alignment is complete (metaphase), a group of proteins that make up the anaphase-promoting complex (APC) is activated. The APC functions in combination with an auxiliary component (either Cdc20 or Cdh1) to label cyclin molecules for destruction, thereby decreasing the concentration of the Cdk–cyclin complex. This initiates a second irreversible transition, Finish, and the chromatids are pulled to opposite poles of the spindle (anaphase). Thus, Start is caused by an explosive increase in the concentration of the Cdk–cyclin complex, while Finish is caused by the degradation of cyclin by APC and the resultant fall in Cdk–cyclin levels.

How do these reactions result in switch-like behavior between G_1 and S-G_2-M, and alternating high and low cyclin concentrations? The switch arises because of the mutually antagonistic interactions between Cdk–cyclin and APC–Cdh1. Not only does APC–Cdh1 inhibit Cdk activity by degrading cyclin, Cdk–cyclin in its turn inhibits APC–Cdh1 activity by phosphorylating Cdh1. Because of this mutual antagonism, the cell

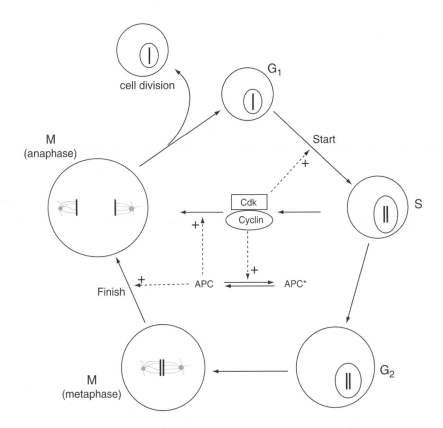

Figure 10.8 Schematic diagram of the primary chemical reactions of the cell cycle. APC* denotes the inactive form of APC. Adapted from Tyson and Novak (2001), Fig. 1.

can have either low Cdk–cyclin activity and high APC–Cdh1 activity (i.e., G_1), or high Cdk–cyclin activity and low APC–Cdh1 activity (S-G_2-M).

To construct a simple model of this reaction scheme we write down differential equations for the concentrations of cyclin (in this case, Cyclin B, called CycB) and unphosphorylated Cdh1, expressed in arbitrary units. The concentration of Cdk does not appear directly in the model because it is assumed to be present in excess. Because CycB binds tightly to Cdk, the concentration of the Cdk–CycB complex is determined solely by the concentration of CycB. Similarly, the activity of the APC–Cdh1 complex is determined by the concentration of Cdh1.

For each of these reactions we use relatively simple kinetics, either obeying mass action, or following a Michaelis–Menten saturating rate function (Chapter 1); more details on the model contruction are given after we present the equations. Thus,

$$\frac{d[\text{CycB}]}{dt} = k_1 m - (k_2' + k_2''[\text{Cdh1}])[\text{CycB}], \tag{10.34}$$

$$\frac{d[\text{Cdh1}]}{dt} = \frac{(k_3' + k_3''A)(1 - [\text{Cdh1}])}{J_3 + 1 - [\text{Cdh1}]} - \frac{k_4[\text{CycB}][\text{Cdh1}]}{J_4 + [\text{Cdh1}]}. \tag{10.35}$$

All the various k's and J's are positive constants, as are A and m. There are a number of important things to note about these equations:

1. CycB is degraded at the intrinsic rate k_2', but is also degraded by APC–Cdh1, with rate constant k_2''. Conversely, Cdh1 is phosphorylated by Cdk–CycB following a Michaelis–Menten saturating rate function. Thus there is mutual inhibition between CycB and Cdh1.

2. The rate of production of Cdh1 is dependent on the concentration of the phosphorylated form, which is [Cdh1]$_{total}$ - [Cdh1]. Since units are arbitrary, we set [Cdh1]$_{total}$ = 1.

3. The rate of production of CycB is dependent on the parameter m, which represents the cell mass. This is a crucial assumption. How can a rate constant be dependent on the mass of a cell? Although there must be some way in which cell mass controls the kinetics of the cell cycle (because the cell cycle and cell growth are closely coupled, as discussed above) the exact mechanisms by which this occurs are unknown. Of course, it is possible to imagine how it might occur; as the cell mass increases, the ratio of cytoplasmic mass to nuclear mass increases. If a protein is made in the cytoplasm but then moves to the nucleus, the greater the cytoplasmic/nuclear volume ratio, the faster the buildup of this protein in the nucleus. However, such explanations remain speculative. In this simple model we assume that CycB builds up in the nucleus, and thus its rate of production is an increasing function of cell mass, as in (10.34).

4. The constant A is related to the activity of Cdc20. Recall that Cdc20, like Cdh1, can pair up with APC. One job of the APC–Cdc20 complex is indirectly to activate a phosphatase that activates Cdh1.

Letting x_1 denote [CycB] and x_2 denote [Cdh1], the steady states of (10.34)–(10.35) are given by

$$x_1 = \frac{k_1 m}{k_2' + k_2'' x_2} \tag{10.36}$$

and

$$p = \left(\frac{J_3 + 1 - x_2}{1 - x_2}\right)\left(\frac{x_2}{J_4 + x_2}\right)\left(\frac{k_1}{k_2' + k_2'' x_2}\right), \tag{10.37}$$

where

$$p = \frac{k_3' + k_3'' A}{k_4 m}. \tag{10.38}$$

These solutions are shown in Fig. 10.9 with x_1 plotted as a function of p, using parameter values taken from Table 10.5. However, since A is not specified, the scale on this plot is arbitrary. The easiest way to plot this curve is to view (10.36) and (10.37) as a parametric curve with underlying parameter x_2. For some values of p there are three

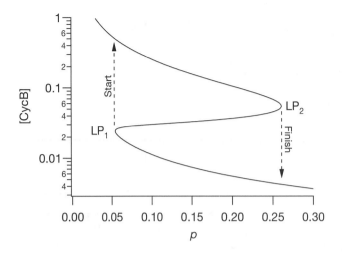

Figure 10.9 Steady states of the two-variable generic model (10.34)–(10.35) plotted against the parameter p.

Table 10.5 Parameters for the six-variable generic cell cycle model (10.34), (10.35), (10.39), (10.40), (10.41), and (10.42). Adapted from Tyson and Novak (2001), Table 1.

Component	Rate constants (min^{-1})	Dimensionless constants
CycB	$k_1 = 0.04,\ k_2' = 0.04,\ k_2'' = 1,\ k_2''' = 1$	$[\mathrm{CycB}]_{\mathrm{threshold}} = 0.05$
Cdh1	$k_3' = 1,\ k_3'' = 10,\ k_4' = 2,\ k_4 = 35$	$J_3 = 0.04,\ J_4 = 0.04$
$\mathrm{Cdc20}_T$	$k_5' = 0.005,\ k_5'' = 0.2,\ k_6 = 0.1$	$J_5 = 0.3,\ n = 4$
Cdc20	$k_7 = 1,\ k_8 = 0.5$	$J_7 = 10^{-3},\ J_8 = 10^{-3}$
IEP	$k_9 = 0.1,\ k_{10} = 0.02$	
m	$\mu = 0.005$	

steady states, for others only one. The two places where the steady states coalesce are limit points labeled LP_1 and LP_2. (Since this is a two-variable system, it is also quite easy to do a phase-plane analysis and to verify that the intermediate steady-state solution is a saddle point, hence unstable. See Exercise 5.)

We can now trace out an approximate cell cycle loop in Fig. 10.9. Note that here neither A nor m has any dynamics; they are both merely increased or decreased as needed. The dynamical system underlying these changes in A and m is discussed in more detail in the next section.

Suppose we start at the lower steady state (i.e., the one corresponding to low $[\mathrm{CycB}]$), to the right of LP_1. As m increases (as the cell grows), p decreases and the cell follows the lower steady state. Eventually p decreases so much that LP_1 is reached, the solution falls off the lower steady state and approaches the high steady state. This corresponds to Start (see Fig. 10.8), at which time the concentration of cyclin B increases explosively. At metaphase the concentration of Cdc20 starts to rise, which corresponds to an increase in the parameter A, which increases p. When p increases too far, the solution can no longer stay on the upper steady state, and falls back down to the lower steady state (Finish). [CycB] thus falls, A decreases again, the cell divides ($m \to m/2$), and the cell begins the cycle over again.

From this point of view, the cell cycle is a hysteresis loop alternating between two branches of steady states.

Activation of APC

In the previous model, A was a parameter that was increased and decreased at will to mimic the activity of APC–Cdc20. Of course, the story is not this simple, since there are several reactions that regulate APC activation, and thus A, which denotes the concentration of Cdc20, is controlled by its own dynamical system. In budding yeast Cdh1 is activated by a phosphatase, which is activated by Cdc20. Furthermore, Cdc20 production is increased by cyclin B. Hence, the explosive increase in [CycB] that occurs at Start leads to an increase in Cdc20, and a subsequent increase in Cdh1. However, when Cdc20 is first made it is not active; it is activated by cyclin B only after a time delay, which is the result of a number of intermediate reaction steps between cyclin B and Cdc20 activation. The exact reactions that cause this delay have not yet been identified, so it is modeled by introducing a fictitious enzyme, IE (intermediate enzyme), with activated form IEP (the P standing for phosphorylation). Thus, in summary, CycB activates IEP, which activates Cdc20, which activates Cdh1, which degrades CycB. This reaction scheme is sketched in Fig. 10.10.

To write the corresponding differential equations, we introduce two new variables; activated IE (IEP) and Cdc20. We also let $[Cdc20]_T$ denote the total concentration of Cdc20, i.e., both the activated and inactivated forms. The rate of production of total Cdc20 is increased by CycB, and thus

$$\frac{d[Cdc20_T]}{dt} = k_5' + k_5'' \frac{[CycB]^n}{J_5^n + [CycB]^n} - k_6[Cdc20_T]. \qquad (10.39)$$

Note that the activation is assumed to follow a simple model of cooperative kinetics, with Hill coefficient n (Chapter 1).

Similarly, Cdc20 is formed from nonactivated Cdc20 (which has concentration $[Cdc20_T] - [Cdc20]$) at a rate that is dependent on [IEP], and is removed in two ways; the same intrinsic degradation rate as $Cdc20_T$, and an additional removal term corresponding to enzymatic conversion of the active form back to the inactive form. Thus,

$$\frac{d[Cdc20]}{dt} = \frac{k_7[IEP]([Cdc20_T] - [Cdc20])}{J_7 + ([Cdc20_T] - [Cdc20])} - \frac{k_8[Cdc20]}{J_8 + [Cdc20]} - k_6[Cdc20]. \qquad (10.40)$$

Finally, we include an equation for [IEP]:

$$\frac{d[IEP]}{dt} = k_9 m[CycB](1 - [IEP]) - k_{10}[IEP]. \qquad (10.41)$$

Note that IEP is activated at a rate that is proportional to [CycB].

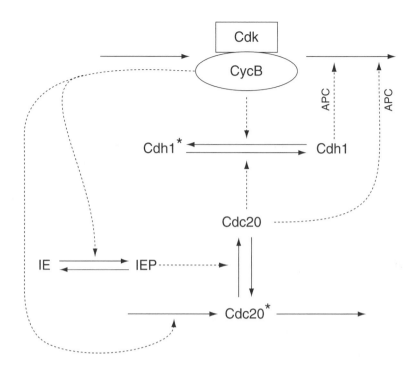

Figure 10.10 Sketch of the reactions involved in the activation of APC. The superscript * indicates an inactive form. IEP denotes the phosphorylated (and active) form of IE. Adapted from Tyson and Novak (2001), Fig. 6.

It remains to specify how to model the growth of the cell. For simplicity, we assume that m grows exponentially, and thus

$$\frac{dm}{dt} = \mu m. \tag{10.42}$$

However, this growth law must be modified to take cell division into account. At Start [CycB] grows explosively, but its subsequent fall is the signal that Finish has occurred and the cell should divide. Thus, we assume that the cell divides in half (i.e., $m \to m/2$) whenever [CycB] falls to some specified low level (in this case 0.05) after Start.

The model now consists of the six differential equations (10.34), (10.35), (10.39), (10.40), (10.41) and (10.42), with $A = $ [Cdc20] in (10.35). In writing these equations we have made a large number of assumptions. Perhaps the most striking of these is that sometimes the kinetics are assumed to follow a first-order law of mass action kinetics, at other times they are assumed to be of Michaelis–Menten form, while at yet other times they are assumed to be cooperative kinetics. Such choices are, in large part, a judgement call on the part of the original modelers, and depend on the available experimental evidence.

The steady states of this model (i.e., steady states holding m fixed) are shown in Fig. 10.11. As in the simpler two-variable model (Fig. 10.9) the curve of steady states

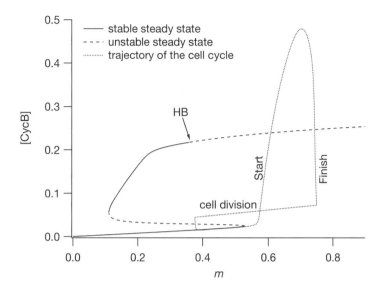

Figure 10.11   Steady states (holding m fixed) of the six-variable generic model and a super-imposed cell cycle trajectory allowing cell growth and division. The parameter values of the model are given in Table 10.5.

has two limit points. However, in this more complicated model the upper steady state becomes unstable via a Hopf bifurcation (which we do not investigate any further, the resultant limit cycles being of no interest at this stage). Thus, when the cell cycle trajectory (shown as the dotted line) falls off the lower branch of steady states (at Start) it cannot end up at the upper steady state, since the upper steady state is unstable. Instead, it loops around the upper steady state, and [CycB] then falls to a low level (Finish) at which time the mass is divided by 2 (cell division), the solution heads back to the lower steady-state branch, and the cycle repeats.

   In this model as it stands, Finish automatically occurs a certain time after Start. Once the trajectory falls off the lower steady-state branch, there is nothing to stop it looping around and initiating Finish when [CycB] falls. In reality there are controls to prevent this from happening if the chromosomes are not aligned properly. However, we omit these controls from this simple model.

A Note About Units

All the concentrations in this model (and the one that follows) are treated as dimensionless. This explains why, in Table 10.5, all the constants k have units of 1/time, and why the J's are dimensionless. Experimentally, it is possible to measure relative concentrations, but it is not possible to measure absolute concentrations. Thus, we assume that there is some scale factor that could be used to scale all the concentrations rendering them dimensionless, although we do not know what the scale factor is. The qualitative behavior of the model remains unchanged by this assumption.

10.3.2 Fission Yeast

Having seen how the cell cycle works in a generic model, we now turn to a more complicated model of the cell cycle of fission yeast (Novak et al., 2001; Tyson et al., 2002). This is a particularly interesting model since the cell cycles of various mutants can be elegantly explained by consideration of the corresponding bifurcation diagrams. A similarly complicated model of the cell cycle of budding yeast is discussed in Chen et al. (2000, 2004), Ciliberto et al. (2003), and Allen et al. (2006), while a model of the mammalian cell cycle is discussed in Novak and Tyson (2004). However, we now need to introduce new names for the major players in the cell cycle. Since the introduction of a list of new names for familiar players has the potential to cause drastic confusion, we urge the reader to pay careful and repeated attention to Table 10.6, where the different names for analogous species are given.

Mitosis-Promoting Factor: MPF

In the generic model, the central player in the cell cycle was the Cdk:cyclin B complex. In fission yeast, the cyclin-dependent kinase is called Cdc2, and the B-type cyclin is called Cdc13 (see Table 10.6. Cdc stands for cell division cycle.). The Cdc2:Cdc13 complex that lies at the heart of the cycle is called *mitosis-promoting factor*, or MPF. As before, Cdc2 is active only when it is bound to the cyclin, Cdc13, and thus MPF is the active species that drives the cell cycle.

To understand the regulation of MPF we need to understand how it is formed, degraded, and inactivated (Figs. 10.12 and 10.13).

- MPF is formed when Cdc13 combines with Cdc2; because Cdc2 is present in excess, the rate of this formation depends only on how much Cdc13 is present. This rate is assumed to depend on the mass of the cell, since Cdc13 can build up inside the nucleus, as discussed previously for CycB.
- The principal degradation pathway is activated by APC, whose auxiliary components in fission yeast are called Ste9 and Slp1 (see Fig. 10.13).

Table 10.6   Cell cycle regulatory proteins.

Generic model	Fission yeast	Role
Cdk	Cdc2	cyclin-dependent kinase
CycB	Cdc13	cyclin
Cdh1	Ste9	APC auxiliary
Cdc20	Slp1	APC auxiliary
IE	IE	intermediate enzyme
	Rum1	inhibitor
	Wee1	tyrosine kinase
	Cdc25	tyrosine phosphatase
	SK	starter kinase
	TF	transcription factor

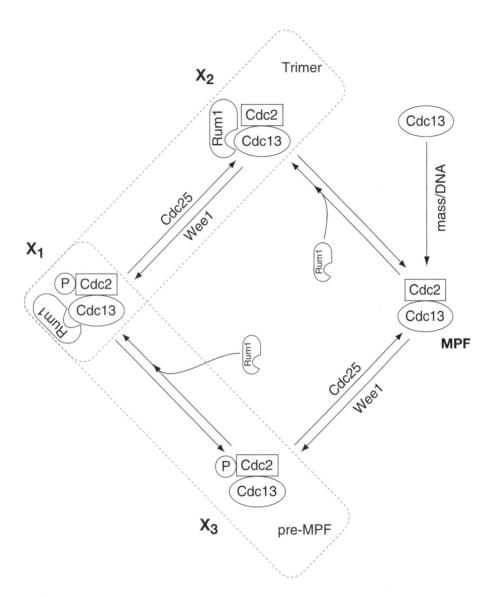

Figure 10.12 Diagram of the states of MPF. In the model of the cell cycle in fission yeast, these states are assumed to be at equilibrium. Each of the states is degraded by a pathway that depends on Slp1 and Ste9, but these are omitted from the diagram for clarity. The circled P denotes a phosphate group.

- MPF can be inactivated in two major ways:

  - It can be phosphorylated by the Wee1 kinase to a protein called preMPF. PreMPF in its turn can be dephosphorylated back to MPF by the phosphorylated form of Cdc25, a tyrosine phosphatase.
  - It can be inhibited by the binding of Rum1, to form an inactive trimer.

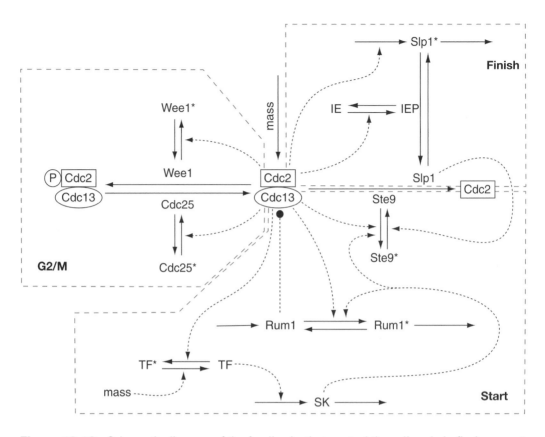

Figure 10.13 Schematic diagram of the feedbacks that control the cell cycle in fission yeast. All the dotted arrows denote activation, while the lines with a solid circle on the end denote inhibition. Adapted from Novak et al. (2001), Fig. 2.

These reactions are summarized in Fig. 10.12. Since the activities of Rum1, Cdc25 and Slp1 are all controlled by MPF, this gives a highly complex series of feedback interactions. We revisit these below. For now, we focus only on the equations modeling the MPF trimer, Cdc13, and Rum1. We follow the presentation of Novak et al. (2001), who constructed the model using the total amount of Cdc13, the total amount of Rum1, and preMPF, as three of the dependent variables. A different choice of dependent variables gives a different version of the same model.

First, we define the total amount of Cdc13, $[\text{Cdc13}_T]$, as

$$[\text{Cdc13}_T] = X_1 + X_2 + X_3 + [\text{MPF}], \tag{10.43}$$

where X_1, X_2, and X_3 refer to the labeled complexes in Fig. 10.12. As usual, X_1 denotes the concentration of X_1. Similarly, we let

$$[\text{preMPF}] = X_1 + X_3 \quad \text{and} \quad [\text{Trimer}] = X_1 + X_2, \tag{10.44}$$

as indicated by the dashed boxes in Fig. 10.12. We also have a conservation equation for Rum1. If we let $[\text{Rum1}_T]$ denote the total amount of Rum1, then

$$[\text{Rum1}_T] = [\text{Rum1}] + [\text{Trimer}] = [\text{Rum1}] + X_1 + X_2. \tag{10.45}$$

We assume that Cdc13, no matter in which state, is degraded at a rate that depends on Slp1 and Ste9, the two components of APC in fission yeast. Thus,

$$\frac{d[\text{Cdc13}_T]}{dt} = k_1 m - (k_2' + k_2''[\text{Ste9}] + k_2'''[\text{Slp1}])[\text{Cdc13}]. \tag{10.46}$$

The equation for preMPF is similar:

$$\frac{d[\text{preMPF}]}{dt} = k_{\text{wee}}([\text{Cdc13}_T] - [\text{preMPF}]) - k_{25}[\text{preMPF}]$$
$$- (k_2' + k_2''[\text{Ste9}] + k_2'''[\text{Slp1}])[\text{preMPF}]. \tag{10.47}$$

Here, k_{wee} and k_{25} are, respectively, the rate constants associated with Wee1 and Cdc25. As described below, they depend on [MPF].

Since k_{wee} and k_{25} depend on [MPF], we need to derive an expression for [MPF] in terms of our chosen dependent variables, $[\text{Cdc13}_T]$, [preMPF] and $[\text{Rum1}_T]$. We begin by deriving an expression for [Trimer]. This we do by assuming that X_1 and X_3 are always in equilibrium, as are X_2 and MPF. Thus

$$[\text{Rum1}][\text{MPF}] = K_d X_2, \tag{10.48}$$
$$[\text{Rum1}]X_3 = K_d X_1, \tag{10.49}$$

where K_d is the equilibrium constant for Rum1 binding. From this, and (10.44), it follows that

$$[\text{Trimer}] = X_1 + X_2$$
$$= \frac{[\text{Rum1}]}{K_d}(X_3 + [\text{MPF}])$$
$$= \frac{1}{K_d}([\text{Rum1}_T] - [\text{Trimer}])([\text{Cdc13}_T] - [\text{Trimer}]). \tag{10.50}$$

This quadratic equation for [Trimer] can be easily solved to give

$$[\text{Trimer}] = \frac{1}{2}(\Sigma - \sqrt{\Sigma^2 - 4[\text{Rum1}_T][\text{Cdc13}_T]}) \tag{10.51}$$
$$= \frac{2[\text{Rum1}_T][\text{Cdc13}_T]}{\Sigma + \sqrt{\Sigma^2 - 4[\text{Rum1}_T][\text{Cdc13}_T]}}, \tag{10.52}$$

where

$$\Sigma = [\text{Cdc13}_T] + [\text{Rum1}_T] + K_d. \tag{10.53}$$

Note that here we have taken the smaller of the two roots, to ensure that $[\text{Trimer}] \to 0$ as $[\text{Cdc13}_T] \to 0$.

With this expression for [Trimer] we can find an equation for [MPF] in terms of the other variables. First, since X_1 and X_3 are assumed to be in equilibrium, we have

$$[\text{preMPF}] = X_1 + X_3 = X_3 \left(1 + \frac{[\text{Rum1}]}{K_d} \right). \tag{10.54}$$

Also, from conservation of Cdc13 (i.e., from (10.43)) we have

$$[\text{Trimer}] + X_3 + [\text{MPF}] = [\text{Cdc13}_T], \tag{10.55}$$

from which, using (10.54), it follows that

$$[\text{MPF}] = [\text{Cdc13}_T] - [\text{Trimer}] - \frac{[\text{preMPF}]K_d}{K_d + [\text{Rum1}_T] - [\text{Trimer}]}. \tag{10.56}$$

The Three Major Sets of Feedbacks

At this point we have differential equations for $[\text{Cdc13}_T]$ and [preMPF], with associated algebraic equations for [Trimer] and MPF. This completes the most complicated part of the model construction. The remainder of the model equations follow in a straightforward manner from the reaction diagram, which is shown in full detail in Fig. 10.13.

This is a complicated reaction diagram; to make sense of it we divide the reactions into three main groups, corresponding to control of Start, Finish, and the G_2/M transition, as indicated by the dashed gray boxes in Fig. 10.13. Each is discussed in turn.

Start This set of feedbacks causes the increase of MPF (i.e., of Cdc2:Cdc13) at Start. There are two mutual inhibition loops. First, MPF inactivates Ste9, thus decreasing the rate of breakdown of MPF. Second, MPF increases the rate of inactivation of Rum1. Since Rum1 inactivates MPF (see Fig. 10.12), this is the second mutual inhibition loop.

Because of these mutual inhibitions, the cell can have either a high concentration of MPF, or a high concentration of Ste9 and Rum1. It cannot have both.

SK denotes a starter kinase that helps begin Start by phosphorylating Rum1 to its inactive state, thus relieving the inhibition on MPF. However, once Start begins, SK must be removed to allow MPF to decrease at Finish. Thus, MPF phosphorylates SK's transcription factor (TF) to its inactive state. Thus, this negative feedback loop works against the two mutual inhibition loops in Start.

Although this scheme appears to be exactly that of the simple generic model, there are some important differences. One important way in which fission yeast differs from the generic model is that, at Start, the increase of MPF is not explosive. Instead, Start is characterized by a precipitous drop in [Ste9], which then allows [MPF] to start increasing. In fission yeast, MPF is held very low during G_1, as shown in Fig. 10.14. At Start, Cdc13 begins to accumulate, but MPF activity remains low because Cdc2 is phosphorylated by Wee1.

S/G_2/M The important reactions are sketched in Fig. 10.13, in a more simplified form than in Fig. 10.12, but including the action of MPF on Wee1 and Cdc25. MPF

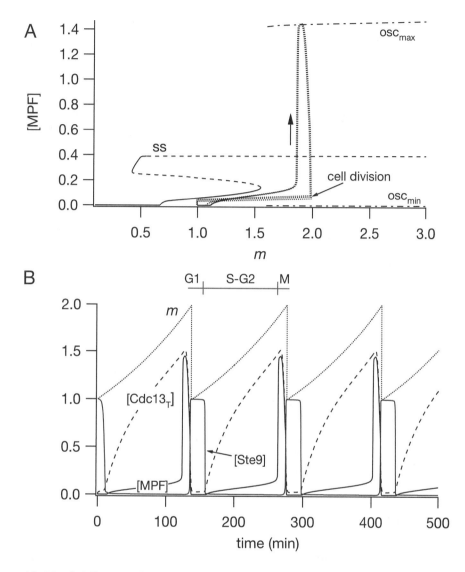

Figure 10.14 A: bifurcation diagram of the model of the cell cycle in fission yeast, using the wild type parameters in Table 10.7. The steady states, labeled ss, are plotted as a function of m, the cell size, with stable branches denoted by solid curves, and unstable branches by dashed curves. A branch of oscillatory solutions exists for larger m. The maximum and minimum of the oscillations are labeled osc_{max} and osc_{min} respectively. A solution of the model for several cell cycles is superimposed (dotted line). B: the solution plotted against time for the same cycles shown in A.

accumulates in the less active, Cdc2-phosphorylated, form, which is sufficient to drive DNA synthesis, but insufficient for mitosis. MPF decreases Wee1 activity by phosphorylating it (Wee1* is the phosphorylated, inactive, form of Wee1). MPF also increases the activity of Cdc25 by phosphorylating it (Cdc25 is the phosphorylated, active, form of Cdc25*). Note that sometimes the phosphorylated state is the inactive state, while at other times it is the active state. Both of these feedbacks are positive.

The cell spends a long time in G2 phase until it grows large enough for the positive feedback loops on Wee1 and Cdc25 to engage and remove the inhibitory phosphate group from Cdc2. The resulting explosive increase in MPF kicks the cell into mitosis. Hence, fission yeast characteristically has a short G_1 phase and a long S/G_2 phase.

Finish This set of feedbacks causes the fast decrease in MPF at Finish (see Fig. 10.11). MPF increases the rate of production of inactive Slp1, which is then activated via the action of IEP, the phosphorylated intermediate enzyme. As in the generic model, the identity of IE is unknown. However, its existence is inferred from the significant delays that occur between an increase in MPF levels and an increase in Slp1. Slp1 increases the rate of MPF breakdown in two different ways; first, by promoting its degradation directly, and second, by activating Ste9, which also inactivates MPF.

It is interesting to note that the Start and Finish feedbacks are, essentially, a fast positive feedback (Start) followed by a slower negative feedback (Finish). This arrangement is seen in many physiological systems, including the Hodgkin–Huxley model of the action potential (Chapter 5) and models of the inositol trisphosphate receptor (Chapter 7).

Cell size affects the model in two different ways. First, we assume that the rate of production of MPF is a function of cell size, as discussed previously. Second, the rate of activation of the transcription factor, TF, is also assumed to be dependent on the cell size (see (10.70)). Thus, as the cell grows, the activity of MPF is increased in two different size-dependent ways. TF affects MPF activity via the production of a starter kinase, SK. SK initiates Start by downregulating Rum1 and Ste9, which allows $[\text{Cdc13}]_T$ to increase and eventually trigger the G2-M transition.

The Model Equations

Our job is now to translate the schematic diagrams in Figs. 10.12 and 10.13 into a set of ordinary differential equations. This has been done already for Cdc13_T and preMPF; it is now necessary to repeat this process for the remaining variables in the model. However, since the equations for the remaining variables are much simpler and easier to derive than those for Cdc13_T and preMPF they are presented here without detailed explanations. Also, so that all the equations are together in a single place, the equations for Cdc13_T and preMPF included here as well.

The model equations are

$$\frac{d[\text{Cdc13}_T]}{dt} = k_1 m - (k_2' + k_2''[\text{Ste9}] + k_2'''[\text{Slp1}])[\text{Cdc13}], \qquad (10.57)$$

$$\frac{d[\text{preMPF}]}{dt} = k_{\text{wee}}([\text{Cdc13}_T] - [\text{preMPF}]) - k_{25}[\text{preMPF}]$$

$$- (k_2' + k_2''[\text{Ste9}] + k_2'''[\text{Slp1}])[\text{preMPF}], \qquad (10.58)$$

$$\frac{d[\text{Ste9}]}{dt} = (k_3' + k_3''[\text{Slp1}])\frac{1 - [\text{Ste9}]}{J_3 + 1 - [\text{Ste9}]}$$
$$- (k_4'[\text{SK}] + k_4[\text{MPF}])\frac{[\text{Ste9}]}{J_4 + [\text{Ste9}]}, \tag{10.59}$$

$$\frac{d[\text{Slp1}_T]}{dt} = k_5' + k_5''\frac{[\text{MPF}]^4}{J_5^4 + [\text{MPF}]^4} - k_6[\text{Slp1}_T], \tag{10.60}$$

$$\frac{d[\text{Slp1}]}{dt} = k_7[\text{IEP}]\frac{[\text{Slp1}_T] - [\text{Slp1}]}{J_7 + [\text{Slp1}_T] - [\text{Slp1}]}$$
$$- k_8\frac{[\text{Slp1}]}{J_8 + [\text{Slp1}]} - k_6[\text{Slp1}], \tag{10.61}$$

$$\frac{d[\text{IEP}]}{dt} = k_9[\text{MPF}]\frac{1 - [\text{IEP}]}{J_9 + 1 - [\text{IEP}]} - k_{10}\frac{[\text{IEP}]}{J_{10} + [\text{IEP}]}, \tag{10.62}$$

$$\frac{d[\text{Rum1}_T]}{dt} = k_{11} - (k_{12} + k_{12}'[\text{SK}] + k_{12}''[\text{MPF}])[\text{Rum1}_T], \tag{10.63}$$

$$\frac{d[\text{SK}]}{dt} = k_{13}[\text{TF}] - k_{14}[\text{SK}], \tag{10.64}$$

where, as before,

$$[\text{Trimer}] = \frac{2[\text{Rum1}_T][\text{Cdc13}_T]}{\Sigma + \sqrt{\Sigma^2 - 4[\text{Rum1}_T][\text{Cdc13}_T]}}, \tag{10.65}$$

$$\Sigma = [\text{Cdc13}_T] + [\text{Rum1}_T] + K_d, \tag{10.66}$$

$$[\text{MPF}] = [\text{Cdc13}_T] - [\text{Trimer}] - \frac{[\text{preMPF}]K_d}{K_d + [\text{Rum1}_T] - [\text{Trimer}]}. \tag{10.67}$$

We also assume that k_{wee} and k_{25} are, respectively, decreasing and increasing Goldbeter–Koshland functions of $[\text{MPF}]$ (Section 1.4.6), and thus

$$k_{\text{wee}} = k_{\text{wee}}' + (k_{\text{wee}}'' - k_{\text{wee}}')G(V_{aw}, V_{iw}[\text{MPF}], J_{aw}, J_{iw}), \tag{10.68}$$
$$k_{25} = k_{25}' + (k_{25}'' - k_{25}')G(V_{a25}[\text{MPF}], V_{i25}, J_{a25}, J_{i25}). \tag{10.69}$$

Finally, the concentration of transcription factor, $[\text{TF}]$, is assumed to be an increasing Goldbeter–Koshland function of the cell mass, m, but a decreasing function of $[\text{MPF}]$, and thus

$$[\text{TF}] = G(k_{15}m, k_{16}' + k_{16}''[\text{MPF}], J_{15}, J_{16}), \tag{10.70}$$

while the mass grows exponentially,

$$\frac{dm}{dt} = \mu m. \tag{10.71}$$

This completes the mathematical description of the reaction diagram in Fig. 10.13. All the parameter values are given in Table 10.7.

Table 10.7   Parameters for the model of the cell cycle in fission yeast (10.34)–(10.42). Adapted from Novak et al. (2001), Table II. After Start occurs, the cell mass is divided by two to mimic cell division when $[CycB] = [CycB]_{threshold}$. All the parameters have units 1/min, except for the J's and K_d, which are dimensionless.

Cdc13
$k_1 = 0.04$, $k_2' = 0.04$, $k_2'' = 1$, $k_2''' = 1$ $[CycB]_{threshold} = 0.05$
Ste9
$k_3' = 1$, $k_3'' = 10$, $k_4' = 2$, $k_4 = 35$ $J_3 = 0.04$, $J_4 = 0.04$
Slp1
$k_5' = 0.005$, $k_5'' = 0.2$, $k_6 = 0.1$, $k_7 = 1$, $k_8 = 0.25$ $J_5 = 0.3$, $J_7 = J_8 = 10^{-3}$
Rum1
$k_{11} = 0.1$, $k_{12} = 0.01$, $k_{12}' = 1$, $k_{12}'' = 3$ $K_d = 0.001$
IEP
$k_9 = 0.1$, $k_{10} = 0.02$
SK
$k_{13} = 0.1$, $k_{14} = 0.1$
TF
$k_{15} = 1.5$, $k_{16}' = 1$, $k_{16}'' = 2$ $J_{15} = 0.01$, $J_{16} = 0.01$
Wee1
$k_{wee}' = 0.15$, $k_{wee}'' = 1.3$, $V_{aw} = 0.25$, $V_{iw} = 1$ $J_{aw} = 0.01$, $J_{iw} = 0.01$
Cdc25
$k_{25}' = 0.05$, $k_{25}'' = 5$, $V_{a25} = 1$, $V_{i25} = 0.25$ $J_{a25} = 0.01$, $J_{i25} = 0.01$
m
$\mu = 0.005$

The Wild Type

The cell cycle in this more complex model has a structure similar to that of the six-variable generic model of Section 10.3.1, and a typical solution is shown in Fig. 10.14. As in the simpler model, the curve of steady states as a function of m is S-shaped. As the cell mass, m, increases, the concentration of MPF also increases until the lower bend of the S is reached, at which point the solution "falls off" the curve of steady states and heads toward the stable oscillation (denoted by the dot-dash line in Fig. 10.14A). The resultant sudden increase in [MPF] pushes the cell into M phase.

Because the stable solution is oscillatory, [MPF] naturally increases and then decreases (Finish), at which point the cell divides, the cell mass is divided by two, and the cycle repeats. A succession of cycles is shown in Fig. 10.14B.

Note that the cell cycle trajectory does not lie exactly on the S-shaped curve of steady states. This is because the cell size is continually changing, never allowing enough time for the solution to reach the steady state that corresponds to a fixed value of m.

The different phases of the cell cycle are shown on the bar in Fig. 10.14B. During G_1, [MPF] is held very low by Ste9. At the Start transition, [Ste9] falls almost to zero, thus allowing [MPF] to increase. This occurs relatively early during the cell cycle, when the cell mass is small. During S/G_2 phase the cell replicates DNA and slowly increases in size until critical mass is reached, whereupon [MPF] increases explosively and the cell is pushed into M phase. Thus, the Start transition here does not correspond to an explosive increase of [MPF], a major point of difference from the simple generic model described earlier. This serves to emphasize the fact that different cell types have different ways of controlling the cell cycle, and, although there are many similarities between cell types, no single mechanism serves as a universal explanation.

Wee1⁻ Cells

One particularly interesting feature of this model is that it can be used to explain the behavior of a number of fission yeast mutants. Here, we discuss only one of these mutants, the wee1⁻ mutant (Sveiczer et al., 2000). Others are discussed in detail in Tyson et al. (2002) and Novak et al. (2001).

The normal fission yeast cell cycle has a short G_1 phase followed by a much longer S/G_2 phase during which most of the cell growth occurs. In the wild type cell, the long S/G_2 phase is caused by the balance between MPF, Wee1 and Cdc25. Because Wee1 inactivates MPF, it is not until [MPF] has increased past a critical threshold that it can inactivate Wee1 and thus allow the explosive growth of [MPF] at the beginning of M phase.

If Wee1 is knocked out (i.e., as in a wee1⁻ mutant), the inactivation of MPF by Wee1 is prevented, or at least greatly decreased. This allows the explosive increase in [MPF] to happen at a lower value of m, and thus the cell enters mitosis when it is much smaller than in the wild type. However, because the cell cycle occurs for smaller cell sizes, and thus lower overall values of [MPF], the G_1 phase is extended, since it takes longer for Ste9 to be overpowered by MPF (which must happen at the end of G_1; see Fig. 10.14B). Hence, wee1⁻ cells have an extended G_1 phase, a shortened S/G_2 phase, and at division are approximately half the size of wild type cells. (The word *wee* is Scottish for small. Paul Nurse originally discovered Wee1 in the early 1970s and coined its name after observing that the absence of the gene made cells divide when they were unusually small.)

The phase plane and a cell cycle of the wee1⁻ mutant are shown in Fig. 10.15. To model the absence of Wee1, the parameter k''_{wee} is decreased to 0.3, thus decreasing the value of k_{wee}, as shown by (10.68). With this change, the S-shaped curve of steady states is shifted to lower values of m, and the solution "falls off" the lower limit point at lower values of m. Thus, mitosis is initiated at lower cell size than in wild type cells. Fig. 10.15B shows that G_1 phase is extended, and that the cell cycle occurs for smaller values of [MPF] than in the wild type.

10.3.3   A Limit Cycle Oscillator in the *Xenopus* Oocyte

There is strong evidence that early embryonic divisions are controlled by a cytoplasmic biochemical limit cycle oscillator. For example, if fertilized (*Xenopus*) frog eggs are enucleated, they continue to exhibit periodic twitches or contractions, as if the cytoplasm continued to generate a signal in the absence of a nucleus. Enucleated sea urchin eggs go a step further by actually dividing a number of times before they notice that they

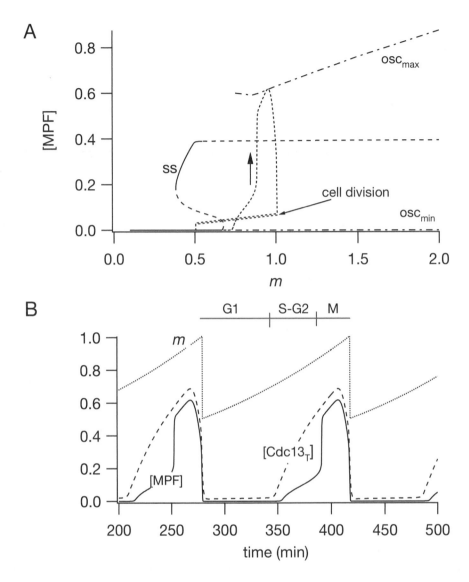

Figure 10.15 A: bifurcation diagram of the wee1$^-$ cell cycle model of fission yeast, using the same parameters in Table 10.7, with $k''_{wee} = 0.3$. The notation is the same as in Fig. 10.14. B: the solution plotted against time for the same cycles shown in A.

contain no genetic material and consequently die. In *Xenopus* the first post-fertilization cell cycle takes approximately one hour, followed by 11 faster cycles (approximately 30 minutes each) during which the cell mass decreases.

These divisions result from a cell cycle oscillator from which some of the checkpoints and controls have been removed. The most detailed model of this oscillator is that of Novak and Tyson (1993a,b; Borisuk and Tyson, 1998; Sha et al., 2003), and this is the model we describe here. A simpler model due to Goldbeter is discussed in

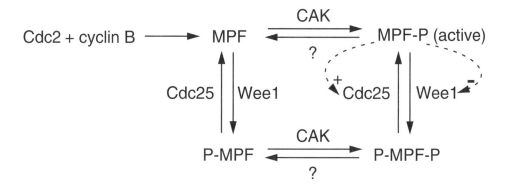

Figure 10.16 Schematic diagram of the regulatory pathway of MPF in *Xenopus* oocytes.

Exercise 7, and a simplified version of the Novak–Tyson model is discussed in Chapter 10 of Fall et al. (2002).

Because frogs are not yeast, the details of the cell cycle in *Xenopus* oocytes are not exactly the same as those of the cell cycle in fission yeast, although there are major similarities.

In fertilized *Xenopus* oocytes, cell division takes place without any cell growth, so the G_1 checkpoint is removed (or inoperable). The MPF that is critical for getting through the G_2 checkpoint is a dimer of Cdc2 and a mitotic cyclin.

The *cdc2* gene encodes a cyclin-dependent protein kinase, Cdc2, which, in combination with B-type cyclins, forms MPF, which induces entry into M phase. The activity of the Cdc2-cyclin B dimer (MPF) is also controlled by phosphorylation at two sites, tyrosine-15 and threonine-167. (Tyrosine and threonine are two of the twenty amino acids that are strung together to form a protein molecule. The number 15 or 167 denotes the location on the protein sequence of Cdc2.) These two sites define four different phosphorylation states. MPF is active when it is phosphorylated at threonine-167 only. The other three phosphorylation states are inactive. Active MPF initiates a chain of reactions that controls mitotic events.

Just as in fission yeast, movement between different phosphorylation states is mediated by Wee1 and Cdc25. Wee1 inactivates MPF by adding a phosphate to the tyrosine-15 site. Cdc25 reverses this by dephosphorylating the tyrosine-15 site. A schematic diagram of this regulation is shown in Fig. 10.16. Cyclin B is synthesized from amino acids and binds with free Cdc2 to form an inactive MPF dimer. The dimer is quickly phosphorylated on threonine-167 (by a protein kinase called CAK) and dephosphorylated at the same site by an unknown enzyme. Simultaneously, Wee1 can phosphorylate the dimer at the tyrosine-15 site, rendering it inactive, and Cdc25 can dephosphorylate the same site. Mitosis is initiated when a sufficient quantity of MPF is active.

This regulation of MPF is one point of divergence with the previous model for fission yeast. The other feedbacks are similar to those we have seen before. First, the active form of MPF regulates the activities of Wee1 and Cdc25 (as in the box labeled G2/M in Fig. 10.13), while the degradation of MPF is controlled by a delayed negative feedback loop through an intermediate enzyme (as in the box labeled Finish in Fig. 10.13). In *Xenopus*, the degradation of MPF is controlled by a ubiquitin-conjugating enzyme (UbE) (which is an outdated name for our friend, APC–Cdc20).

We now have a complete verbal description of a model of the initiation of mitosis. In summary, as cyclin is produced, it combines with Cdc2 to form MPF. MPF is quickly phosphorylated to its active form. Active MPF turns on its own autocatalytic production by activating Cdc25 and inactivating Wee1. By activating UbE, which activates the destruction of cyclin, active MPF also turns on its own destruction, but with a delay, thus completing the cycle.

Of course, this verbal description is incomplete, because there are many other features of M phase control that have not been included. It also does not follow from verbal arguments alone that this model actually controls mitosis in a manner consistent with experimental observations. To check that this model is indeed sufficient to explain some features of the cell cycle, it is necessary to present it in quantitative form (Novak and Tyson, 1993).

The chemical species that must be tracked include the Cdc2 and cyclin monomers, the dimer MPF in its active and inactive states, as well as the four regulatory enzymes Wee1, Cdc25, IE, and UbE in their phosphorylated and unphosphorylated states.

First, cyclin (with concentration y) is produced at a steady rate and is degraded or combines with Cdc2 (with concentration c) to form the MPF dimer (r):

$$\frac{dy}{dt} = k_1[A] - k_2 y - k_3 Yc. \tag{10.72}$$

The MPF dimer can be in one of four phosphorylation states, with phosphate at tyrosine-15 (s), at threonine-167 (concentration m), at both sites (concentration n), or at none (concentration r). The movement among these states is regulated by the enzymes Wee1, Cdc25, CAK, and one unknown enzyme ("?"). Thus,

$$\frac{dr}{dt} = -(k_2 + k_{CAK} + k_{wee})r + k_3 yc + k_? n + k_{25} s, \tag{10.73}$$

$$\frac{ds}{dt} = -(k_2 + k_{CAK} + k_{25})s + k_? n + k_{wee} r, \tag{10.74}$$

$$\frac{dm}{dt} = -(k_2 + k_? + k_{wee})m + k_{CAK} r + k_{25} n, \tag{10.75}$$

$$\frac{dn}{dt} = -(k_2 + k_? + k_{25})n + k_{wee} m + k_{CAK} s. \tag{10.76}$$

Notice that in the above equations, cyclin degradation at rate k_2 is permitted for free cyclin as well as for cyclin that is dimerized with Cdc2. We assume that, if cyclin degrades directly from a phosphorylated dimer, then the phosphate is also immediately removed thus producing free Cdc2. Thus,

$$\frac{dc}{dt} = k_2(r + s + n + m) - k_3 cy. \tag{10.77}$$

These six equations would form a closed system were it not for feedback. Notice that the last equation, (10.77), is redundant, since $m + r + s + n + c = $ constant. The feedback shows up in the nonlinear dependence of rate constants on the enzymes Cdc25, Wee1, IE, and UbE. This is expressed as

$$k_{25} = V'_{25}[\text{Cdc25}] + V''_{25}[\text{Cdc25\_P}], \tag{10.78}$$

$$k_{\text{wee}} = V'_{\text{wee}}[\text{Wee1\_P}] + V''_{\text{wee}}[\text{Wee1}], \tag{10.79}$$

$$k_2 = V'_2[\text{UbE}] + V''_2[\text{UbE}^*]. \tag{10.80}$$

In addition, the active states of Cdc25, Wee1, IE, and UbE are governed by Michaelis–Menten rate laws of the form

$$\frac{d[\text{Cdc25\_P}]}{dt} = \frac{k_a m[\text{Cdc25}]}{K_a + [\text{Cdc25}]} - \frac{k_b[\text{PPase}][\text{Cdc25\_P}]}{K_b + [\text{Cdc25\_P}]}, \tag{10.81}$$

$$\frac{d[\text{Wee1\_P}]}{dt} = \frac{k_e m[\text{Wee1}]}{K_e + [\text{Wee1}]} - \frac{k_f[\text{PPase}][\text{Wee1\_P}]}{K_f + [\text{Wee1\_P}]}, \tag{10.82}$$

$$\frac{d[\text{IE\_P}]}{dt} = \frac{k_g m[\text{IE}]}{K_g + [\text{IE}]} - \frac{k_h[\text{PPase}][\text{IE\_P}]}{K_h + [\text{IE\_P}]}, \tag{10.83}$$

$$\frac{d[\text{UbE}^*]}{dt} = \frac{k_c[\text{IE\_P}][\text{UbE}]}{K_c + [\text{UbE}]} - \frac{k_d[\text{IE}_{\text{anti}}][\text{UbE}^*]}{K_d + [\text{UbE}^*]}. \tag{10.84}$$

The quantities with _P attached correspond to phosphorylated forms of the enzyme quantity. The total amounts of each of these enzymes are assumed to be constant, so equations for the inactive forms are not necessary. PPase denotes a phosphatase that dephosphorylates Cdc25_P.

This forms a complete model with nine differential equations having eight Michaelis–Menten parameters and eighteen rate constants. There are two ways to gain an understanding of the behavior of this system of differential equations: by numerical simulation using reasonable parameter values, or by approximating by a smaller system of equations and studying the simpler system by analytical means.

The parameter values used by Novak and Tyson to simulate *Xenopus* oocyte extracts are shown in Tables 10.8 and 10.9.

While numerical simulation of these nine differential equations is not difficult, to gain an understanding of the basic behavior of the model it is convenient to make some simplifying assumptions. Suppose k_{CAK} is large and $k_?$ is small, as experiments suggest. Then the phosphorylation of Cdc2 on threonine-167 occurs immediately

---

Table 10.8 Michaelis–Menten constants for the cell cycle model of Novak and Tyson (1993).

$K_a/[\text{Cdc25}_{\text{total}}] = 0.1$	$K_b/[\text{Cdc25}_{\text{total}}] = 0.1$
$K_c/[\text{UbE}_{\text{total}}] = 0.01$	$K_d/[\text{UbE}_{\text{total}}] = 0.01$
$K_e/[\text{Wee1}_{\text{total}}] = 0.3$	$K_f/[\text{Wee1}_{\text{total}}] = 0.3$
$K_g/[\text{IE}_{\text{total}}] = 0.01$	$K_h/[\text{IE}_{\text{total}}] = 0.01$

Table 10.9 Rate constants for the cell cycle model of Novak and Tyson (1993).

$k_1[A]/[Cdc2_{total}]$	= 0.01	$k_3[Cdc2_{total}]$	= 1.0
$V_2'[UbE_{total}]$	= 0.015 (0.03)	$V_2''[UbE_{total}]$	= 1.0
$V_{25}'[Cdc25_{total}]$	= 0.1	$V_{25}''[Cdc25_{total}]$	= 2.0
$V_{wee}'[Wee1_{total}]$	= 0.1	$V_{wee}''[Wee1_{total}]$	= 1.0
k_{CAK}	= 0.25	$k_?$	= 0.25
$k_a[Cdc2_{total}]/[Cdc25_{total}]$	= 1.0	$k_b[PPase]/[Cdc25_{total}]$	= 0.125
$k_c[IE_{total}]/[UbE_{total}]$	= 0.1	$k_d[IE_{anti}]/[UbE_{total}]$	= 0.095
$k_e[Cdc2_{total}]/[Wee1_{total}]$	= 1.33	$k_f[PPase]/[Wee1_{total}]$	= 0.1
$k_g[Cdc2_{total}]/[IE_{total}]$	= 0.65	$k_h[PPase]/[IE_{total}]$	= 0.087

after formation of the MPF dimer. This allows us to ignore the quantities r and s. Next we assume that the activities of the regulatory enzymes, (10.78)–(10.80), can be approximated by

$$k_{wee} = \text{constant}, \tag{10.85}$$

$$k_2 = k_2' + k_2''m^2, \tag{10.86}$$

$$k_{25} = k_{25}' + k_{25}''m^2. \tag{10.87}$$

This leaves only three equations for the three unknowns y (free cyclin), m (active MPF), and Cdc2 monomer (q) as follows:

$$\frac{dy}{dt} = k_1 - k_2 y - k_3 yq, \tag{10.88}$$

$$\frac{dm}{dt} = k_3 yq - k_2 m + k_{25}n - k_{wee}m, \tag{10.89}$$

$$\frac{dq}{dt} = -k_3 yq + k_2(m + n), \tag{10.90}$$

where $m + n + q = c$ is the total Cdc2. It follows that the total cyclin $l = y + m + n$ satisfies the differential equation

$$\frac{dl}{dt} = k_1 - k_2 l. \tag{10.91}$$

Any three of these four equations describe the behavior of the system. However, in the limit that k_3 is large compared to other rate constants, the system can be further reduced to a two-variable system for which phase-plane analysis is applicable. With $v = k_3 y$,

$$\frac{dv}{dt} + k_2 v = k_3(k_1 - qv), \tag{10.92}$$

so that $qv = k_1$ to leading order. If k_1 is small, then y is small as well, so that (10.89) becomes

$$\frac{dm}{dt} = k_1 - k_2 m + k_{25}(l - m) - k_{wee}m. \tag{10.93}$$

The two equations (10.91) and (10.93) form a closed system that can be studied using the phase portrait in the (l, m) plane. In this approximation, $q = c - l$. The nullclines are described by the equations

$$\frac{dl}{dt} = 0: \qquad l = \frac{k_1}{k_2(m)}, \tag{10.94}$$

and

$$\frac{dm}{dt} = 0: \qquad l = \frac{k_{wee}m + k_2(m)m - k_1}{k_{25}(m)} + m, \tag{10.95}$$

which are plotted in Fig. 10.17.

For these parameter values, there is a unique unstable steady-state solution surrounded by a limit cycle oscillation. By adjusting parameters, one can have a stable fixed point on the leftmost branch of the N-shaped curve corresponding to the G_2 checkpoint, or one can have a stable fixed point on the rightmost branch, yielding an M phase checkpoint. Here a possible control parameter is k_{wee}, since increasing k_{wee} makes the m nullcline move up. Thus, increasing k_{wee} creates a G_2 checkpoint on the leftmost branch of the N-shaped curve.

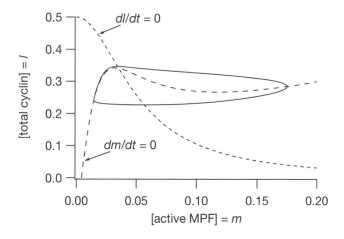

Figure 10.17 Solution trajectory and nullclines for the system of equations (10.91), (10.93). Parameter values are $k_1 = 0.004$, $k_{wee} = 0.9$, $k_2 = 0.008$, $k_2' = 3.0$, $k_{25} = 0.03$, $k_{25}' = 50.0$.

10.3.4   Conclusion

We are a long way from a complete understanding of the biochemistry of the cell cycle. Here we have seen a few of the main players, although there are major portions of the cell cycle that are still a mystery. With the modern development of biochemistry, many similar stories relating to regulation of cell function are being unfolded. It is likely that in the coming years, many of the details of this story will change and many new details will be added. However, the basic modeling process of turning a verbal description into a mathematical model and the consequent analysis of the model will certainly remain an important tool to aid our understanding of these complicated and extremely important processes. Furthermore, as the details of the stories become more complicated (as they are certain to do), mathematical analysis will become even more important in helping us understand how these processes work.

10.4   EXERCISES

1. Suppose that the production of an enzyme is turned on by m molecules of the enzyme according to

$$G + mP \underset{k_-}{\overset{k_+}{\rightleftarrows}} X,$$

where G is the inactive state of the gene and X is the active state of the gene. Suppose that mRNA is produced when the gene is in the active state and the enzyme is produced by mRNA and is degraded at some linear rate. Find a system of differential equations governing the behavior of mRNA and enzyme. Give a phase portrait analysis of this system and show that it has a "switch-like" behavior.

2. For a simplified lac operon model, assume that external lactose is converted directly into allolactose, and take

$$\frac{dA}{dt} = \alpha_L P \frac{L_e}{K_{Le} + L_e} - \beta_A B \frac{A}{K_A + A} - \gamma_A A, \tag{10.96}$$

eliminating (10.20) from the model.

Give a graphical steady-state analysis of the system of equations by plotting the null-clines $\frac{dM}{dt} = 0$ and $\frac{dA}{dt} = 0$ in the M, A phase plane (this is not a true phase plane, since it is a four-variable system.)

3. Find the steady-state solutions of the Goodwin model (10.21)–(10.22) with $n = 3$, and with $k_1 = k_2 = k_3$. Under what conditions is the steady-state solution unstable, and under what conditions is the instability an oscillatory instability? Find conditions on the parameters for a Hopf bifurcation.

4. A modification of the Goodwin model that does not require such a high level of cooperativity was formulated by Bliss et al. (1982) and is given by

$$\frac{dx_1}{dt} = \frac{a}{1 + x_3} - b_1 x_1, \tag{10.97}$$

$$\frac{dx_2}{dt} = b_1 x_1 - b_2 x_2, \tag{10.98}$$

$$\frac{dx_3}{dt} = b_2 x_2 - \frac{c x_3}{K + x_3}. \tag{10.99}$$

Determine the stability of the steady-state solution and find any Hopf bifurcation points in the special case that $K = 1$, $b_1 = b_2 < c$, and $a = c\left(\sqrt{\frac{c}{b_1}} - 1\right)$.

5. Sketch the phase portrait for the two-variable generic cell cycle model given by (10.34) and (10.35).

6. Instead of using (10.54), the original presentation of the model of Novak et al. (2001) used the expression

$$[\text{MPF}] = \frac{([\text{Cdc}13_T] - [\text{preMPF}])([\text{Cdc}13_T] - [\text{Trimer}])}{[\text{Cdc}13_T]}. \tag{10.100}$$

Show that these two expressions for [MPF] are the same. Hint: Make it easier by defining $c_T = [\text{Cdc}13_T]$, $T = [\text{Trimer}]$, $r = [\text{Rum1}]$, $M = [\text{MPF}]$, and $p = [\text{preMPF}]$. There are a number of ways to show this. One way is to start by

$$c_T = c_T + p - p = (c_T - p) + (X_1 + X_3). \tag{10.101}$$

Then show that $X_1 + X_3 = (c_T - p)X_3/M$, and factor out $c_T - p$ to get the result.

7. Goldbeter (1996) developed and studied a minimal cascade model of the mitotic oscillator. The model assumes that cyclin B is synthesized at a constant rate and activates Cdc25 kinase. The activated Cdc25 kinase in turn activates Cdc2 kinase (M), and the activated Cdc2 kinase is inactivated by the kinase Wee1. There is also a cyclin protease X that is activated by Cdc2 kinase and inactivated by an additional phosphatase. The differential equations for this reaction scheme (shown in Fig. 10.18) are

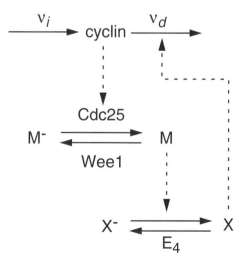

Figure 10.18 Diagram for the minimal cascade mitosis model of Goldbeter (1996).

Table 10.10 Parameter values for the Goldbeter minimal mitotic cycle model.

$K_1 = 0.1$	$V_{M1} = 0.5$ min^{-1}
$K_2 = 0.1$	$V_2 = 0.167$ min^{-1}
$K_3 = 0.1$	$V_{M3} = 0.2$ min^{-1}
$K_4 = 0.1$	$V_4 = 0.1$ min^{-1}
$v_i = 0.023$ μM/min	$v_d = 0.1$ μM/min
$K_c = 0.3$ μM	$K_d = 0.02$ μM
$k_d = 3.33 \times 10^{-3}$ min^{-1}	

$$\frac{dc}{dt} = v_i - v_d x \frac{c}{K_d + c} - k_d c, \tag{10.102}$$

$$\frac{dm}{dt} = V_1 \frac{1-m}{K_1 + 1 - m} - V_2 \frac{m}{K_2 + m}, \tag{10.103}$$

$$\frac{dx}{dt} = V_3 \frac{1-x}{K_3 + 1 - X} - V_4 \frac{x}{K_4 + X}, \tag{10.104}$$

and

$$V_1 = V_{M1} \frac{c}{K_c + c}, \qquad V_3 = V_{M3} m, \tag{10.105}$$

where c denotes the cyclin B concentration, and m and x denote the fractions of active Cdc2 kinase and of active cyclin protease, respectively. The parameters v_i and v_d are the constant rate of cyclin synthesis and maximum rate of cyclin degradation by protease X, achieved at $x = 1$. The parameters K_d and K_c denote Michaelis–Menten constants for cyclin degradation and cyclin activation, while k_d is the rate of nonspecific degradation of cyclin. The remaining parameters V_i and K_i, $i = 1, \ldots, 4$, are the effective maximum rate and Michaelis–Menten constants for each of the four enzymes Cdc25, Wee1, Cdc2, and the protease phosphatase (E_4), respectively. Typical parameter values are shown in Table 10.10. Simulate this system of equations to show that there is a stable limit cycle oscillation for this model. What is the period of oscillation?

Appendix: Units and Physical Constants

Quantity	Name	Symbol	Units
Amount	mole	mol	
Electric charge	coulomb	C	
Mass	gram	g	
Temperature	kelvin	K	
Time	second	s	
Length	meter	m	
Force	newton	N	$kg \cdot m \cdot s^{-2}$
Energy	joule	J	$N \cdot m$
Pressure	pascal	Pa	$N \cdot m^{-2}$
Capacitance	farad	F	$A \cdot s \cdot V^{-1}$
Resistance	ohm	Ω	$V \cdot A^{-1}$
Electric current	ampere	A	$C \cdot s^{-1}$
Conductance	siemen	S	$A \cdot V^{-1} = \Omega^{-1}$
Potential difference	volt	V	$N \cdot m \cdot C^{-1}$
Concentration	molar	M	$mol \cdot L^{-1}$
Atomic Mass	dalton	D	$g\, N_A^{-1}$

Unit Scale Factors

Name	Prefix	Scale factor
femto	f	$\times 10^{-15}$
pico	p	$\times 10^{-12}$
nano	n	$\times 10^{-9}$
micro	μ	$\times 10^{-6}$
milli	m	$\times 10^{-3}$
centi	c	$\times 10^{-2}$
deci	d	$\times 10^{-1}$
kilo	k	$\times 10^{3}$
mega	M	$\times 10^{6}$
giga	G	$\times 10^{9}$

Physical Constant	Symbol	Value
Boltzmann's constant	k	1.381×10^{-23} J $\cdot$ K^{-1}
Planck's constant	h	6.626×10^{-34} J $\cdot$ s
Avogadro's number	N_A	6.02257×10^{23} mol^{-1}
unit charge	q	1.6×10^{-19} C
gravitational constant	g	9.78049 m/s^2
Faraday's constant	F	9.649×10^{4}C $\cdot$ mol^{-1}
permittivity of free space	ϵ_0	8.854×10^{-12} F/m
universal gas constant	R	8.315 J mol^{-1} $\cdot$ K^{-1}
atmosphere	atm	1.01325×10^{5} N $\cdot$ m^{-2}
insulin unit	U	$\frac{1}{24000}$ g

Other Useful Identities and Quantities

1 atm = 760 mmHg

$R = kN_A$

$F = qN_A$

pH $= -\log_{10}[H^+]$ with $[H^+]$ in moles per liter

273.15 K = 0°C (ice point)

$T_{\text{Kelvin}} = T_{\text{centigrade}} - 273.15$

$T_{\text{Farenheit}} = \frac{9}{5}T_{\text{centigrade}} + 32$

$\frac{RT}{F} = 25.8$ mV at 27°C

density of pure water at 4°C = 1 gm/cm^3

C_m; capacitance of the cell membrane ≈ 1 μF/cm^2

Liter; 1 liter = 10^{-3} m^3

ϵ; dielectric constant for water = 80.4 ϵ_0

Lumen: 1 lm = quantity of light emitted by $\frac{1}{60}$ cm^2 surface area of pure platinum at its melting temperature (1770°C), within a solid angle of 1 steradian.

References

Adair, G. S. (1925) A critical study of the direct method of measuring osmotic pressure of hemoglobin, *Proceedings of the Royal Society of London A*. 108: 627–637.

Agre, P., L. S. King, M. Yasui, W. B. Guggino, O. P. Ottersen, Y. Fujiyoshi, A. Engel and S. Nielsen (2002) Aquaporin water channels–from atomic structure to clinical medicine, *Journal of Physiology*. 542: 3–16.

Aharon, S., H. Parnas and I. Parnas (1994) The magnitude and significance of Ca^{2+} domains for release of neurotransmitter, *Bulletin of Mathematical Biology*. 56: 1095–1119.

Akin, E. and H. M. Lacker (1984) Ovulation control: the right number or nothing, *Journal of Mathematical Biology*. 20: 113–132.

Albers, R. W., S. Fahn and G. J. Koval (1963) The role of sodium ions in the activation of electrophorus electric organ adenosine triphosphatase, *Proceedings of the National Academy of Sciences USA*. 50: 474–481.

Alberts, B., D. Bray, J. Lewis, M. Raff, K. Roberts and J. D. Watson (1994) *Molecular Biology of the Cell (Third Edition)*: Garland Publishing, Inc., New York, London.

Aldrich, R. W., D. P. Corey and C. F. Stevens (1983) A reinterpretation of mammalian sodium channel gating based on single channel recording, *Nature*. 306: 436–441.

Allbritton, N. L., T. Meyer and L. Stryer (1992) Range of messenger action of calcium ion and inositol 1,4,5-trisphosphate, *Science*. 258: 1812–1815.

Allen, N. A., K. C. Chen, C. A. Shaffer, J. J. Tyson and L. T. Watson (2006) Computer evaluation of network dynamics models with application to cell cycle control in budding yeast, *IEE Proceedings Systems Biology*. 153: 13–21.

Allescher, H. D., K. Abraham-Fuchs, R. E. Dunkel and M. Classen (1998) Biomagnetic 3-dimensional spatial and temporal characterization of electrical activity of human stomach, *Digestive Diseases and Sciences*. 43: 683–693.

Alt, W. and D. A. Lauffenberger (1987) Transient behavior of a chemotaxis system modelling certain types of tissue inflammation, *Journal of Mathematical Biology*. 24: 691–722.

Alvarez, W. C. (1922) The electrogastrogram and what it shows, *Journal of the American Medical Association*. 78: 1116–1119.

Anand, M., K. Rajagopal and K. R. Rajagopal (2003) A model incorporating some of the mechanical and biochemical factors underlying clot formation and dissolution in flowing blood, *Journal of Theoretical Medicine*. 5: 183–218.

Anand, M., K. Rajagopal and K. R. Rajagopal (2005) A model for the formation and lysis of blood clots, *Pathophysiology of Haemostasis and Thrombosis*. 34: 109–120.

Anderson, C., N. Trayanova and K. Skouibine (2000) Termination of spiral waves with biphasic shocks: role of virtual electrode polarization, *Journal of Cardiovascular Electrophysiology*. 11: 1386–1396.

Anliker, M., R. L. Rockwell and E. Ogden (1971a) Nonlinear analysis of flow pulses and shock waves in arteries. Part I: derivation and properties of mathematical model, *Zeitschrift für Angewandte Mathematik und Physik*. 22: 217–246.

Anliker, M., R. L. Rockwell and E. Ogden (1971b) Nonlinear analysis of flow pulses and shock waves in arteries. Part II: parametric study related to clinical problems., *Zeitschrift für Angewandte Mathematik und Physik*. 22: 563–581.

Apell, H. J. (1989) Electrogenic properties of the Na,K pump, *Journal of Membrane Biology*. 110: 103–114.

Apell, H. J. (2004) How do P-type ATPases transport ions?, *Bioelectrochemistry*. 63: 149–156.

Armstrong, C. M. (1981) Sodium channels and gating currents, *Physiological Reviews*. 61: 644–683.

Armstrong, C. M. and F. Bezanilla (1973) Currents related to movement of the gating particles of the sodium channels, *Nature*. 242: 459–461.

Armstrong, C. M. and F. Bezanilla (1974) Charge movement associated with the opening and closing of the activation gates of the Na channels, *Journal of General Physiology*. 63: 533–552.

Armstrong, C. M. and F. Bezanilla (1977) Inactivation of the sodium channel II Gating current experiments, *Journal of General Physiology*. 70: 567–590.

Arnold, V. I. (1983) *Geometric Methods in the Theory of Ordinary Differential Equations*: Springer-Verlag, New York.

Aronson, D. G. and H. F. Weinberger (1975) *Nonlinear diffusion in population genetics, combustion and nerve pulse propagation*: Springer-Verlag, New York.

Arrhenius, S. (1889) On the reaction velocity of the inversion of cane sugar by acids, *Zeitschrift fr physikalische Chemie*. 4: 226 ff.

Asdell, S. A. (1946) *Patterns of Mammalian Reproduction*: Comstock Publishing Company, New York.

Ashmore, J. F. and D. Attwell (1985) Models for electrical tuning in hair cells, *Proceedings of the Royal Society of London B*. 226: 325–344.

Ashmore, J. F., G. S. Geleoc and L. Harbott (2000) Molecular mechanisms of sound amplification in the mammalian cochlea, *Proceedings of the National Academy of Sciences USA*. 97: 11759–11764.

Ataullakhanov, F. I., Y. V. Krasotkina, V. I. Sarbash, R. I. Volkova, E. I. Sinauridse and A. Y. Kondratovich (2002a) Spatio-temporal dynamics of blood coagulation and pattern formation. An experimental study., *International Journal of Bifurcation and Chaos*. 12: 1969–1983.

Ataullakhanov, F. I., V. I. Zarnitsina, A. V. Pokhilko, A. I. Lobanov and O. L. Morozova (2002b) Spatio-temporal dynamics of blood coagulation and pattern formation. A theoretical approach., *International Journal of Bifurcation and Chaos*. 12: 1985–2002.

Atri, A., J. Amundson, D. Clapham and J. Sneyd (1993) A single-pool model for intracellular calcium oscillations and waves in the *Xenopus laevis* oocyte, *Biophysical Journal*. 65: 1727–1739.

Atwater, I., C. M. Dawson, A. Scott, G. Eddlestone and E. Rojas (1980) The nature of the oscillatory behavior in electrical activity for pancreatic β-cell, *Hormone and Metabolic Research*. 10 (suppl.): 100–107.

Atwater, I. and J. Rinzel (1986) *The β-cell bursting pattern and intracellular calcium*. In: Ionic Channels in Cells and Model Systems, Ed: R. Latorre, Plenum Press, New York, London.

Bélair, J., M. C. Mackey and J. M. Mahaffy (1995) Age-structured and two-delay models for erythropoiesis, *Mathematical Biosciences*. 128: 317–346.

Backx, P. H., P. P. d. Tombe, J. H. K. V. Deen, B. J. M. Mulder and H. E. D. J. t. Keurs (1989) A model of propagating calcium-induced calcium release mediated by calcium diffusion, *Journal of General Physiology*. 93: 963–977.

Ball, F. G., Y. Cai, J. B. Kadane and A. O'Hagan (1999) Bayesian inference for ion-channel gating mechanisms directly from single-channel recordings, using Markov chain Monte Carlo, *Proceedings of the Royal Society of London A*. 455: 2879–2932.

Bar, M. and M. Eiswirth (1993) Turbulence due to spiral breakup in a continuous excitable medium, *Physical Review E*. 48: 1635–1637.

Barcilon, V. (1992) Ion flow through narrow membrane channels: Part I, *SIAM Journal on Applied Mathematics*. 52: 1391–1404.

Barcilon, V., D. P. Chen and R. S. Eisenberg (1992) Ion flow through narrow membrane channels: Part II, *SIAM Journal on Applied Mathematics*. 52: 1405–1425.

Barkley, D. (1994) Euclidean symmetry and the dynamics of rotating spiral waves, *Physical Review Letters*. 72: 164–167.

Barlow, H. B. (1972) Single units and sensation: a neuron doctrine for perceptual psychology?, *Perception*. 1: 371–394.

Barlow, H. B., R. Fitzhugh and S. W. Kuffler (1957) Change of organization in the receptive fields of the cat's retina during dark adaptation, *Journal of Physiology*. 137: 338–354.

Barlow, H. B. and W. R. Levick (1965) The mechanism of directionally selective units in rabbit's retina, *Journal of Physiology*. 178: 477–504.

Batchelor, G. K. (1967) *An Introduction to Fluid Dynamics*: Cambridge University Press, Cambridge.

Batzel, J. J., F. Kappel, D. Schneditz and H. T. Tran (2007) *Cardiovascular and Respiratory System: Modeling, Analysis and Control*: SIAM, Philadelphia.

Batzel, J. J. and H. T. Tran (2000a) Stability of the human respiratory control system. I. Analysis of a two-dimensional delay state-space model, *Journal of Mathematical Biology*. 41: 45–79.

Batzel, J. J. and H. T. Tran (2000b) Stability of the human respiratory control system. II. Analysis of a three-dimensional delay state-space model, *Journal of Mathematical Biology*. 41: 80–102.

Batzel, J. J. and H. T. Tran (2000c) Modeling instability in the control system for human respiration: applications to infant non-REM sleep, *Applied Mathematics and Computation*. 110: 1–51.

Baylor, D. A. and A. L. Hodgkin (1974) Changes in time scale and sensitivity in turtle photo-receptors, *Journal of Physiology*. 242: 729–758.

Baylor, D. A., A. L. Hodgkin and T. D. Lamb (1974a) The electrical response of turtle cones to flashes and steps of light, *Journal of Physiology*. 242: 685–727.

Baylor, D. A., A. L. Hodgkin and T. D. Lamb (1974b) Reconstruction of the electrical responses of turtle cones to flashes and steps of light, *Journal of Physiology*. 242: 759–791.

Baylor, D. A., T. D. Lamb and K. W. Yau (1979) Responses of retinal rods to single photons, *Journal of Physiology*. 288: 613–634.

Baylor, D. A., G. Matthews and K. W. Yau (1980) Two components of electrical dark noise in toad retinal rod outer segments, *Journal of Physiology*. 309: 591–621.

Beck, J. S., R. Laprade and J.-Y. Lapointe (1994) Coupling between transepithelial Na transport and basolateral K conductance in renal proximal tubule, *American Journal of Physiology — Renal, Fluid and Electrolyte Physiology*. 266: F517–527.

Beeler, G. W. and H. J. Reuter (1977) Reconstruction of the action potential of ventricular myocardial fibers, *Journal of Physiology*. 268: 177–210.

Begenisich, T. B. and M. D. Cahalan (1980) Sodium channel permeation in squid axons I: reversal potential experiments, *Journal of Physiology*. 307: 217–242.

Beltrami, E. and J. Jesty (1995) Mathematical analysis of activation thresholds in enzyme-catalyzed positive feedbacks: application to the feedbacks of blood coagulation, *Proceedings of the National Academy of Sciences USA*. 92: 8744–8748.

Ben-Tal, A. (2006) Simplified models for gas exchange in the human lungs, *Journal of Theoretical Biology*. 238: 474–495.

Ben-Tal, A. and J. C. Smith (2008) A model for control of breathing in mammals: coupling neural dynamics to peripheral gas exchange and transport, *Journal of Theoretical Biology*. 251: 480–497.

Bennett, M. R., L. Farnell and W. G. Gibson (2004) The facilitated probability of quantal secretion within an array of calcium channels of an active zone at the amphibian neuromuscular junction, *Biophysical Journal*. 86: 2674–2690.

Bennett, M. R., L. Farnell and W. G. Gibson (2005) A quantitative model of purinergic junctional transmission of calcium waves in astrocyte networks, *Biophysical Journal*. 89: 2235–2250.

Bennett, M. R., L. Farnell, W. G. Gibson and P. Dickens (2007) Mechanisms of calcium sequestration during facilitation at active zones of an amphibian neuromuscular junction, *Journal of Theoretical Biology*. 247: 230–241.

Bensoussan, A., J.-L. Lions and G. Papanicolaou (1978) *Asymptotic Analysis for Periodic Structures*: North-Holland, Amsterdam, New York.

Bentele, K. and M. Falcke (2007) Quasi-steady approximation for ion channel currents, *Biophysical Journal*. 93: 2597–2608.

Bergman, R. N. (1989) Toward physiological understanding of glucose tolerance: minimal-model approach, *Diabetes*. 38: 1512–1527.

Bergman, R. N., Y. Z. Ider, C. R. Bowden and C. Cobelli (1979) Quantitative estimation of insulin sensitivity, *American Journal of Physiology — Endocrinology and Metabolism*. 236: E667–677.

Bergstrom, R. W., W. Y. Fujimoto, D. C. Teller and C. D. Haën (1989) Oscillatory insulin secretion in perifused isolated rat islets, *American Journal of Physiology — Endocrinology and Metabolism*. 257: E479–485.

Berman, N., H.-F. Chou, A. Berman and E. Ipp (1993) A mathematical model of oscillatory insulin secretion, *American Journal of Physiology — Regulatory, Integrative and Comparative Physiology*. 264: R839–851.

Berne, R. M. and M. N. Levy, Eds. (1993) *Physiology (Third Edition)*. Mosby Year Book, St. Louis.

Berne, R. M. and M. N. Levy, Eds. (1998) *Physiology (Fourth Edition)*. Mosby Year Book, St. Louis.

Bernstein, J. (1902) Untersuchungen zur Thermodynamik der bioelektrischen Ströme. Erster Theil., *Pflügers Archive*. 82: 521–562.

Berridge, M. J. (1997) Elementary and global aspects of calcium signalling, *Journal of Physiology*. 499: 291–306.

Berridge, M. J., M. D. Bootman and H. L. Roderick (2003) Calcium signalling: dynamics, homeostasis and remodelling, *Nature Reviews Molecular Cell Biology*. 4: 517–529.

Berridge, M. J. and A. Galione (1988) Cytosolic calcium oscillators, *FASEB Journal*. 2: 3074–3082.

Bers, D. M. (2001) *Excitation–Contraction Coupling and Cardiac Contractile Force (Second edition)*: Kluwer, New York.

Bers, D. M. (2002) Cardiac excitation–contraction coupling, *Nature*. 415: 198–205.

Bers, D. M. and E. Perez-Reyes (1999) Ca channels in cardiac myocytes: structure and function in Ca influx and intracellular Ca release, *Cardiovascular Research*. 42: 339–360.

Bertram, R., M. J. Butte, T. Kiemel and A. Sherman (1995) Topological and phenomenological classification of bursting oscillations, *Bulletin of Mathematical Biology*. 57: 413–439.

Bertram, R., M. Egli, N. Toporikova and M. E. Freeman (2006b) A mathematical model for the mating-induced prolactin rhythm of female rats, *American Journal of Physiology — Endocrinology and Metabolism*. 290: E573–582.

Bertram, R., M. Gram Pedersen, D. S. Luciani and A. Sherman (2006a) A simplified model for

mitochondrial ATP production, *Journal of Theoretical Biology*. 243: 575–586.

Bertram, R., L. Satin, M. Zhang, P. Smolen and A. Sherman (2004) Calcium and glycolysis mediate multiple bursting modes in pancreatic islets, *Biophysical Journal*. 87: 3074–3087.

Bertram, R., L. S. Satin, M. G. Pedersen, D. S. Luciani and A. Sherman (2007b) Interaction of glycolysis and mitochondrial respiration in metabolic oscillations of pancreatic islets, *Biophysical Journal*. 92: 1544–1555.

Bertram, R. and A. Sherman (2004a) A calcium-based phantom bursting model for pancreatic islets, *Bulletin of Mathematical Biology*. 66: 1313–1344.

Bertram, R. and A. Sherman (2004b) Filtering of calcium transients by the endoplasmic reticulum in pancreatic beta-cells, *Biophysical Journal*. 87: 3775–3785.

Bertram, R., A. Sherman and L. S. Satin (2007a) Metabolic and electrical oscillations: partners in controlling pulsatile insulin secretion, *American Journal of Physiology — Endocrinology and Metabolism*. 293: E890–900.

Bertram, R., A. Sherman and E. F. Stanley (1996) Single-domain/bound calcium hypothesis of transmitter release and facilitation, *Journal of Neurophysiology*. 75: 1919–1931.

Beuter, A., L. Glass, M. C. Mackey and M. S. Titcombe (2003) *Nonlinear Dynamics in Physiology and Medicine*: Springer-Verlag, New York.

Bezprozvanny, I., J. Watras and B. E. Ehrlich (1991) Bell-shaped calcium-response curves of Ins(1,4,5)P$_3$- and calcium-gated channels from endoplasmic reticulum of cerebellum, *Nature*. 351: 751–754.

Bier, M., B. Teusink, B. N. Kholodenko and H. V. Westerhoff (1996) Control analysis of glycolytic oscillations, *Biophysical Chemistry*. 62: 15–24.

Blakemore, C., Ed. (1990) *Vision: Coding and Efficiency*. Cambridge University Press, Cambridge, UK.

Blatow, M., A. Caputi, N. Burnashev, H. Monyer and A. Rozov (2003) Ca^{2+} buffer saturation underlies paired pulse facilitation in calbindin-D28k-containing terminals, *Neuron*. 38: 79–88.

Bliss, R. D., P. R. Painter and A. G. Marr (1982) Role of feedback inhibition in stabilizing the classical operon, *Journal of Theoretical Biology*. 97: 177–193.

Blum, J. J., D. D. Carr and M. C. Reed (1992) Theoretical analysis of lipid transport in sciatic nerve, *Biochimica et Biophysica Acta*. 1125: 313–320.

Blum, J. J. and M. C. Reed (1985) A model for fast axonal transport, *Cell Motility*. 5: 507–527.

Blum, J. J. and M. C. Reed (1989) A model for slow axonal transport and its application to neurofilamentous neuropathies, *Cell Motility and the Cytoskeleton*. 12: 53–65.

Bluman, G. W. and H. C. Tuckwell (1987) Methods for obtaining analytical solutions for Rall's model neuron, *Journal of Neuroscience Methods*. 20: 151–166.

Bogumil, R. J., M. Ferin, J. Rootenberg, L. Speroff and R. L. Van De Wiele (1972) Mathematical studies of the human menstrual cycle. I. formulation of a mathematical model., *Journal of Clinical Endocrinology and Metabolism*. 35: 126–156.

Bohr, C., K. A. Hasselbach and A. Krogh (1904) Über einen in biologischen Beziehung wichtigen Einfluss, den die Kohlen-sauerspannung des Blutes auf dessen Sauerstoffbindung übt., *Skand Arch Physiol*. 15: 401–412.

Boitano, S., E. R. Dirksen and M. J. Sanderson (1992) Intercellular propagation of calcium waves mediated by inositol trisphosphate, *Science*. 258: 292–294.

Bootman, M., E. Niggli, M. Berridge and P. Lipp (1997a) Imaging the hierarchical Ca^{2+} signalling system in HeLa cells, *Journal of Physiology*. 499: 307–314.

Bootman, M. D., M. J. Berridge and P. Lipp (1997b) Cooking with calcium: the recipes for composing global signals from elementary events, *Cell*. 91: 367–373.

Borghans, J. A. M., R. J. De Boer and L. A. Segel (1996) Extending the quasi-steady state approximation by changing variables, *Bulletin of Mathematical Biology*. 58: 43–63.

Borghans, J. M., G. Dupont and A. Goldbeter (1997) Complex intracellular calcium oscillations. A theoretical exploration of possible mechanisms, *Biophysical Chemistry*. 66: 25–41.

Borisuk, M. T. and J. J. Tyson (1998) Bifurcation analysis of a model of mitotic control in frog eggs, *Journal of Theoretical Biology*. 195: 69–85.

Borst, A. (2000) Models of motion detection, *Nature Neuroscience*. 3 Suppl: 1168.

Borst, A. and M. Egelhaaf (1989) Principles of visual motion detection, *Trends in Neurosciences*. 12: 297–306.

Bowen, J. R., A. Acrivos and A. K. Oppenheim (1963) Singular perturbation refinement to quasi-steady state approximation in chemical kinetics, *Chemical Engineering Science*. 18: 177–188.

Boyce, W. E. and R. C. DiPrima (1997) *Elementary Differential Equations and Boundary Value Problems (Sixth Edition)*: John Wiley and Sons, New York.

Boyd, I. A. and A. R. Martin (1956) The end-plate potential in mammalian muscle, *Journal of Physiology*. 132: 74–91.

Brabant, G., K. Prank and C. Schöfl (1992) Pulsatile patterns in hormone secretion, *Trends in Endocrinology and Metabolism*. 3: 183–190.

Bradshaw, L. A., S. H. Allos, J. P. Wikswo, Jr. and W. O. Richards (1997) Correlation and comparison of magnetic and electric detection of small intestinal electrical activity, *American Journal of Physiology — Gastrointestinal and Liver Physiology*. 272: G1159–1167.

Bradshaw, L. A., J. K. Ladipo, D. J. Staton, J. P. Wikswo, Jr. and W. O. Richards (1999) The human vector magnetogastrogram and magnetoenterogram, *IEEE Transactions in Biomedical Engineering*. 46: 959–970.

Braun, M. (1993) *Differential Equations and their Applications (Fourth Edition)*: Springer-Verlag, New York.

Briggs, G. E. and J. B. S. Haldane (1925) A note on the kinematics of enzyme action, *Biochemical Journal*. 19: 338–339.

Brink, P. R. and S. V. Ramanan (1985) A model for the diffusion of fluorescent probes in the septate giant axon of earthworm: axoplasmic diffusion and junctional membrane permeability, *Biophysical Journal*. 48: 299–309.

Britton, N. F. (1986) *Reaction–Diffusion Equations and their Applications to Biology*: Academic Press, London.

Brodie, S. E., B. W. Knight and F. Ratliff (1978a) The response of the Limulus retina to moving stimuli: a prediction by Fourier synthesis, *Journal of General Physiology*. 72: 129–166.

Brodie, S. E., B. W. Knight and F. Ratliff (1978b) The spatiotemporal transfer function of the Limulus lateral eye, *Journal of General Physiology*. 72: 167–202.

Brokaw, C. J. (1976) Computer simulation of movement-generating cross-bridges, *Biophysical Journal*. 16: 1013–1027.

Brown, B. H., H. L. Duthie, A. R. Horn and R. H. Smallwood (1975) A linked oscillator model of electrical activity of human small intestine, *American Journal of Physiology*. 229: 384–388.

Brown, D., A. E. Herbison, J. E. Robinson, R. W. Marrs and G. Leng (1994) Modelling the luteinizing hormone-releasing hormone pulse generator, *Neuroscience*. 63: 869–879.

Brown, D., E. A. Stephens, R. G. Smith, G. Li and G. Leng (2004) Estimation of parameters for a mathematical model of growth hormone secretion, *Journal of Neuroendocrinology*. 16: 936–946.

Bugrim, A., R. Fontanilla, B. B. Eutenier, J. Keizer and R. Nuccitelli (2003) Sperm initiate a Ca^{2+} wave in frog eggs that is more similar to Ca^{2+} waves initiated by IP_3 than by Ca^{2+}, *Biophysical Journal*. 84: 1580–1590.

Bungay, S. D., P. A. Gentry and R. D. Gentry (2003) A mathematical model of lipid-mediated thrombin generation, *Mathematical Medicine and Biology*. 20: 105–129.

Burger, H. C. and J. B. Van Milaan (1946) Heart-vector and leads, *British Heart Journal*. 8: 157–161.

Burger, H. C. and J. B. Van Milaan (1947) Heart-vector and leads. Part II, *British Heart Journal*. 9: 154–160.

Burger, H. C. and J. B. Van Milaan (1948) Heart-vector and leads. Part III: geometrical representation, *British Heart Journal*. 10: 229–233.

Bursztyn, L., O. Eytan, A. J. Jaffa and D. Elad (2007) Modeling myometrial smooth muscle contraction, *Annals of the New York Academy of Sciences*. 1101: 110–138.

Burton, W. K., N. Cabrera and F. C. Frank (1951) The growth of crystals and the equilibrium structure of their surfaces, *Philosophical Transactions of the Royal Society of London A*. 243: 299–358.

Butera, R. J., Jr., J. Rinzel and J. C. Smith (1999a) Models of respiratory rhythm generation in the pre-Bötzinger complex. I. Bursting pacemaker neurons, *Journal of Neurophysiology*. 82: 382–397.

Butera, R. J., Jr., J. Rinzel and J. C. Smith (1999b) Models of respiratory rhythm generation in the pre-Bötzinger complex. II. Populations Of coupled pacemaker neurons, *Journal of Neurophysiology*. 82: 398–415.

Cajal, S. R. (1893) Sur les ganglions et plexus nerveux de lintestin, *Comptes Rendus des Seances de la Société de Biologie et de ses Filiales*. 45: 217–223.

Cajal, S. R. (1911) *Histologie du Systéme Nerveux de l'Homme et des Vertébrés*: Maloine, Paris.

Callamaras, N., J. S. Marchant, X. P. Sun and I. Parker (1998) Activation and co-ordination of $InsP_3$-mediated elementary Ca^{2+} events during global Ca^{2+} signals in Xenopus oocytes, *Journal of Physiology*. 509: 81–91.

Čamborová, P., P. Hubka, I. Šulková and I. Hulín (2003) The pacemaker activity of interstitial cells of Cajal and gastric electrical activity, *Physiological Research*. 52: 275–284.

Campbell, D. T. and B. Hille (1976) Kinetic and pharmacological properties of the sodium channel of frog skeletal muscle, *Journal of General Physiology*. 67: 309–323.

Carafoli, E. (2002) Calcium signaling: a tale for all seasons, *Proceedings of the National Academy of Sciences USA*. 99: 1115–1122.

Carafoli, E., L. Santella, D. Branca and M. Brini (2001) Generation, control, and processing of cellular calcium signals, *Critical Reviews in Biochemistry and Molecular Biology*. 36: 107–260.

Carpenter, G. (1977) A geometric approach to singular perturbation problems with applications to nerve impulse equations, *Journal of Differential Equations*. 23: 335–367.

Carpenter, G. A. and S. Grossberg (1981) Adaptation and transmitter gating in vertebrate photoreceptors, *Journal of Theoretical Neurobiology*. 1: 1–42.

Cartwright, M. and M. Husain (1986) A model for the control of testosterone secretion, *Journal of Theoretical Biology*. 123: 239–250.

Castellan, G. W. (1971) *Physical Chemistry (Second Edition)*: Addison-Wesley, Reading,

Casten, R. G., H. Cohen and P. A. Lagerstrom (1975) Perturbation analysis of an approximation to the Hodgkin–Huxley theory, *Quarterly of Applied Mathematics*. 32: 365–402.

Caulfield, J. B. and T. K. Borg (1979) The collagen network of the heart, *Laboratory Investigation*. 40: 364–372.

Caumo, A., R. N. Bergman and C. Cobelli (2000) Insulin sensitivity from meal tolerance tests in normal subjects: a minimal model index, *Journal of Clinical Endocrinology and Metabolism*. 85: 4396–4402.

Chávez-Ross, A., S. Franks, H. D. Mason, K. Hardy and J. Stark (1997) Modelling the control of ovulation and polycystic ovary syndrome, *Journal of Mathematical Biology*. 36: 95–118.

Chadwick, R. (1980) *Studies in cochlear mechanics*. In: Mathematical Modeling of the Hearing Process Lecture Notes in Biomathematics, 43, Ed: M. H. Holmes and L. A. Rubenfeld, Springer-Verlag, Berlin, Heidelberg, New York.

Chadwick, R. S., A. Inselberg and K. Johnson (1976) Mathematical model of the cochlea. II: results and conclusions, *SIAM Journal on Applied Mathematics*. 30: 164–179.

Champneys, A. R., V. Kirk, E. Knobloch, B. E. Oldeman and J. Sneyd (2007) When Shil'nikov meets Hopf in excitable systems, *SIAM Journal on Applied Dynamical Systems*. 6: 663–693.

Changeux, J. P. (1965) The control of biochemical reactions, *Scientific American*. 212: 36–45.

Chapman, R. A. and C. H. Fry (1978) An analysis of the cable properties of frog ventricular myocardium, *Journal of Physiology*. 283: 263–282.

Charles, A. (1998) Intercellular calcium waves in glia, *Glia*. 24: 39–49.

Charles, A. C., J. E. Merrill, E. R. Dirksen and M. J. Sanderson (1991) Intercellular signaling in glial cells: calcium waves and oscillations in response to mechanical stimulation and glutamate, *Neuron*. 6: 983–992.

Charles, A. C., C. C. G. Naus, D. Zhu, G. M. Kidder, E. R. Dirksen and M. J. Sanderson (1992) Intercellular calcium signaling via gap junctions in glioma cells, *Journal of Cell Biology*. 118: 195–201.

Charnock, J. S. and R. L. Post (1963) Studies of the mechanism of cation transport. I. The preparation and properties of a cation-stimulated adenosine-triphosphatase from guinea pig kidney cortex, *Australian Journal of Experimental Biology and Medical Science*. 41: 547–560.

Chay, T. R. (1986) On the effect of the intracellular calcium-sensitive K^+ channel in the bursting pancreatic β-cell, *Biophysical Journal*. 50: 765–777.

Chay, T. R. (1987) The effect of inactivation of calcium channels by intracellular Ca^{2+} ions in the bursting pancreatic β-cell, *Cell Biophysics*. 11: 77–90.

Chay, T. R. (1996a) Modeling slowly bursting neurons via calcium store and voltage-independent calcium current, *Neural Computation*. 8: 951–978.

Chay, T. R. (1996b) Electrical bursting and luminal calcium oscillation in excitable cell models, *Biological Cybernetics*. 75: 419–431.

Chay, T. R. (1997) Effects of extracellular calcium on electrical bursting and intracellular and luminal calcium oscillations in insulin secreting pancreatic beta-cells, *Biophysical Journal*. 73: 1673–1688.

Chay, T. R. and D. L. Cook (1988) Endogenous bursting patterns in excitable cells, *Mathematical Biosciences*. 90: 139–153.

Chay, T. R. and H. S. Kang (1987) *Multiple oscillatory states and chaos in the endogeneous activity of excitable cells: pancreatic β-cell as an example*. In: Chaos in Biological Systems, Ed: H. Degn, A. V. Holden and L. F. Olsen, Plenum Press, New York.

Chay, T. R. and J. Keizer (1983) Minimal model for membrane oscillations in the pancreatic β-cell, *Biophysical Journal*. 42: 181–190.

Cheer, A., R. Nuccitelli, G. F. Oster and J.-P. Vincent (1987) Cortical waves in vertebrate eggs I: the activation waves, *Journal of Theoretical Biology*. 124: 377–404.

Chen, D. and R. Eisenberg (1993) Charges, currents, and potentials in ionic channels of one conformation, *Biophysical Journal*. 64: 1405–1421.

Chen, D. P., V. Barcilon and R. S. Eisenberg (1992) Constant fields and constant gradients in open ionic channels, *Biophysical Journal*. 61: 1372–1393.

Chen, K. C., L. Calzone, A. Csikasz-Nagy, F. R. Cross, B. Novak and J. J. Tyson (2004) Integrative analysis of cell cycle control in budding yeast, *Molecular Biology of the Cell*. 15: 3841–3862.

Chen, K. C., A. Csikasz-Nagy, B. Gyorffy, J. Val, B. Novak and J. J. Tyson (2000) Kinetic analysis of a molecular model of the budding yeast cell cycle, *Molecular Biology of the Cell*. 11: 369–391.

Chen, L. and M. Q. Meng (1995) Compact and scattered gap junctions in diffusion mediated cell-cell communication, *Journal of Theoretical Biology*. 176: 39–45.

Chen, P. S., P. D. Wolf, E. G. Dixon, N. D. Danieley, D. W. Frazier, W. M. Smith and R. E. Ideker (1988) Mechanism of ventricular vulnerability to single premature stimuli in open-chest dogs, *Circulation Research*. 62: 1191–1209.

Cheng, H., M. Fill, H. Valdivia and W. J. Lederer (1995) Models of Ca^{2+} release channel adaptation, *Science*. 267: 2009–2010.

Cheng, H., W. J. Lederer and M. B. Cannell (1993) Calcium sparks: elementary events underlying excitation–contraction coupling in heart muscle, *Science*. 262: 740–744.

Cherniack, N. S. and G. S. Longobardo (2006) Mathematical models of periodic breathing and their usefulness in understanding cardiovascular and respiratory disorders, *Experimental Physiology*. 91: 295–305.

Cheyne, J. (1818) A case of apoplexy, in which the fleshy part of the heart was converted into fat, *Dublin Hospital Reports*. 2: 216.

Christ, G. J., P. R. Brink and S. V. Ramanan (1994) Dynamic gap junctional communication: a delimiting model for tissue responses, *Biophysical Journal*. 67: 1335–1344.

Ciliberto, A., B. Novak and J. J. Tyson (2003) Mathematical model of the morphogenesis checkpoint in budding yeast, *Journal of Cell Biology*. 163: 1243–1254.

Civan, M. M. and R. J. Bookman (1982) Transepithelial Na^+ transport and the intracellular fluids: a computer study, *Journal of Membrane Biology*. 65: 63–80.

Civan, M. M. and R. J. Podolsky (1966) Contraction kinetics of striated muscle fibres following quick changes in load, *Journal of Physiology*. 184: 511–534.

Clément, F., M. A. Gruet, P. Monget, M. Terqui, E. Jolivet and D. Monniaux (1997) Growth kinetics of the granulosa cell population in ovarian follicles: an approach by mathematical modelling, *Cell Proliferation*. 30: 255–270.

Clément, F., D. Monniaux, J. Stark, K. Hardy, J. C. Thalabard, S. Franks and D. Claude (2001) Mathematical model of FSH-induced cAMP production in ovarian follicles, *American Journal of Physiology — Endocrinology and Metabolism*. 281: E35–53.

Clément, F., D. Monniaux, J. C. Thalabard and D. Claude (2002) Contribution of a mathematical modelling approach to the understanding of the ovarian function, *Comptes Rendus des Seances de la Société de Biologie et de ses Filiales*. 325: 473–485.

Clancy, C. E. and Y. Rudy (1999) Linking a genetic defect to its cellular phenotype in a cardiac arrhythmia, *Nature*. 400: 566–569.

Clancy, C. E. and Y. Rudy (2001) Cellular consequences of HERG mutations in the long QT syndrome: precursors to sudden cardiac death, *Cardiovascular Research*. 50: 301–313.

Clapham, D. (1995) Intracellular calcium–replenishing the stores, *Nature*. 375: 634–635.

Clark, A. J. (1933) *The Mode of Action of Drugs on Cells*: Edward Arnold and Co., London.

Clark, L. H., P. M. Schlosser and J. F. Selgrade (2003) Multiple stable periodic solutions in a model for hormonal control of the menstrual cycle, *Bulletin of Mathematical Biology*. 65: 157–173.

Clifford, C. W. and M. R. Ibbotson (2003) Fundamental mechanisms of visual motion detection: models, cells and functions, *Progress in Neurobiology*. 68: 409–437.

Cobbs, W. H. and E. N. Pugh (1987) Kinetics and components of the flash photocurrent of isolated retinal rods of the larval salamander *Ambystoma Tigrinum*, *Journal of Physiology*. 394: 529–572.

Coddington, E. A. and N. Levinson (1984) *Ordinary Differential Equations*: Robert E. Krieger Publishing Company, Malabar, Florida.

Cohen, M. A. and J. A. Taylor (2002) Short-term cardiovascular oscillations in man: measuring and modelling the physiologies, *Journal of Physiology*. 542: 669–683.

Cole, K. S., H. A. Antosiewicz and P. Rabinowitz (1955) Automatic computation of nerve excitation, *SIAM Journal on Applied Mathematics*. 3: 153–172.

Cole, K. S. and H. J. Curtis (1940) Electric impedance of the squid giant axon during activity, *Journal of General Physiology*. 22: 649–670.

Colegrove, S. L., M. A. Albrecht and D. D. Friel (2000) Quantitative analysis of mitochondrial Ca^{2+} uptake and release pathways in sympathetic neurons. Reconstruction of the recovery after depolarization-evoked $[Ca^{2+}]_i$ elevations, *Journal of General Physiology*. 115: 371–388.

Colijn, C. and M. C. Mackey (2005a) A mathematical model of hematopoiesis–I. Periodic chronic myelogenous leukemia, *Journal of Theoretical Biology*. 237: 117–132.

Colijn, C. and M. C. Mackey (2005b) A mathematical model of hematopoiesis: II. Cyclical neutropenia, *Journal of Theoretical Biology*. 237: 133–146.

Colli-Franzone, P., L. Guerri and S. Rovida (1990) Wavefront propagation in an activation model of the anisotropic cardiac tissue: asymptotic analysis and numerical simulations, *Journal of Mathematical Biology*. 28: 121–176.

Colli-Franzone, P., L. Guerri and B. Taccardi (1993) Spread of excitation in a myocardial volume: simulation studies in a model of anisotropic ventricular muscle activated by a point stimulation, *Journal of Cardiovascular Electrophysiology*. 4: 144–160.

Colquhoun, D. (1994) *Practical analysis of single-channel records*. In: Microelectrode Techniques: The Plymouth Workshop Handbook (Second Edition), Ed: D. Ogden, Mill Hill, London.

Colquhoun, D. (2006) The quantitative analysis of drug–receptor interactions: a short history, *Trends in Pharmacological Sciences*. 27: 149–157.

Colquhoun, D. and A. G. Hawkes (1977) Relaxation and fluctuations of membrane currents that flow through drug-operated channels, *Proceedings of the Royal Society of London B*. 199: 231–262.

Colquhoun, D. and A. G. Hawkes (1981) On the stochastic properties of single ion channels, *Proceedings of the Royal Society of London B*. 211: 205–235.

Colquhoun, D. and A. G. Hawkes (1982) On the stochastic properties of bursts of single ion channel openings and of clusters of bursts, *Philosophical Transactions of the Royal Society of London B*. 300: 1–59.

Colquhoun, D. and A. G. Hawkes (1994) *The interpretation of single channel recordings*. In: Microelectrode Techniques: The Plymouth Workshop Handbook (Second Edition), Ed: D. Ogden, Mill Hill, London.

Coombes, S. (2001) The effect of ion pumps on the speed of travelling waves in the fire-diffuse-fire model of Ca^{2+} release, *Bulletin of Mathematical Biology*. 63: 1–20.

Coombes, S. and P. C. Bressloff (2003) Saltatory waves in the spike-diffuse-spike model of active dendritic spines, *Physical Review Letters*. 91: 028102.

Coombes, S., R. Hinch and Y. Timofeeva (2004) Receptors, sparks and waves in a fire-diffuse-fire framework for calcium release, *Progress in Biophysics and Molecular Biology*. 85: 197–216.

Coombes, S. and Y. Timofeeva (2003) Sparks and waves in a stochastic fire-diffuse-fire model of Ca^{2+} release, *Physical Review E — Statistical, Nonlinear and Soft Matter Physics*. 68: 021915.

Cornell-Bell, A. H., S. M. Finkbeiner, M. S. Cooper and S. J. Smith (1990) Glutamate induces calcium waves in cultured astrocytes: long-range glial signaling, *Science*. 247: 470–473.

Cornish-Bowden, A. and R. Eisenthal (1974) Statistical considerations in the estimation of enzyme kinetic parameters by the direct linear plot and other methods, *Biochemical Journal*. 139: 721–730.

Costa, K. D., P. J. Hunter, J. M. Rogers, J. M. Guccione, L. K. Waldman and A. D. McCulloch (1996a) A three-dimensional finite element method for large elastic deformations of ventricular myocardium:

I–Cylindrical and spherical polar coordinates, *Journal of Biomechanical Engineering*. 118: 452–463.

Costa, K. D., P. J. Hunter, J. S. Wayne, L. K. Waldman, J. M. Guccione and A. D. McCulloch (1996b) A three-dimensional finite element method for large elastic deformations of ventricular myocardium: II–Prolate spheroidal coordinates, *Journal of Biomechanical Engineering*. 118: 464–472.

Courant, R. and D. Hilbert (1953) *Methods of Mathematical Physics*: Wiley-Interscience, New York.

Courtemanche, M., J. P. Keener and L. Glass (1993) Instabilities of a propagating pulse in a ring of excitable media, *Physical Review Letters*. 70: 2182–2185.

Courtemanche, M., J. P. Keener and L. Glass (1996) A delay equation representation of pulse circulation on a ring in excitable media, *SIAM Journal on Applied Mathematics*. 56: 119–142.

Courtemanche, M., R. J. Ramirez and S. Nattel (1998) Ionic mechanisms underlying human atrial action potential properties: insights from a mathematical model, *American Journal of Physiology — Heart and Circulatory Physiology*. 275: H301–321.

Courtemanche, M. and A. T. Winfree (1991) Re-entrant rotating waves in a Beeler–Reuter-based model of 2-dimensional cardiac electrical activity, *International Journal of Bifurcation and Chaos*. 1: 431–444.

Cox, S. J. (2004) Estimating the location and time course of synaptic input from multi-site potential recordings, *Journal of Computational Neuroscience*. 17: 225–243.

Crawford, A. C. and R. Fettiplace (1981) An electrical tuning mechanism in turtle cochlear hair cells, *Journal of Physiology*. 312: 377–412.

Curran, P. F. and J. R. MacIntosh (1962) A model system for biological water transport, *Nature*. 193: 347–348.

Cuthbertson, K. S. R. and T. R. Chay (1991) Modelling receptor-controlled intracellular calcium oscillators, *Cell Calcium*. 12: 97–109.

Cytrynbaum, E. and J. P. Keener (2002) Stability conditions for the traveling pulse: Modifying the restitution hypothesis, *Chaos*. 12: 788–799.

Dörlochter, M. and H. Stieve (1997) The Limulus ventral photoreceptor: light response and the role of calcium in a classic preparation, *Progress in Neurobiology*. 53: 451–515.

Dallos, P., C. D. Geisler, J. W. Matthews, M. A. Ruggero and C. R. Steele, Eds. (1990)

The Mechanics and Biophysics of Hearing. Lecture Notes in Biomathematics, 87: Springer-Verlag, Berlin, Heidelberg, New York.

Dallos, P., A. N. Popper and R. R. Fay, Eds. (1996) *The Cochlea*. Springer-Verlag, New York.

Daly, S. J. and R. I. Normann (1985) Temporal information processing in cones: effects of light adaptation on temporal summation and modulation, *Vision Research*. 25: 1197–1206.

Danø, S., P. G. Sorensen and F. Hynne (1999) Sustained oscillations in living cells, *Nature*. 402: 320–322.

Dani, J. A. and D. G. Levitt (1990) Diffusion and kinetic approaches to describe permeation in ionic channels, *Journal of Theoretical Biology*. 146: 289–301.

Danielsen, M. and J. T. Ottesen (2001) Describing the pumping heart as a pressure source, *Journal of Theoretical Biology*. 212: 71–81.

Dargan, S. L. and I. Parker (2003) Buffer kinetics shape the spatiotemporal patterns of IP_3-evoked Ca^{2+} signals, *Journal of Physiology*. 553: 775–788.

Davis, B. O., N. Holtz and J. C. Davis (1985) *Conceptual Human Physiology*: C.E. Merrill Pub. Co., Columbus.

Davis, R. C., L. Garafolo and F. P. Gault (1957) An exploration of abdominal potential, *Journal of Comparative and Physiological Psychology*. 52: 519–523.

Dawis, S. M., R. M. Graeff, R. A. Heyman, T. F. Walseth and N. D. Goldberg (1988) Regulation of cyclic GMP metabolism in toad photoreceptors, *Journal of Biological Chemistry*. 263: 8771–8785.

Dawson, D. C. (1992) *Water transport: principles and perspectives*. In: The Kidney: Physiology and Pathophysiology, Ed: D. W. Seldin and G. Giebisch, Raven Press, New York.

Dawson, D. C. and N. W. Richards (1990) Basolateral K conductance: role in regulation of NaCl absorption and secretion, *American Journal of Physiology — Cell Physiology*. 259: C181–195.

Dayan, P. and L. F. Abbott (2001) *Theoretical Neuroscience: Computational and Mathematical Modeling of Neural Systems*: MIT Press, Cambridge, MA.

de Beus, A. M., T. L. Fabry and H. M. Lacker (1993) A gastric acid secretion model, *Biophysical Journal*. 65: 362–378.

De Gaetano, A. and O. Arino (2000) Mathematical modelling of the intravenous glucose tolerance test, *Journal of Mathematical Biology*. 40: 136–168.

de Schutter, E., Ed. (2000) *Computational Neuroscience: Realistic Modeling for Experimentalists*. CRC Press, Boca Raton, Florida.

de Vries, G. (1995) *Analysis of models of bursting electrical activity in pancreatic beta cells. PhD Thesis, Department of Mathematics*: University of British Columbia, Vancouver.

De Young, G. W. and J. Keizer (1992) A single pool IP_3-receptor based model for agonist stimulated Ca^{2+} oscillations, *Proceedings of the National Academy of Sciences USA*. 89: 9895–9899.

deBoer, R. W., J. M. Karemaker and J. Strackee (1987) Hemodynamic fluctuations and baroreflex sensitivity in humans: a beat-to-beat model, *American Journal of Physiology — Heart and Circulatory Physiology*. 253: H680–689.

DeFronzo, R. A., J. D. Tobin and R. Andres (1979) Glucose clamp technique: a method for quantifying insulin secretion and resistance, *American Journal of Physiology — Endocrinology and Metabolism*. 237: E214–223.

del Castillo, J. and B. Katz (1954) Quantal components of the end-plate potential, *Journal of Physiology*. 124: 560–573.

del Castillo, J. and B. Katz (1957) Interaction at end-plate receptors between different choline derivatives, *Proceedings of the Royal Society of London B*. 146: 369–381.

Del Negro, C. A., C. G. Wilson, R. J. Butera, H. Rigatto and J. C. Smith (2002) Periodicity, mixed-mode oscillations, and quasiperiodicity in a rhythm-generating neural network, *Biophysical Journal*. 82: 206–214.

Demer, L. L., C. M. Wortham, E. R. Dirksen and M. J. Sanderson (1993) Mechanical stimulation induces intercellular calcium signalling in bovine aortic endothelial cells, *American Journal of Physiology — Heart and Circulatory Physiology*. 33: H2094–2102.

Deonier, R. C., S. Tavaré and M. S. Waterman (2004) *Computational Genome Analysis: An Introduction*: Springer-Verlag, New York.

Destexhe, A., Z. F. Mainen and T. J. Sejnowski (1998) *Kinetic models of synaptic transmission*. In: Methods in Neuronal Modeling, Ed: C. Koch and I. Segev, MIT Press, Cambridge, MA.

Detwiler, P. B. and A. L. Hodgkin (1979) Electrical coupling between cones in the turtle retina, *Journal of Physiology*. 291: 75–100.

DeVries, G. W., A. I. Cohen, O. H. Lowry and J. A. Ferendelli (1979) Cyclic nucleotides in the cone-dominant ground squirrel retina, *Experimental Eye Research*. 29: 315–321.

Diamant, N. E. and A. Bortoff (1969) Nature of the intestinal slow-wave frequency gradient, *American Journal of Physiology*. 216: 301–307.

Diamant, N. E., P. K. Rose and E. J. Davison (1970) Computer simulation of intestinal slow-wave frequency gradient, *American Journal of Physiology*. 219: 1684–1690.

Diamond, J. M. and W. H. Bossert (1967) Standing-gradient osmotic flow. A mechanism for coupling of water and solute transport in epithelia, *Journal of General Physiology*. 50: 2061–2083.

DiFrancesco, D. and D. Noble (1985) A model of cardiac electrical activity incorporating ionic pumps and concentration changes, *Philosophical Transactions of the Royal Society of London B*. 307: 353–398.

Dixon, M. and E. C. Webb (1979) *Enzymes (Third Edition)*: Academic Press, New York.

Doan, T., A. Mendez, P. B. Detwiler, J. Chen and F. Rieke (2006) Multiple phosphorylation sites confer reproducibility of the rod's single-photon responses, *Science*. 313: 530–533.

Dockery, J. D. and J. P. Keener (1989) Diffusive effects on dispersion in excitable media, *SIAM Journal on Applied Mathematics*. 49: 539–566.

Dodd, A. N., J. Love and A. A. Webb (2005a) The plant clock shows its metal: circadian regulation of cytosolic free Ca^{2+}, *Trends in Plant Science*. 10: 15–21.

Dodd, A. N., N. Salathia, A. Hall, E. Kevei, R. Toth, F. Nagy, J. M. Hibberd, A. J. Millar and A. A. Webb (2005b) Plant circadian clocks increase photosynthesis, growth, survival, and competitive advantage, *Science*. 309: 630–633.

Doedel, E. (1986) *Software for continuation and bifurcation problems in ordinary differential equations*: California Institute of Technology, Pasadena.

Dolmetsch, R. E., K. Xu and R. S. Lewis (1998) Calcium oscillations increase the efficiency and specificity of gene expression, *Nature*. 392: 933–936.

Domijan, M., R. Murray and J. Sneyd (2006) Dynamic probing of the mechanisms underlying calcium oscillations, *Journal of Nonlinear Science*. 16: 483–506.

Duffy, M. R., N. F. Britton and J. D. Murray (1980) Spiral wave solutions of practical reaction–diffusion systems, *SIAM Journal on Applied Mathematics*. 39: 8–13.

Dunlap, J. C. (1998) Circadian rhythms; an end in the beginning, *Science*. 280: 1548–1549.

Dunlap, J. C. (1999) Molecular bases for circadian clocks, *Cell*. 96: 271–290.

Dupont, G. and C. Erneux (1997) Simulations of the effects of inositol 1,4,5-trisphosphate 3-kinase and 5-phosphatase activities on Ca^{2+} oscillations, *Cell Calcium*. 22: 321–331.

Dupont, G., O. Koukoui, C. Clair, C. Erneux, S. Swillens and L. Combettes (2003) Ca^{2+} oscillations in hepatocytes do not require the modulation of $InsP_3$ 3-kinase activity by Ca^{2+}, *FEBS Letters*. 534: 101–105.

Dupont, G., T. Tordjmann, C. Clair, S. Swillens, M. Claret and L. Combettes (2000) Mechanism of receptor-oriented intercellular calcium wave propagation in hepatocytes, *FASEB Journal*. 14: 279–289.

Durand, D. (1984) The somatic shunt cable model for neurons, *Biophysical Journal*. 46: 645–653.

Durrett, R. (2002) *Probability Models for DNA Sequence Evolution*: Springer Verlag, New York.

Eason, J. and N. Trayanova (2002) Phase singularities and termination of spiral wave reentry, *Journal of Cardiovascular Electrophysiology*. 13: 672–679.

Eatock, R. A. (2000) Adaptation in hair cells, *Annual Review of Neuroscience*. 23: 285–314.

Eaton, W. A., E. R. Henry, J. Hofrichter and A. Mozzarelli (1999) Is cooperative oxygen binding by hemoglobin really understood?, *Nature Structural Biology*. 6: 351–358.

Ebihara, L. and E. A. Johnson (1980) Fast sodium current in cardiac muscle, a quantitative description, *Biophysical Journal*. 32: 779–790.

Echenim, N., D. Monniaux, M. Sorine and F. Clément (2005) Multi-scale modeling of the follicle selection process in the ovary, *Mathematical Biosciences*. 198: 57–79.

Edelstein-Keshet, L. (1988) *Mathematical Models in Biology*: McGraw-Hill, New York.

Efimov, I. R., F. Aguel, Y. Cheng, B. Wollenzier and N. Trayanova (2000a) Virtual electrode polarization in the far field: implications for external defibrillation, *American Journal of Physiology — Heart and Circulatory Physiology*. 279: H1055–1070.

Efimov, I. R., Y. Cheng, D. R. Van Wagoner, T. Mazgalev and P. J. Tchou (1998) Virtual electrode-induced phase singularity: a basic mechanism of defibrillation failure, *Circulation Research*. 82: 918–925.

Efimov, I. R., R. A. Gray and B. J. Roth (2000b) Virtual electrodes and deexcitation: new insights into fibrillation induction and defibrillation, *Journal of Cardiovascular Electrophysiology*. 11: 339–353.

Eguiluz, V. M., M. Ospeck, Y. Choe, A. J. Hudspeth and M. O. Magnasco (2000) Essential nonlinearities in hearing, *Physical Review Letters*. 84: 5232–5235.

Einstein, A. (1906) Eine neue Bestimmung der Moleküldimensionen, *Annals of Physics*. 19: 289.

Eisenberg, E. and L. E. Greene (1980) The relation of muscle biochemistry to muscle physiology, *Annual Review of Physiology*. 42: 293–309.

Eisenberg, E. and T. L. Hill (1978) A cross-bridge model of muscle contraction, *Progress in Biophysics and Molecular Biology*. 33: 55–82.

Eisenthal, R. and A. Cornish-Bowden (1974) A new graphical method for estimating enzyme kinetic parameters, *Biochemical Journal*. 139: 715–720.

Elston, T., H. Wang and G. Oster (1998) Energy transduction in ATP synthase, *Nature*. 391: 510–513.

Elston, T. C. (2000) Models of post-translational protein translocation, *Biophysical Journal*. 79: 2235–2251.

Elston, T. C. and C. S. Peskin (2000) The role of protein flexibility in molecular motor function: coupled diffusion in a tilted periodic potential, *SIAM Journal on Applied Mathematics*. 60: 842–867.

Endo, M., M. Tanaka and Y. Ogawa (1970) Calcium-induced release of calcium from the sarcoplasmic reticulum of skinned skeletal muscle fibres, *Nature*. 228: 34–36.

Engel, E., A. Peskoff, G. L. Kauffman and M. I. Grossman (1984) Analysis of hydrogen ion concentration in the gastric gel mucus layer, *American Journal of Physiology — Gastrointestinal and Liver Physiology*. 247: G321–338.

Ermentrout, B. (2002) *Simulating, Analyzing, and Animating Dynamical Systems: A Guide to Xppaut for Researchers and Students*: SIAM, Philadelphia.

Ermentrout, G. B. and N. Kopell (1984) Frequency plateaus in a chain of weakly coupled oscillators, *SIAM Journal on Mathematical Analysis*. 15: 215–237.

Ewens, W. J. and G. Grant (2005) *Statistical Methods in Bioinformatics: An Introduction*: Springer Verlag, New York.

Eyring, H., R. Lumry and J. W. Woodbury (1949) Some applications of modern rate theory to physiological systems, *Record of Chemical Progress*. 10: 100–114.

Fabiato, A. (1983) Calcium-induced release of calcium from the cardiac sarcoplasmic reticulum, *American Journal of Physiology — Cell Physiology*. 245: C1–14.

Faddy, M. J. and R. G. Gosden (1995) A mathematical model of follicle dynamics in the human ovary, *Human Reproduction*. 10: 770–775.

Fain, G. and H. R. Matthews (1990) Calcium and the mechanism of light adaptation in vertebrate photoreceptors, *Trends in Neuroscience*. 13: 378–384.

Fain, G. L. (1976) Sensitivity of toad rods: Dependence on wave-length and background illumination, *Journal of Physiology*. 261: 71–101.

Fain, G. L., H. R. Matthews, M. C. Cornwall and Y. Koutalos (2001) Adaptation in vertebrate photoreceptors, *Physiological Reviews*. 81: 117–151.

Fajmut, A., M. Brumen and S. Schuster (2005) Theoretical model of the interactions between Ca^{2+}, calmodulin and myosin light chain kinase, *FEBS Letters*. 579: 4361–4366.

Falcke, M. (2003a) Buffers and oscillations in intracellular Ca^{2+} dynamics, *Biophysical Journal*. 84: 28–41.

Falcke, M. (2003b) On the role of stochastic channel behavior in intracellular Ca^{2+} dynamics, *Biophysical Journal*. 84: 42–56.

Falcke, M. (2004) Reading the patterns in living cells–the physics of Ca^{2+} signaling, *Advances in Physics*. 53: 255–440.

Falcke, M., J. L. Hudson, P. Camacho and J. D. Lechleiter (1999) Impact of mitochondrial Ca^{2+} cycling on pattern formation and stability, *Biophysical Journal*. 77: 37–44.

Falcke, M., M. Or-Guil and M. Bar (2000) Dispersion gap and localized spiral waves in a model for intracellular Ca^{2+} dynamics, *Physical Review Letters*. 84: 4753–4756.

Fall, C. P., E. S. Marland, J. M. Wagner and J. J. Tyson, Eds. (2002) *Computational Cell Biology*: Springer-Verlag, New York.

Farkas, I., D. Helbing and T. Vicsek (2002) Mexican waves in an excitable medium, *Nature*. 419: 131–132.

Fast, V. G. and A. G. Kleber (1993) Microscopic conduction in cultured strands of neonatal rat heart cells measured with voltage-sensitive dyes, *Circulation Research*. 73: 914–925.

Fatt, P. and B. Katz (1952) Spontaneous subthreshold activity at motor nerve endings, *Journal of Physiology*. 117: 109–128.

Feldman, J. L. and C. A. Del Negro (2006) Looking for inspiration: new perspectives on respiratory rhythm, *Nature Reviews Neuroscience*. 7: 232–242.

Felmy, F., E. Neher and R. Schneggenburger (2003) The timing of phasic transmitter release is Ca^{2+}-dependent and lacks a direct influence of presynaptic membrane potential, *Proceedings of the National Academy of Sciences USA*. 100: 15200–15205.

Fenn, W. O., H. Rahn and A. B. Otis (1946) A theoretical study of the composition of the alveolar air at altitude, *American Journal of Physiology*. 146: 637–653.

Field, G. D. and F. Rieke (2002) Mechanisms regulating variability of the single photon responses of mammalian rod photoreceptors, *Neuron*. 35: 733–747.

Fields, R. D. and B. Stevens-Graham (2002) New insights into neuron–glia communication, *Science*. 298: 556–562.

Fife, P. (1979) *Mathematical Aspects of Reacting and Diffusing Systems*: Springer-Verlag, Berlin.

Fife, P. C. and J. B. McLeod (1977) The approach of solutions of nonlinear diffusion equations to travelling front solutions, *Archive for Rational Mechanics and Analysis*. 65: 335–361.

Fill, M. and J. A. Copello (2002) Ryanodine receptor calcium release channels, *Physiological Reviews*. 82: 893–922.

Fill, M., A. Zahradnikova, C. A. Villalba-Galea, I. Zahradnik, A. L. Escobar and S. Gyorke (2000) Ryanodine receptor adaptation, *Journal of General Physiology*. 116: 873–882.

Finkelstein, A. and C. S. Peskin (1984) Some unexpected consequences of a simple physical mechanism for voltage-dependent gating in biological membranes, *Biophysical Journal*. 46: 549–558.

Fishman, M. A. and A. S. Perelson (1993) Modeling T cell-antigen presenting cell interactions, *Journal of Theoretical Biology*. 160: 311–342.

Fishman, M. A. and A. S. Perelson (1994) Th1/Th2 cross regulation, *Journal of Theoretical Biology*. 170: 25–56.

FitzHugh, R. (1960) Thresholds and plateaus in the Hodgkin–Huxley nerve equations, *Journal of General Physiology*. 43: 867–896.

FitzHugh, R. (1961) Impulses and physiological states in theoretical models of nerve membrane, *Biophysical Journal*. 1: 445–466.

FitzHugh, R. (1969) *Mathematical models of excitation and propagation in nerve*. In:

Biological Engineering, Ed: H. P. Schwan, McGraw-Hill, New York.

Fletcher, H. (1951) On the dynamics of the cochlea, *Journal of the Acoustical Society of America*. 23: 637–645.

Foerster, P., S. Muller and B. Hess (1989) Critical size and curvature of wave formation in an excitable chemical medium, *Proceedings of the National Academy of Sciences USA*. 86: 6831–6834.

Fogelson, A. L. (1992) Continuum models of platelet aggregation: Formulation and mechanical properties, *SIAM Journal on Applied Mathematics*. 52: 1089–1110.

Fogelson, A. L. and R. D. Guy (2004) Platelet–wall interactions in continuum models of platelet thrombosis: formulation and numerical solution, *Mathematical Medicine and Biology*. 21: 293–334.

Fogelson, A. L. and A. L. Kuharsky (1998) Membrane binding-site density can modulate activation thresholds in enzyme systems, *Journal of Theoretical Biology*. 193: 1–18.

Fogelson, A. L. and N. Tania (2005) Coagulation under flow: the influence of flow-mediated transport on the initiation and inhibition of coagulation, *Pathophysiology of Haemostasis and Thrombosis*. 34: 91–108.

Fogelson, A. L. and R. S. Zucker (1985) Presynaptic calcium diffusion from various arrays of single channels, *Biophysical Journal*. 48: 1003–1017.

Forger, D. B. and C. S. Peskin (2003) A detailed predictive model of the mammalian circadian clock, *Proceedings of the National Academy of Sciences USA*. 100: 14806–14811.

Forger, D. B. and C. S. Peskin (2004) Model based conjectures on mammalian clock controversies, *Journal of Theoretical Biology*. 230: 533–539.

Forger, D. B. and C. S. Peskin (2005) Stochastic simulation of the mammalian circadian clock, *Proceedings of the National Academy of Sciences USA*. 102: 321–324.

Forti, S., A. Menini, G. Rispoli and V. Torre (1989) Kinetics of phototransduction in retinal rods of the newt *Triturus Cristatus*, *Journal of Physiology*. 419: 265–295.

Fowler, A. C. and G. P. Kalamangalam (2000) The role of the central chemoreceptor in causing periodic breathing, *IMA Journal of Mathematics Applied in Medicine and Biology*. 17: 147–167.

Fowler, A. C. and G. P. Kalamangalam (2002) Periodic breathing at high altitude, *IMA*

Journal of Mathematics Applied in Medicine and Biology. 19: 293–313.

Fowler, A. C., G. P. Kalamangalam and G. Kember (1993) A mathematical analysis of the Grodins model of respiratory control, *IMA Journal of Mathematics Applied in Medicine and Biology*. 10: 249–280.

Fowler, A. C. and M. J. McGuinness (2005) A delay recruitment model of the cardiovascular control system, *Journal of Mathematical Biology*. 51: 508–526.

Fox, R. F. and Y. Lu (1994) Emergent collective behavior in large numbers of globally coupled independently stochastic ion channels, *Physical Review E*. 49: 3421–3431.

Frank, O. (1899) Die Grundform des Arteriellen Pulses (see translation by Sagawa et al., 1990), *Zeitschrift für Biologie*. 37: 483–526.

Frankel, M. L. and G. I. Sivashinsky (1987) On the nonlinear diffusive theory of curved flames, *Journal de Physique (Paris)*. 48: 25–28.

Frankel, M. L. and G. I. Sivashinsky (1988) On the equation of a curved flame front, *Physica D*. 30: 28–42.

Frankenhaeuser, B. (1960a) Quantitative description of sodium currents in myelinated nerve fibres of *Xenopus laevis*, *Journal of Physiology*. 151: 491–501.

Frankenhaeuser, B. (1960b) Sodium permeability in toad nerve and in squid nerve, *Journal of Physiology*. 152: 159–166.

Frankenhaeuser, B. (1963) A quantitative description of potassium currents in myelinated nerve fibres of *Xenopus laevis*, *Journal of Physiology*. 169: 424–430.

Frazier, D. W., P. D. Wolf and R. E. Ideker (1988) Electrically induced reentry in normal myocardium–evidence of a phase singularity, *PACE — Pacing and Clinical Electrophysiology*. 11: 482.

Frazier, D. W., P. D. Wolf, J. M. Wharton, A. S. Tang, W. M. Smith and R. E. Ideker (1989) Stimulus-induced critical point. Mechanism for electrical initiation of reentry in normal canine myocardium, *Journal of Clinical Investigation*. 83: 1039–1052.

Fredkin, D. R. and J. A. Rice (1992) Bayesian restoration of single-channel patch clamp recordings, *Biometrics*. 48: 427–448.

Frenzen, C. L. and P. K. Maini (1988) Enzyme kinetics for a two-step enzymic reaction with comparable initial enzyme-substrate ratios, *Journal of Mathematical Biology*. 26: 689–703.

Friel, D. (1995) $[Ca^{2+}]_i$ oscillations in sympathetic neurons: an experimental test of a theoretical model, *Biophysical Journal*. 68: 1752–1766.

Friel, D. (2000) Mitochondria as regulators of stimulus-evoked calcium signals in neurons., *Cell Calcium*. 28: 307–316.

Fye, W. B. (1986) Carl Ludwig and the Leipzig Physiological Institute: "a factory of new knowledge", *Circulation*. 74: 920–928.

Gardiner, C. W. (2004) *Handbook of Stochastic Methods for Physics, Chemistry and the Natural Sciences (Third Edition)*: Springer-Verlag, New York.

Garfinkel, A., Y. H. Kim, O. Voroshilovsky, Z. Qu, J. R. Kil, M. H. Lee, H. S. Karagueuzian, J. N. Weiss and P. S. Chen (2000) Preventing ventricular fibrillation by flattening cardiac restitution, *Proceedings of the National Academy of Sciences USA*. 97: 6061–6066.

Gatti, R. A., W. A. Robinson, A. S. Denaire, M. Nesbit, J. J. McCullogh, M. Ballow and R. A. Good (1973) Cyclic leukocytosis in chronic myelogenous leukemia, *Blood*. 41: 771–782.

Gerhardt, M., H. Schuster and J. J. Tyson (1990) A cellular automaton model of excitable media including curvature and dispersion, *Science*. 247: 1563–1566.

Gilkey, J. C., L. F. Jaffe, E. B. Ridgway and G. T. Reynolds (1978) A free calcium wave traverses the activating egg of the medaka, *Oryzias latipes*, *Journal of Cell Biology*. 76: 448–466.

Givelberg, E. (2004) Modeling elastic shells immersed in fluid, *Communications on Pure and Applied Mathematics*. 57: 283–309.

Givelberg, E. and J. Bunn (2003) A comprehensive three-dimensional model of the cochlea, *Journal of Computational Physics*. 191: 377–391.

Glass, L., A. L. Goldberger, M. Courtemanche and A. Shrier (1987) Nonlinear dynamics, chaos and complex cardiac arrhythmias, *Proceedings of the Royal Society of London A*. 413: 9–26.

Glass, L. and D. Kaplan (1995) *Understanding Nonlinear Dynamics*: Springer-Verlag, New York.

Glass, L. and M. C. Mackey (1988) *From Clocks to Chaos*: Princeton University Press, Princeton.

Glynn, I. M. (2002) A hundred years of sodium pumping, *Annual Review of Physiology*. 64: 1–18.

Goel, P., J. Sneyd and A. Friedman (2006) Homogenization of the cell cytoplasm: the calcium bidomain equations, *SIAM Journal on*

Multiscale Modeling and Simulation. 5: 1045–1062.

Gold, T. (1948) Hearing. II. The physical basis of the action of the cochlea, *Proceedings of the Royal Society of London B*. 135: 492–498.

Goldberger, A. L. and E. Goldberger (1994) *Clinical Electrocardiography: a Simplified Approach (Fifth Edition)*: Mosby, St. Louis.

Goldbeter, A. (1995) A model for circadian oscillations in the Drosophila period protein (PER), *Proceedings of the Royal Society of London B*. 261: 319–324.

Goldbeter, A. (1996) *Biochemical Oscillations and Cellular Rhythms: the Molecular Bases of Periodic and Chaotic Behaviour*: Cambridge University Press, Cambridge.

Goldbeter, A., G. Dupont and M. J. Berridge (1990) Minimal model for signal-induced Ca^{2+} oscillations and for their frequency encoding through protein phosphorylation, *Proceedings of the National Academy of Sciences USA*. 87: 1461–1465.

Goldbeter, A. and D. E. Koshland, Jr. (1981) An amplified sensitivity arising from covalent modification in biological systems, *Proceedings of the National Academy of Sciences USA*. 78: 6840–6844.

Goldbeter, A. and R. Lefever (1972) Dissipative structures for an allosteric model; application to glycolytic oscillations, *Biophysical Journal*. 12: 1302–1315.

Gomatam, J. and P. Grindrod (1987) Three dimensional waves in excitable reaction–diffusion systems, *Journal of Mathematical Biology*. 25: 611–622.

Goodner, C. J., B. C. Walike, D. J. Koerker, J. W. Ensinck, A. C. Brown, E. W. Chideckel, J. Palmer and L. Kalnasy (1977) Insulin, glucagon, and glucose exhibit synchronous, sustained oscillations in fasting monkeys, *Science*. 195: 177–179.

Goodwin, B. C. (1965) Oscillatory behavior in enzymatic control processes, *Advances in Enzyme Regulation*. 3: 425–438.

Gordon, A. M., A. F. Huxley and F. J. Julian (1966) The variation in isometric tension with sarcomere length in vertebrate muscle fibres, *Journal of Physiology*. 184: 170–192.

Graham, N. (1989) *Visual Pattern Analyzers*: Oxford University Press, New York.

Greenstein, J. L., R. Hinch and R. L. Winslow (2006) Mechanisms of excitation–contraction coupling in an integrative model of the cardiac ventricular myocyte, *Biophysical Journal*. 90: 77–91.

Greenstein, J. L. and R. L. Winslow (2002) An integrative model of the cardiac ventricular myocyte incorporating local control of Ca^{2+} release, *Biophysical Journal*. 83: 2918–2945.

Griffith, J. S. (1968a) Mathematics of cellular control processes. I. Negative feedback to one gene, *Journal of Theoretical Biology*. 20: 202–208.

Griffith, J. S. (1968b) Mathematics of cellular control processes. II. Positive feedback to one gene, *Journal of Theoretical Biology*. 20: 209–216.

Griffith, J. S. (1971) *Mathematical Neurobiology*: Academic Press, London.

Grindrod, P. (1991) *Patterns and Waves: the Theory and Application of Reaction-Diffusion Equations*: Clarendon Press, Oxford.

Grodins, F. S., J. Buell and A. J. Bart (1967) Mathematical analysis and digital simulation of the respiratory control system, *Journal of Applied Physiology*. 22: 260–276.

Grubelnik, V., A. Z. Larsen, U. Kummer, L. F. Olsen and M. Marhl (2001) Mitochondria regulate the amplitude of simple and complex calcium oscillations, *Biophysical Chemistry*. 94: 59–74.

Grzywacz, N. M., P. Hillman and B. W. Knight (1992) Response transfer functions of Limulus ventral photoreceptors: interpretation in terms of transduction mechanisms, *Biological Cybernetics*. 66: 429–435.

Guccione, J. M. and A. D. McCulloch (1993) Mechanics of active contraction in cardiac muscle: Part I–Constitutive relations for fiber stress that describe deactivation, *Journal of Biomechanical Engineering*. 115: 72–81.

Guccione, J. M., L. K. Waldman and A. D. McCulloch (1993) Mechanics of active contraction in cardiac muscle: Part II–Cylindrical models of the systolic left ventricle, *Journal of Biomechanical Engineering*. 115: 82–90.

Guckenheimer, J. and P. Holmes (1983) *Nonlinear Oscillations, Dynamical Systems, and Bifurcations of Vector Fields*: Springer-Verlag, New York, Heidelberg, Berlin.

Guevara, M. R. and L. Glass (1982) Phase locking, period doubling bifurcations and chaos in a mathematical model of a periodically driven oscillator: a theory for the entrainment of biological oscillators and the generation of cardiac dysrhythmias, *Journal of Mathematical Biology*. 14: 1–23.

Guneroth, W. G. (1965) *Pediatric Electrocardiography*: W.B. Saunders Co., Philadelphia.

Guo, J. S. and J. C. Tsai (2006) The asymptotic behavior of solutions of the buffered bistable system, *Journal of Mathematical Biology*. 53: 179–213.

Guyton, A. C. (1963) *Circulatory Physiology: Cardiac Output and its Regulation*: W.B. Saunders, Philadelphia.

Guyton, A. C. and J. E. Hall (1996) *Textbook of Medical Physiology (Ninth Edition)*: W.B. Saunders, Philadelphia.

Höfer, T. (1999) Model of intercellular calcium oscillations in hepatocytes: synchronization of heterogeneous cells, *Biophysical Journal*. 77: 1244–1256.

Höfer, T., H. Nathansen, M. Lohning, A. Radbruch and R. Heinrich (2002) GATA-3 transcriptional imprinting in Th2 lymphocytes: a mathematical model, *Proceedings of the National Academy of Sciences USA*. 99: 9364–9368.

Höfer, T., A. Politi and R. Heinrich (2001) Intercellular Ca^{2+} wave propagation through gap-junctional Ca^{2+} diffusion: a theoretical study, *Biophysical Journal*. 80: 75–87.

Höfer, T., L. Venance and C. Giaume (2002) Control and plasticity of intercellular calcium waves in astrocytes: a modeling approach, *Journal of Neuroscience*. 22: 4850–4859.

Haberman, R. (2004) *Applied Partial Differential Equations : with Fourier Series and Boundary Value Problems*: Pearson Prentice Hall, Upper Saddle River, N.J.

Hai, C. M. and R. A. Murphy (1989a) Ca^{2+}, crossbridge phosphorylation, and contraction, *Annual Review of Physiology*. 51: 285–298.

Hai, C. M. and R. A. Murphy (1989b) Cross-bridge dephosphorylation and relaxation of vascular smooth muscle, *American Journal of Physiology — Cell Physiology*. 256: C282–287.

Hajnóczky, G. and A. P. Thomas (1997) Minimal requirements for calcium oscillations driven by the IP_3 receptor, *Embo Journal*. 16: 3533–3543.

Hale, J. K. and H. Koçak (1991) *Dynamics and Bifurcations*: Springer-Verlag, New York, Berlin, Heidelberg.

Hall, G. M., S. Bahar and D. J. Gauthier (1999) Prevalence of rate-dependent behaviors in cardiac muscle, *Physical Review Letters*. 82: 2995–2998.

Hamer, R. D. (2000a) Computational analysis of vertebrate phototransduction: combined quantitative and qualitative modeling of dark- and light-adapted responses in amphibian rods, *Visual Neuroscience*. 17: 679–699.

Hamer, R. D. (2000b) Analysis of Ca^{++}-dependent gain changes in PDE activation in vertebrate rod phototransduction, *Molecular Vision*. 6: 265–286.

Hamer, R. D., S. C. Nicholas, D. Tranchina, T. D. Lamb and J. L. Jarvinen (2005) Toward a unified model of vertebrate rod phototransduction, *Visual Neuroscience*. 22: 417–436.

Hamer, R. D., S. C. Nicholas, D. Tranchina, P. A. Liebman and T. D. Lamb (2003) Multiple steps of phosphorylation of activated rhodopsin can account for the reproducibility of vertebrate rod single-photon responses, *Journal of General Physiology*. 122: 419–444.

Hamer, R. D. and C. W. Tyler (1995) Phototransduction: modeling the primate cone flash response, *Visual Neuroscience*. 12: 1063–1082.

Hamill, O. P., A. Marty, E. Neher, B. Sakmann and F. J. Sigworth (1981) Improved patch-clamp techniques for high-resolution current recording from cells and cell-free membrane patches, *Pflügers Arch*. 391: 85–100.

Hargrave, P. A., K. P. Hoffman and U. B. Kaupp, Eds. (1992) *Signal Transduction in Photoreceptor Cells*: Springer-Verlag, Berlin.

Harootunian, A. T., J. P. Kao and R. Y. Tsien (1988) Agonist-induced calcium oscillations in depolarized fibroblasts and their manipulation by photoreleased $Ins(1,4,5)P_3$, Ca^{++}, and Ca^{++} buffer, *Cold Spring Harbor Symposia on Quantitative Biology*. 53: 935–943.

Hartline, H. K. and B. W. Knight, Jr. (1974) The processing of visual information in a simple retina, *Annals of the New York Academy of Sciences*. 231: 12–18.

Hartline, H. K. and F. Ratliff (1957) Inhibitory interaction of receptor units in the eye of Limulus, *Journal of General Physiology*. 40: 357–376.

Hartline, H. K. and F. Ratliff (1958) Spatial summation of inhibitory influences in the eye of Limulus, and the mutual interaction of receptor units, *Journal of General Physiology*. 41: 1049–1066.

Hartline, H. K., H. G. Wagner and F. Ratliff (1956) Inhibition in the eye of Limulus, *Journal of General Physiology*. 39: 651–673.

Hassenstein, B. and W. Reichardt (1956) Systemtheorische analyse der zeit-, reihenfolgen- und vorzeichenauswertung bei der bewegungsperzeption des rüsselkäfers

Chlorophanus, *Zeitschrift für Naturforschung B*. 11: 513–524.

Hastings, S., J. Tyson and D. Webster (1977) Existence of periodic solutions for negative feedback cellular control systems, *Journal of Differential Equations*. 25: 39–64.

Hastings, S. P. (1975) The existence of progressive wave solutions to the Hodgkin–Huxley equations, *Archive for Rational Mechanics and Analysis*. 60: 229–257.

Haurie, C., D. C. Dale and M. C. Mackey (1998) Cyclical neutropenia and other periodic hematological disorders: a review of mechanisms and mathematical models, *Blood*. 92: 2629–2640.

Haurie, C., D. C. Dale, R. Rudnicki and M. C. Mackey (2000) Modeling complex neutrophil dynamics in the grey collie, *Journal of Theoretical Biology*. 204: 505–519.

Haurie, C., R. Person, D. C. Dale and M. C. Mackey (1999) Hematopoietic dynamics in grey collies, *Experimental Hematology*. 27: 1139–1148.

Hearn, T., C. Haurie and M. C. Mackey (1998) Cyclical neutropenia and the peripheral control of white blood cell production, *Journal of Theoretical Biology*. 192: 167–181.

Hearon, J. Z. (1948) The kinetics of blood coagulation, *Bulletin of Mathematical Biology*. 10: 175–186.

Heineken, F. G., H. M. Tsuchiya and R. Aris (1967) On the mathematical status of the pseudo-steady state hypothesis of biochemical kinetics, *Mathematical Biosciences*. 1: 95–113.

Helmholtz, H. L. F. (1875) *On the Sensations of Tone as a Physiological Basis for the Theory of Music*: Longmans, Green and Co., London.

Henriquez, C. S. (1993) Simulating the electrical behavior of cardiac tissue using the bidomain model, *CRC Critical Reviews in Biomedical Engineering*. 21: 1–77.

Henriquez, C. S., A. L. Muzikant and C. K. Smoak (1996) Anisotropy, fiber curvature, and bath loading effects on activation in thin and thick cardiac tissue preparations: simulations in a three-dimensional bidomain model, *Journal of Cardiovascular Electrophysiology*. 7: 424–444.

Hering, E. (1869) Uber den einfluss der atmung auf den kreislauf i. Uber athenbewegungen des gefasssystems., *Sitzungsberichte Kaiserlich Akad Wissenschaft Mathemat-Naturwissenschaft Classe*. 60: 829–856.

Hess, B. and A. Boiteux (1973) *Substrate Control of Glycolytic Oscillations*. In: Biological and Biochemical Oscillators, Ed: B. Chance, E. K. Pye, A. K. Ghosh and B. Hess, Academic Press, New York.

Hilgemann, D. W. (2004) New insights into the molecular and cellular workings of the cardiac Na^+/Ca^{2+} exchanger, *American Journal of Physiology — Cell Physiology*. 287: C1167–1172.

Hill, A. V. (1938) The heat of shortening and the dynamic constants of muscle, *Proceedings of the Royal Society of London B*. 126: 136–195.

Hill, T. L. (1974) Theoretical formalism for the sliding filament model of contraction of striated muscle. Part I., *Progress in Biophysics and Molecular Biology*. 28: 267–340.

Hill, T. L. (1975) Theoretical formalism for the sliding filament model of contraction of striated muscle. Part II., *Progress in Biophysics and Molecular Biology*. 29: 105–159.

Hille, B. (2001) *Ionic Channels of Excitable Membranes (Third Edition)*: Sinauer, Sunderland, MA.

Hille, B. and W. Schwartz (1978) Potassium channels as multi-ion single-file pores, *Journal of General Physiology*. 72: 409–442.

Himmel, D. M. and T. R. Chay (1987) Theoretical studies on the electrical activity of pancreatic β-cells as a function of glucose, *Biophysical Journal*. 51: 89–107.

Hindmarsh, J. L. and R. M. Rose (1982) A model of the nerve impulse using two first order differential equations, *Nature*. 296: 162–164.

Hindmarsh, J. L. and R. M. Rose (1984) A model of neuronal bursting using three coupled first order differential equations, *Proceedings of the Royal Society of London B*. 221: 87–102.

Hirose, K., S. Kadowaki, M. Tanabe, H. Takeshima and M. Iino (1999) Spatiotemporal dynamics of inositol 1,4,5-trisphosphate that underlies complex Ca^{2+} mobilization patterns, *Science*. 284: 1527–1530.

Hirsch, M. W., C. C. Pugh and M. Shub (1977) *Invariant Manifolds*: Springer-Verlag, New York.

Hirsch, M. W. and S. Smale (1974) *Differential Equations, Dynamical Systems and Linear Algebra*: Academic Press, New York.

Hirst, G. D. and F. R. Edwards (2004) Role of interstitial cells of Cajal in the control of gastric motility, *Journal of Pharmacological Sciences*. 96: 1–10.

Hirst, G. D. and S. M. Ward (2003) Interstitial cells: involvement in rhythmicity and neural control of gut smooth muscle, *Journal of Physiology*. 550: 337–346.

Hodgkin, A. L. (1976) Chance and design in electrophysiology: an informal account of certain experiments on nerve carried out between 1934 and 1952, *Journal of Physiology*. 263: 1–21.

Hodgkin, A. L. and A. F. Huxley (1952a) Currents carried by sodium and potassium ions through the membrane of the giant axon of *Loligo*, *Journal of Physiology*. 116: 449–472.

Hodgkin, A. L. and A. F. Huxley (1952b) The components of membrane conductance in the giant axon of *Loligo*, *Journal of Physiology*. 116: 473–496.

Hodgkin, A. L. and A. F. Huxley (1952c) The dual effect of membrane potential on sodium conductance in the giant axon of *Loligo*, *Journal of Physiology*. 116: 497–506.

Hodgkin, A. L. and A. F. Huxley (1952d) A quantitative description of membrane current and its application to conduction and excitation in nerve, *Journal of Physiology*. 117: 500–544.

Hodgkin, A. L., A. F. Huxley and B. Katz (1952) Measurement of current-voltage relations in the membrane of the giant axon of *Loligo*, *Journal of Physiology*. 116: 424–448.

Hodgkin, A. L. and B. Katz (1949) The effect of sodium ions on the electrical activity of the giant axon of the squid, *Journal of Physiology*. 108: 37–77.

Hodgkin, A. L. and R. D. Keynes (1955) The potassium permeability of a giant nerve fibre, *Journal of Physiology*. 128: 61–88.

Hodgkin, A. L. and B. J. Nunn (1988) Control of light-sensitive current in salamander rods, *Journal of Physiology*. 403: 439–471.

Hodgkin, A. L. and W. A. H. Rushton (1946) The electrical constants of a crustacean nerve fibre, *Proceedings of the Royal Society of London B*. 133: 444–479.

Hodgson, M. E. A. and P. J. Green (1999) Bayesian choice among Markov models of ion channels using Markov chain Monte Carlo, *Proceedings of the Royal Society of London A*. 455: 3425–3448.

Hofer, A. M. and E. M. Brown (2003) Extracellular calcium sensing and signalling, *Nature Reviews Molecular Cell Biology*. 4: 530–538.

Holmes, M. H. (1980a) An analysis of a low-frequency model of the cochlea, *Journal of the Acoustical Society of America*. 68: 482–488.

Holmes, M. H. (1980b) Low frequency asymptotics for a hydroelastic model of the cochlea, *SIAM Journal on Applied Mathematics*. 38: 445–456.

Holmes, M. H. (1982) A mathematical model of the dynamics of the inner ear, *Journal of Fluid Mechanics*. 116: 59–75.

Holmes, M. H. (1995) *Introduction to Perturbation Methods*: Springer-Verlag, New York.

Holstein-Rathlou, N. H. (1993) Oscillations and chaos in renal blood flow control, *Journal of the American Society of Nephrology*. 4: 1275–1287.

Holstein-Rathlou, N. H. and P. P. Leyssac (1987) Oscillations in the proximal intratubular pressure: a mathematical model, *American Journal of Physiology — Renal, Fluid and Electrolyte Physiology*. 252: F560–572.

Holstein-Rathlou, N. H. and D. J. Marsh (1989) Oscillations of tubular pressure, flow, and distal chloride concentration in rats, *American Journal of Physiology — Renal, Fluid and Electrolyte Physiology*. 256: F1007–1014.

Holstein-Rathlou, N. H. and D. J. Marsh (1990) A dynamic model of the tubuloglomerular feedback mechanism, *American Journal of Physiology — Renal, Fluid and Electrolyte Physiology*. 258: F1448–1459.

Holstein-Rathlou, N. H. and D. J. Marsh (1994) A dynamic model of renal blood flow autoregulation, *Bulletin of Mathematical Biology*. 56: 411–429.

Holstein-Rathlou, N. H., A. J. Wagner and D. J. Marsh (1991) Tubuloglomerular feedback dynamics and renal blood flow autoregulation in rats, *American Journal of Physiology — Renal, Fluid and Electrolyte Physiology*. 260: F53–68.

Hood, D. C. (1998) Lower-level visual processing and models of light adaptation, *Annual Review of Psychology*. 49: 503–535.

Hood, D. C. and M. A. Finkelstein (1986) *Sensitivity to light*. In: Handbook of Perception and Human Performance, Volume 1: Sensory Processes and Perception (Chapter 5), Ed: K. R. Boff, L. Kaufman and J. P. Thomas, Wiley, New York.

Hooks, D. A., K. A. Tomlinson, S. G. Marsden, I. J. LeGrice, B. H. Smaill, A. J. Pullan and P. J. Hunter (2002) Cardiac microstructure: implications for electrical propagation and defibrillation in the heart, *Circulation Research*. 91: 331–338.

Hoppensteadt, F. C. and J. P. Keener (1982) Phase locking of biological clocks, *Journal of Mathematical Biology*. 15: 339–349.

Hoppensteadt, F. C. and C. S. Peskin (2001) *Modeling and Simulation in Medicine and the Life Sciences*: Springer-Verlag, New York.

Houart, G., G. Dupont and A. Goldbeter (1999) Bursting, chaos and birhythmicity originating from self-modulation of the inositol 1,4,5-trisphosphate signal in a model for intracellular Ca^{2+} oscillations, *Bulletin of Mathematical Biology*. 61: 507–530.

Howard, J. (2001) *Mechanics of Motor Proteins and the Cytoskeleton*: Sinauer Associates, Sunderland, MA.

Howard, J., A. J. Hudspeth and R. D. Vale (1989) Movement of microtubules by single kinesin molecules, *Nature*. 342: 154–158.

Hudspeth, A. (1997) Mechanical amplification of stimuli by hair cells, *Current Opinion in Neurobiology*. 7: 480–486.

Hudspeth, A. J. (1985) The cellular basis of hearing: the biophysics of hair cells, *Science*. 230: 745–752.

Hudspeth, A. J. (1989) How the ear's works work, *Nature*. 341: 397–404.

Hudspeth, A. J. (2005) How the ear's works work: mechanoelectrical transduction and amplification by hair cells, *Comptes Rendus des Seances de la Société de Biologie et de ses Filiales*. 328: 155–162.

Hudspeth, A. J., Y. Choe, A. D. Mehta and P. Martin (2000) Putting ion channels to work: mechanoelectrical transduction, adaptation, and amplification by hair cells, *Proceedings of the National Academy of Sciences USA*. 97: 11765–11772.

Hudspeth, A. J. and P. G. Gillespie (1994) Pulling springs to tune transduction: adaptation by hair cells, *Neuron*. 12: 1–9.

Hudspeth, A. J. and R. S. Lewis (1988a) Kinetic analysis of voltage- and ion-dependent conductances in saccular hair cells of the bull-frog, *Rana Catesbeiana*, *Journal of Physiology*. 400: 237–274.

Hudspeth, A. J. and R. S. Lewis (1988b) A model for electrical resonance and frequency tuning in saccular hair cells of the bull-frog, *Rana Catesbeiana*, *Journal of Physiology*. 400: 275–297.

Huertas, M. A. and G. D. Smith (2007) The dynamics of luminal depletion and the stochastic gating of Ca^{2+}-activated Ca^{2+} channels and release sites, *Journal of Theoretical Biology*. 246: 332–354.

Hunter, P. J. (1995) Myocardial constitutive laws for continuum mechanics models of the heart, *Advances in Experimental Medicine and Biology*. 382: 303–318.

Hunter, P. J., A. D. McCulloch and H. E. ter Keurs (1998) Modelling the mechanical properties of cardiac muscle, *Progress in Biophysics and Molecular Biology*. 69: 289–331.

Hunter, P. J., A. J. Pullan and B. H. Smaill (2003) Modeling total heart function, *Annual Review of Biomedical Engineering*. 5: 147–177.

Huntsman, L. L., E. O. Attinger and A. Noordergraaf (1978) *Metabolic autoregulation of blood flow in skeletal muscle*. In: Cardiovascular System Dynamics, Ed: J. Baan, A. Noordergraaf and J. Raines, MIT Press, Cambridge MA.

Huxley, A. F. (1957) Muscle structure and theories of contraction, *Progress in Biophysics*. 7: 255–318.

Huxley, A. F. and R. M. Simmons (1971) Proposed mechanism of force generation in striated muscle, *Nature*. 233: 533–538.

Iacobas, D. A., S. O. Suadicani, D. C. Spray and E. Scemes (2006) A stochastic two-dimensional model of intercellular Ca^{2+} wave spread in glia, *Biophysical Journal*. 90: 24–41.

Imredy, J. P. and D. T. Yue (1994) Mechanism of Ca^{2+}-sensitive inactivation of L-type Ca^{2+} channels, *Neuron*. 12: 1301–1318.

Inselberg, A. and R. S. Chadwick (1976) Mathematical model of the cochlea. I: formulation and solution, *SIAM Journal on Applied Mathematics*. 30: 149–163.

Irving, M., J. Maylie, N. L. Sizto and W. K. Chandler (1990) Intracellular diffusion in the presence of mobile buffers: application to proton movement in muscle, *Biophysical Journal*. 57: 717–721.

Iyer, A. N. and R. A. Gray (2001) An experimentalist's approach to accurate localization of phase singularities during reentry, *Annals of Biomedical Engineering*. 29: 47–59.

Izhikevich, E. M. (2000) Neural excitability, spiking and bursting, *International Journal of Bifurcation and Chaos*. 10: 1171–1266.

Izu, L. T., W. G. Wier and C. W. Balke (2001) Evolution of cardiac calcium waves from stochastic calcium sparks, *Biophysical Journal*. 80: 103–120.

Jack, J. J. B., D. Noble and R. W. Tsien (1975) *Electric Current Flow in Excitable Cells*: Oxford University Press, Oxford.

Jacob, F. and J. Monod (1961) Genetic regulatory mechanisms in the synthesis of proteins, *Journal of Molecular Biology*. 3: 318–356.

Jacob, F., D. Perrin, C. Sanchez and J. Monod (1960) L'opéron : groupe de gène à expression par un opérateur, *Comptes Rendus des*

Sceances de la Société de Biologie et de ses Filiales. 250: 1727–1729.

Jaffrin, M.-Y. and C. G. Caro (1995) *Biological flows*: Plenum Press, New York.

Jafri, M. S. and J. Keizer (1995) On the roles of Ca^{2+} diffusion, Ca^{2+} buffers and the endoplasmic reticulum in IP_3-induced Ca^{2+} waves, *Biophysical Journal*. 69: 2139–2153.

Jafri, M. S., J. J. Rice and R. L. Winslow (1998) Cardiac Ca^{2+} dynamics: the roles of ryanodine receptor adaptation and sarcoplasmic reticulum load, *Biophysical Journal*. 74: 1149–1168.

Jahnke, W., C. Henze and A. T. Winfree (1988) Chemical vortex dynamics in three-dimensional excitable media, *Nature*. 336: 662–665.

Jahnke, W. and A. T. Winfree (1991) A survey of spiral-wave behaviors in the Oregonator model, *International Journal of Bifurcation and Chaos*. 1: 445–466.

Jakobsson, E. (1980) Interactions of cell volume, membrane potential, and membrane transport parameters, *American Journal of Physiology — Cell Physiology*. 238: C196–206.

Janeway, C. A., P. Travers, M. Walport and M. Shlomchik (2001) *Immunobiology: The Immune System in Health and Disease (Fifth Edition)*: Garland Publishing, New York.

Jelić, S., Z. Čupić and L. Kolar-Anić (2005) Mathematical modeling of the hypothalamic–pituitary–adrenal system activity, *Mathematical Biosciences*. 197: 173–187.

Jesty, J., E. Beltrami and G. Willems (1993) Mathematical analysis of a proteolytic positive-feedback loop: dependence of lag time and enzyme yields on the initial conditions and kinetic parameters, *Biochemistry*. 32: 6266–6274.

Jewell, B. R. and D. R. Wilkie (1958) An analysis of the mechanical components in frog's striated muscle, *Journal of Physiology*. 143: 515–540.

Jones, C. K. R. T. (1984) Stability of the traveling wave solutions of the FitzHugh–Nagumo system, *Transactions of the American Mathematical Society*. 286: 431–469.

Jones, K. C. and K. G. Mann (1994) A model for the tissue factor pathway to thrombin. II. A mathematical simulation, *Journal of Biological Chemistry*. 269: 23367–23373.

Joseph, I. M., Y. Zavros, J. L. Merchant and D. Kirschner (2003) A model for integrative study of human gastric acid secretion, *Journal of Applied Physiology*. 94: 1602–1618.

Julian, F. J. (1969) Activation in a skeletal muscle contraction model with a modification for insect fibrillar muscle, *Biophysical Journal*. 9: 547–570.

Julien, C. (2006) The enigma of Mayer waves: Facts and models, *Cardiovascular Research*. 70: 12–21.

Jung, P., A. Cornell-Bell, K. S. Madden and F. Moss (1998) Noise-induced spiral waves in astrocyte syncytia show evidence of self-organized criticality, *Journal of Neurophysiology*. 79: 1098–1101.

Jung, P., A. Cornell-Bell, F. Moss, S. Kadar, J. Wang and K. Showalter (1998) Noise sustained waves in subexcitable media: From chemical waves to brain waves, *Chaos*. 8: 567–575.

Just, A. (2006) Mechanisms of renal blood flow autoregulation: dynamics and contributions, *American Journal of Physiology — Regulatory Integrative and Comparative Physiology*. 292: 1–17.

Kang, T. M. and D. W. Hilgemann (2004) Multiple transport modes of the cardiac Na^+/Ca^{2+} exchanger, *Nature*. 427: 544–548.

Kaplan, W. (1981) *Advanced Engineering Mathematics*: Addison-Wesley, Reading, MA.

Karma, A. (1993) Spiral breakup in model equations of action potential propagation in cardiac tissue, *Physical Review Letters*. 71: 1103–1106.

Karma, A. (1994) Electrical alternans and spiral wave breakup in cardiac tissue, *Chaos*. 4: 461–472.

Katz, B. and R. Miledi (1968) The role of calcium in neuromuscular facilitation, *Journal of Physiology*. 195: 481–492.

Keener, J. P. (1980a) Waves in excitable media, *SIAM Journal on Applied Mathematics*. 39: 528–548.

Keener, J. P. (1980b) Chaotic behavior in piecewise continuous difference equations, *Transactions of the American Mathematical Society*. 261: 589–604.

Keener, J. P. (1981) On cardiac arrhythmias: AV conduction block, *Journal of Mathematical Biology*. 12: 215–225.

Keener, J. P. (1983) Analog circuitry for the van der Pol and FitzHugh–Nagumo equation, *IEEE Transactions on Systems, Man and Cybernetics*. SMC-13: 1010–1014.

Keener, J. P. (1986) A geometrical theory for spiral waves in excitable media, *SIAM Journal on Applied Mathematics*. 46: 1039–1056.

Keener, J. P. (1987) Propagation and its failure in coupled systems of discrete excitable cells, *SIAM Journal on Applied Mathematics*. 47: 556–572.

Keener, J. P. (1988) The dynamics of three dimensional scroll waves in excitable media, *Physica D*. 31: 269–276.

Keener, J. P. (1991a) An eikonal–curvature equation for action potential propagation in myocardium, *Journal of Mathematical Biology*. 29: 629–651.

Keener, J. P. (1991b) The effects of discrete gap junctional coupling on propagation in myocardium, *Journal of Theoretical Biology*. 148: 49–82.

Keener, J. P. (1992) The core of the spiral, *SIAM Journal on Applied Mathematics*. 52: 1372–1390.

Keener, J. P. (1994) Symmetric spirals in media with relaxation kinetics and two diffusing species, *Physica D*. 70: 61–73.

Keener, J. P. (1998) *Principles of Applied Mathematics, Transformation and Approximation (Second Edition)*: Perseus Books, Cambridge, Massachusetts.

Keener, J. P. (2000a) Homogenization and propagation in the bistable equation, *Physica D*. 136: 1–17.

Keener, J. P. (2000b) Propagation of waves in an excitable medium with discrete release sites, *SIAM Journal on Applied Mathematics*. 61: 317–334.

Keener, J. P. (2004) The topology of defibrillation, *Journal of Theoretical Biology*. 230: 459–473.

Keener, J. P. (2006) Stochastic calcium oscillations, *Mathematical Medicine and Biology*. 23: 1–25.

Keener, J. P. and E. Cytrynbaum (2003) The effect of spatial scale of resistive inhomogeneity on defibrillation of cardiac tissue, *Journal of Theoretical Biology*. 223: 233–248.

Keener, J. P. and L. Glass (1984) Global bifurcations of a periodically forced oscillator, *Journal of Mathematical Biology*. 21: 175–190.

Keener, J. P., F. C. Hoppensteadt and J. Rinzel (1981) Integrate and fire models of nerve membrane response to oscillatory input, *SIAM Journal on Applied Mathematics*. 41: 503–517.

Keener, J. P. and A. V. Panfilov (1995) *Three-dimensional propagation in the heart: the effects of geometry and fiber orientation on propagation in myocardium*. In: Cardiac Electrophysiology From Cell to Bedside, Ed: D. P. Zipes and J. Jalife, Saunders, Philadelphia PA.

Keener, J. P. and A. V. Panfilov (1996) A biophysical model for defibrillation of cardiac tissue, *Biophysical Journal*. 71: 1335–1345.

Keener, J. P. and A. V. Panfilov (1997) *The effects of geometry and fibre orientation on propagation and extracellular potentials in myocardium*. In: Computational Biology of the Heart, Ed: A. V. Panfilov and A. V. Holden, John Wiley and Sons, New York.

Keener, J. P. and J. J. Tyson (1986) Spiral waves in the Belousov–Zhabotinsky reaction, *Physica D*. 21: 307–324.

Keener, J. P. and J. J. Tyson (1992) The dynamics of scroll waves in excitable media, *SIAM Review*. 34: 1–39.

Keizer, J. and G. DeYoung (1994) Simplification of a realistic model of IP_3-induced Ca^{2+} oscillations, *Journal of Theoretical Biology*. 166: 431–442.

Keizer, J. and L. Levine (1996) Ryanodine receptor adaptation and Ca^{2+}-induced Ca^{2+} release-dependent Ca^{2+} oscillations, *Biophysical Journal*. 71: 3477–3487.

Keizer, J. and G. Magnus (1989) ATP-sensitive potassium channel and bursting in the pancreatic beta cell, *Biophysical Journal*. 56: 229–242.

Keizer, J. and G. D. Smith (1998) Spark-to-wave transition: saltatory transmission of calcium waves in cardiac myocytes, *Biophysical Chemistry*. 72: 87–100.

Keizer, J., G. D. Smith, S. Ponce-Dawson and J. E. Pearson (1998) Saltatory propagation of Ca^{2+} waves by Ca^{2+} sparks, *Biophysical Journal*. 75: 595–600.

Keizer, J. and P. Smolen (1991) Bursting electrical activity in pancreatic β-cells caused by Ca^{2+} and voltage-inactivated Ca^{2+} channels, *Proceedings of the National Academy of Sciences USA*. 88: 3897–3901.

Keller, E. F. and L. A. Segel (1971) Models for chemotaxis, *Journal of Theoretical Biology*. 30: 225–234.

Kernevez, J.-P. (1980) *Enzyme Mathematics*: North-Holland Publishing Company, Amsterdam, New York.

Kessler, D. A. and R. Kupferman (1996) Spirals in excitable media: the free-boundary limit with diffusion, *Physica D*. 97: 509–516.

Kevorkian, J. (2000) *Partial Differential Equations: Analytical Solution Techniques*: Springer, New York.

Kevorkian, J. and J. D. Cole (1996) *Multiple Scale and Singular Perturbation Methods*: Springer-Verlag, New York.

Khoo, M. C., R. E. Kronauer, K. P. Strohl and A. S. Slutsky (1982) Factors inducing periodic breathing in humans: a general model, *Journal of Applied Physiology*. 53: 644–659.

Kidd, J. F., K. E. Fogarty, R. A. Tuft and P. Thorn (1999) The role of Ca^{2+} feedback in shaping InsP$_3$-evoked Ca^{2+} signals in mouse pancreatic acinar cells, *Journal of Physiology*. 520: 187–201.

Kim, W. T., M. G. Rioult and A. H. Cornell-Bell (1994) Glutamate-induced calcium signaling in astrocytes, *Glia*. 11: 173–184.

Klingauf, J. and E. Neher (1997) Modeling buffered Ca^{2+} diffusion near the membrane: implications for secretion in neuroendocrine cells, *Biophysical Journal*. 72: 674–690.

Kluger, Y., Z. Lian, X. Zhang, P. E. Newburger and S. M. Weissman (2004) A panorama of lineage-specific transcription in hematopoiesis, *Bioessays*. 26: 1276–1287.

Knepper, M. A. and F. C. Rector, Jr. (1991) *Urinary concentration and dilution*. In: The Kidney (4th edition) Volume 1, Ed: B. M. Brenner and F. C. Rector, Jr., Saunders, Philadelphia.

Knight, B. W. (1972) Dynamics of encoding a population of neurons, *Journal of General Physiology*. 59: 734–766.

Knight, B. W., J. I. Toyoda and F. A. Dodge, Jr. (1970) A quantitative description of the dynamics of excitation and inhibition in the eye of Limulus, *Journal of General Physiology*. 56: 421–437.

Knisley, S. B., T. F. Blitchington, B. C. Hill, A. O. Grant, W. M. Smith, T. C. Pilkington and R. E. Ideker (1993) Optical measurements of transmembrane potential changes during electric field stimulation of ventricular cells, *Circulation Research*. 72: 255–270.

Knobil, E. (1981) Patterns of hormonal signals and hormone action, *New England Journal of Medicine*. 305: 1582–1583.

Knox, B. E., P. N. Devreotes, A. Goldbeter and L. A. Segel (1986) A molecular mechanism for sensory adaptation based on ligand-induced receptor modification, *Proceedings of the National Academy of Sciences USA*. 83: 2345–2349.

Koch, C. (1999) *Biophysics of Computation: Information Processing in Single Neurons*: Oxford University Press, Oxford.

Koch, C. and I. Segev, Eds. (1998) *Methods in Neuronal Modeling: From Ions to Networks (Third Edition)*. MIT Press, Cambridge, MA.

Koch, K.-W. and L. Stryer (1988) Highly cooperative feedback control of retinal rod guanylate cyclase by calcium ions, *Nature*. 334: 64–66.

Koefoed-Johnsen, V. and H. H. Ussing (1958) The nature of the frog skin potential, *Acta Physiologica Scandinavica*. 42: 298–308.

Koenigsberger, M., R. Sauser, J. L. Beny and J. J. Meister (2006) Effects of arterial wall stress on vasomotion, *Biophysical Journal*. 91: 1663–1674.

Koenigsberger, M., R. Sauser, M. Lamboley, J. L. Beny and J. J. Meister (2004) Ca^{2+} dynamics in a population of smooth muscle cells: modeling the recruitment and synchronization, *Biophysical Journal*. 87: 92–104.

Koenigsberger, M., R. Sauser and J. J. Meister (2005) Emergent properties of electrically coupled smooth muscle cells, *Bulletin of Mathematical Biology*. 67: 1253–1272.

Kohler, H.-H. and K. Heckman (1979) Unidirectional fluxes in saturated single-file pores of biological and artificial membranes I: pores containing no more than one vacancy, *Journal of Theoretical Biology*. 79: 381–401.

Kopell, N. and G. B. Ermentrout (1986) Subcellular oscillations and bursting, *Mathematical Biosciences*. 78: 265–291.

Kopell, N. and L. N. Howard (1973) Plane wave solutions to reaction–diffusion equations, *Studies in Applied Mathematics*. 52: 291–328.

Koshland, D. E., Jr. and K. Hamadani (2002) Proteomics and models for enzyme cooperativity, *Journal of Biological Chemistry*. 277: 46841–46844.

Koshland, D. E., Jr., G. Nemethy and D. Filmer (1966) Comparison of experimental binding data and theoretical models in proteins containing subunits, *Biochemistry*. 5: 365–385.

Koutalos, Y., K. Nakatani and K. W. Yau (1995) The cGMP–phosphodiesterase and its contribution to sensitivity regulation in retinal rods, *Journal of General Physiology*. 106: 891–921.

Kramers, H. A. (1940) Brownian motion in a field of force and the diffusion model of chemical reactions, *Physica*. 7: 284–304.

Krane, D. E. and M. L. Raymer (2003) *Fundamental Concepts of Bioinformatics*: Benjamin Cummings, San Francisco.

Krassowska, W., T. C. Pilkington and R. E. Ideker (1987) Periodic conductivity as a mechanism for cardiac stimulation and defibrillation, *IEEE Transactions in Biomedical Engineering*. 34: 555–560.

Krausz, H. I. and K.-I. Naka (1980) Spatiotemporal testing and modeling of catfish retinal neurons, *Biophysical Journal*. 29: 13–36.

Kreyszig, E. (1994) *Advanced Engineering Mathematics (Seventh Edition)*: John Wiley and Sons, New York.

Kucera, J. P., S. Rohr and Y. Rudy (2002) Localization of sodium channels in intercalated disks modulates cardiac conduction, *Circulation Research*. 91: 1176–1182.

Kuffler, S. W. (1953) Discharge patterns and functional organization of the mammalian retina, *Journal of Neurophysiology*. 16: 37–68.

Kuffler, S. W. (1973) The single-cell approach in the visual system and the study of receptive fields, *Investigative Ophthalmology*. 12: 794–813.

Kuffler, S. W., J. G. Nicholls and R. Martin (1984) *From Neuron to Brain (Second Edition)*: Sinaeur Associates, Sunderland, MA.

Kuramoto, Y. and T. Tsuzuki (1976) Persistent propagation of concentration waves in dissipative media far from thermal equilibrium, *Progress of Theoretical Physics*. 55: 356–369.

Kuramoto, Y. and T. Yamada (1976) Pattern formation in oscillatory chemical reactions, *Progress of Theoretical Physics*. 56: 724–740.

Läuger, P. (1973) Ion transport through pores: a rate-theory analysis, *Biochimica et Biophysica Acta*. 311: 423–441.

Lacker, H. M. (1981) Regulation of ovulation number in mammals: a follicle interaction law that controls maturation, *Biophysical Journal*. 35: 433–454.

Lacker, H. M. and C. S. Peskin (1981) *Control of ovulation number in a model of ovarian follicular maturation*. In: Lectures on Mathematics in the Life Sciences, Ed: S. Childress, American Mathematical Society, Providence.

Lacker, H. M. and C. S. Peskin (1986) A mathematical method for unique determination of cross-bridge properties from steady-state mechanical and energetic experiments on macroscopic muscle, *Lectures on Mathematics in the Life Sciences*. 16: 121–153.

Lacy, A. H. (1967) The unit of insulin, *Diabetes*. 16: 198–200.

Laidler, K. J. (1969) *Theories of Chemical Reaction Rates*: McGraw-Hill, New York.

Lamb, T. D. (1981) The involvement of rod photoreceptors in dark adaptation, *Vision Research*. 21: 1773–1782.

Lamb, T. D. and E. N. Pugh (1992) A quantitative account of the activation steps involved in phototransduction in amphibian photoreceptors, *Journal of Physiology*. 449: 719–758.

Lamb, T. D. and E. J. Simon (1977) Analysis of electrical noise in turtle cones, *Journal of Physiology*. 272: 435–468.

Landy, M. S. and J. A. Movshon, Eds. (1991) *Computational Models of Visual Processing*: MIT Press, Cambridge, MA.

Lane, D. C., J. D. Murray and V. S. Manoranjan (1987) Analysis of wave phenomena in a morphogenetic mechanochemical model and an application to post-fertilisation waves on eggs, *IMA Journal of Mathematics Applied in Medicine and Biology*. 4: 309–331.

Lange, R. L. and H. H. Hecht (1962) The mechanism of Cheyne–Stokes respiration, *Journal of Clinical Investigation*. 41: 42–52.

Langer, G. A. and A. Peskoff (1996) Calcium concentration and movement in the diadic cleft space of the cardiac ventricular cell, *Biophysical Journal*. 70: 1169–1182.

Lapointe, J. Y., M. Gagnon, S. Poirier and P. Bissonnette (2002) The presence of local osmotic gradients can account for the water flux driven by the Na^+-glucose cotransporter, *Journal of Physiology*. 542: 61–62.

Layton, A. T. and H. E. Layton (2002) A numerical method for renal models that represent tubules with abrupt changes in membrane properties, *Journal of Mathematical Biology*. 45: 549–567.

Layton, A. T. and H. E. Layton (2003) A region-based model framework for the rat urine concentrating mechanism, *Bulletin of Mathematical Biology*. 65: 859–901.

Layton, A. T. and H. E. Layton (2005a) A region-based mathematical model of the urine concentrating mechanism in the rat outer medulla. I. Formulation and base-case results, *American Journal of Physiology — Renal, Fluid and Electrolyte Physiology*. 289: F1346–1366.

Layton, A. T. and H. E. Layton (2005b) A region-based mathematical model of the urine concentrating mechanism in the rat outer medulla. II. Parameter sensitivity and tubular inhomogeneity, *American Journal of Physiology — Renal, Fluid and Electrolyte Physiology*. 289: F1367–1381.

Layton, A. T., T. L. Pannabecker, W. H. Dantzler and H. E. Layton (2004) Two modes for concentrating urine in rat inner medulla, *American Journal of Physiology — Renal, Fluid and Electrolyte Physiology*. 287: F816–839.

Layton, H. E., E. B. Pitman and M. A. Knepper (1995a) A dynamic numerical method for models of the urine concentrating mechanism, *SIAM Journal on Applied Mathematics*. 55: 1390–1418.

Layton, H. E., E. B. Pitman and L. C. Moore (1991) Bifurcation analysis of TGF-mediated oscillations in SNGFR, *American Journal of Physiology — Renal, Fluid and Electrolyte Physiology*. 261: F904–919.

Layton, H. E., E. B. Pitman and L. C. Moore (1995b) Instantaneous and steady-state gains in the tubuloglomerular feedback system, *American Journal of Physiology — Renal, Fluid and Electrolyte Physiology*. 268: F163–174.

Layton, H. E., E. B. Pitman and L. C. Moore (1997) Spectral properties of the tubuloglomerular feedback system, *American Journal of Physiology — Renal, Fluid and Electrolyte Physiology*. 273: F635–649.

Layton, H. E., E. B. Pitman and L. C. Moore (2000) Limit-cycle oscillations and tubuloglomerular feedback regulation of distal sodium delivery, *American Journal of Physiology — Renal, Fluid and Electrolyte Physiology*. 278: F287–301.

LeBeau, A. P., A. B. Robson, A. E. McKinnon, R. A. Donald and J. Sneyd (1997) Generation of action potentials in a mathematical model of corticotrophs, *Biophysical Journal*. 73: 1263–1275.

LeBeau, A. P., A. B. Robson, A. E. McKinnon and J. Sneyd (1998) Analysis of a reduced model of corticotroph action potentials, *Journal of Theoretical Biology*. 192: 319–339.

Lechleiter, J. and D. Clapham (1992) Molecular mechanisms of intracellular calcium excitability in *X. laevis* oocytes, *Cell*. 69: 283–294.

Lechleiter, J., S. Girard, D. Clapham and E. Peralta (1991a) Subcellular patterns of calcium release determined by G protein-specific residues of muscarinic receptors, *Nature*. 350: 505–508.

Lechleiter, J., S. Girard, E. Peralta and D. Clapham (1991b) Spiral calcium wave propagation and annihilation in *Xenopus laevis* oocytes, *Science*. 252: 123–126.

Leloup, J. C. and A. Goldbeter (1998) A model for circadian rhythms in Drosophila incorporating the formation of a complex between the PER and TIM proteins, *Journal of Biological Rhythms*. 13: 70–87.

Leloup, J. C. and A. Goldbeter (2003) Toward a detailed computational model for the mammalian circadian clock, *Proceedings of the National Academy of Sciences USA*. 100: 7051–7056.

Leloup, J. C. and A. Goldbeter (2004) Modeling the mammalian circadian clock: sensitivity analysis and multiplicity of oscillatory mechanisms, *Journal of Theoretical Biology*. 230: 541–562.

Lenbury, Y. and P. Pornsawad (2005) A delay-differential equation model of the feedback-controlled hypothalamus-pituitary-adrenal axis in humans, *Mathematical Medicine and Biology*. 22: 15–33.

Lesser, M. B. and D. A. Berkley (1972) Fluid mechanics of the cochlea. Part I., *Journal of Fluid Mechanics*. 51: 497–512.

Levine, I. N. (2002) *Physical Chemistry (Fifth Edition)*: McGraw-Hill, Tokyo.

Levine, S. N. (1966) Enzyme amplifier kinetics, *Science*. 152: 651–653.

Lew, V. L., H. G. Ferreira and T. Moura (1979) The behaviour of transporting epithelial cells. I. Computer analysis of a basic model, *Proceedings of the Royal Society of London B*. 206: 53–83.

Lewis, M. (2005) The *lac* repressor, *CR Biologies*. 328: 521–548.

Lewis, T. J. and J. P. Keener (2000) Wave-blocking in excitable media due to regions of depressed excitability, *SIAM Journal on Applied Mathematics*. 61: 293–396.

Leyssac, P. P. and L. Baumbach (1983) An oscillating intratubular pressure response to alterations in Henle loop flow in the rat kidney, *Acta Physiologica Scandinavica*. 117: 415–419.

Li, W., J. Llopis, M. Whitney, G. Zlokarnik and R. Y. Tsien (1998) Cell-permeant caged $InsP_3$ ester shows that Ca^{2+} spike frequency can optimize gene expression, *Nature*. 392: 936–941.

Li, Y.-X. and A. Goldbeter (1989) Frequency specificity in intercellular communication: influence of patterns of periodic signaling on target cell responsiveness, *Biophysical Journal*. 55: 125–145.

Li, Y.-X., J. Keizer, S. S. Stojilkovic and J. Rinzel (1995) Ca^{2+} excitability of the ER membrane: an explanation for IP_3-induced Ca^{2+} oscillations, *American Journal of Physiology — Cell Physiology*. 269: C1079–1092.

Li, Y.-X. and J. Rinzel (1994) Equations for $InsP_3$ receptor-mediated $[Ca^{2+}]$ oscillations derived from a detailed kinetic model: a Hodgkin–Huxley like formalism, *Journal of Theoretical Biology*. 166: 461–473.

Li, Y.-X., J. Rinzel, J. Keizer and S. S. Stojilkovic (1994) Calcium oscillations in pituitary gonadotrophs: comparison of experiment and theory, *Proceedings of the National Academy of Sciences USA*. 91: 58–62.

Li, Y.-X., S. S. Stojilkovic, J. Keizer and J. Rinzel (1997) Sensing and refilling calcium stores in an excitable cell, *Biophysical Journal*. 72: 1080–1091.

Lighthill, J. (1975) *Mathematical Biofluiddynamics*: SIAM, Philadelphia, PA.

Lin, C. C. and L. A. Segel (1988) *Mathematics Applied to Deterministic Problems in the Natural Sciences*: SIAM, Philadelphia.

Lin, S. C. and D. E. Bergles (2004) Synaptic signaling between neurons and glia, *Glia*. 47: 290–298.

Liu, B.-Z. and G.-M. Deng (1991) An improved mathematical model of hormone secretion in the hypothalamo–pituitary–gonadal axis in man, *Journal of Theoretical Biology*. 150: 51–58.

Llinás, R., I. Z. Steinberg and K. Walton (1976) Presynaptic calcium currents and their relation to synaptic transmission: voltage clamp study in squid giant synapse and theoretical model for the calcium gate, *Proceedings of the National Academy of Sciences USA*. 73: 2918–2922.

Loeb, J. N. and S. Strickland (1987) Hormone binding and coupled response relationships in systems dependent on the generation of secondary mediators, *Molecular Endocrinology*. 1: 75–82.

Lombard, W. P. (1916) The Life and Work of Carl Ludwig, *Science*. 44: 363–375.

Longobardo, G., C. J. Evangelisti and N. S. Cherniack (2005) Introduction of respiratory pattern generators into models of respiratory control, *Respiratory Physiology and Neurobiology*. 148: 285–301.

Longtin, A. and J. G. Milton (1989) Modelling autonomous oscillations in the human pupil light reflex using non-linear delay-differential equations, *Bulletin of Mathematical Biology*. 51: 605–624.

Loo, D. D., E. M. Wright and T. Zeuthen (2002) Water pumps, *Journal of Physiology*. 542: 53–60.

Lugosi, E. and A. T. Winfree (1988) Simulation of wave propagation in three dimensions using Fortran on the Cyber 205, *Journal of Computational Chemistry*. 9: 689–701.

Luo, C. H. and Y. Rudy (1991) A model of the ventricular cardiac action potential; depolarization, repolarization and their interaction, *Circulation Research*. 68: 1501–1526.

Luo, C. H. and Y. Rudy (1994a) A dynamic model of the cardiac ventricular action potential; I: Simulations of ionic currents and concentration changes, *Circulation Research*. 74: 1071–1096.

Luo, C. H. and Y. Rudy (1994b) A dynamic model of the cardiac ventricular action potential; II: Afterdepolarizations, triggered activity and potentiation, *Circulation Research*. 74: 1097–1113.

Lytton, J., M. Westlin, S. E. Burk, G. E. Shull and D. H. MacLennan (1992) Functional comparisons between isoforms of the sarcoplasmic or endoplasmic reticulum family of calcium pumps, *Journal of Biological Chemistry*. 267: 14483–14489.

MacGregor, D. J. and G. Leng (2005) Modelling the hypothalamic control of growth hormone secretion, *Journal of Neuroendocrinology*. 17: 788–803.

Mackey, M. C. (1978) Unified hypothesis for the origin of aplastic anemia and periodic hematopoiesis, *Blood*. 51: 941–956.

Mackey, M. C. (1979) Periodic auto-immune hemolytic anemia: an induced dynamical disease, *Bulletin of Mathematical Biology*. 41: 829–834.

Mackey, M. C. and L. Glass (1977) Oscillation and chaos in physiological control systems, *Science*. 197: 287–289.

Mackey, M. C., C. Haurie and J. Bélair (2003) *Cell replication and control*. In: Nonlinear Dynamics in Physiology and Medicine, Ed: A. Beuter, L. Glass, M. C. Mackey and M. S. Titcombe, Springer-Verlag, New York.

Mackey, M. C. and J. G. Milton (1987) Dynamical diseases, *Annals of the New York Academy of Sciences*. 504: 16–32.

Mackey, M. C., M. Santillán and N. Yildirim (2004) Modeling operon dynamics: the tryptophan and lactose operons as paradigms, *Comptes Rendus des Seances de la Société de Biologie et de ses Filiales*. 327: 211–224.

Macknight, A. D. C. (1988) Principles of cell volume regulation, *Renal Physiology and Biochemistry*. 3–5: 114–141.

MacLennan, D. H. and E. G. Kranias (2003) Phospholamban: a crucial regulator of cardiac contractility, *Nature Reviews Molecular Cell Biology*. 4: 566–577.

MacLennan, D. H., W. J. Rice and N. M. Green (1997) The mechanism of Ca^{2+} transport by sarco(endo)plasmic reticulum Ca^{2+}-ATPases, *Journal of Biological Chemistry*. 272: 28815–28818.

Madsen, M. F., S. Dano and P. G. Sorensen (2005) On the mechanisms of glycolytic oscillations in yeast, *FEBS Journal*. 272: 2648–2660.

Maginu, K. (1985) Geometrical characteristics associated with stability and bifurcations of periodic travelling waves in reaction–diffusion equations, *SIAM Journal on Applied Mathematics*. 45: 750–774.

Magleby, K. L. and C. F. Stevens (1972) A quantitative description of end-plate currents, *Journal of Physiology*. 223: 173–197.

Magnus, G. and J. Keizer (1997) Minimal model of beta-cell mitochondrial Ca^{2+} handling, *American Journal of Physiology — Cell Physiology*. 273: C717–733.

Magnus, G. and J. Keizer (1998a) Model of beta-cell mitochondrial calcium handling and electrical activity. I. Cytoplasmic variables, *American Journal of Physiology — Cell Physiology*. 274: C1158–1173.

Magnus, G. and J. Keizer (1998b) Model of beta-cell mitochondrial calcium handling and electrical activity. II. Mitochondrial variables, *American Journal of Physiology — Cell Physiology*. 274: C1174–1184.

Mallik, R. and S. P. Gross (2004) Molecular motors: strategies to get along, *Current Biology*. 14: R971–982.

Manley, G. A. (2001) Evidence for an active process and a cochlear amplifier in nonmammals, *Journal of Neurophysiology*. 86: 541–549.

Marchant, J., N. Callamaras and I. Parker (1999) Initiation of IP$_3$-mediated Ca^{2+} waves in Xenopus oocytes, *Embo Journal*. 18: 5285–5299.

Marchant, J. S. and I. Parker (2001) Role of elementary Ca^{2+} puffs in generating repetitive Ca^{2+} oscillations, *Embo Journal*. 20: 65–76.

Marchant, J. S. and C. W. Taylor (1998) Rapid activation and partial inactivation of inositol trisphosphate receptors by inositol trisphosphate, *Biochemistry*. 37: 11524–11533.

Marhl, M., T. Haberichter, M. Brumen and R. Heinrich (2000) Complex calcium oscillations and the role of mitochondria and cytosolic proteins., *Biosystems*. 57: 75–86.

Mari, A. (2002) Mathematical modeling in glucose metabolism and insulin secretion, *Current Opinion in Clinical Nutrition and Metabolic Care*. 5: 495–501.

Mari, A., G. Pacini, E. Murphy, B. Ludvik and J. J. Nolan (2001) A model-based method for assessing insulin sensitivity from the oral glucose tolerance test, *Diabetes Care*. 24: 539–548.

Mariani, L., M. Lohning, A. Radbruch and T. Hofer (2004) Transcriptional control networks of cell differentiation: insights from helper T lymphocytes, *Progress in Biophysics and Molecular Biology*. 86: 45–76.

Marland, E. (1998) *The Dynamics of the Sarcomere*: PhD Thesis, Department of Mathematics, University of Utah, Salt Lake City.

Martin, P., A. D. Mehta and A. J. Hudspeth (2000) Negative hair-bundle stiffness betrays a mechanism for mechanical amplification by the hair cell, *Proceedings of the National Academy of Sciences USA*. 97: 12026–12031.

Matveev, V., R. Bertram and A. Sherman (2006) Residual bound Ca^{2+} can account for the effects of Ca^{2+} buffers on synaptic facilitation, *Journal of Neurophysiology*. 96: 3389–3397.

Matveev, V., A. Sherman and R. S. Zucker (2002) New and corrected simulations of synaptic facilitation, *Biophysical Journal*. 83: 1368–1373.

Matveev, V., R. S. Zucker and A. Sherman (2004) Facilitation through buffer saturation: constraints on endogenous buffering properties, *Biophysical Journal*. 86: 2691–2709.

Mayer, S. (1877) Studien zur physiologie des herzens und der blutgefasse. V. Uber spontane blutdruckschwankungen, *Sitzungsberichte Kaiserlich Akad Wissenschaft Mathemat-Naturwissenschaft Classe*. 74: 281–307.

McAllister, R. E., D. Noble and R. W. Tsien (1975) Reconstruction of the electrical activity of cardiac Purkinje fibres, *Journal of Physiology*. 251: 1–59.

McCulloch, A., L. Waldman, J. Rogers and J. Guccione (1992) Large-scale finite element analysis of the beating heart, *Critical Reviews in Biomedical Engineering*. 20: 427–449.

McCulloch, A. D. (1995) *Cardiac biomechanics*. In: Biomedical Engineering Handbook: The Electrical Engineering Handbook Series, Ed: J. D. Branzino, CRC Press, Boca Raton.

McCulloch, A. D. and G. Paternostro (2005) Cardiac systems biology, *Annals of the New York Academy of Sciences*. 1047: 283–295.

McDonald, D. A. (1974) *Blood Flow in Arteries (Second Edition)*: Arnold, London.

McKean, H. P. (1970) Nagumo's equation, *Advances in Mathematics*. 4: 209–223.

McKenzie, A. and J. Sneyd (1998) On the formation and breakup of spiral waves of calcium, *International Journal of Bifurcation and Chaos*. 8: 2003–2012.

McLachlan, R. I., N. L. Cohen, K. D. Dahl, W. J. Bremner and M. R. Soules (1990) Serum inhibin levels during the periovulatory interval in normal women: relationships with sex steroid and gonadotrophin levels, *Clinical Endocrinology*. 32: 39–48.

McNaughton, P. A. (1990) Light response of vertebrate photoreceptors, *Physiological Reviews*. 70: 847–883.

McQuarrie, D. A. (1967) *Stochastic Approach to Chemical Kinetics*: Methuen and Co., London.

Meinrenken, C. J., J. G. Borst and B. Sakmann (2003) Local routes revisited: the space and time dependence of the Ca^{2+} signal for phasic transmitter release at the rat calyx of Held, *Journal of Physiology*. 547: 665–689.

Meyer, T. and L. Stryer (1988) Molecular model for receptor-stimulated calcium spiking, *Proceedings of the National Academy of Sciences USA*. 85: 5051–5055.

Meyer, T. and L. Stryer (1991) Calcium spiking, *Annual Review of Biophysics and Biophysical Chemistry*. 20: 153–174.

Michaelis, L. and M. I. Menten (1913) Die Kinetik der Invertinwirkung, *Biochemische Zeitschrift*. 49: 333–369.

Miftakhov, R. N., G. R. Abdusheva and J. Christensen (1999a) Numerical simulation of motility patterns of the small bowel. 1. formulation of a mathematical model, *Journal of Theoretical Biology*. 197: 89–112.

Miftakhov, R. N., G. R. Abdusheva and J. Christensen (1999b) Numerical simulation of motility patterns of the small bowel. II. Comparative pharmacological validation of a mathematical model, *Journal of Theoretical Biology*. 200: 261–290.

Mijailovich, S. M., J. P. Butler and J. J. Fredberg (2000) Perturbed equilibria of myosin binding in airway smooth muscle: bond-length distributions, mechanics, and ATP metabolism, *Biophysical Journal*. 79: 2667–2681.

Milhorn, H. T., Jr. and P. E. Pulley, Jr. (1968) A theoretical study of pulmonary capillary gas exchange and venous admixture, *Biophysical Journal*. 8: 337–357.

Miller, R. N. and J. Rinzel (1981) The dependence of impulse propagation speed on firing frequency, dispersion, for the Hodgkin–Huxley model, *Biophysical Journal*. 34: 227–259.

Milton, J. (2003) *Pupil light reflex: delays and oscillations*. In: Nonlinear Dynamics in Physiology and Medicine, Ed: A. Beuter, L. Glass, M. C. Mackey and M. S. Titcombe, Springer-Verlag, New York.

Milton, J. G. and M. C. Mackey (1989) Periodic haematological diseases: mystical entities or dynamical disorders?, *Journal of the Royal College of Physicians of London*. 23: 236–241.

Mines, G. R. (1914) On circulating excitations in heart muscle and their possible relation to tachycardia and fibrillation, *Transactions of the Royal Society of Canada*. 4: 43–53.

Minorsky, N. (1962) *Nonlinear Oscillations*: Van Nostrand, New York.

Miura, R. M. (1981) *Nonlinear waves in neuronal cortical structures*. In: Nonlinear Phenomena in Physics and Biology, Ed: R. H. Enns, B. L. Jones, R. M. Miura and S. S. Rangnekar, Plenum Press, New York.

Moe, G. K., W. C. Rheinboldt and J. A. Abildskov (1964) A computer model of atrial fibrillation, *American Heart Journal*. 67: 200–220.

Mogilner, A., T. C. Elston, H. Wang and G. Oster (2002) *Molecular motors: examples*. In: Computational Cell Biology, Ed: C. P. Fall, E. S. Marland, J. M. Wagner and J. J. Tyson, Springer-Verlag, New York.

Mogilner, A. and G. Oster (1996) Cell motility driven by actin polymerization, *Biophysical Journal*. 71: 3030–3045.

Mogilner, A. and G. Oster (1999) The polymerization ratchet model explains the force–velocity relation for growing microtubules, *European Journal of Biophysics*. 28: 235–242.

Mogilner, A. and G. Oster (2003) Force generation by actin polymerization II: the elastic ratchet and tethered filaments, *Biophysical Journal*. 84: 1591–1605.

Monod, J., J. Wyman and J. P. Changeux (1965) On the nature of allosteric transition: A plausible model, *Journal of Molecular Biology*. 12: 88–118.

Morley, A. (1979) Cyclic hemopoiesis and feedback control, *Blood Cells*. 5: 283–296.

Morris, C. and H. Lecar (1981) Voltage oscillations in the barnacle giant muscle fiber, *Biophysical Journal*. 35: 193–213.

Mount, D. W. (2001) *Bioinformatics: Sequence and Genome Analysis*: Cold Spring Harbor Laboratory Press, Cold Spring Harbor, N.Y.

Mountcastle, V. B., Ed. (1974) *Medical Physiology (Thirteenth Edition)*: C.V. Mosby Co., Saint Louis.

Murphy, R. A. (1994) What is special about smooth muscle? The significance of covalent crossbridge regulation, *FASEB Journal*. 8: 311–318.

Murray, J. D. (1971) On the molecular mechanism of facilitated oxygen diffusion by haemoglobin and myoglobin, *Proceedings of the Royal Society of London B*. 178: 95–110.

Murray, J. D. (1984) *Asymptotic Analysis*: Springer-Verlag, New York.

Murray, J. D. (2002) *Mathematical Biology (Third Edition)*: Springer-Verlag, New York.

Murray, J. D. and J. Wyman (1971) Facilitated diffusion: the case of carbon monoxide, *Journal of Biological Chemistry*. 246: 5903–5906.

Nagumo, J., S. Arimoto and S. Yoshizawa (1964) An active pulse transmission line simulating nerve axon, *Proceedings of the Institute of Radio Engineers*. 50: 2061–2070.

Naka, K. I. and W. A. Rushton (1966) S-potentials from luminosity units in the retina of fish (Cyprinidae), *Journal of Physiology*. 185: 587–599.

Nakatani, K., T. Tamura and K. W. Yau (1991) Light adaptation in retinal rods of the rabbit and two other nonprimate mammals, *Journal of General Physiology*. 97: 413–435.

Nakayama, K. (1985) Biological image motion processing: a review, *Vision Research*. 25: 625–660.

Naraghi, M., T. H. Muller and E. Neher (1998) Two-dimensional determination of the cellular Ca^{2+} binding in bovine chromaffin cells, *Biophysical Journal*. 75: 1635–1647.

Naraghi, M. and E. Neher (1997) Linearized buffered Ca^{2+} diffusion in microdomains and its implications for calculation of $[Ca^{2+}]$ at the mouth of a calcium channel, *Journal of Neuroscience*. 17: 6961–6973.

Nash, M. S., K. W. Young, R. A. Challiss and S. R. Nahorski (2001) Intracellular signalling. Receptor-specific messenger oscillations, *Nature*. 413: 381–382.

Nasmyth, K. (1995) Evolution of the cell cycle, *Philosophical Transactions of the Royal Society of London B*. 349: 271–281.

Nasmyth, K. (1996) Viewpoint: putting the cell cycle in order, *Science*. 274: 1643–1645.

Nathanson, M. H., A. D. Burgstahler, A. Mennone, M. B. Fallon, C. B. Gonzalez and J. C. Saez (1995) Ca^{2+} waves are organized among hepatocytes in the intact organ, *American Journal of Physiology — Gastrointestinal and Liver Physiology*. 269: G167–171.

Nedergaard, M. (1994) Direct signaling from astrocytes to neurons in cultures of mammalian brain cells, *Science*. 263: 1768–1771.

Neher, E. (1998a) Usefulness and limitations of linear approximations to the understanding of Ca^{++} signals, *Cell Calcium*. 24: 345–357.

Neher, E. (1998b) Vesicle pools and Ca^{2+} microdomains: new tools for understanding their roles in neurotransmitter release, *Neuron*. 20: 389–399.

Nelsen, T. S. and J. C. Becker (1968) Simulation of the electrical and mechanical gradient of the small intestine, *American Journal of Physiology*. 214: 749–757.

Nesheim, M. E., R. P. Tracy and K. G. Mann (1984) "Clotspeed," a mathematical simulation of the functional properties of prothrombinase, *Journal of Biological Chemistry*. 259: 1447–1453.

Nesheim, M. E., R. P. Tracy, P. B. Tracy, D. S. Boskovic and K. G. Mann (1992) Mathematical simulation of prothrombinase, *Methods in Enzymology*. 215: 316–328.

Nesher, R. and E. Cerasi (2002) Modeling phasic insulin release: immediate and time-dependent effects of glucose, *Diabetes*. 51 Suppl 1: S53–59.

Neu, J. C. (1979) Chemical waves and the diffusive coupling of limit cycle oscillators, *SIAM Journal on Applied Mathematics*. 36: 509–515.

Neu, J. C. and W. Krassowska (1993) Homogenization of syncytial tissues, *Critical Reviews in Biomedical Engineering*. 21: 137–199.

Nicholls, J. G., A. R. Martin and B. G. Wallace (1992) *From Neuron to Brain (Third Edition)*: Sinauer Associates, Inc., Sunderland, MA.

Niederer, S. A., P. J. Hunter and N. P. Smith (2006) A quantitative analysis of cardiac myocyte relaxation: a simulation study, *Biophysical Journal*. 90: 1697–1722.

Nielsen, K., P. G. Sørensen and F. Hynne (1997) Chaos in glycolysis, *Journal of Theoretical Biology*. 186: 303–306.

Nielsen, P. M. F., I. J. LeGrice and B. H. Smaill (1991) A mathematical model of geometry and fibrous structure of the heart, *American Journal of Physiology — Heart and Circulatory Physiology*. 260: H1365–1378.

Nikonov, S., N. Engheta and E. N. Pugh, Jr. (1998) Kinetics of recovery of the dark-adapted salamander rod photoresponse, *Journal of General Physiology*. 111: 7–37.

Nobili, R., F. Mammano and J. Ashmore (1998) How well do we understand the cochlea?, *Trends in Neuroscience*. 21: 159–167.

Noble, D. (1962) A modification of the Hodgkin–Huxley equations applicable to Purkinje fiber action and pacemaker potential, *Journal of Physiology*. 160: 317–352.

Noble, D. (2002a) Modelling the heart: insights, failures and progress, *Bioessays*. 24: 1155–1163.

Noble, D. (2002b) Modeling the heart–from genes to cells to the whole organ, *Science*. 295: 1678–1682.

Noble, D. and S. J. Noble (1984) A model of sino-atrial node electrical activity using a modification of the DiFrancesco–Noble (1984) equations, *Proceedings of the Royal Society of London B*. 222: 295–304.

Nolasco, J. B. and R. W. Dahlen (1968) A graphic method for the study of alternation in cardiac action potentials, *Journal of Applied Physiology*. 25: 191–196.

Norman, R. A. and I. Perlman (1979) The effects of background illumination on the photoresponses of red and green cones, *Journal of Physiology*. 286: 491–507.

Novak, B., Z. Pataki, A. Ciliberto and J. J. Tyson (2001) Mathematical model of the cell division cycle of fission yeast, *Chaos*. 11: 277–286.

Novak, B. and J. J. Tyson (1993a) Numerical analysis of a comprehensive model of M-phase control in *Xenopus* oocyte extracts and intact embryos, *Journal of Cell Science*. 106: 1153–1168.

Novak, B. and J. J. Tyson (1993b) Modeling the cell division cycle: M phase trigger oscillations and size control, *Journal of Theoretical Biology*. 165: 101–134.

Novak, B. and J. J. Tyson (2004) A model for restriction point control of the mammalian cell cycle, *Journal of Theoretical Biology*. 230: 563–579.

Nowak, M. A. and R. M. May (2000) *Virus Dynamics: Mathematical Principles of Immunology and Virology*: Oxford University Press, Oxford.

Nowycky, M. C. and M. J. Pinter (1993) Time courses of calcium and calcium-bound buffers

following calcium influx in a model cell, *Biophysical Journal*. 64: 77–91.

Nuccitelli, R., D. L. Yim and T. Smart (1993) The sperm-induced Ca^{2+} wave following fertilization of the Xenopus egg requires the production of Ins(1,4,5)P$_3$, *Developmental Biology*. 158: 200–212.

Nunemaker, C. S., R. Bertram, A. Sherman, K. Tsaneva-Atanasova, C. R. Daniel and L. S. Satin (2006) Glucose modulates $[Ca^{2+}]_i$ oscillations in pancreatic islets via ionic and glycolytic mechanisms, *Biophysical Journal*. 91: 2082–2096.

O'Neill, P. V. (1983) *Advanced Engineering Mathematics*: Wadsworth, Belmont CA.

Ohta, T., M. Mimura and R. Kobayashi (1989) Higher dimensional localized patterns in excitable media, *Physica D*. 34: 115–144.

Olufsen, M. S., A. Nadim and L. A. Lipsitz (2002) Dynamics of cerebral blood flow regulation explained using a lumped parameter model, *American Journal of Physiology — Regulatory, Integrative, and Comparative Physiology*. 282: R611–622.

Olufsen, M. S., C. S. Peskin, W. Y. Kim, E. M. Pedersen, A. Nadim and J. Larsen (2000) Numerical simulation and experimental validation of blood flow in arteries with structured-tree outflow conditions, *Annals of Biomedical Engineering*. 28: 1281–1299.

Orrenius, S., B. Zhivotovsky and P. Nicotera (2003) Regulation of cell death: the calcium–apoptosis link, *Nature Reviews Molecular Cell Biology*. 4: 552–565.

Ortoleva, P. and J. Ross (1973) Phase waves in oscillatory chemical reactions, *Journal of Chemical Physics*. 58: 5673–5680.

Ortoleva, P. and J. Ross (1974) On a variety of wave phenomena in chemical reactions, *Journal of Chemical Physics*. 60: 5090–5107.

Osher, S. and J. A. Sethian (1988) Fronts propagating with curvature-dependent speed: algorithms based on Hamilton–Jacobi formulations, *Journal of Computational Physics*. 79: 12–49.

Otani, N. F. and R. F. Gilmour, Jr. (1997) Memory models for the electrical properties of local cardiac systems, *Journal of Theoretical Biology*. 187: 409–436.

Othmer, H. G. (1976) The qualitative dynamics of a class of biochemical control circuits, *Journal of Mathematical Biology*. 3: 53–78.

Ottesen, J. T. (1997) Modelling of the baroreflex-feedback mechanism with time-delay, *Journal of Mathematical Biology*. 36: 41–63.

Ottesen, J. T., M. S. Olufsen and J. K. Larsen (2004) *Applied Mathematical Models in Human Physiology*: SIAM, Philadelphia.

Ozbudak, E. M., M. Thattai, H. N. Lim, B. I. Shraiman and A. Van Oudenaarden (2004) Multistability in the lactose utilization network of Escherichia coli, *Nature*. 427: 737–740.

Pace, N., E. Strajman and E. L. Walker (1950) Acceleration of carbon monoxide elimination in man by high pressure oxygen, *Science*. 111: 652–654.

Panfilov, A. V. and P. Hogeweg (1995) Spiral break-up in a modified FitzHugh–Nagumo model, *Physics Letters A*. 176: 295–299.

Panfilov, A. V. and A. V. Holden (1990) Self-generation of turbulent vortices in a two-dimensional model of cardiac tissue, *Physics Letters A*. 151: 23–26.

Panfilov, A. V. and J. P. Keener (1995) Re-entry in an anatomical model of the heart, *Chaos, Solitons and Fractals*. 5: 681–689.

Papoulis, A. (1962) *The Fourier Integral and its Applications*: McGraw-Hill, New York.

Parker, I., J. Choi and Y. Yao (1996b) Elementary events of InsP$_3$-induced Ca^{2+} liberation in Xenopus oocytes: hot spots, puffs and blips, *Cell Calcium*. 20: 105–121.

Parker, I. and Y. Yao (1996) Ca^{2+} transients associated with openings of inositol trisphosphate-gated channels in Xenopus oocytes, *Journal of Physiology*. 491: 663–668.

Parker, I., Y. Yao and V. Ilyin (1996a) Fast kinetics of calcium liberation induced in *Xenopus* oocytes by photoreleased inositol trisphosphate, *Biophysical Journal*. 70: 222–237.

Parker, I., W. J. Zang and W. G. Wier (1996c) Ca^{2+} sparks involving multiple Ca^{2+} release sites along Z-lines in rat heart cells, *Journal of Physiology*. 497: 31–38.

Parnas, H., G. Hovav and I. Parnas (1989) Effect of Ca^{2+} diffusion on the time course of neurotransmitter release, *Biophysical Journal*. 55: 859–874.

Parnas, H. and L. A. Segel (1980) A theoretical explanation for some effects of calcium on the facilitation of neurotransmitter release, *Journal of Theoretical Biology*. 84: 3–29.

Parnas, H., J. C. Valle-Lisboa and L. A. Segel (2002) Can the Ca^{2+} hypothesis and the Ca^{2+}-voltage hypothesis for neurotransmitter release be reconciled?, *Proceedings of the National Academy of Sciences USA*. 99: 17149–17154.

Pate, E. (1997) *Mathematical modeling of muscle crossbridge mechanics*. In: Case Studies in Mathematical Biology, Ed: H. Othmer, F. Adler, M. Lewis and J. Dallon, Prentice Hall, Upper Saddle River, New Jersey.

Pate, E. and R. Cooke (1989) A model of crossbridge action: the effects of ATP, ADP and Pi, *Journal of Muscle Research and Cell Motility*. 10: 181–196.

Pate, E. and R. Cooke (1991) Simulation of stochastic processes in motile crossbridge systems, *Journal of Muscle Research and Cell Motility*. 12: 376–393.

Patlak, J. (1991) Molecular kinetics of voltage-dependent Na$^+$ channels., *Physiological Reviews*. 71: 1047–1080.

Patneau, D. K. and M. L. Mayer (1991) Kinetic analysis of interactions between kainate and AMPA: evidence for activation of a single receptor in mouse hippocampal neurons, *Neuron*. 6: 785–798.

Patton, R. J. and D. A. Linkens (1978) Hodgkin–Huxley type electronic modelling of gastrointestinal electrical activity, *Medical and Biological Engineering and Computing*. 16: 195–202.

Pauling, L. (1935) The oxygen equilibrium of hemoglobin and its structural interpretation, *Proceedings of the National Academy of Sciences USA*. 21: 186–191.

Pauwelussen, J. P. (1981) Nerve impulse propagation in a branching nerve system: a simple model, *Physica D*. 4: 67–88.

Payne, S. and C. Stephens (2005) The response of the cross-bridge cycle model to oscillations in intracellular calcium: A mathematical analysis, *Conference proceedings: Annual International Conference of the IEEE Engineering in Medicine and Biology Society*. 7: 7305–7308.

Pearson, J. E. and S. Ponce-Dawson (1998) Crisis on skid row, *Physica A*. 257: 141–148.

Pedley, T. J. (1980) *The Fluid Mechanics of Large Blood Vessels*: Cambridge University Press, Cambridge.

Pelce, P. and J. Sun (1991) Wave front interaction in steadily rotating spirals, *Physica D*. 48: 353–366.

Pepperberg, D. R., M. C. Cornwall, M. Kahlert, K. P. Hofmann, J. Jin, G. J. Jones and H. Ripps (1992) Light-dependent delay in the falling phase of the retinal rod photoresponse, *Visual Neuroscience*. 8: 9–18.

Perelson, A. S. (2002) Modelling viral and immune system dynamics, *Nature Reviews Immunology*. 2: 28–36.

Perlman, I. and R. A. Normann (1998) Light adaptation and sensitivity controlling mechanisms in vertebrate photoreceptors, *Progress in Retinal and Eye Research*. 17: 523–563.

Pernarowski, M. (1994) Fast subsystem bifurcations in a slowly varying Liénard system exhibiting bursting, *SIAM Journal on Applied Mathematics*. 54: 814–832.

Pernarowski, M., R. M. Miura and J. Kevorkian (1991) *The Sherman–Rinzel–Keizer model for bursting electrical activity in the pancreatic β-cell*. In: Differential Equations Models in Biology, Epidemiology and Ecology, Ed: S. Busenberg and M. Martelli, Springer-Verlag, New York.

Pernarowski, M., R. M. Miura and J. Kevorkian (1992) Perturbation techniques for models of bursting electrical activity in pancreatic β-cells, *SIAM Journal on Applied Mathematics*. 52: 1627–1650.

Perutz, M. F. (1970) Stereochemistry of cooperative effects in haemoglobin, *Nature*. 228: 726–739.

Perutz, M. F., W. Bolton, R. Diamond, H. Muirhead and H. Watson (1964) Structure of haemoglobin. An X-ray examination of reduced horse haemoglobin, *Nature*. 203: 687–690.

Peskin, C. S. (1975) *Mathematical Aspects of Heart Physiology*: Courant Institute of Mathematical Sciences Lecture Notes, New York.

Peskin, C. S. (1976) *Partial Differential Equations in Biology*: Courant Institute of Mathematical Sciences Lecture Notes, New York.

Peskin, C. S. (1981) Lectures on mathematical aspects of physiology, *AMS Lectures in Applied Mathematics*. 19: 38–69.

Peskin, C. S. (1991) *Mathematical Aspects of Neurophysiology*: Courant Institute of Mathematical Sciences Lecture Notes, New York.

Peskin, C. S. (2002) The immersed boundary method, *Acta Numerica*. 11: 479–517.

Peskin, C. S. and D. M. McQueen (1989) A three-dimensional computational method for blood flow in the heart. I. Immersed elastic fibers in a viscous incompressible fluid., *Journal of Computational Physics*. 81: 372–405.

Peskin, C. S. and D. M. McQueen (1992) Cardiac fluid dynamics, *Critical Reviews in Biomedical Engineering*. 20: 451–459.

Peskin, C. S., G. M. Odell and G. F. Oster (1993) Cellular motions and thermal fluctuations: the Brownian ratchet, *Biophysical Journal*. 65: 316–324.

Peskin, C. S. and G. Oster (1995) Coordinated hydrolysis explains the mechanical behavior of kinesin, *Biophysical Journal*. 68: 202S–210.

Peskin, C. S. and G. F. Oster (1995) Force production by depolymerizing microtubules: load–velocity curves and run-pause statistics, *Biophysical Journal*. 69: 2268–2276.

Peskoff, A. and G. A. Langer (1998) Calcium concentration and movement in the ventricular cardiac cell during an excitation–contraction cycle, *Biophysical Journal*. 74: 153–174.

Peskoff, A., J. A. Post and G. A. Langer (1992) Sarcolemmal calcium binding sites in heart: II. Mathematical model for diffusion of calcium released from the sarcoplasmic reticulum into the diadic region, *Journal of Membrane Biology*. 129: 59–69.

Peterson, L. C. and B. P. Bogert (1950) A dynamical theory of the cochlea, *Journal of the Acoustical Society of America*. 22: 369–381.

Pickles, J. O. (1982) *An Introduction to the Physiology of Hearing*: Academic Press, London.

Pitman, E. B., R. M. Zaritski, K. J. Kesseler, L. C. Moore and H. E. Layton (2004) Feedback-mediated dynamics in two coupled nephrons, *Bulletin of Mathematical Biology*. 66: 1463–1492.

Plant, R. E. (1981) Bifurcation and resonance in a model for bursting nerve cells, *Journal of Mathematical Biology*. 11: 15–32.

Podolsky, R. J. and A. C. Nolan (1972) *Cross-bridge properties derived from physiological studies of frog muscle fibres*. In: Contractility of Muscle Cells and Related Processes, Ed: R. J. Podolsky, Prentice Hall, Englewood Cliffs, NJ.

Podolsky, R. J. and A. C. Nolan (1973) *Muscle Contraction Transients, Cross-Bridge Kinetics and the Fenn Effect.*: 37th Cold Spring Harbor Symposium of Quantitative Biology, Cold Spring Harbor, New York.

Podolsky, R. J., A. C. Nolan and S. A. Zaveler (1969) Cross-bridge properties derived from muscle isotonic velocity transients, *Proceedings of the National Academy of Sciences USA*. 64: 504–511.

Politi, A., L. D. Gaspers, A. P. Thomas and T. Hofer (2006) Models of IP_3 and Ca^{2+} oscillations: frequency encoding and identification of underlying feedbacks, *Biophysical Journal*. 90: 3120–3133.

Pollack, G. H. (1976) Intercellular coupling in the atrioventricular node and other tissues of the rabbit heart, *Journal of Physiology*. 255: 275–298.

Ponce-Dawson, S., J. Keizer and J. E. Pearson (1999) Fire-diffuse-fire model of dynamics of intracellular calcium waves, *Proceedings of the National Academy of Sciences USA*. 96: 6060–6063.

Preston, G. M., T. P. Carroll, W. B. Guggino and P. Agre (1992) Appearance of water channels in *Xenopus* oocytes expressing red cell CHIP28 protein, *Science*. 256: 385–387.

Pries, A. R. and T. W. Secomb (2000) Microcirculatory network structures and models, *Annals of Biomedical Engineering*. 28: 916–921.

Pries, A. R. and T. W. Secomb (2005) Control of blood vessel structure: insights from theoretical models, *American Journal of Physiology — Heart and Circulatory Physiology*. 288: H1010–1015.

Pries, A. R., T. W. Secomb and P. Gaehtgens (1996) Biophysical aspects of blood flow in the microvasculature, *Cardiovascular Research*. 32: 654–667.

Pugh, E. N. and T. D. Lamb (1990) Cyclic GMP and calcium: messengers of excitation and adaptation in vertebrate photoreceptors, *Vision Research*. 30: 1923–1948.

Pullan, A., L. Cheng, R. Yassi and M. Buist (2004) Modelling gastrointestinal bioelectric activity, *Progress in Biophysics and Molecular Biology*. 85: 523–550.

Qian, H. (2000) The mathematical theory of molecular motor movement and chemomechanical energy transduction, *Journal of Mathematical Chemistry*. 27: 219–234.

Qu, Z., A. Garfinkel, P. S. Chen and J. N. Weiss (2000) Mechanisms of discordant alternans and induction of reentry in simulated cardiac tissue, *Circulation*. 102: 1664–1670.

Röttingen, J. and J. G. Iversen (2000) Ruled by waves? Intracellular and intercellular calcium signalling, *Acta Physiologica Scandinavica*. 169: 203–219.

Rahn, H. (1949) A concept of mean alveolar air and the ventilation–bloodflow relationships during pulmonary gas exchange, *American Journal of Physiology*. 158: 21–30.

Rall, W. (1957) Membrane time constant of motoneurons, *Science*. 126: 454.

Rall, W. (1959) Branching dendritic trees and motoneuron membrane resistivity, *Experimental Neurology*. 2: 491–527.

Rall, W. (1960) Membrane potential transients and membrane time constant of motoneurons, *Experimental Neurology*. 2: 503–532.

Rall, W. (1969) Time constants and electrotonic length of membrane cylinders and neurons, *Biophysical Journal*. 9: 1483–1508.

Rall, W. (1977) *Core conductor theory and cable properties of neurons*. In: Handbook of Physiology The Nervous System I, Ed: J. M. Brookhart and V. B. Mountcastle, American Physiological Society, Bethesda, MD.

Ramamoorthy, S., N. V. Deo and K. Grosh (2007) A mechano-electro-acoustical model for the cochlea: response to acoustic stimuli, *Journal of the Acoustical Society of America*. 121: 2758–2773.

Ramanan, S. V. and P. R. Brink (1990) Exact solution of a model of diffusion in an infinite chain or monolayer of cells coupled by gap junctions, *Biophysical Journal*. 58: 631–639.

Ramirez, J. M. and D. W. Richter (1996) The neuronal mechanisms of respiratory rhythm generation, *Current Opinion in Neurobiology*. 6: 817–825.

Rand, R. H. and P. J. Holmes (1980) Bifurcation of periodic motions in two weakly coupled van der Pol oscillators, *Journal of Non-linear Mechanics*. 15: 387–399.

Ranke, O. F. (1950) Theory of operation of the cochlea: A contribution to the hydrodynamics of the cochlea, *Journal of the Acoustical Society of America*. 22: 772–777.

Rapp, P. E. (1975) A theoretical investigation of a large class of biochemical oscillations, *Mathematical Biosciences*. 25: 165–188.

Rapp, P. E. (1976) Mathematical techniques for the study of oscillations in biochemical control loops, *Bulletin of the Institute of Mathematics and its Applications*. 12: 11–21.

Rapp, P. E. and M. J. Berridge (1977) Oscillations in calcium–cyclic AMP control loops form the basis of pacemaker activity and other high frequency biological rhythms, *Journal of Theoretical Biology*. 66: 497–525.

Ratliff, F. (1961) *Inhibitory interaction and the detection and enhancement of contours*. In: Sensory Communication, Ed: W. A. Rosenblith, MIT Press, Cambridge, MA.

Ratliff, F. and H. K. Hartline (1959) The responses of Limulus optic nerve fibers to patterns of illumination on the receptor mosaic, *Journal of General Physiology*. 42: 1241–1255.

Ratliff, F., B. W. Knight, Jr., F. A. Dodge, Jr. and H. K. Hartline (1974) Fourier analysis of dynamics of excitation and inhibition in the eye of Limulus: amplitude, phase and distance, *Vision Research*. 14: 1155–1168.

Rauch, J. and J. Smoller (1978) Qualitative theory of the FitzHugh–Nagumo equations, *Advances in Mathematics*. 27: 12–44.

Reed, M. C. and J. J. Blum (1986) Theoretical analysis of radioactivity profiles during fast axonal transport: effects of deposition and turnover, *Cell Motility and the Cytoskeleton*. 6: 620–627.

Reeve, E. B. and A. C. Guyton, Eds. (1967) *Physical Bases of Circulatory Transport: Regulation and Exchange*: W.B. Saunders, Philadelphia.

Reichardt, W. (1961) *Autocorrelation, a principle for the evaluation of sensory information by the central nervous system*. In: Sensory Communication, Ed: W. A. Rosenblith, Cambridge, MA.

Reijenga, K. A., H. V. Westerhoff, B. N. Kholodenko and J. L. Snoep (2002) Control analysis for autonomously oscillating biochemical networks, *Biophysical Journal*. 82: 99–108.

Reimann, P. (2002) Brownian motors: Noisy transport far from equilibrium, *Physics Reports*. 361: 57–265.

Reuss, L. and B. H. Hirst (2002) Water transport controversies–an overview, *Journal of Physiology*. 542: 1–2.

Rhode, W. S. (1984) Cochlear mechanics, *Annual Review of Physiology*. 46: 231–246.

Richter, D. W. (1996) *Neural regulation of respiration: rhythmogenesis and afferent control*. In: Comprehensive Human Physiology, Ed: R. Gregor and U. Windhorst, Springer-Verlag, Berlin.

Ridgway, E. B., J. C. Gilkey and L. F. Jaffe (1977) Free calcium increases explosively in activating medaka eggs, *Proceedings of the National Academy of Sciences USA*. 74: 623–627.

Rieke, F. and D. A. Baylor (1998a) Origin of reproducibility in the responses of retinal rods to single photons, *Biophysical Journal*. 75: 1836–1857.

Rieke, F. and D. A. Baylor (1998b) Single-photon detection by rod cells of the retina, *Reviews of Modern Physics*. 70: 1027–1036.

Riley, R. L. and A. Cournand (1949) " Ideal" alveolar air and the analysis of ventilation–perfusion relationships in the lungs, *Journal of Applied Physiology*. 1: 825–847.

Riley, R. L. and A. Cournand (1951) Analysis of factors affecting partial pressures of oxygen and carbon dioxide in gas and blood of lungs; theory, *Journal of Applied Physiology*. 4: 77–101.

Rinzel, J. (1978) On repetitive activity in nerve, *Federation Proceedings*. 37: 2793–2802.

Rinzel, J. (1985) *Bursting oscillations in an excitable membrane model*. In: Ordinary and Partial Differential Equations, Ed: B. D. Sleeman and R. J. Jarvis, Springer-Verlag, New York.

Rinzel, J. (1987) *A formal classification of bursting mechanisms in excitable systems*. In: Mathematical Topics in Population Biology, Morphogenesis, and Neurosciences, Lecture Notes in Biomathematics, Vol 71, Ed: E. Teramoto and M. Yamaguti, Springer-Verlag, Berlin.

Rinzel, J. (1990) Electrical excitability of cells, theory and experiment: review of the Hodgkin–Huxley foundation and an update, *Bulletin of Mathematical Biology*. 52: 5–23.

Rinzel, J. and J. P. Keener (1983) Hopf bifurcation to repetitive activity in nerve, *SIAM Journal on Applied Mathematics*. 43: 907–922.

Rinzel, J. and J. B. Keller (1973) Traveling wave solutions of a nerve conduction equation, *Biophysical Journal*. 13: 1313–1337.

Rinzel, J. and Y. S. Lee (1986) *On different mechanisms for membrane potential bursting*. In: Nonlinear Oscillations in Biology and Chemistry, Lecture Notes in Biomathematics, Vol 66, Ed: H. G. Othmer, Springer-Verlag, New York.

Rinzel, J. and Y. S. Lee (1987) Dissection of a model for neuronal parabolic bursting, *Journal of Mathematical Biology*. 25: 653–675.

Rinzel, J. and K. Maginu (1984) *Kinematic analysis of wave pattern formation in excitable media*. In: Non-equilibrium Dynamics in Chemical Systems, Ed: A. Pacault and C. Vidal, Springer-Verlag, Berlin.

Robb-Gaspers, L. D. and A. P. Thomas (1995) Coordination of Ca^{2+} signaling by intercellular propagation of Ca^{2+} waves in the intact liver, *Journal of Biological Chemistry*. 270: 8102–8107.

Roberts, D. and A. M. Scher (1982) Effect of tissue anisotropy on extracellular potential fields in canine myocardium *in situ*, *Circulation Research*. 50: 342–351.

Robertson-Dunn, B. and D. A. Linkens (1974) A mathematical model of the slow-wave electrical activity of the human small intestine, *Medical and Biological Engineering*. 12: 750–758.

Robinson, T. F., L. Cohen-Gould and S. M. Factor (1983) Skeletal framework of mammalian heart muscle. Arrangement of inter- and pericellular connective tissue structures, *Laboratory Investigation*. 49: 482–498.

Rodieck, R. W. (1965) Quantitative analysis of cat retinal ganglion cell response to visual stimuli, *Vision Research*. 5: 583–601.

Rooney, T. A. and A. P. Thomas (1993) Intracellular calcium waves generated by Ins(1,4,5)P$_3$-dependent mechanisms, *Cell Calcium*. 14: 674–690.

Roper, P., J. Callaway and W. Armstrong (2004) Burst initiation and termination in phasic vasopressin cells of the rat supraoptic nucleus: a combined mathematical, electrical, and calcium fluorescence study, *Journal of Neuroscience*. 24: 4818–4831.

Roper, P., J. Callaway, T. Shevchenko, R. Teruyama and W. Armstrong (2003) AHP's, HAP's and DAP's: how potassium currents regulate the excitability of rat supraoptic neurones, *Journal of Computational Neuroscience*. 15: 367–389.

Rorsman, P. and G. Trube (1986) Calcium and delayed potassium currents in mouse pancreatic β-cells under voltage clamp conditions, *Journal of Physiology*. 375: 531–550.

Roth, B. J. (1992) How the anisotropy of the intracellular and extracellular conductivities influences stimulation of cardiac muscle, *Journal of Mathematical Biology*. 30: 633–646.

Roughton, F. J. W., E. C. DeLand, J. C. Kernohan and J. W. Severinghaus (1972) *Some recent studies of the oxyhaemoglobin dissociation curve of human blood under physiological conditions and the fitting of the Adair equation to the standard curve*. In: Oxygen Affinity of Hemoglobin and Red Cell Acid Base States, Ed: M. Rorth and P. Astrup, Academic Press, New York.

Roy, D. R., H. E. Layton and R. L. Jamison (1992) *Countercurrent mechanism and its regulation*. In: The Kidney: Physiology and Pathophysiology, Ed: D. W. Seldin and G. Giebisch, Raven Press, New York.

Rubinow, S. I. (1973) *Mathematical Problems in the Biological Sciences*: SIAM, Philadelphia.

Rubinow, S. I. (1975) *Introduction to Mathematical Biology*: John Wiley and Sons, New York.

Rubinow, S. I. and M. Dembo (1977) The facilitated diffusion of oxygen by hemoglobin and myoglobin, *Biophysical Journal*. 18: 29–42.

Ruggero, M. A. (1992) Responses to sound of the basilar membrane of the mammalian cochlea, *Current Opinion in Neurobiology*. 2: 449–456.

Rushmer, R. F. (1976) *Structure and Function of the Cardiovascular System (Second Edition)*: W.B. Saunders Co., Philadelphia.

Rybak, I. A., J. F. Paton and J. S. Schwaber (1997a) Modeling neural mechanisms for genesis of respiratory rhythm and pattern. I. Models of respiratory neurons, *Journal of Neurophysiology*. 77: 1994–2006.

Rybak, I. A., J. F. Paton and J. S. Schwaber (1997b) Modeling neural mechanisms for genesis of respiratory rhythm and pattern. II. Network models of the central respiratory pattern generator, *Journal of Neurophysiology*. 77: 2007–2026.

Rybak, I. A., N. A. Shevtsova, J. F. Paton, T. E. Dick, W. M. St-John, M. Morschel and M. Dutschmann (2004) Modeling the ponto-medullary respiratory network, *Respiratory Physiology and Neurobiology*. 143: 307–319.

Sabah, N. H. and R. A. Spangler (1970) Repetitive response of the Hodgkin–Huxley model for the squid giant axon, *Journal of Theoretical Biology*. 29: 155–171.

Sachs, F., F. Qin and P. Palade (1995) Models of Ca^{2+} release channel adaptation, *Science*. 267: 2010–2011.

Sagawa, K., R. K. Lie and J. Schaefer (1990) Translation of Otto Frank's Paper "Die Grundform des Arteriellen Pulses", Zeitschrift für Biologie, 37:483-526 (1899), *Journal of Molecular and Cellular Cardiology*. 22: 253–277.

Sagawa, K., H. Suga and K. Nakayama (1978) *Instantaneous pressure–volume ratio of the left ventricle versus instantaneous force–length relation of papillary muscle*. In: Cardiovascular System Dynamics, Ed: J. Baan, A. Noordergraaf and J. Raines, MIT Press, Cambridge, MA.

Sakmann, B. and E. Neher (1995) *Single-Channel Recording (Second Edition)*: Plenum Press, New York.

Sala, F. and A. Hernández-Cruz (1990) Calcium diffusion modeling in a spherical neuron: relevance of buffering properties, *Biophysical Journal*. 57: 313–324.

Sanders, K. M. (1996) A case for interstitial cells of Cajal as pacemakers and mediators of neurotransmission in the gastrointestinal tract, *Gastroenterology*. 111: 492–515.

Sanderson, M. J., A. C. Charles, S. Boitano and E. R. Dirksen (1994) Mechanisms and function of intercellular calcium signaling, *Molecular and Cellular Endocrinology*. 98: 173–187.

Sanderson, M. J., A. C. Charles and E. R. Dirksen (1990) Mechanical stimulation and intercellular communication increases intracellular Ca^{2+} in epithelial cells, *Cell Regulation*. 1: 585–596.

Santillán, M. and M. C. Mackey (2001a) Dynamic behavior in mathematical models of the tryptophan operon, *Chaos*. 11: 261–268.

Santillán, M. and M. C. Mackey (2001b) Dynamic regulation of the tryptophan operon: a modeling study and comparison with experimental data, *Proceedings of the National Academy of Sciences USA*. 98: 1364–1369.

Santillán, M. and M. C. Mackey (2004) Influence of catabolite repression and inducer exclusion on the bistable behavior of the lac operon, *Biophysical Journal*. 86: 1282–1292.

Sarna, S. K. (1989) *In vivo myoelectric activity: methods, analysis and interpretation*. In: Handbook of Physiology. Section 6: The Gastrointestinal System, Ed: S. G. Schultz, J. D. Wood and B. B. Rauner, American Physiological Society, Bethesda, Maryland.

Sarna, S. K., E. E. Daniel and Y. J. Kingma (1971) Simulation of the slow wave electrical activity of small intestine, *American Journal of Physiology*. 221: 166–175.

Sarty, G. E. and R. A. Pierson (2005) An application of Lacker's mathematical model for the prediction of ovarian response to superstimulation, *Mathematical Biosciences*. 198: 80–96.

Schiefer, A., G. Meissner and G. Isenberg (1995) Ca^{2+} activation and Ca^{2+} inactivation of canine reconstituted cardiac sarcoplasmic reticulum Ca^{2+}-release channels, *Journal of Physiology*. 489: 337–348.

Schlosser, P. M. and J. F. Selgrade (2000) A model of gonadotropin regulation during the menstrual cycle in women: qualitative features, *Environmental Health Perspectives*. 108 Suppl 5: 873–881.

Schmitz, S., H. Franke, J. Brusis and H. E. Wichmann (1993) Quantification of the cell kinetic effects of G-CSF using a model of human granulopoiesis, *Experimental Hematology*. 21: 755–760.

Schmitz, S., M. Loeffler, J. B. Jones, R. D. Lange and H. E. Wichmann (1990) Synchrony of bone marrow proliferation and maturation as the origin of cyclic haemopoiesis, *Cell and Tissue Kinetics*. 23: 425–442.

Schultz, S. G. (1981) Homocellular regulatory mechanisms in sodium-transporting epithelia: avoidance of extinction by "flush-through", *American Journal of Physiology — Renal, Fluid and Electrolyte Physiology*. 241: F579–590.

Schumaker, M. F. and R. MacKinnon (1990) A simple model for multi-ion permeation, *Biophysical Journal*. 58: 975–984.

Schuster, S., M. Marhl and T. Höfer (2002) Modelling of simple and complex calcium oscillations. From single-cell responses to intercellular signalling, *European Journal of Biochemistry*. 269: 1333–1355.

Schwartz, N. B. (1969) A model for the regulation of ovulation in the rat, *Recent Progress in Hormone Research*. 25: 1–53.

Segel, I. H. (1975) *Enzyme Kinetics: Behavior and Analysis of Rapid Equilibrium and Steady-State Enzyme Systems*: John Wiley & Sons. Republished in the Wiley Classics Library Edition, 1993. Wiley, Hoboken, New Jersey.

Segel, L. and A. Goldbeter (1994) Scaling in biochemical kinetics: dissection of a relaxation oscillator, *Journal of Mathematical Biology*. 32: 147–160.

Segel, L. A. (1970) Standing-gradient flows driven by active solute transport, *Journal of Theoretical Biology*. 29: 233–250.

Segel, L. A. (1977) *Mathematics Applied to Continuum Mechanics*: MacMillan, New York.

Segel, L. A. (1988) On the validity of the steady state assumption of enzyme kinetics, *Bulletin of Mathematical Biology*. 50: 579–593.

Segel, L. A., I. Chet and Y. Henis (1977) A simple quantitative assay for bacterial motility, *Journal of General Microbiology*. 98: 329–337.

Segel, L. A., A. Goldbeter, P. N. Devreotes and B. E. Knox (1986) A mechanism for exact sensory adaptation based on receptor modification, *Journal of Theoretical Biology*. 120: 151–179.

Segel, L. A. and A. S. Perelson (1992) Plasmid copy number control: a case study of the quasi-steady state assumption, *Journal of Theoretical Biology*. 158: 481–494.

Segel, L. A. and M. Slemrod (1989) The quasi-steady state assumption: a case study in perturbation, *SIAM Review*. 31: 446–447.

Segev, I., J. Rinzel and G. M. Shepherd (1995) *The Theoretical Foundation of Dendritic Function*: MIT Press, Cambridge, MA.

Sel'kov, E. E. (1968) Self-oscillations in glycolysis, *European Journal of Biochemistry*. 4: 79–86.

Selgrade, J. F. and P. M. Schlosser (1999) A model for the production of ovarian hormones during the menstrual cycle, *Fields Institute Communications*. 21: 429–446.

Selivanov, V. A., F. Ichas, E. L. Holmuhamedov, L. S. Jouaville, Y. V. Evtodienko and J. P. Mazat (1998) A model of mitochondrial Ca^{2+}-induced Ca^{2+} release simulating the Ca^{2+} oscillations and spikes generated by mitochondria, *Biophysical Chemistry*. 72: 111–121.

Sha, W., J. Moore, K. Chen, A. D. Lassaletta, C. S. Yi, J. J. Tyson and J. C. Sible (2003) Hysteresis drives cell-cycle transitions in Xenopus laevis egg extracts, *Proceedings of the National Academy of Sciences USA*. 100: 975–980.

Shannon, T. R., F. Wang, J. Puglisi, C. Weber and D. M. Bers (2004) A mathematical treatment of integrated Ca dynamics within the ventricular myocyte, *Biophysical Journal*. 87: 3351–3371.

Shapley, R. M. and C. Enroth-Cugell (1984) *Visual adaptation and retinal gain controls*. In: Progress in Retinal Research, Ed: N. Osborne and G. Chader, Pergamon Press, London.

Sheetz, M. P. and J. A. Spudich (1983) Movement of myosin-coated fluorescent beads on actin cables in vitro, *Nature*. 303: 31–35.

Shen, P. and R. Larter (1995) Chaos in intracellular Ca^{2+} oscillations in a new model for non-excitable cells, *Cell Calcium*. 17: 225–232.

Sherman, A. (1994) Anti-phase, asymmetric and aperiodic oscillations in excitable cells–I. coupled bursters, *Bulletin of Mathematical Biology*. 56: 811–835.

Sherman, A. and J. Rinzel (1991) Model for synchronization of pancreatic β-cells by gap junction coupling, *Biophysical Journal*. 59: 547–559.

Sherman, A., J. Rinzel and J. Keizer (1988) Emergence of organized bursting in clusters of pancreatic β-cells by channel sharing, *Biophysical Journal*. 54: 411–425.

Shorten, P. R., A. B. Robson, A. E. McKinnon and D. J. Wall (2000) CRH-induced electrical activity and calcium signalling in pituitary corticotrophs, *Journal of Theoretical Biology*. 206: 395–405.

Shorten, P. R. and D. J. Wall (2000) A Hodgkin–Huxley model exhibiting bursting oscillations, *Bulletin of Mathematical Biology*. 62: 695–715.

Shotkin, L. M. (1974a) A model for LH levels in the recently-castrated adult rat and its comparison with experiment, *Journal of Theoretical Biology*. 43: 1–14.

Shotkin, L. M. (1974b) A model for the effect of daily injections of gonadal hormones on LH levels in recently-castrated adult rats and its comparison with experiment, *Journal of Theoretical Biology*. 43: 15–28.

Shuai, J. W. and P. Jung (2002a) Optimal intracellular calcium signaling, *Physical Review Letters*. 88: 068102.

Shuai, J. W. and P. Jung (2002b) Stochastic properties of Ca^{2+} release of inositol 1,4,5-trisphosphate receptor clusters, *Biophysical Journal*. 83: 87–97.

Shuai, J. W. and P. Jung (2003) Optimal ion channel clustering for intracellular calcium signaling, *Proceedings of the National Academy of Sciences USA*. 100: 506–510.

Shuttleworth, T. J. (1999) What drives calcium entry during $[Ca^{2+}]_i$ oscillations?–challenging the capacitative model, *Cell Calcium*. 25: 237–246.

Siebert, W. M. (1974) Ranke revisited–a simple short-wave cochlear model, *Journal of the Acoustical Society of America*. 56: 594–600.

Simon, S. M., C. S. Peskin and G. F. Oster (1992) What drives the translocation of proteins?, *Proceedings of the National Academy of Sciences USA*. 89: 3770–3774.

Sinha, S. (1988) Theoretical study of tryptophan operon: application in microbial technology, *Biotechnology and Bioengineering*. 31: 117–124.

Smart, J. L. and J. A. McCammon (1998) Analysis of synaptic transmission in the neuromuscular junction using a continuum finite element model, *Biophysical Journal*. 75: 1679–1688.

Smith, G. D. (1996) Analytical steady-state solution to the rapid buffering approximation near an open Ca^{2+} channel, *Biophysical Journal*. 71: 3064–3072.

Smith, G. D., L. Dai, R. M. Miura and A. Sherman (2001) Asymptotic analysis of buffered calcium diffusion near a point source, *SIAM Journal on Applied Mathematics*. 61: 1816–1838.

Smith, G. D., J. E. Keizer, M. D. Stern, W. J. Lederer and H. Cheng (1998) A simple numerical model of calcium spark formation and detection in cardiac myocytes, *Biophysical Journal*. 75: 15–32.

Smith, G. D., J. E. Keizer, M. D. Stern, W. J. Lederer and H. Cheng (1998) A simple numerical model of calcium spark formation and detection in cardiac myocytes, *Biophysical Journal*. 75: 15–32.

Smith, G. D., J. Wagner and J. Keizer (1996) Validity of the rapid buffering approximation near a point source of calcium ions, *Biophysical Journal*. 70: 2527–2539.

Smith, J. C., A. P. Abdala, H. Koizumi, I. A. Rybak and J. F. Paton (2007) Spatial and functional architecture of the Mammalian brain stem respiratory network: a hierarchy of three oscillatory mechanisms, *Journal of Neurophysiology*. 98: 3370–3387.

Smith, J. C., H. H. Ellenberger, K. Ballanyi, D. W. Richter and J. L. Feldman (1991) Pre-Bötzinger complex: a brainstem region that may generate respiratory rhythm in mammals, *Science*. 254: 726–729.

Smith, J. M. and R. J. Cohen (1984) Simple finite element model accounts for wide range of cardiac dysrhythmias, *Proceedings of the National Academy of Sciences USA*. 81: 233–237.

Smith, N. P. and E. J. Crampin (2004) Development of models of active ion transport for whole-cell modelling: cardiac sodium–potassium pump as a case study, *Progress in Biophysics and Molecular Biology*. 85: 387–405.

Smith, W. R. (1980) Hypothalamic regulation of pituitary secretion of luteinizing hormone–II. Feedback control of gonadotropin secretion, *Bulletin of Mathematical Biology*. 42: 57–78.

Smith, W. R. (1983) Qualitative mathematical models of endocrine systems, *American Journal of Physiology — Regulatory, Integrative and Comparative Physiology*. 245: R473–477.

Smolen, P. (1995) A model for glycolytic oscillations based on skeletal muscle phosphofructokinase kinetics, *Journal of Theoretical Biology*. 174: 137–148.

Smolen, P. and J. Keizer (1992) Slow voltage inactivation of Ca^{2+} currents and bursting mechanisms for the mouse pancreatic β-cell, *Journal of Membrane Biology*. 127: 9–19.

Smoller, J. (1994) *Shock Waves and Reaction-Diffusion Equations (Second Edition)*: Springer-Verlag, New York.

Sneyd, J., Ed. (2005) *Tutorials in Mathematical Biosciences II: Mathematical Modeling of Calcium Dynamics and Signal Transduction*. Springer-Verlag, New York.

Sneyd, J. and A. Atri (1993) Curvature dependence of a model for calcium wave propagation, *Physica D*. 65: 365–372.

Sneyd, J., A. C. Charles and M. J. Sanderson (1994) A model for the propagation of intercellular calcium waves, *American Journal of Physiology — Cell Physiology*. 266: C293–302.

Sneyd, J., P. D. Dale and A. Duffy (1998) Traveling waves in buffered systems: applications to calcium waves, *SIAM Journal on Applied Mathematics*. 58: 1178–1192.

Sneyd, J. and J. F. Dufour (2002) A dynamic model of the type-2 inositol trisphosphate receptor, *Proceedings of the National Academy of Sciences USA*. 99: 2398–2403.

Sneyd, J. and M. Falcke (2005) Models of the inositol trisphosphate receptor, *Progress in Biophysics and Molecular Biology*. 89: 207–245.

Sneyd, J., M. Falcke, J. F. Dufour and C. Fox (2004a) A comparison of three models of the inositol trisphosphate receptor, *Progress in Biophysics and Molecular Biology*. 85: 121–140.

Sneyd, J., J. Keizer and M. J. Sanderson (1995b) Mechanisms of calcium oscillations and waves: a quantitative analysis, *FASEB Journal*. 9: 1463–1472.

Sneyd, J. and J. Sherratt (1997) On the propagation of calcium waves in an inhomogeneous medium, *SIAM Journal on Applied Mathematics*. 57: 73–94.

Sneyd, J. and D. Tranchina (1989) Phototransduction in cones: an inverse problem in enzyme kinetics, *Bulletin of Mathematical Biology*. 51: 749–784.

Sneyd, J., K. Tsaneva-Atanasova, V. Reznikov, Y. Bai, M. J. Sanderson and D. I. Yule (2006) A method for determining the dependence of calcium oscillations on inositol trisphosphate oscillations, *Proceedings of the National Academy of Sciences USA*. 103: 1675–1680.

Sneyd, J., K. Tsaneva-Atanasova, D. I. Yule, J. L. Thompson and T. J. Shuttleworth (2004b) Control of calcium oscillations by membrane fluxes, *Proceedings of the National Academy of Sciences USA*. 101: 1392–1396.

Sneyd, J., B. Wetton, A. C. Charles and M. J. Sanderson (1995a) Intercellular calcium waves mediated by diffusion of inositol trisphosphate: a two-dimensional model, *American Journal of Physiology — Cell Physiology*. 268: C1537–1545.

Sneyd, J., M. Wilkins, A. Strahonja and M. J. Sanderson (1998) Calcium waves and oscillations driven by an intercellular gradient of inositol (1,4,5)-trisphosphate, *Biophysical Chemistry*. 72: 101–109.

Sobie, E. A., K. W. Dilly, J. dos Santos Cruz, W. J. Lederer and M. S. Jafri (2002)

Termination of cardiac Ca^{2+} sparks: an investigative mathematical model of calcium-induced calcium release, *Biophysical Journal*. 83: 59–78.

Soeller, C. and M. B. Cannell (1997) Numerical simulation of local calcium movements during L-type calcium channel gating in the cardiac diad, *Biophysical Journal*. 73: 97–111.

Soeller, C. and M. B. Cannell (2002a) Estimation of the sarcoplasmic reticulum Ca^{2+} release flux underlying Ca^{2+} sparks, *Biophysical Journal*. 82: 2396–2414.

Soeller, C. and M. B. Cannell (2002b) A Monte Carlo model of ryanodine receptor gating in the diadic cleft of cardiac muscle, *Biophysical Journal*. 82: 76a.

Soeller, C. and M. B. Cannell (2004) Analysing cardiac excitation–contraction coupling with mathematical models of local control, *Progress in Biophysics and Molecular Biology*. 85: 141–162.

Spach, M. S., W. T. Miller, 3rd, D. B. Geselowitz, R. C. Barr, J. M. Kootsey and E. A. Johnson (1981) The discontinuous nature of propagation in normal canine cardiac muscle. Evidence for recurrent discontinuities of intracellular resistance that affect the membrane currents, *Circulation Research*. 48: 39–54.

Spilmann, L. and J. S. Werner, Eds. (1990) *Visual Perception: The Neurophysiological Foundations*: Academic Press, London.

Spitzer, V., M. J. Ackerman, A. L. Scherzinger and R. M. Whitlock (1996) The visible human male: a technical report, *Journal of the American Medical Informatics Association*. 3: 118–130.

Spudich, J. A., S. J. Kron and M. P. Sheetz (1985) Movement of myosin-coated beads on oriented filaments reconstituted from purified actin, *Nature*. 315: 584–586.

Stakgold, I. (1998) *Green's Functions and Boundary Value Problems*: Wiley, New York.

Starmer, C. F., A. R. Lancaster, A. A. Lastra and A. O. Grant (1992) Cardiac instability amplified by use-dependent Na channel blockade, *American Journal of Physiology — Heart and Circulatory Physiology*. 262: H1305–1310.

Starmer, C. F., A. A. Lastra, V. V. Nesterenko and A. O. Grant (1991) Proarrhythmic response to sodium channel blockade. Theoretical model and numerical experiments, *Circulation*. 84: 1364–1377.

Steele, C. R. (1974) Behavior of the basilar membrane with pure-tone excitation, *Journal of the Acoustical Society of America*. 55: 148–162.

Steele, C. R. and L. Tabor (1979a) Comparison of WKB and finite difference calculations for a two-dimensional cochlear model, *Journal of the Acoustical Society of America*. 65: 1001–1006.

Steele, C. R. and L. Tabor (1979b) Comparison of WKB calculations and experimental results for three-dimensional cochlear models, *Journal of the Acoustical Society of America*. 65: 1007–1018.

Stephenson, J. L. (1972) Concentration of the urine in a central core model of the counterflow system, *Kidney International*. 2: 85–94.

Stephenson, J. L. (1992) *Urinary concentration and dilution: models*. In: Handbook of Physiology. Section 8: Renal Physiology, Ed: E. E. Windhager, American Physiological Society, Bethesda, Maryland.

Stern, M. D. (1992) Buffering of calcium in the vicinity of a channel pore, *Cell Calcium*. 13: 183–192.

Stern, M. D. (1992) Theory of excitation–contraction coupling in cardiac muscle, *Biophysical Journal*. 63: 497–517.

Stern, M. D., G. Pizarro and E. Ríos (1997) Local control model of excitation–contraction coupling in skeletal muscle, *Journal of General Physiology*. 110: 415–440.

Stern, M. D., L. S. Song, H. Cheng, J. S. Sham, H. T. Yang, K. R. Boheler and E. Rios (1999) Local control models of cardiac excitation–contraction coupling. A possible role for allosteric interactions between ryanodine receptors, *Journal of General Physiology*. 113: 469–489.

Stiles, J. R. and T. M. Bartol (2000) *Monte Carlo methods for simulating realistic synaptic microphysiology using MCell*. In: Computational Neuroscience: Realistic Modeling for Experimentalists, Ed: E. D. Schutter, CRC Press, New York.

Stojilkovic, S. S., J. Reinhart and K. J. Catt (1994) Gonadotropin-releasing hormone receptors: structure and signal transduction pathways, *Endocrine Reviews*. 15: 462–499.

Stoker, J. J. (1950) *Nonlinear Vibrations*: Interscience, New York.

Stokes, W. (1854) *The Diseases of the Heart and Aorta*: Hodges and Smith, Dublin.

Strang, G. (1986) *Introduction to Applied Mathematics*: Wellesley-Cambridge Press, Wellesley, MA.

Streeter, D. D. J. (1979) *Gross morphology and fiber geometry of the heart*. In: Handbook of Physiology. Section 2: The Cardiovascular System, Volume I: The Heart, Ed: American Physiological Society, Bethesda, MD.

Strieter, J., J. L. Stephenson, L. G. Palmer and A. M. Weinstein (1990) Volume-activated chloride permeability can mediate cell volume regulation in a mathematical model of a tight epithelium, *Journal of General Physiology*. 96: 319–344.

Strogatz, S. H. (1994) *Nonlinear Dynamics and Chaos*: Addison-Wesley, Reading, MA.

Stryer, L. (1986) Cyclic GMP cascade of vision, *Annual Review of Neuroscience*. 9: 87–119.

Stryer, L. (1988) *Biochemistry (Third Edition)*: W.H. Freeman, New York.

Sturis, J., K. S. Polonsky, E. Mosekilde and E. V. Cauter (1991) Computer model for mechanisms underlying ultradian oscillations of insulin and glucose, *American Journal of Physiology — Endocrinology and Metabolism*. 260: E801–809.

Sturis, J., A. J. Scheen, R. Leproult, K. S. Polonsky and E. van Cauter (1995) 24-hour glucose profiles during continuous or oscillatory insulin infusion. Demonstration of the functional significance of ultradian insulin oscillations, *Journal of Clinical Investigation*. 95: 1464–1471.

Sun, X. P., N. Callamaras, J. S. Marchant and I. Parker (1998) A continuum of InsP$_3$-mediated elementary Ca^{2+} signalling events in Xenopus oocytes, *Journal of Physiology*. 509: 67–80.

Sveiczer, A., A. Csikasz-Nagy, B. Gyorffy, J. J. Tyson and B. Novak (2000) Modeling the fission yeast cell cycle: quantized cycle times in *wee1$^-$ cdc25Δ* mutant cells, *Proceedings of the National Academy of Sciences USA*. 97: 7865–7870.

Swillens, S., P. Champeil, L. Combettes and G. Dupont (1998) Stochastic simulation of a single inositol 1,4,5-trisphosphate-sensitive Ca^{2+} channel reveals repetitive openings during "blip-like" Ca^{2+} transients, *Cell Calcium*. 23: 291–302.

Swillens, S., G. Dupont, L. Combettes and P. Champeil (1999) From calcium blips to calcium puffs: theoretical analysis of the requirements for interchannel communication, *Proceedings of the National Academy of Sciences USA*. 96: 13750–13755.

Swillens, S. and D. Mercan (1990) Computer simulation of a cytosolic calcium oscillator, *Biochemical Journal*. 271: 835–838.

Taccardi, B., R. L. Lux and P. R. Erschler (1992) Effect of myocardial fiber direction on 3-dimensional shape of excitation wavefront and associated potential distributions in ventricular walls, *Circulation*. 86 (Suppl. I): 752.

Tai, K., S. D. Bond, H. R. MacMillan, N. A. Baker, M. J. Holst and J. A. McCammon (2003) Finite element simulations of acetylcholine diffusion in neuromuscular junctions, *Biophysical Journal*. 84: 2234–2241.

Takaki, M. (2003) Gut pacemaker cells: the interstitial cells of Cajal (ICC), *Journal of Smooth Muscle Research*. 39: 137–161.

Tameyasu, T. (2002) Simulation of Ca^{2+} release from the sarcoplasmic reticulum with three-dimensional sarcomere model in cardiac muscle, *Japanese Journal of Physiology*. 52: 361–369.

Tamura, T., K. Nakatani and K.-W. Yau (1991) Calcium feedback and sensitivity regulation in primate rods, *Journal of General Physiology*. 98: 95–130.

Tang, Y., T. Schlumpberger, T. Kim, M. Lueker and R. S. Zucker (2000) Effects of mobile buffers on facilitation: experimental and computational studies, *Biophysical Journal*. 78: 2735–2751.

Tang, Y. and J. L. Stephenson (1996) Calcium dynamics and homeostasis in a mathematical model of the principal cell of the cortical collecting tubule, *Journal of General Physiology*. 107: 207–230.

Tang, Y., J. L. Stephenson and H. G. Othmer (1996) Simplification and analysis of models of calcium dynamics based on IP$_3$-sensitive calcium channel kinetics, *Biophysical Journal*. 70: 246–263.

Taniguchi, K., S. Kaya, K. Abe and S. Mardh (2001) The oligomeric nature of Na/K-transport ATPase, *Journal of Biochemistry*. 129: 335–342.

Tawhai, M. H. and K. S. Burrowes (2003) Developing integrative computational models of pulmonary structure, *Anatomical record Part B, New anatomist*. 275: 207–218.

Tawhai, M. H., P. Hunter, J. Tschirren, J. Reinhardt, G. McLennan and E. A. Hoffman (2004) CT-based geometry analysis and finite element models of the human and ovine bronchial tree, *Journal of Applied Physiology*. 97: 2310–2321.

Tawhai, M. H., M. P. Nash and E. A. Hoffman (2006) An imaging-based computational approach to model ventilation distribution and soft-tissue deformation in the ovine lung, *Academic Radiology*. 13: 113–120.

Taylor, C. W. (1998) Inositol trisphosphate receptors: Ca^{2+}-modulated intracellular Ca^{2+} channels, *Biochimica et Biophysica Acta*. 1436: 19–33.

Taylor, W. R., S. He, W. R. Levick and D. I. Vaney (2000) Dendritic computation of direction selectivity by retinal ganglion cells, *Science*. 289: 2347–2350.

ten Tusscher, K. H., D. Noble, P. J. Noble and A. V. Panfilov (2004) A model for human ventricular tissue, *American Journal of Physiology — Heart and Circulatory Physiology*. 286: H1573–1589.

Thomas, A. P., G. S. Bird, G. Hajnoczky, L. D. Robb-Gaspers and J. W. Putney, Jr. (1996) Spatial and temporal aspects of cellular calcium signaling, *FASEB Journal*. 10: 1505–1517.

Thomas, D., P. Lipp, S. C. Tovey, M. J. Berridge, W. Li, R. Y. Tsien and M. D. Bootman (2000) Microscopic properties of elementary Ca^{2+} release sites in non-excitable cells, *Current Biology*. 10: 8–15.

Tolić, I. M., E. Mosekilde and J. Sturis (2000) Modeling the insulin–glucose feedback system: the significance of pulsatile insulin secretion, *Journal of Theoretical Biology*. 207: 361–375.

Topor, Z. L., M. Pawlicki and J. E. Remmers (2004) A computational model of the human respiratory control system: responses to hypoxia and hypercapnia, *Annals of Biomedical Engineering*. 32: 1530–1545.

Tordjmann, T., B. Berthon, M. Claret and L. Combettes (1997) Coordinated intercellular calcium waves induced by noradrenaline in rat hepatocytes: dual control by gap junction permeability and agonist, *Embo Journal*. 16: 5398–5407.

Tordjmann, T., B. Berthon, E. Jacquemin, C. Clair, N. Stelly, G. Guillon, M. Claret and L. Combettes (1998) Receptor-oriented intercellular calcium waves evoked by vasopressin in rat hepatocytes, *Embo Journal*. 17: 4695–4703.

Torre, V., S. Forti, A. Menini and M. Campani (1990) Model of phototransduction in retinal rods, *Cold Spring Harbor Symposia in Quantitative Biology*. 55: 563–573.

Tosteson, D. C. and J. F. Hoffman (1960) Regulation of cell volume by active cation transport in high and low potassium sheep red cells, *Journal of General Physiology*. 44: 169–194.

Tranchina, D., J. Gordon and R. Shapley (1984) Retinal light adaptation–evidence for a feedback mechanism, *Nature*. 310: 314–316.

Tranchina, D., J. Sneyd and I. D. Cadenas (1991) Light adaptation in turtle cones: testing and analysis of a model for phototransduction, *Biophysical Journal*. 60: 217–237.

Tranquillo, R. and D. Lauffenberger (1987) Stochastic models of leukocyte chemosensory movement, *Journal of Mathematical Biology*. 25: 229–262.

Traube, L. (1865) Ueber periodische thatigkeits-aeusserungen des vasomotorischen und hemmungs-nervencentrums, *Medizin Wissenschaft*. 56: 881–885.

Troy, J. B. and T. Shou (2002) The receptive fields of cat retinal ganglion cells in physiological and pathological states: where we are after half a century of research, *Progress in Retinal and Eye Research*. 21: 263–302.

Troy, W. C. (1976) Bifurcation phenomena in FitzHugh's nerve conduction equations, *Journal of Mathematical Analysis and Applications*. 54: 678–690.

Troy, W. C. (1978) The bifurcation of periodic solutions in the Hodgkin–Huxley equations, *Quarterly of Applied Mathematics*. 36: 73–83.

Tsai, J.-C. and J. Sneyd (2005) Existence and stability of traveling waves in buffered systems, *SIAM Journal on Applied Mathematics*. 66: 1675–1680.

Tsai, J.-C. and J. Sneyd (2007a) Are buffers boring?: uniqueness and asymptotical stability of traveling wave fronts in the buffered bistable system, *Journal of Mathematical Biology*. 54: 513–553.

Tsai, J. C. and J. Sneyd (2007b) Traveling waves in the discrete fast buffered bistable system, *Journal of Mathematical Biology*. 55: 605–652.

Tsaneva-Atanasova, K., D. I. Yule and J. Sneyd (2005) Calcium oscillations in a triplet of pancreatic acinar cells, *Biophysical Journal*. 88: 1535–1551.

Tsaneva-Atanasova, K., C. L. Zimliki, R. Bertram and A. Sherman (2006) Diffusion of calcium and metabolites in pancreatic islets: killing oscillations with a pitchfork, *Biophysical Journal*. 90: 3434–3446.

Tse, A., F. W. Tse, W. Almers and B. Hille (1993) Rhythmic exocytosis stimulated by GnRH-induced calcium oscillations in rat gonadotropes, *Science*. 260: 82–84.

Tuckwell, H. C. (1988) *Introduction to Theoretical Neurobiology*: Cambridge University Press, Cambridge.

Tuckwell, H. C. and R. M. Miura (1978) A mathematical model for spreading cortical depression, *Biophysical Journal*. 23: 257–276.

Tung, L. (1978) *A bi-domain model for describing ischemic myocardial D-C potentials. Ph.D. Thesis*: MIT, Cambridge, MA.

Tyson, J. J., A. Csikasz-Nagy and B. Novak (2002) The dynamics of cell cycle regulation, *Bioessays*. 24: 1095–1109.

Tyson, J. J. and P. C. Fife (1980) Target patterns in a realistic model of the Belousov–Zhabotinskii reaction, *Journal of Chemical Physics*. 73: 2224–2237.

Tyson, J. J., C. I. Hong, C. D. Thron and B. Novak (1999) A simple model of circadian rhythms based on dimerization and proteolysis of PER and TIM, *Biophysical Journal*. 77: 2411–2417.

Tyson, J. J. and J. P. Keener (1988) Singular perturbation theory of traveling waves in excitable media, *Physica D*. 32: 327–361.

Tyson, J. J. and B. Novak (2001) Regulation of the eukaryotic cell cycle: molecular antagonism, hysteresis, and irreversible transitions, *Journal of Theoretical Biology*. 210: 249–263.

Tyson, J. J. and H. G. Othmer (1978) The dynamics of feedback control circuits in biochemical pathways, *Progress in Theoretical Biology*. 5: 1–62.

Ullah, G., P. Jung and A. H. Cornell-Bell (2006) Anti-phase calcium oscillations in astrocytes via inositol (1,4,5)-trisphosphate regeneration, *Cell Calcium*. 39: 197–208.

Urban, B. W. and S. B. Hladky (1979) Ion transport in the simplest single file pore, *Biochimica et Biophysica Acta*. 554: 410–429.

Ursino, M. (1998) Interaction between carotid baroregulation and the pulsating heart: a mathematical model, *American Journal of Physiology — Heart and Circulatory Physiology*. 275: H1733–1747.

Ursino, M. (1999) A mathematical model of the carotid baroregulation in pulsating conditions, *IEEE Transactions in Biomedical Engineering*. 46: 382–392.

Ussing, H. H. (1949) The distinction by means of tracers between active transport and diffusion, *Acta Physiologica Scandinavica*. 19: 43–56.

Ussing, H. H. (1982) Volume regulation of frog skin epithelium, *Acta Physiologica Scandinavica*. 114: 363–369.

Uyeda, T. Q., S. J. Kron and J. A. Spudich (1990) Myosin step size. Estimation from slow sliding movement of actin over low densities of heavy meromyosin, *Journal of Molecular Biology*. 214: 699–710.

van der Pol, B. and J. van der Mark (1928) The heartbeat considered as a relaxation oscillation, and an electrical model of the heart, *Philosophical Magazine*. 6: 763–775.

van Kampen, N. G. (2007) *Stochastic Processes in Physics and Chemistry (Third Edition)*: Elsevier, Amsterdam.

van Milligen, B. P., P. D. Bons, B. A. Carreras and R. Sanchez (2006) On the applicability of Fick's law to diffusion in inhomogeneous systems, *European Journal of Physics*. 26: 913–925.

Vandenberg, C. A. and F. Bezanilla (1991) A sodium channel gating model based on single channel, macroscopic ionic, and gating currents in the squid giant axon, *Biophysical Journal*. 60: 1511–1533.

Vaney, D. I. and W. R. Taylor (2002) Direction selectivity in the retina, *Current Opinion in Neurobiology*. 12: 405–410.

Vesce, S., P. Bezzi and A. Volterra (1999) The active role of astrocytes in synaptic transmission, *Cellular and Molecular Life Sciences*. 56: 991–1000.

Vielle, B. (2005) Mathematical analysis of Mayer waves, *Journal of Mathematical Biology*. 50: 595–606.

Vodopick, H., E. M. Rupp, C. L. Edwards, F. A. Goswitz and J. J. Beauchamp (1972) Spontaneous cyclic leukocytosis and thrombocytosis in chronic granulocytic leukemia, *New England Journal of Medicine*. 286: 284–290.

von Békésy, V. (1960) *Experiments in Hearing*: McGraw-Hill, New York. Reprinted in 1989 by the Acoustical Society of America.

von Euler, C. (1980) Central pattern generation during breathing, *Trends in Neuroscience*. 3: 275–277.

Wagner, J. and J. Keizer (1994) Effects of rapid buffers on Ca^{2+} diffusion and Ca^{2+} oscillations, *Biophysical Journal*. 67: 447–456.

Wagner, J., Y.-X. Li, J. Pearson and J. Keizer (1998) Simulation of the fertilization Ca^{2+} wave in *Xenopus laevis* eggs, *Biophysical Journal*. 75: 2088–2097.

Waldo, A. L., A. J. Camm, et al. (1996) Effect of d-sotalol on mortality in patients with left ventricular dysfunction after recent and remote myocardial infarction. The SWORD

Investigators. Survival With Oral d-Sotalol, *Lancet*. 348: 7–12.

Wang, H. and G. Oster (1998) Energy transduction in the F1 motor of ATP synthase, *Nature*. 396: 279–282.

Wang, X.-J. and J. Rinzel (1995) *Oscillatory and bursting properties of neurons*. In: The Handbook of Brain Theory and Neural Networks, Ed: M. Arbib, MIT Press, Cambridge, MA.

Wang, I., A.Z. Politi, N. Tania, Y. Bai, M.J. Sanderson and J. Sneyd (2008) A mathematical model of airway and pulmonary arteriole smooth muscle, *Biophysical Journal*. 94: 2053–2064.

Watanabe, M., N. F. Otani and R. F. Gilmour, Jr. (1995) Biphasic restitution of action potential duration and complex dynamics in ventricular myocardium, *Circulation Research*. 76: 915–921.

Watanabe, M. A., F. H. Fenton, S. J. Evans, H. M. Hastings and A. Karma (2001) Mechanisms for discordant alternans, *Journal of Cardiovascular Electrophysiology*. 12: 196–206.

Waterman, M. S. (1995) *Introduction to Computational Biology: Maps, Sequences, and Genomes*: Chapman and Hall, Boca Raton, FL.

Webb, S. E. and A. L. Miller (2003) Calcium signalling during embryonic development, *Nature Reviews Molecular Cell Biology*. 4: 539–551.

Weber, E. H. (1834) *De pulsu, resorptione, auditu et tactu annotationes anatomicæ, et physiologicæ. Author's summary: Ueber den Tastsinn*, Arch. Anat. u. Physiol., *1835, 152*: Leipzig.

Weinstein, A. (1992) Analysis of volume regulation in an epithelial cell model, *Bulletin of Mathematical Biology*. 54: 537–561.

Weinstein, A. (1996) Coupling of entry to exit by peritubular K^+ permeability in a mathematical model of rat proximal tubule, *American Journal of Physiology — Renal, Fluid and Electrolyte Physiology*. 271: F158–168.

Weinstein, A. M. (1994) Mathematical models of tubular transport, *Annual Review of Physiology*. 56: 691–709.

Weinstein, A. M. (1998a) A mathematical model of the inner medullary collecting duct of the rat: pathways for Na and K transport, *American Journal of Physiology — Renal, Fluid and Electrolyte Physiology*. 274: F841–855.

Weinstein, A. M. (1998b) A mathematical model of the inner medullary collecting duct of the

rat: acid/base transport, *American Journal of Physiology — Renal, Fluid and Electrolyte Physiology*. 274: F856–867.

Weinstein, A. M. (2000) A mathematical model of the outer medullary collecting duct of the rat, *American Journal of Physiology — Renal, Fluid and Electrolyte Physiology*. 279: F24–45.

Weinstein, A. M. (2003) Mathematical models of renal fluid and electrolyte transport: acknowledging our uncertainty, *American Journal of Physiology — Renal, Fluid and Electrolyte Physiology*. 284: F871–884.

Weinstein, A. M. and J. L. Stephenson (1981) Coupled water transport in standing gradient models of the lateral intercellular space, *Biophysical Journal*. 35: 167–191.

Weiss, J. N., A. Karma, Y. Shiferaw, P. S. Chen, A. Garfinkel and Z. Qu (2006) From pulsus to pulseless: the saga of cardiac alternans, *Circulation Research*. 98: 1244–1253.

Wenckebach, K. F. (1904) *Arrhythmia of the Heart: a Physiological and Clinical Study*: Green, Edinburgh.

West, J. B. (1985) *Ventilation/Blood Flow and Gas Exchange*: Blackwell Scientific Publications, Oxford.

West, J. B. (2004) Understanding pulmonary gas exchange: ventilation–perfusion relationships, *American Journal of Physiology — Lung, Cellular and Molecular Physiology*. 287: L1071–1072.

Wetsel, W. C., M. M. Valenca, I. Merchenthaler, Z. Liposits, F. J. Lopez, R. I. Weiner, P. L. Mellon and A. Negro-Vilar (1992) Intrinsic pulsatile secretory activity of immortalized luteinizing hormone-releasing hormone-secreting neurons, *Proceedings of the National Academy of Sciences USA*. 89: 4149–4153.

White, D. C. S. and J. Thorson (1975) *The Kinetics of Muscle Contraction*: Pergamon Press. Originally published in Progress in Biophysics and Molecular Biology, volume 27, 1973, Oxford, New York.

White, J. B., G. P. Walcott, A. E. Pollard and R. E. Ideker (1998) Myocardial discontinuities: a substrate for producing virtual electrodes that directly excite the myocardium by shocks, *Circulation*. 97: 1738–1745.

Whiteley, J. P., D. J. Gavaghan and C. E. Hahn (2001) Modelling inert gas exchange in tissue and mixed-venous blood return to the lungs, *Journal of Theoretical Biology*. 209: 431–443.

Whiteley, J. P., D. J. Gavaghan and C. E. Hahn (2002) Mathematical modelling of oxygen transport to tissue, *Journal of Mathematical Biology*. 44: 503–522.

Whiteley, J. P., D. J. Gavaghan and C. E. Hahn (2003a) Periodic breathing induced by arterial oxygen partial pressure oscillations, *Mathematical Medicine and Biology*. 20: 205–224.

Whiteley, J. P., D. J. Gavaghan and C. E. Hahn (2003b) Mathematical modelling of pulmonary gas transport, *Journal of Mathematical Biology*. 47: 79–99.

Whitham, G. B. (1974) *Linear and Nonlinear Waves*: Wiley-Interscience, New York.

Whitlock, G. G. and T. D. Lamb (1999) Variability in the time course of single photon responses from toad rods: termination of rhodopsin's activity, *Neuron*. 23: 337–351.

Wichmann, H. E., M. Loeffler and S. Schmitz (1988) A concept of hemopoietic regulation and its biomathematical realization, *Blood Cells*. 14: 411–429.

Wiener, N. and A. Rosenblueth (1946) The mathematical formulation of the problem of conduction of impulses in a network of connected excitable elements, specifically in cardiac muscle, *Archivos del Instituto de Cardiologia de Mexico*. 16: 205–265.

Wier, W. G., T. M. Egan, J. R. Lopez-Lopez and C. W. Balke (1994) Local control of excitation–contraction coupling in rat heart cells, *Journal of Physiology*. 474: 463–471.

Wierschem, K. and R. Bertram (2004) Complex bursting in pancreatic islets: a potential glycolytic mechanism, *Journal of Theoretical Biology*. 228: 513–521.

Wiggins, S. (2003) *Introduction to Applied Nonlinear Dynamical Systems and Chaos*: Springer-Verlag, New York.

Wikswo, J. P., Jr., S. F. Lin and R. A. Abbas (1995) Virtual electrodes in cardiac tissue: a common mechanism for anodal and cathodal stimulation, *Biophysical Journal*. 69: 2195–2210.

Wildt, L., A. Häusler, G. Marshall, J. S. Hutchison, T. M. Plant, P. E. Belchetz and E. Knobil (1981) Frequency and amplitude of gonadotropin-releasing hormone stimulation and gonadotropin secretion in the Rhesus monkey, *Endocrinology*. 109: 376–385.

Willems, G. M., T. Lindhout, W. T. Hermens and H. C. Hemker (1991) Simulation model for thrombin generation in plasma, *Haemostasis*. 21: 197–207.

Williams, M. M. (1990) *Hematology*: McGraw-Hill, New York.

Winfree, A. T. (1967) Biological rhythms and the behavior of populations of coupled oscillators, *Journal of Theoretical Biology*. 16: 15–42.

Winfree, A. T. (1972) Spiral waves of chemical activity, *Science*. 175: 634–636.

Winfree, A. T. (1973) Scroll-shaped waves of chemical activity in three dimensions, *Science*. 181: 937–939.

Winfree, A. T. (1974) Rotating chemical reactions, *Scientific American*. 230: 82–95.

Winfree, A. T. (1980) *The Geometry of Biological Time*: Springer-Verlag, New York.

Winfree, A. T. (1987) *When Time Breaks Down*: Princeton University Press, Princeton, NJ.

Winfree, A. T. (1991) Varieties of spiral wave behavior: an experimentalist's approach to the theory of excitable media, *Chaos*. 1: 303–334 .

Winfree, A. T. and S. H. Strogatz (1983a) Singular filaments organize chemical waves in three dimensions: 1. Geometrically simple waves, *Physica D*. 8: 35–49.

Winfree, A. T. and S. H. Strogatz (1983b) Singular filaments organize chemical waves in three dimensions: 2. twisted waves, *Physica D*. 9: 65–80.

Winfree, A. T. and S. H. Strogatz (1983c) Singular filaments organize chemical waves in three dimensions: 3. knotted waves, *Physica D*. 9: 333–345.

Winfree, A. T. and S. H. Strogatz (1984) Singular filaments organize chemical waves in three dimensions: 4. wave taxonomy, *Physica D*. 13: 221–233.

Winslow, R. L., R. Hinch and J. L. Greenstein (2005) *Mechanisms and models of cardiac excitation–contraction coupling*. In: Tutorials in Mathematical Biosciences II: Mathematical Modeling of Calcium Dynamics and Signal Transduction, Ed: J. Sneyd, Springer-Verlag, New York.

Wittenberg, J. B. (1966) The molecular mechanism of haemoglobin-facilitated oxygen diffusion, *Journal of Biological Chemistry*. 241: 104–114.

Wong, P., S. Gladney and J. D. Keasling (1997) Mathematical model of the lac operon: inducer exclusion, catabolite repression, and diauxic growth on glucose and lactose, *Biotechnology Progress*. 13: 132–143.

Woodbury, J. W. (1971) *Eyring rate theory model of the current–voltage relationship of ion channels in excitable membranes*. In: Chemical Dynamics: Papers in Honor of Henry Eyring, Ed: J. Hirschfelder, John Wiley and Sons, Inc., New York.

Wyman, J. (1966) Facilitated diffusion and the possible role of myoglobin as a transport mechanism, *Journal of Biological Chemistry*. 241: 115–121.

Wyman, R. J. (1977) Neural generation of breathing rhythm, *Annual Review of Physiology*. 39: 417–448.

Yamada, W. M. and R. S. Zucker (1992) Time course of transmitter release calculated from simulations of a calcium diffusion model, *Biophysical Journal*. 61: 671–682.

Yanagida, E. (1985) Stability of fast travelling pulse solutions of the FitzHugh–Nagumo equation, *Journal of Mathematical Biology*. 22: 81–104.

Yanagihara, K., A. Noma and H. Irisawa (1980) Reconstruction of sino-atrial node pacemaker potential based on voltage clamp experiments, *Japanese Journal of Physiology*. 30: 841–857.

Yao, Y., J. Choi and I. Parker (1995) Quantal puffs of intracellular Ca^{2+} evoked by inositol trisphosphate in Xenopus oocytes, *Journal of Physiology*. 482: 533–553.

Yates, A., R. Callard and J. Stark (2004) Combining cytokine signalling with T-bet and GATA-3 regulation in Th1 and Th2 differentiation: a model for cellular decision-making, *Journal of Theoretical Biology*. 231: 181–196.

Yehia, A. R., D. Jeandupeux, F. Alonso and M. R. Guevara (1999) Hysteresis and bistability in the direct transition from 1:1 to 2:1 rhythm in periodically driven single ventricular cells, *Chaos*. 9: 916–931.

Yildirim, N. and M. C. Mackey (2003) Feedback regulation in the lactose operon: a mathematical modeling study and comparison with experimental data, *Biophysical Journal*. 84: 2841–2851.

Yildirim, N., M. Santillán, D. Horike and M. C. Mackey (2004) Dynamics and bistability in a reduced model of the lac operon, *Chaos*. 14: 279–292.

Young, K. W., M. S. Nash, R. A. Challiss and S. R. Nahorski (2003) Role of Ca^{2+} feedback on single cell inositol 1,4,5-trisphosphate oscillations mediated by G-protein-coupled receptors, *Journal of Biological Chemistry*. 278: 20753–20760.

Young, R. C. (1997) A computer model of uterine contractions based on action potential propagation and intercellular calcium waves, *Obstetrics and Gynecology*. 89: 604–608.

Young, R. C. and R. O. Hession (1997) Paracrine and intracellular signaling mechanisms of

calcium waves in cultured human uterine myocytes, *Obstetrics and Gynecology*. 90: 928–932.

Yule, D. I., E. Stuenkel and J. A. Williams (1996) Intercellular calcium waves in rat pancreatic acini: mechanism of transmission, *American Journal of Physiology — Cell Physiology*. 271: C1285–1294.

Zahradnikova, A. and I. Zahradnik (1996) A minimal gating model for the cardiac calcium release channel, *Biophysical Journal*. 71: 2996–3012.

Zarnitsina, V. I., A. V. Pokhilko and F. I. Ataullakhanov (1996a) A mathematical model for the spatio-temporal dynamics of intrinsic pathway of blood coagulation. I. The model description, *Thrombosis Research*. 84: 225–236.

Zarnitsina, V. I., A. V. Pokhilko and F. I. Ataullakhanov (1996b) A mathematical model for the spatio-temporal dynamics of intrinsic pathway of blood coagulation. II. Results, *Thrombosis Research*. 84: 333–344.

Zeuthen, T. (2000) Molecular water pumps, *Reviews of Physiology, Biochemistry and Pharmacology*. 141: 97–151.

Zhang, M., P. Goforth, R. Bertram, A. Sherman and L. Satin (2003) The Ca^{2+} dynamics of isolated mouse beta-cells and islets: implications for mathematical models, *Biophysical Journal*. 84: 2852–2870.

Zhou, X., S. B. Knisley, W. M. Smith, D. Rollins, A. E. Pollard and R. E. Ideker (1998) Spatial changes in the transmembrane potential during extracellular electric stimulation, *Circulation Research*. 83: 1003–1014.

Zhou, Z. and E. Neher (1993) Mobile and immobile calcium buffers in bovine adrenal chromaffin cells, *Journal of Physiology*. 469: 245–273.

Zigmond, S. H. (1977) Ability of polymorphonuclear leukocytes to orient in gradients of chemotactic factors, *Journal of Cell Biology*. 75: 606–616.

Zigmond, S. H., H. I. Levitsky and B. J. Kreel (1981) Cell polarity: an examination of its behavioral expression and its consequences for polymorphonuclear leukocyte chemotaxis, *Journal of Cell Biology*. 89: 585–592.

Zinner, B. (1992) Existence of traveling wavefront solutions for the discrete Nagumo equation, *Journal of Differential Equations*. 96: 1–27.

Zipes, D. P. and J. Jalife (1995) *Cardiac Electrophysiology; From Cell to Bedside*

(Second Edition): W. B. Saunders Co., Philadelphia.

Zucker, R. S. and A. L. Fogelson (1986) Relationship between transmitter release and presynaptic calcium influx when calcium enters through discrete channels, *Proceedings of the National Academy of Sciences USA*. 83: 3032–3036.

Zucker, R. S. and L. Landò (1986) Mechanism of transmitter release: voltage hypothesis and calcium hypothesis, *Science*. 231: 574–579.

Zucker, R. S. and W. G. Regehr (2002) Short-term synaptic plasticity, *Annual Review of Physiology*. 64: 355–405.

Zwislocki, J. (1953) Review of recent mathematical theories of cochlear dynamics, *Journal of the Acoustical Society of America*. 25: 743–751.

Index

Printed in the United States of America